Human Impacts on Weather and Climate
Second Edition

This new edition of *Human Impacts on Weather and Climate* examines the scientific debates surrounding anthropogenic impacts on the Earth's climate and presents the most recent theories, data, and modeling studies. The book discusses the concepts behind deliberate human attempts to modify the weather through cloud seeding, as well as inadvertent modification of weather and climate on regional and global scales through the emission of aerosols and gases and change in land-use. The natural variability of weather and climate greatly complicates our ability to determine a clear cause-and-effect relationship to human activity. The authors examine the strengths and weaknesses of the various hypotheses regarding human impacts on global climate in simple and accessible terms.

Like the first edition, this fully revised new edition will be a valuable resource for undergraduate and graduate courses in atmospheric and environmental science, and will also appeal to policy-makers and general readers interested in how humans are affecting the global climate.

WILLIAM COTTON is a Professor in the Department of Atmospheric Science at Colorado State University. He is a Fellow of the American Meteorological Society and the Cooperative Institute for Research in the Atmosphere (CIRA).

ROGER PIELKE Sr. is a Senior Research Associate in the Department of Atmospheric and Oceanic Sciences, Senior Research Scientist at the Cooperative Institute for Research in Environmental Sciences at the University of Colorado–Boulder, and an Emeritus Professor of Atmospheric Science at Colorado State University. He is also a Fellow of the American Geophysical Union and of the American Meteorological Society.

HUMAN IMPACTS ON WEATHER AND CLIMATE

Second Edition

WILLIAM R. COTTON
Colorado State University

and

ROGER A. PIELKE SR.
University of Colorado at Boulder

CAMBRIDGE
UNIVERSITY PRESS

CAMBRIDGE UNIVERSITY PRESS
Cambridge, New York, Melbourne, Madrid, Cape Town, Singapore, São Paulo, Delhi

Cambridge University Press
The Edinburgh Building, Cambridge CB2 8RU, UK

Published in the United States of America by Cambridge University Press, New York

www.cambridge.org
Information on this title: www.cambridge.org/9780521840866

First published 2007
Reprinted 2008

Printed in the United Kingdom at the University Press, Cambridge

A catalog record for this publication is available from the British Library

ISBN 978-0-521-84086-6 hardback
ISBN 978-0-521-60056-9 paperback

Contents

Acknowledgments

The study of human impacts on weather and climate continues to be a high-interest topic area, not only among scientists but also the public. Our second edition has continued to build on our funded research studies from the National Science Foundation, the National Aeronautics and Space Administration, the Environmental Protection Agency, the Department of Defense, the National Oceanic and Atmospheric Administration, and the United States Geological Survey. Our numerous research collaborators at the Natural Resource Ecology Laboratory and Civil Engineering at Colorado State University have continued to provide valuable insight on this subject. Over our multidecadal career, the fundamental insights into weather and climate provided by our education at the Pennsylvania State University have become increasingly recognized. We also want to recognize the perspective on these subjects, and science in general, that Robert and Joanne Simpson have provided us in our careers. Their mentorship and philosophy of research, of course, is but one of their many seminal accomplishments.

Roger Pielke would like to thank everyone who contributed to compiling the information in Tables 6.2 and 11.2 especially Roni Avissar, Richard Betts, Gordon Bonan, Lahouari Bounoua, Rafael Bras, Chris Castro, Will Cheng, Martin Claussen, Bob Dickinson, Paul Dirmeyer, Han Dolman, Elfatih Eltahir, Jon Foley, Pavel Kabat, George Kallos, Axel Kleidon, Curtis Marshall, Pat Michaels, Nicole Mölders, Udaysankar Nair, Andy Pitman, Adriana Beltran-Przekurat, Rick Raddatz, Chris Rozoff, J. Marshall Shepherd, Lou Steyaert, and Yongkang Xue. In addition, Roger would like to thank Dr. Adriana Beltrán for her assistance with figures in this edition.

As is always the case, Dallas Staley's editorial leadership and Brenda Thompson's assistance in completing the book has been invaluable and is very much appreciated.

Part I

The rise and fall of the science of weather modification by cloud seeding

In Part I we examine human attempts at purposely modifying weather and climate. We also trace the history of the science of weather modification by cloud seeding describing its scientific basis and the rise and fall of funding of weather modification scientific programs, particularly in the United States.

1

The rise of the science of weather modification by cloud seeding

Throughout history and probably prehistory man has sought to modify weather by a variety of means. Many primitive tribes have employed witch doctors or medicine men to bring clouds and rainfall during periods of drought and to drive away rain clouds during flooding episodes. Numerous examples exist where modern man has shot cannons, fired rockets, rung bells, etc. in attempts to modify the weather (Changnon and Ivens, 1981).

It was Schaefer's (1948a) discovery in 1946 that the introduction of dry ice into a freezer containing cloud droplets cooled well below 0 °C (what we call supercooled droplets) resulted in the formation of ice crystals, that launched us into the modern age of the science of weather modification.[1] Working for the General Electric Research Laboratory under the direction of Irving Langmuir on a project investigating ways to combat aircraft icing, Schaefer learned to form a supercooled cloud by blowing moist air into a home freezer unit lined with black velvet. He noted that at temperatures as cold as −23 °C, ice crystals failed to form in the cloud. Introducing a variety of substances in the cloud failed to convert the cloud to ice crystals. It was only after a piece of dry ice was lowered into the cloud that thousands of twinkling ice crystals could be seen in the light beam passing through the chamber. He subsequently showed that only small grains of dry ice or even a needle cooled in liquid air could trigger the nucleation of millions of ice crystals.

Motivated by Schaefer's discovery, Vonnegut (1947), also a researcher at the General Electric Research Laboratory, began a systematic search through chemical tables for materials that have a crystallographic structure similar to ice. He hypothesized that such a material would serve as an artificial ice nucleus. It was well known at that time that under ordinary conditions, the formation (or nucleation) of ice crystals required the presence of a foreign substance called a

---

[1] A summary of this early work is given in Havens *et al.* (1978).

nucleus or mote that would promote their formation. For some time European researchers such as A. Wegener, T. Bergeron, and W. Findeisen had hypothesized that the presence of supercooled droplets in clouds indicated a scarcity of ice-forming nuclei in the atmosphere. It was believed that the dry ice in Schaefer's experiment cooled the air to such a low temperature that nucleation took place without an available nuclei; the process is referred to as *homogeneous nucleation.* Vonnegut's search through the chemical tables revealed three substances which had the desired crystallographic similarity to ice: lead iodide, silver iodide, and antimony. Dispersal of a powder of these substances in a cold box had little effect. Vonnegut then decided to produce a smoke of these substances by vaporizing the material, and as it condensed a smoke of very small crystals of the material was created. Vonnegut found that a smoke of silver iodide particles produced numerous ice crystals in the cold box at temperatures warmer than $-20\,^{\circ}$C similar to dry ice in Schaefer's experiment.

The stage was now set to attempt to introduce dry ice or silver iodide smoke into real supercooled clouds and observe the impact on those clouds. Again, the background of previous research by the Europeans (Wegener, Bergeron, and Findeisen) was important for this stage. They showed that ice crystals once formed in a supercooled cloud could grow very rapidly by deposition of vapor onto them at the expense of supercooled cloud droplets. This is due to the fact that the saturation vapor pressure with respect to ice is lower than the saturation vapor pressure with respect to water at temperatures colder than zero degrees centigrade. As shown in Fig. 1.1, the supersaturation with respect to ice increases linearly with decreasing temperature below 0 °C for a water-saturated cloud. Thus an ice crystal nucleated in a cloud that is water saturated finds itself in an environment which is supersaturated with respect to ice and can thereby grow rapidly by deposition of vapor. As vapor is deposited on the growing ice crystals the vapor in the cloud is depleted, and the cloud vapor pressure lowers to below water saturation. Thus cloud droplets evaporate providing a reservoir of water vapor for growing ice crystals. The ice crystals, therefore, grow at the expense of the cloud droplets.

It was thus hypothesized that the insertion of dry ice or silver iodide in a supercooled cloud would initiate the formation of ice crystals, which in turn would grow by vapor deposition into ice crystals. Precipitation could be artificially initiated in such clouds.

Langmuir (1953) calculated theoretically the number of ice crystals that would form from dry ice pellets of a given size. He also predicted that the latent heat released as the ice crystals grew by vapor deposition would warm the seeded part of the cloud, causing upward motion and turbulence which would disperse the

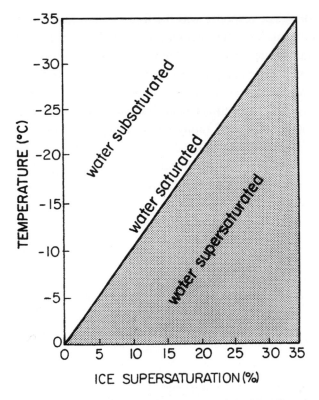

Figure 1.1 Supersaturation with respect to ice as a function of temperature for a water-saturated cloud. The shaded area represents a water-supersaturated cloud. From Cotton and Anthes (1989).

mist of ice crystals created by seeding over a large volume of the unseeded part of the cloud.

On November 13, 1946, Schaefer (1948b) dropped about 1.4 kg of dry ice pellets from an aircraft flying over a supercooled stratus cloud near Schenectady, New York. Similar to the laboratory cold box experiments, the cloud rapidly converted to ice crystals which fell out as snow beneath the stratus deck. This, as well as a number of other exploratory seeding experiments, led to the formation of Project Cirrus.

1.1 Project Cirrus

Under Project Cirrus, Langmuir and Schaefer performed a number of exploratory cloud seeding experiments including seeding of cirrus clouds, supercooled stratus clouds, cumulus clouds, and even hurricanes. Supercooled stratus clouds yielded the clearest response to seeding. A variety of aircraft patterns were flown over the stratus clouds while dropping dry ice. Patterns included L-shaped, race track, and Greek gammas. The response was the formation of holes in the clouds whose shape mirrored the aircraft flight pattern (see Fig. 1.2).

Seeding of supercooled cumulus clouds produced more controversial results. Dry ice and silver iodide seeding experiments were carried out at a variety of locations with the most comprehensive experiments being over New Mexico. Based on four seeding operations near Albuquerque, New Mexico, Langmuir claimed that seeding produced rainfall over a quarter of the area of the state of New Mexico. He concluded that "The odds in favor of this conclusion as compared to the rain was due to natural causes are millions to one." Langmuir was even more enthusiastic about the consequences of silver iodide seeding over New Mexico. The explosive growth of a cumulonimbus cloud and the heavy rainfall near Albuquerque and Santa Fe were attributed to the direct results of ground-based silver iodide seeding. In fact Langmuir concluded that nearly all the rainfall that occurred over New Mexico on the dry ice seeding day and the silver iodide seeding day were the result of seeding.

One of the most controversial experiments performed during Project Cirrus was the periodic seeding experiment. In this experiment a ground-based silver iodide generator was operated on a 7-day periodic schedule with the generator

Figure 1.2 Race track pattern approximately 20 miles long produced by dropping crushed dry ice from an airplane. The safety-pin-like loop at the near end of the pattern resulted when the dry ice dispenser was inadvertently left running as the airplane began climbing to attain altitude from which to photograph results. From Havens *et al.* (1978). Photo courtesy of Dr. Vincent Schaefer.

being operated 8 hours a day on Tuesday, Wednesday, and Thursday and turned off the rest of the week. A total of 1000 g of silver iodide was used per week and the experiment was carried out from December 1949 to the middle of 1951. The analysis of precipitation and other weather records over the Ohio River basin and other regions to the east of New Mexico revealed a highly significant 7-day periodicity. Langmuir and his colleagues were convinced that this periodicity in the rainfall records was a direct result of their seeding in New Mexico. Other scientists were not so convinced (Lewis, 1951; Wahl, 1951; Wexler, 1951; Brier, 1955; Byers, 1974). They showed that large-amplitude 7-day periodicities in rainfall and other meteorological variables, though not common, had occurred during the period 1899–1951. Thus they felt the rainfall periodicity was due to *natural variability* rather than to a direct consequence of cloud seeding.

Convinced that cloud seeding was a miraculous cure to all of nature's evils, Langmuir and his colleagues carried out a trial seeding experiment of a hurricane with the hope of altering the course of the storm or reducing its intensity. On October 10, 1947, a hurricane was seeded off the east coast of the United States. About 102 kg of dry ice was dropped in clouds in the storm. Due to logistical reasons, the eyewall region and the dominate spiral band were not seeded. Observers interpreted visual observations of snow showers as evidence that seeding had some effect on cloud structure. Following seeding, the hurricane changed direction from a northeasterly to a westerly course, crossing the coast into Georgia. The change in course may have been a result of the storm's interaction with the larger-scale flow field. Nonetheless, General Electric Corporation became the target of lawsuits for damage claims associated with the hurricane.

While the main focus of research during Project Cirrus was the dry ice and silver iodide seeding of supercooled clouds, some theoretical and experimental effort was directed toward stimulated rain formation in non-freezing clouds or what we will refer to as warm clouds. In 1948, Langmuir (1948) published his theoretical study of rain formation by chain reaction. According to his theory, once a few raindrops grew by colliding and coalescing with smaller drops to such a size that they would break up, the fragments they produced would serve as embryos for further growth by collection. The smaller-sized embryos would then ascend in the cloud updrafts while growing by collection and also break up creating more raindrop embryos. Langmuir hypothesized that insertion of only a few raindrops in a cloud could infect the cloud with raindrops through the chain-reaction process. Some attempts were made to initiate rain in warm clouds by water-drop seeding in Puerto Rico, though no suitable clouds were found. Subsequently Braham *et al.* (1957) and others at the University of Chicago demonstrated that one could initiate rainfall by water-drop seeding. This experiment will be discussed more fully in a later section.

In summary, Project Cirrus launched the United States and much of the world into the age of cloud seeding. The impact of this project on the science of cloud seeding, cloud physics research, and the entire field of atmospheric science was similar to the effects of the launching of Sputnik on the United States aerospace industry.

2

The glory years of weather modification

2.1 Introduction

The exploratory cloud seeding experiments performed by Langmuir, Schaefer, and Project Cirrus personnel fueled a new era in weather modification research as well as basic research in the microphysics of precipitation processes, cloud dynamics, and small-scale weather systems, in general. At the same time commercial cloud seeding companies sprung up worldwide practicing the art of cloud seeding to enhance and suppress rainfall, dissipate fog, and decrease hail damage. Armed with only rudimentary knowledge of the physics of clouds and the meteorology of small-scale weather systems, these weather modification practitioners sought to alleviate all the symptoms of undesirable weather by prescribing cloud seeding medication. The prevailing view was "cloud seeding is good!"

Scientists were now faced with the major challenge of proving that cloud seeding did indeed result in the enhancement of precipitation or produce some other desired response, as well as unravel the intricate web of physical processes responsible for both natural and artificially stimulated rainfall. We, therefore, entered the era where scientists had to get down in the trenches and sift through every little piece of physical evidence to unravel the mysteries of cloud microphysics and precipitation processes.

As the science of weather modification developed, two schools of cloud seeding methodology emerged. One school embraced what is called the *static mode* of seeding while the other is called the *dynamic mode* of seeding. In the next few sections, we will review these two approaches including the application of cloud seeding to hail suppression, hurricane modification, and precipitation enhancement in warm clouds.

2.2 The static mode of cloud seeding

We have seen that the pioneering experiments of Schaefer and Langmuir suggested that the introduction of dry ice or silver iodide into supercooled clouds could initiate a precipitation process. The underlying concept behind the *static mode* of cloud seeding is that natural clouds are deficient in ice nuclei. (For an excellent, more technical review of static seeding, see Silverman (1986).)

As a result many clouds contain an abundance of supercooled liquid water which represents an underutilized water resource. Supercooled clouds are thus viewed to be inefficient in precipitation formation, where precipitation efficiency is defined as the ratio of the rainfall rate or flux of rainfall on the ground to the flux of water substance entering the base of a cloud. The major focus of the static mode of cloud seeding is to increase the precipitation efficiency of a cloud or cloud system.

In its simplest form the static mode of cloud seeding was based on the Bergeron–Findeisen concept in which ice crystals nucleated either naturally or through seeding in a water-saturated supercooled cloud will grow by vapor deposition at the expense of cloud droplets. Figure 2.1 illustrates schematically the Bergeron–Findeisen process. Seeding therefore can convert a naturally inefficient cloud containing supercooled cloud droplets into a precipitating cloud in which the precipitation is in the form of vapor-grown ice crystals or raindrops formed from melted ice crystals. The "seedability" of a cloud is thus primarily a function of the availability of supercooled water. Because laboratory cloud chambers predicted that natural ice nuclei concentrations increased exponentially with the degree of supercooling (i.e., degrees colder than 0 °C) and because the amount of water vapor available for condensation increases with temperature, it was generally believed that the availability of supercooled water was greatest at warm temperatures, or between 0 °C and −20 °C.

Cloud seeding experiments and research on the basic physics of clouds during the 1950s through the early 1980s revealed that this simple concept of static

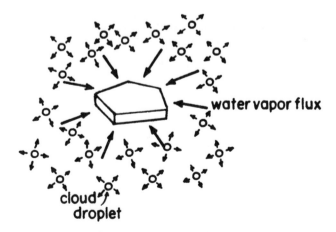

Figure 2.1 Schematic illustration of the Bergeron–Findeisen process.

seeding is only applicable to a limited range of clouds. It was found that in many supercooled clouds, the primary natural precipitation process was not growth of ice crystals by vapor deposition but growth of precipitation by collision and coalescence, or collection (see Fig. 2.2). It was found that clouds containing relatively low concentrations of cloud condensation nuclei (CCN) were more likely to produce rain by collision and coalescence among cloud droplets than clouds containing high concentrations of CCN. If a cloud condenses a given amount of supercooled liquid water, then a cloud containing low CCN concentrations will produce fewer cloud droplets than a cloud containing high CCN concentrations. As a result, in a cloud containing fewer cloud droplets, the droplets will be bigger on the average and fall faster than a cloud containing numerous, slowly settling cloud droplets. Because some of the bigger cloud droplets will settle through a population of smaller droplets more readily in a cloud containing low CCN concentrations, a cloud containing low CCN concentrations is more likely to initiate a precipitation process by collision and coalescence among cloud droplets than a cloud with a high CCN concentration. Generally clouds forming in a maritime airmass have lower concentrations of CCN than clouds forming in continental regions, often differing by an order of magnitude or more, and in polluted air masses the CCN concentrations can be 40 times that found in a clean maritime airmass.

It was also found that clouds having relatively warm cloud base temperatures were richer in liquid water content than clouds having cold cloud base temperatures. This is because the saturation vapor pressure increases exponentially with temperature. As a result clouds with warm cloud base temperatures have much more water vapor entering cloud base available to be condensed in the upper

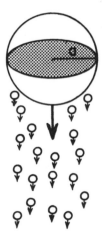

Figure 2.2 Illustration of growth of a drop by colliding and coalescing with smaller, slower-settling cloud droplets. From Cotton (1990).

levels of the cloud than a cloud with cold base temperatures. What this means is that clouds forming in a maritime airmass with low CCN concentrations and having warm cloud base temperatures have a high potential of being very efficient natural rain producers by collision and coalescence of cloud droplets.

The collision and coalescence process is not limited to just liquid drops colliding with liquid drops. Once ice crystals become large enough and begin to settle through a cloud of small supercooled droplets, the ice crystals can grow by collecting those droplets as they rapidly fall through a population of cloud droplets to form what we call *rimed* ice crystals or graupel particles (see Fig. 2.3). Frozen raindrops can also readily collide with supercooled cloud droplets to form hailstones or large graupel particles. The larger the liquid water content in clouds, the more likely that precipitation will form by one of the above collection mechanisms. Therefore, natural clouds can be far more efficient precipitation producers than would be expected from the simple concept of precipitation formation primarily by vapor growth of ice crystals.

Research during the same period revealed that laboratory ice nucleus counters were not always good predictors of ice crystal concentrations. Observations of ice crystal concentrations showed that in many clouds the observed ice crystal concentrations exceeded estimates of ice crystal concentrations by four to five orders of magnitude! The greatest discrepancies between observed ice crystal concentrations and concentrations diagnosed from ice nucleus counters occurred in clouds with relatively warm cloud top temperatures (i.e., warmer than $-10\,°C$) and those having significant concentrations of heavily rimed ice particles such as graupel and frozen raindrops. These are the clouds that contain relatively high liquid water contents and/or an active collection process. In other words, clouds that are warm-based and maritime are most likely to contain much higher ice crystal concentrations than ice nuclei concentrations. On the other hand, clouds in which ice crystal growth by vapor deposition prevails and in which riming

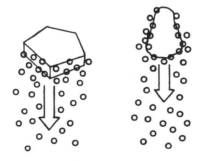

Figure 2.3 Riming of ice crystals or graupel particles.

is modest generally exhibit ice crystal concentrations comparable to ice nuclei concentrations.

The reasons for the discrepancy between ice crystal concentrations and ice nuclei concentrations are not fully understood today. Some researchers concluded from observational studies that temperature has little influence on the ice crystal concentrations (Hobbs and Rangno, 1985). Instead, it is argued that the droplet size distribution in clouds has the major controlling influence on ice crystal concentrations.

In recent years several laboratory experiments have revealed that under certain cloud conditions, ice crystal concentrations can be greatly enhanced by an ice multiplication process (Hallett and Mossop, 1974; Mossop and Hallett, 1974). The laboratory studies suggest that over the temperature range $-3\,°C$ to $-8\,°C$, copious quantities of secondary ice crystals are produced when an ice crystal or graupel particle collects or rimes supercooled cloud droplets. The secondary production of ice crystals is greatest when the supercooled cloud droplet population contains a significant number of large cloud droplets ($r > 12\,\mu m$). Figure 2.4 illustrates the rime-splinter secondary ice crystal production process. The presence of large cloud droplets would be greatest in clouds that are warm-based and maritime. Moreover, warm-based maritime clouds are more likely to contain supercooled raindrops which, when frozen, can serve as active sites for riming growth and secondary particle production. Thus, the Hallett–Mossop rime-splinter process is consistent with many field observations which suggest that clouds that are

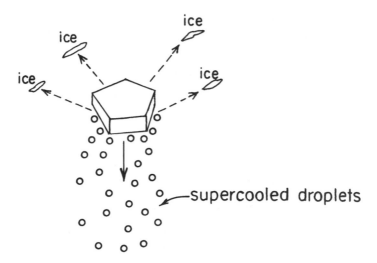

Figure 2.4 Illustration of secondary ice particle production by ice particle collection of supercooled cloud droplets at temperatures between $-4\,°C$ to $-8\,°C$. From Cotton (1990).

maritime and warm-based are more likely to contain ice crystal concentrations greatly in excess of ice nuclei concentrations.

The rime-splinter secondary ice crystal production process may not explain all the observations of high ice crystal concentrations relative to ice nuclei estimates but it is consistent with many of them. Still other processes not understood at this time may be operating in some cases of observed high ice crystal concentrations relative to ice nuclei concentrations.

The implication of these physical studies is that the "window of opportunity" for precipitation enhancement by cloud seeding is much smaller than was originally thought. Clouds that are warm-based and maritime have a high natural potential for producing precipitation. On the other hand, clouds that are cold-based and continental have reduced natural potential for precipitation formation and, hence, the opportunity for precipitation enhancement by cloud seeding is much greater, although the total water available would be less than in a warm-based cloud.

This is consistent with the results of field experiments testing the static seeding hypothesis. The Israeli I and II Experiments were quite successful in producing positive yields of precipitation in seeded clouds (Gagin and Neumann, 1981). The clouds that were seeded over Israel had relatively cold bases (5–8 °C) and were generally continental such that there was little evidence of a vigorous warm rain process or the presence of large quantities of heavily rimed graupel particles. Other cloud seeding experiments were not so fortunate and either no effects of seeding or even decreases in precipitation were inferred (Tukey *et al.*, 1978; Kerr, 1982). Presumably the opportunities for vigorous warm rain processes and secondary ice particle formation were greater in those clouds. In those clouds, seeding could not compete effectively with natural precipitation formation processes or natural precipitation processes masked the seeding effects so that they could not be separated from the *natural variability* of precipitation.

A number of observational and theoretical studies have also suggested that there is a cold temperature "window of opportunity" as well. Studies of both orographic and convective clouds have suggested that clouds colder than −25 °C have sufficiently large concentrations of natural ice crystals that seeding can either have no effect or even reduce precipitation (Grant and Elliot, 1974; Gagin and Neumann, 1981; Gagin *et al.*, 1985; Grant, 1986). It is possible that seeding such cold clouds could reduce precipitation by creating so many ice crystals that they compete for the limited supply of water vapor and result in numerous, slowly settling ice crystals which evaporate before reaching the ground. Such clouds are said to be *overseeded*.

There are also indications that there is a warm temperature limit to seeding effectiveness (Grant and Elliot, 1974; Gagin and Neumann, 1981; Cooper and Lawson, 1984). This is believed to be due to the low efficiency of ice crystal production by silver iodide at temperatures approaching $-4\,°C$ and to the slow rates of ice crystal vapor deposition growth at warm temperatures. Thus there appears to be a "temperature window" of about $-10\,°C$ to $-25\,°C$ where clouds respond favorably to seeding (i.e., exhibit seedability).

There also seems to be a "time window" of opportunity for seeding in many clouds; especially convective clouds. It is well known that the life cycle of convective clouds is significantly affected by the entrainment of dry environmental air. As dry environmental air is entrained into cumuli, cloud droplets formed in moist updraft air are evaporated, causing cooling that then forms downdrafts which terminate the life cycle of the cumuli. It was found during the HIPLEX-1 Experiment (Cooper and Lawson, 1984) that the timescale for which sufficient supercooled cloud water was available for seeding in ordinary cumuli was less than 14 minutes. In towering cumuli and small cumulonimbi, the timescale is not so limited by entrainment as in smaller cumuli, but those clouds are more likely to produce precipitation naturally thus competing for the available cloud water. The time window of opportunity in larger cumuli is therefore much more variable and uncertain since it depends not only on dynamic timescales which control entrainment, but also on the timescales of natural precipitation formation.

Physical studies and inferences drawn from statistical seeding experiments suggest there exists a more limited window of opportunity for precipitation enhancement by the static mode of cloud seeding than originally thought. The window of opportunity for cloud seeding appears to be limited to:

(1) clouds that are relatively cold-based and continental;
(2) clouds having top temperatures in the range $-10\,°C$ to $-25\,°C$;
(3) a timescale limited by the availability of significant supercooled water before depletion by entrainment and natural precipitation processes.

This limited scope of opportunities for rainfall enhancement by the static mode of cloud seeding that has emerged in recent years may explain why some cloud seeding experiments have been successful while others have yielded inferred reductions in rainfall from seeded clouds or no effect. A successful experiment in one region does not guarantee that seeding in another region will be successful unless all environmental conditions are replicated as well as the methodology of seeding. This, of course, is highly unlikely.

We must also recognize that implementing a seeding experiment or operational program that operates only in the above listed windows of opportunity is extremely difficult and costly. It means that in a field setting we must forecast the top temperatures of clouds to assure that they fall within the $-10\,°C$ to $-25\,°C$ temperature window. We must determine the extent to which clouds are maritime and warm-based, versus continental and cold-based, or the likelihood that clouds will naturally contain broad droplet spectra and an active warm-rain process. Because there are no routine measurements of cloud condensation nuclei spectra nor forecast models with a demonstrated skill in predicting the particular modes of precipitation formation, the potential of successfully implementing a seeding strategy in the field in which consideration is given to the natural widths of cloud droplet spectra and to the natural modes of precipitation formation is not very good. Furthermore, consideration of a time window complicates implementation of an operational seeding strategy even more. Seeding material would have to be targeted at the right time in a cumulus cloud before either entrainment depletes the available supercooled water or natural precipitation processes deplete the available liquid water. This would require airborne delivery of seeding material, which is expensive, and a prediction of the timescales of liquid water availability in clouds of differing types.

The success of a cloud seeding experiment or operation, therefore, requires a cloud forecast skill that is far greater than currently in use. As a result, such experiments or operations are at the mercy of the *natural variability* of clouds. The impact of *natural variability* may be reduced in some regions where the local climatology favors clouds that are in the appropriate temperature windows and are more continental. The time window will still exist, however, and this will yield uncertainty to the results unless the field personnel are particularly skillful in selecting suitable clouds.

Orographic clouds are less susceptible to a time window as they are steady clouds and offer a greater opportunity for successful precipitation enhancement than cumulus clouds. A "time window" of a different type does exist for orographic clouds which is related to the time it takes a parcel of air to condense to form supercooled liquid water and ascend to the mountain crest. If winds are weak, there may be sufficient time for natural precipitation processes to occur efficiently. Stronger winds may not allow efficient natural precipitation processes but seeding may speed up precipitation formation. Even stronger winds may not provide enough time for seeded ice crystals to grow to precipitation before being blown over the mountain crest and evaporating in the sinking subsaturated air to the lee of the mountain. A time window related to the ambient winds, however, is much easier to assess in a field setting than the time window in cumulus clouds.

In summary, the static mode of cloud seeding has been shown to cause the expected alterations in cloud microstructure including increased concentrations of ice crystals, reductions of supercooled liquid water content, and more rapid production of precipitation elements in both cumuli (Cooper and Lawson, 1984) and orographic clouds (Reynolds and Dennis, 1986; Reynolds, 1988; Super and Boe, 1988; Super and Heimbach, 1988; Super *et al.*, 1988). The documentation of increases in precipitation on the ground due to static seeding of cumuli, however, has been far more elusive with the Israeli experiment (Gagin and Neumann, 1981) providing the strongest evidence that static seeding of cold-based, continental cumuli can cause significant increases of precipitation on the ground. The evidence that orographic clouds can cause significant increases in snowpack is far more compelling, particularly in the more continental and cold-based orographic clouds (Mielke *et al.*, 1981; Super and Heimbach, 1988).

But even these conclusions have been brought into question. The Climax I and II wintertime orographic cloud seeding experiments (Grant and Mielke, 1967; Chappell *et al.*, 1971; Mielke *et al.*, 1971, 1981) are generally acknowledged by the scientific community (National Academy of Sciences, 1975; Tukey *et al.*, 1978) for providing the strongest evidence that seeding those clouds can significantly increase precipitation. Nonetheless, Rangno and Hobbs (1987, 1993) question both the randomization techniques and the quality of data collected during those experiments and conclude that the Climax II experiment failed to confirm that precipitation can be increased by cloud seeding in the Colorado Rockies. Even so, Rangno and Hobbs (1987) did show that precipitation may have been increased by about 10% in the combined Climax I and II experiments. This should be compared, however, to the original analyses by Grant *et al.* (1969) and Mielke *et al.* (1970, 1971) which indicated greater than 100% increase in precipitation on seeded days for Climax I and 24% for Climax II. Subsequently, Mielke (1995) explained a number of the criticisms made by Rangno and Hobbs regarding the statistical design of the experiments, in particular the randomization procedures, the quality and selection of target and control data, and the use of 500 mb temperature as a partitioning criteria. It is clear that the design, implementation, and analysis of this experiment was a learning process not only for meteorologists but statisticians as well.

The results of the many reanalyses of the Climax I and II experiments have clearly "watered down" the overall magnitude of the possible increases in precipitation in wintertime orographic clouds. Furthermore, they have revealed that many of the concepts that were the basis of the experiments are far too simplified compared to what we know today. Furthermore, many of the cloud systems seeded were not simple "blanket-type orographic clouds" but were part of major wintertime cyclonic storms that pass through the region. As such, there was a greater opportunity for ice multiplication processes and riming processes to be operative in those storms, making them less susceptible to cloud seeding.

Another problem with ground-based seeding of winter orographic clouds is that it depends on boundary layer transport and dispersion of the seeding material. Often the generators are located in valleys in order to facilitate access and maintenance of the generators. There strong inversions can trap the seeding material in the valleys preventing it from becoming entrained into the clouds in the higher terrain. Chemical analysis of chemical tracers in snow that are concurrently released with the seeding material (Warburton *et al.*, 1985, 1995a,b) have revealed that most of the precipitation falling in the targets during seeded periods does not contain seeding material and thereby suggests that seeding did not impact those sites, assuming that the absence of the seeding chemical in the snowfall can be used for making such a deduction. This problem can be minimized by having more generators and using modern remotely operated telemetered generators, placing the generators out of the valleys in higher terrain.

Two other randomized orographic cloud seeding experiments, the Lake Almanor Experiment (Mooney and Lunn, 1969) and the Bridger Range Experiment (BRE) as reported by Super and Heimbach (1983) and Super (1986) suggested positive results. However, these particular experiments used high-elevation silver iodide generators, which increases the chance that the silver iodide plumes get into the supercooled clouds. Moreover, both experiments provided physical measurements that support the statistical results (Super, 1974; Super and Heimbach, 1983, 1988). Using trace chemistry analysis of snowfall for the Lake Almanor project, Warburton *et al.* (1995a) found particularly good agreement with earlier statistical suggestions of seeding-induced snowfall enhancement with cold westerly flow. They concluded that failure to produce positive statistical results with southerly flow cases was likely related to seeding mistargeting of the seeded material. These two randomized experiments strongly suggest that higher-elevation seeding in mountainous terrain can produce meaningful seasonal snowfall increases.

We noted above, that the strongest evidence of significant precipitation increases by static seeding of cumulus clouds came from the Israeli I and II experiments. Even these experiments have come under attack by Rangno and Hobbs (1995). From their reanalysis of both the Israeli I and II experiments, they argue that the appearance of seeding-caused increases in rainfall in the Israeli I experiment was due to "lucky draws" or a Type I statistical error. Furthermore, they argued that during Israeli II, naturally heavy rainfall over a wide region encompassing the north target area gave the appearance that seeding caused increases in rainfall over the north target area. At the same time, lower natural rainfall in the region encompassing the south target area gave the appearance that seeding decreased rainfall over that target area.

Rosenfeld and Farbstein (1992) suggested that the differences in seeding effects between the north and south target areas during Israeli II is the result of the incursion of desert dust into the cloud systems. They argue that the desert dust contains more active natural ice nuclei and that they can also serve as coalescence embryos enhancing collision and coalescence among droplets. Together, the dust can make the clouds more efficient rain-producers and less amenable to cloud seeding.

We argued above that the "apparent" success of the Israeli seeding experiments was due to the fact that they are more susceptible to precipitation enhancement by cloud seeding. This is because numerous studies (Gagin, 1971, 1975, 1986; Gagin and Neumann, 1974) have shown that the clouds over Israel are continental having cloud droplet concentrations of about $1000\,cm^{-3}$ and that ice particle concentrations are generally small until cloud top temperatures are colder than $-14\,°C$. Furthermore, there is little evidence for ice particle multiplication processes operating in those clouds.

Rangno and Hobbs (1995) also reported on observations of clouds over Israel containing large supercooled droplets and quite high ice crystal concentrations at relatively warm temperatures. In addition, Levin *et al.* (1996) presented evidence of active ice multiplication processes in Israeli clouds. This further erodes the perception that the clouds over Israel were as susceptible to seeding as originally thought. Naturally, the Rangno and Hobbs (1995) paper generated quite a large reaction in the weather modification community. The March issue of the *Journal of Applied Meteorology* contained a series of comments and replies related to their paper (Ben-Zvi, 1997; Dennis and Orville, 1997; Rangno and Hobbs, 1997a,b,c,d; Rosenfeld, 1997; Woodley, 1997). These comments and responses clarify many of the issues raised by Rangno and Hobbs (1995). Nonetheless, the image of what was originally thought of as the best example of the potential for precipitation enhancement of cumulus clouds by static seeding has become considerably tarnished.

Ryan and King (1997) presented a comprehensive overview of over 47 years of cloud seeding experiments in Australia. These studies almost exclusively focused on the static seeding concept. In this water-limited country, cloud seeding has been considered as a potentially important contributor to water management. As a result their review included discussions of the overall benefits/costs to various regions.

Over 14 cloud seeding experiments were conducted covering much of southeastern, western, and central Australia as well as the island of Tasmania. Ryan and King (1997) concluded that static seeding over the plains of Australia is not effective. They argue that for orographic stratiform clouds, there is strong statistical evidence that cloud seeding increased rainfall, perhaps by as much as 30% over Tasmania when cloud top temperatures are between $-10\,°C$ and $-12\,°C$ in southwesterly airflow. The evidence that cloud seeding had similar effects in orographic clouds over the mainland of southeastern Australia is much weaker. This is somewhat surprising from a physical point of view since the clouds over Tasmania are maritime. As such one would expect the opportunities for warm-cloud collision and coalescence precipitation processes to be fairly large. Furthermore, in those maritime clouds ice multiplication processes should be operative; especially when embedded cumuliform cloud elements are present. Thus natural ice crystal concentrations should be competitive with concentrations expected from static seeding, especially in the $-10\,°C$ to $-12\,°C$ temperature range. If the results of the Tasmanian experiments are real, benefit/cost analysis suggests that seeding has a gain of about 13/1. This is viewed as a real gain to hydrologic energy production.

It is clear, however, that we still do not have the ability to produce statistically significant increases in surface precipitation from all supercooled cumuli or orographic clouds. At the very least we conclude that we do not yet have the ability to discriminate seeding-induced increases in surface precipitation from the background "noise" created by the high *natural variability* of surface precipitation for many cloud systems. The stronger evidence for positive seeding effects on orographic clouds versus cumuli is due in large measure to the lower *natural variability* of wintertime precipitation in orographic clouds than in summertime cumuli.

2.3 The dynamic mode of cloud seeding

2.3.1 Introduction

We have seen that the fundamental concept of the static mode of cloud seeding is that precipitation can be increased in clouds by enhancing their precipitation efficiency. While alterations in the dynamics or air motion in clouds due to latent heat release of growing ice particles, redistribution of condensed water, and evaporation of precipitation is inevitable with static mode seeding, it is not the primary aim of the strategy. By contrast, the focus of the *dynamic mode* of cloud seeding is to enhance the vertical air currents in clouds and thereby vertically process more water through the clouds resulting in increased precipitation. (For an excellent, more technical review of dynamic seeding see Orville (1986).) In this section we examine the concepts behind the dynamic mode strategy and discuss the physical/statistical evidence supporting the concept.

2.3.2 Fundamental concepts

We noted earlier that Langmuir postulated that the latent heat released as ice crystals grow by vapor deposition would warm the seeded part of a cloud and cause upward motion and turbulence. The concept is a simple one. As ice crystals grow by vapor deposition a phase transition takes place in which water vapor molecules deposit on an ice crystal lattice. During the phase transformation the latent heat of sublimation, 2.83×10^6 J kg^{-1}, is released, warming the immediate environment of the ice crystals. If the cloud contains cloud droplets, however, the growth of ice crystals causes the lowering of the cloud saturation pressure below water saturation, resulting in the evaporation of cloud droplets to restore the cloud to water saturation. The evaporation of cloud droplets absorbs the latent heat of vaporization or 2.50×10^6 J kg^{-1}, resulting in a net warming of the cloud of 0.33×10^6 J kg^{-1} for the vapor deposited on the ice crystals. Only if all condensed liquid water is evaporated and deposited on ice crystals will the cloud experience the full warming effects of the latent heat of sublimation.

Moreover, if supercooled cloud droplets or raindrops freeze by contacting an ice crystal or ice nuclei, the phase transformation from liquid to ice will release the latent heat of fusion or 0.33×10^6 J kg^{-1} of water frozen. In some instances, so much supercooled water may freeze that the cloud can become subsaturated with respect to ice causing the sublimation of ice crystals and partially negating the positive heat released by freezing.

Why is this heating important to clouds? Many clouds such as cumulus clouds are buoyancy-driven. When a small volume of air, which we shall call an air parcel, becomes warmer than its environment it expands and displaces a volume of environmental air equal to the weight of the warm air. According to Archimedes' principle, the warmed air will be buoyed up with a force that is equal to the weight of the displaced environmental air. This upward-directed buoyancy force will then accelerate a cloud parcel upwards similar to the upward acceleration one can experience in a hot air balloon when the air inside the balloon is heated with a propane burner. The simple addition of heat to atmospheric air parcels, however, does not guarantee that the air will become buoyant.

The buoyancy of a cloud is determined not only by how warm a cloud is with respect to its environment, but also by how much water is condensed in a cloud. Condensed water produces negative buoyancy, such that a cloud that is warmer than its environment can actually become negatively buoyant due to the load of condensed water it must carry. One consequence of a precipitation process is that it unloads the upper portions of a cloud from its burden of condensed water (see Fig. 2.5a). Unleashed from its burden of condensed water, the top of the cloud can penetrate deeper into the atmosphere. Of course, the water that settles from the upper part of the cloud transfers the burden of condensed water to lower levels, causing a weakening of updrafts or formation of downdrafts at lower levels.

Once the raindrops settle into the subsaturated, subcloud layer, they begin to evaporate. Evaporation of the raindrops absorbs latent heat from the surrounding air, thereby cooling the air. The denser, evaporatively chilled air sinks towards the surface, spreading horizontally as it approaches the ground (see Fig. 2.5). The dense, horizontally spreading air undercuts the warm, moist air, often elevating it to the lifting condensation level (LCL) and perhaps the level of free convection (LFC). Thus, the settling of raindrops below cloud base can promote

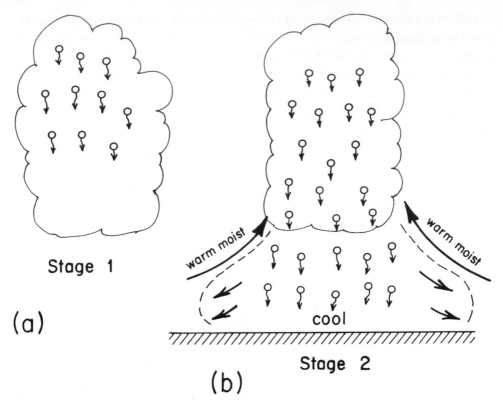

Figure 2.5 (a) Illustration of droplets settling from the upper levels of a cloud, thus reducing the amount of liquid water content or water-loading burden on the cloud. (b) Illustration of the formation of an evaporatively chilled layer near the surface which can lift surrounding moist air sometimes to the lifting condensation level (LCL) and level of free convection (LFC). From Cotton (1990).

the development of new cumulus clouds or help sustain existing ones by causing lifting of warm, moist air into the cloud base level.

Because towering cumulus clouds are taller than fair-weather cumulus clouds, they often extend to heights that are colder than 0 °C, or the freezing level. Before significant precipitation occurs, these clouds are called *cumulus congestus*. Ice particles can therefore form by either the freezing of supercooled drops or by nucleation on ice nuclei (IN). As far as the overall behavior of a cumulus cloud is concerned, the important consequence of droplet freezing and vapor deposition growth of ice crystals is that additional latent heat is added to the cloudy air. The latent heat liberated during the freezing and vapor deposition growth of ice particles therefore contributes to the buoyancy of the cloud, giving the cloud a boost in its vertical ascent. As a result, towering cumulus clouds often exhibit *explosive* vertical development once ice phase precipitation processes take place.

The taller cumulus clouds typically produce more rainfall and perturb the stably stratified environment more, thus producing gravity waves which may impact the development of other cumulus clouds (see Cotton and Anthes, 1989).

An important step in the transition of cumulus congestus clouds to thunderstorms or cumulonimbus clouds is the merger of a number of neighboring towering cumulus clouds. Figure 2.6 illustrates the merger of two cumulus clouds due to the interaction of low-level, cool outflows from neighboring clouds. As the merger process proceeds, a "bridge" of smaller cumuli forms between the two clouds. The bridge of clouds eventually deepens and fills the gap between the clouds, resulting in wider and often taller clouds. Clouds resulting from the merger generally produce larger rainfall rates, last longer, and are bigger, so that the volume of rainfall from merged clouds is sometimes a factor of ten or more greater than the sum of the rain volumes from similar, non-merged clouds (Simpson *et al.*, 1980).

There are many factors which influence the merger of cumulus clouds. We know that merger takes place more frequently in regions where there exists convergence of warm, moist air at low levels on a scale greater than the individual cumulus clouds. The convergence of warm, moist air provides the fuel necessary to sustain convection on the scale of the merged system.

Merger is often accompanied by the explosive growth of at least one of the neighboring clouds. Explosive growth of a cumulus cloud, perhaps due to the release of additional latent heat from the growth of ice particles, generally results in greater precipitation which, in turn, causes stronger subcloud cooling and outflow. Also, the more vigorously growing clouds create a region of low pressure beneath their bases, which draws warm, moist air into the cloud base, and perhaps along with it, draws in neighboring cumulus clouds. Moreover, explosively rising

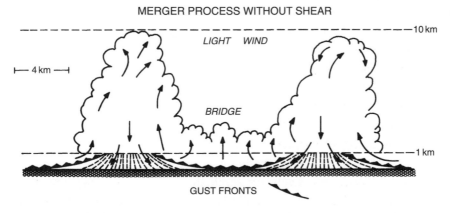

Figure 2.6 Schematic illustration relating downdraft interaction to bridging and merger in case of light wind and weak shear. From Simpson *et al.* (1980).

cumulus clouds perturb the stably stratified, surrounding environment, triggering gravity waves which can enhance the growth of some clouds and weaken others.

One may ask, if the latent heat released by freezing supercooled drops is only about one-eighth the latent heat released during the condensation of an equivalent mass of vapor onto droplets, why are we interested in its impact on cloud growth? The reason is that at cold temperatures where the ice phase becomes prevalent, the saturation vapor pressure with respect to water is relatively small and varies more slowly with temperature. As a result, as a cloud volume rises and becomes colder, the amount of water available to be condensed in a cloud and correspondingly the latent heat released becomes less and less. Moreover, unless the cloud is raining heavily, the water vapor that has condensed in the cloud to form water drops at warmer temperatures is available in large amounts for freezing. If this stored water is then frozen by seeding or spontaneously through natural ice nucleation processes, the cloud will experience a boost in buoyancy at precisely those levels where the latent heat liberated during condensation is lessened. In addition, since at colder temperatures the saturation vapor pressure becomes small in magnitude, the differences between environmental vapor pressures and the saturation vapor pressure in the cloudy region become smaller. As a result, entrainment of dry environmental air into the cloud causes less evaporative cooling and the consequences of entrainment are less of a brake on cloud vertical development. Thus the artificial stimulation of the ice phase in a cloud by seeding can cause a boost in the buoyancy of a cloud that is less likely to be destroyed by entrainment of dry environmental air.

All these factors must be considered when estimating whether or not the latent heat released by freezing or vapor deposition growth of ice crystals created by seeding will boost a cloud upwards in the atmosphere. We must examine the local environment or each individual sounding in the neighborhood of a cloud to see if it will support deep convection and if natural cloud vertical growth will be limited by a stable layer of inversion or by the effects of entrainment. Will the cloud experience a sufficient boost in buoyancy when seeded to overcome the effects of entrainment or a stable inversion layer so that its vertical growth will be enhanced? To answer these and other questions about cloud behavior, we must simulate the behavior of natural and seeded clouds on a computer.

The computer simulation of clouds involves the use of a mathematical or numerical model. Such a model simulates a cloud by solving or integrating a prescribed set of equations numerically on a computer. The earliest cloud models (and those used most extensively for simulating dynamic seeding) are based on the hypothesis that clouds behave similarly to buoyant laboratory thermals or jets (see Fig. 2.7). The laboratory studies suggest that thermals or jets are primarily buoyancy-driven, and that the rate of rise can be mathematically described in

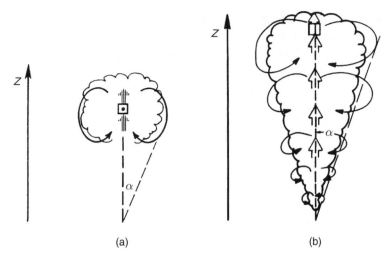

Figure 2.7 (a) Schematic view of the "bubble" or "thermal" model of lateral entrainment in cumuli. (b) Schematic view of the "steady-state jet" model of lateral entrainment in cumuli. Here Z represents distance (height) from the thermal source and α represents the half-angle of spread of the bubble or plume. From Cotton and Anthes (1989).

terms of the cloud buoyancy, vertical momentum, and the rate at which buoyancy is eroded as dry, cooler environmental air is entrained into the bubble or jet. Different entrainment laws were hypothesized for jets and thermals based on laboratory tank calibrations.

Application of these models to atmospheric clouds involves the use of a thermodynamic energy equation along with the vertical rise rate equation. The models are typically initialized with a local atmospheric sounding of temperature, relative humidity, and winds, as well as prescribing some initial ascent at an estimated or prescribed cloud base height. As illustrated by the square in Fig. 2.7a and b, a small parcel of air is then integrated upward while calculating the changes in cloud buoyancy and rise rate due to condensation of vapor, freezing of raindrops, and vapor deposition on ice crystals as well as removal of condensate products by precipitation. The calculations are terminated when the modeled cloud loses all positive buoyancy. To simulate the effects of dynamic seeding, the calculations are first done for a *natural* cloud in which natural ice nucleation processes are simulated, and then they are repeated for a seeded cloud in which enhanced ice particle nucleation is simulated for an assumed amount of seeding material. The difference in height between *natural* and *seeded* clouds is defined as the dynamic seeding potential or *seedability* of clouds that develop in such an environment.

Application of such models to the semi-tropical and tropical atmosphere often resulted in seedability predictions of 2–3 km, while in midlatitudes the predicted height changes due to seeding are generally less though some days exhibit large values of predicted explosive growth. Simple models such as these were used to support field experiments by predicting the potential for obtaining significant increases in cloud growth on a given day. They have been also used for identifying how effective seeding actually was. Figure 2.8 illustrates an example of values of seedability predicted versus the observed heights of both seeded and unseeded clouds. These results strongly suggest there is a significant difference between the heights of seeded versus unseeded clouds, at least in the semi-tropics.

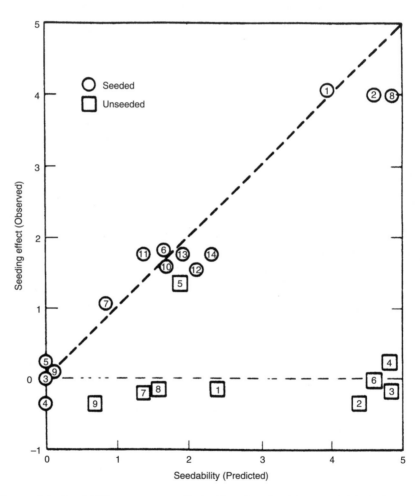

Figure 2.8 Seedability versus seeding effect for the 14 seeded (circles) and 9 control (squares) clouds studied in 1965. Note that seeded clouds lie mainly along a straight line with slope 1 (seeding effect is close to seedability), while control clouds lie mainly along a straight horizontal line (showing little or no seeding effect regardless of magnitude of seedability). Units of each axis are in kilometers. From Simpson *et al.* (1967).

The higher cloud top heights do not necessarily mean that the desired goal of greater rainfall on the ground has been achieved. It is generally well known that in a population of natural clouds, taller clouds produce more rain on the average (Gagin *et al.*, 1985). Because seeded clouds are altered microphysically, it does not necessarily follow that taller seeded clouds rain more. Some limited exploratory field experiments have been conducted that suggest that seeding clouds for dynamic effects can increase rainfall (Woodley, 1970). Woodley speculated that the seeded clouds were larger, longer lasting, and processed more moisture than their unseeded counterparts resulting in an increase in precipitation. Extensive area-wide, randomized statistical experiments have not been able to confirm the earlier exploratory studies (Dennis *et al.*, 1975; Woodley *et al.*, 1982a, 1983; Barnston *et al.*, 1983; Meitín *et al.*, 1984). The reasons for the failure of the confirmatory seeding experiments are not fully known but they may be due to: (1) the simple model relating increased cloud growth to enhanced surface rainfall may not work for all clouds and in some environments (i.e., certain wind shear profiles, some mesoscale weather regimes); (2) large *natural variability* of rainfall over fixed targets of large areal extent and inadequate models (physical or statistical) to account for that variability; and (3) the size of the sample of clouds seeded and not seeded was not large enough to accommodate the *natural variability* in rainfall (i.e., a single, heavy rainfall day swamped the natural rainfall statistics).

Lacking in the dynamic seeding research is an identification of the hypothesized chain of physical processes that lead to enhanced rainfall on the ground over a target region. Observations in clouds seeded for dynamic effects showed that seeding did indeed glaciate the clouds (convert the cloud from liquid to primarily ice) (Sax, 1976; Sax *et al.*, 1979; Sax and Keller, 1980; Hallett, 1981). The one-dimensional models clearly predict that artificial glaciation of a cloud should result in increased vertical development of the cloud. Those one-dimensional models, however, cannot simulate the consequences of increased vertical growth. A chain of physical responses to dynamic seeding has been hypothesized (Woodley *et al.*, 1982b) that includes:

(1) pressure falls beneath the seeded cloud towers and convergence of unstable air in the cloud will as a result develop;
(2) downdrafts are enhanced;
(3) new towers will therefore form;
(4) the cloud will widen;
(5) the likelihood that the new cloud will merge with neighboring clouds will therefore increase;
(6) increased moist air is processed by the cloud to form rain.

Few of these hypothesized responses to dynamic seeding have been observationally documented in any systematic way. A few exploratory attempts to identify some of the hypothesized links in the chain of responses were attempted using multiple Doppler radars, but they were largely unsuccessful since they occurred at a time that multiple Doppler technology was still in its infancy. Likewise, two- and three-dimensional numerical prediction models were applied to simulate dynamic effects. These models have the potential for simulating pressure perturbations caused by seeding throughout the cloud, as well as the formation of downdrafts, new towers, cloud merger, and increased rainfall. Only a few attempts were made to simulate dynamic seeding with multidimensional cloud models (Orville and Chen, 1982; Levy and Cotton, 1984) but these simulations did not produce the hypothesized sequence of responses including enhanced rainfall on the ground. This could have been due to the inadequacies of the models at that time, or to the fact that the soundings selected were not ideal for dynamic responses, or the chain of hypothesized events did not occur. More research is needed to determine which is indeed the case.

In recent years the dynamic seeding strategy has been applied to Thailand and West Texas. Results from exploratory dynamic seeding experiments over west Texas have been reported by Rosenfeld and Woodley (1989, 1993). Analysis of the seeding of 183 convective cells suggests that seeding increased the maximum height of the clouds by 7%, the areas of the cells by 43%, the durations by 36%, and the rain volumes of the cells by 130%. Overall the results are encouraging but such small increases in vertical development of the clouds is hardly consistent with earlier exploratory seeding experiments.

As a result of their experience in Texas, Rosenfeld and Woodley (1993) proposed an altered conceptual model of dynamic seeding as follows:

(1) NONSEEDED STAGES

(i) *Cumulus growth stage*
The freezing of supercooled raindrops plays a major role in the revised dynamic seeding conceptual model. Therefore, a suitable cloud is one that has a warm base and a vigorous updraft that is strong enough to carry any raindrops that are formed in the updraft above the $0\,°C$ isotherm level. Such a cloud has a vast reservoir of latent heat that is available to be tapped by natural processes or by seeding.

(ii) *Supercooled rain stage*
At this stage a significant amount of supercooled cloud and rainwater exists between the $0°$ and the $-10\,°C$ levels, which is a potential energy source for future cloud growth.

A cloud with active warm rain processes but a weak updraft will lose most of the water from its upper regions in the form of rain before growing into the supercooled region. Therefore, only a small amount of water remains in the supercooled region for the conversion to ice. Such a cloud has no dynamic seeding potential.

(iii) *The cloud-top rain-out stage*

If the updraft is not strong enough to sustain the rain in the supercooled region until it freezes naturally, most of it will fall back toward the warmer parts of the cloud without freezing. The supercooled water that remains will ultimately glaciate. The falling rain will load the updraft and eventually suppress it, cutting off the supply of moisture and heat to the upper regions of the cloud, thus terminating its vertical growth. This is a common occurrence in warm rain showers from cumulus clouds.

(iv) *The downdraft stage*

At this stage, the rain and its associated downdraft reach the surface, resulting in a short-lived rain shower and gust front.

(v) *The dissipation stage*

The rain shower, downdraft, and convergence near the gust front weaken during this stage, lending no support for the continued growth of secondary clouds, which may have been triggered by the downdraft and its gust front.

(2) SEEDED STAGES

(i) *Cumulus growth and supercooled rain*

These stages are the same for the seeded sequence as they are for natural processes.

(ii) *The glaciation stage*

The freezing of the supercooled rain and cloud water near the cloud top at this stage may occur either naturally or be induced artificially by glaciogenic seeding. This conceptual model is equally valid for both cases.

The required artificial glaciation is accomplished at this stage through intensive, on-top seeding of the updraft region of a vigorous supercooled cloud tower using a glaciogenic agent (e.g., AgI). The seeding rapidly converts most of the supercooled water to ice during the cloud's growth phase. The initial effect is the formation of numerous small ice crystals and frozen raindrops.

This rapid conversion of water to ice releases fusion heat—faster and greater for the freezing of raindrops—which acts to increase tower buoyancy and updraft and, potentially, its top height. The magnitude of the added buoyancy is modified by the depositional heating or cooling that may occur during the adjustment to ice saturation; see Orville and Hubbard (1973). Entrainment is likely enhanced in conjunction with the invigorated cloud circulation.

The frozen water drops continue to grow as graupel as they accrete any remaining supercooled liquid water in the seeded volume and/or when they fall into regions of high supercooled liquid water content. These graupel particles will grow faster and stay aloft longer because their growth rate per unit mass is larger and their terminal fall velocity is smaller than water drops of comparable mass. This will cause the tower to retain more precipitation mass in it upper portions. Some or all of the increase cloud buoyancy from seeding will be needed to overcome the increased precipitation load.

If the buoyancy cannot compensate for the increased loading, however, the cloud will be destroyed by the downdraft that contains the ice mass. The downdraft will be augmented further by cooling from the melting of the ice hydrometeors just below the freezing level.

The retention of the precipitation mass in the cloud's upper portions delays the formation of the precipitation-induced downdraft and the resultant disruption of the updraft circulation beneath the precipitation mass. This delay allows more time for the updraft to feed additional moisture into the growing cloud.

(iii) *The unloading stage*

The greater precipitation mass in the upper portion of the tower eventually moves downward along with the evaporatively cooled air that was entrained from the drier environment during the tower's growth phase. When the precipitation descends through the updraft, it suppresses the updraft. If the invigorated pulse of convection has had increased residence time in regions of light to moderate wind shear, however, the precipitation-induced downdraft may form adjacent to the updraft, forming an enhanced updraft-downdraft couplet. This unloading of the updraft may allow the cloud a second surge of growth to cumulonimbus stature.

When the ice mass reaches the melting level, some of the heat released in the updraft during the glaciation process is reclaimed as cooling in the downdraft. This downrush of precipitation and cooled air enhances the downdraft and the resulting outflow beneath the tower.

(iv) *The downdraft and merger stage*

The precipitation beneath the cloud tower is enhanced when the increased water mass reaches the surface. In addition, the enhancement of the downdraft increases the convergence at its gust front.

(v) *The mature cumulonimbus stage*

The enhanced convergence acts to stimulate more neighboring cloud growth, some of which will also produce precipitation, leading to an expansion of the cloud system and its conversion to a fully developed cumulonimbus system.

When this process is applied to one or more suitable towers residing within a convective cell as viewed by radar, greater cell area, duration, and rainfall are the result. Increased echo-top height is a likely but not a necessary outcome of the seeding, depending on how much of the seeding-induced buoyancy is needed to overcome the increased precipitation loading.

(vi) *The convective complex stage*

When seeding is applied to towers within several neighboring cells, increased cell merging and growth will result, producing a small mesoscale convective system and greater overall rainfall.

This is an idealized sequence of events. Dissipation may follow the glaciation stage or at any subsequent stage if the required conditions are not present.

Figure 2.9 illustrates their revised conceptual model of dynamic seeding. This conceptual model differs from the earlier one in that it emphasizes the conversion of liquid water into graupel particles which fall slower and grow faster than water drops of comparable mass. The seeding-induced graupel particles will reside in the cloud updraft longer and achieve greater size than a population of water drops in a similar unseeded cloud. They explain the lack of enhanced vertical development of the seeded clouds to increased precipitation mass loading. The enhanced thermal

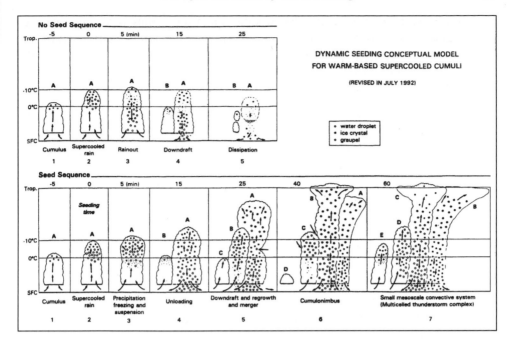

Figure 2.9 Diagrammatic illustration of the dynamic seeding conceptual model for warm-based supercooled cumuli. SFC, surface. Revised as of July 1992. From Rosenfeld and Woodley (1993).

buoyancy of the cloud due to seeding-induced ice phase conversion, they argue, is offset by the increased mass loading which results in only modest increases in updraft strength and cloud top height.

This new concept emphasizes that rapid conversion of supercooled liquid water into graupel must take place in the seeded plume. As such, it is limited to rather warm-based, maritime clouds having a broad cloud droplet distribution and supercooled raindrops. Numerous modeling studies have shown that the speed of conversion of supercooled liquid water to ice is facilitated by the presence of supercooled raindrops (Cotton, 1972a,b; Koenig and Murray, 1976; Scott and Hobbs, 1977; Lamb *et al.*, 1981). The supercooled raindrops readily collect the ice crystals nucleated by the seeding agent and freeze. The frozen raindrops then collect cloud droplets becoming low-density graupel particles if the liquid water content of the cloud is low or modest, or become high-density hailstones if the liquid water contents are rather large.

Rosenfeld and Woodley (1993) argue that the retention of the increased ice mass in the form of graupel is an important new aspect of their dynamic seeding conceptual model. This may delay the formation of a downdraft and allows more time for further growth of the cloud. The eventual unloading of the enhanced water mass, they argue, is favorable for subsequent regeneration of the cloud by the downdraft-induced gust fronts leading to larger, longer-lived cells.

As pointed out by Silverman (2001), however, application of the revised hypothesis in Thailand (Woodley *et al.*, 1999a,b) did not yield a statistically significant increase in rainfall, and the rainfall and enhanced downdraft presumably produced by it did not appear to be delayed (Woodley *et al.*, 1999b). Moreover, the differences between seed and no-seed rain volumes increased with time after 2 hours and reached a maximum at 8 hours. Such a long timescale for a response to seeding, if real, will require further modifications to the seeding hypothesis.

Analysis of an operational cloud seeding program in Texas by Woodley and Rosenfeld (2004) provides some optimism that dynamic seeding increases rainfall. By defining floating targets with NEXRAD radar data for lifetimes from first echo to the disappearance of all echoes, they superimpose the track and seeding actions of the project seeder aircraft and objectively define seeded (S) and non-seeded (NS) analysis units. They estimated that seeding increased rainfall by 50% and volumetric rainfall by 3000 acre feet. These responses were found outside of the operational target area and within 2 hours following seeding.

In summary, the concept of dynamic seeding is a physically plausible hypothesis that offers the opportunity to increase rainfall by much larger amounts than simply enhancing the precipitation efficiency of a cloud. It is a much more complex hypothesis, however, requiring greater quantitative understanding of the behavior of cumulus clouds and their interaction with each other, with the boundary layer, and with larger-scale weather systems. Fundamental research in dynamic seeding essentially terminated at the time that new cloud observing tools and multidimensional cloud models became available to the community. Systematic application of these tools to the study of dynamic seeding would help in evaluating if the hypothesized chain of events is possible and perhaps the conditions under which such responses are likely to occur.

2.4 Modification of warm clouds

2.4.1 Introduction

Attempts to augment precipitation by cloud seeding has not been limited to supercooled clouds. In tropical and semi-tropical regions, in particular, clouds too shallow to extend into freezing levels are prevalent. During drought periods, if any clouds form at all, they are generally warm, non-precipitating clouds. The motivation is therefore strong to develop techniques for extracting rainfall from non-ice-phase clouds.

In this section we review the basic physical concepts governing the formation of precipitation in warm clouds. We then discuss the various concepts for enhancing precipitation in warm clouds and describe the physical/statistical experiments and modeling studies attempting to refine and develop warm cloud seeding strategies.

2.4.2 Basic physical concepts of precipitation formation in warm clouds

The process of precipitation formation in warm clouds begins with the nucleation of cloud droplets on hygroscopic aerosol particles. These are airborne dust particles that are normally quite small (i.e., less than $1\,\mu m$ in diameter) and have a natural affinity for water vapor. Examples are salts such as ammonium sulfate and sodium chloride particles. A rising volume of cloud-free air will cool dry adiabatically resulting in an increase in relative humidity. At relative humidities greater than about 78%, hygroscopic particles begin taking on water vapor and swell in size. Eventually the relative humidity will exceed 100% and we say the cloud is *supersaturated*. In a supersaturated environment the hygroscopic aerosol particles (CCN) will allow deposition of water vapor on the particle surface eventually forming a cloud droplet. As long as supersaturation in a cloud is maintained (generally by continued ascent and adiabatic cooling) vapor will deposit on the surface of the droplet, allowing the cloud droplet to grow bigger (see Fig. 2.10). Because vapor deposition occurs on the surface of cloud droplets, the rate of growth (in terms of droplet radius) diminishes as the droplet gets bigger. This is because the surface-to-volume ratio of a nearly spherical droplet gets less the bigger the droplet. That is, the surface area of the droplet diminishes relative to the amount of vapor mass that must be added to the droplet to make it expand. Cloud droplets therefore grow quite rapidly to a size of 4 to $12\,\mu m$ in radius, but then grow very slowly to radii exceeding $20\,\mu m$. In fact, for a droplet to grow by vapor deposition to raindrop size ($1000\,\mu m$ or $1\,mm$) in a smooth ascending

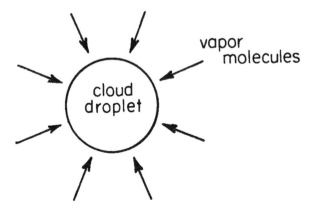

Figure 2.10 Schematic illustration of a droplet growing by deposition of vapor molecules.

updraft takes more than a day. Vapor deposition is, therefore, not the major process causing the formation of warm rain.

Instead, warm rain formation is dominated by collision and coalescence among droplets. Collision and coalescence refers to the process in which a large drop settling through the air at a high terminal velocity overtakes a small drop with a smaller settling velocity. Figure 2.2 illustrates a large drop that is settling through a population of smaller droplets. The large drop sweeps out a cross-sectional area (πa^2) indicated by the shaded region. Thus the larger the drop relative to a smaller drop, the greater will be its fall velocity through the air relative to a smaller drop, and the greater the cross-sectional area the large drop will sweep out. Likewise the more numerous the smaller drops and the more mass they collectively have (i.e., the higher the liquid water content or mass of water per unit mass of air), the faster will the larger drop grow by colliding and coalescing with smaller droplets.

Unfortunately the problem is a bit more complicated since not all the drops in front of the larger drop shown in Fig. 2.2 actually collide and coalesce with the larger droplet. As illustrated in Fig. 2.11, streamlines of the air flowing about a drop are quite curved near the drop surface. Some smaller drops near the center line of the big drop depart slightly from the airflow streamline and make an impact with the larger drop. On the other hand, drops further from the center line of the bigger drop also follow the airflow streamline, departing from it only slightly, and are not able to collide with the larger drop. The likelihood that a small drop will cross streamlines and impact the larger drop is greater the larger the smaller drop. This likelihood of a collision is also greater if the larger drop falls faster through the air.

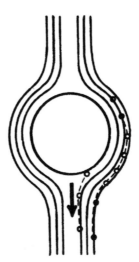

Figure 2.11 Schematic of streamlines of airflow around a falling droplet.

Moreover, not all the drops that make impact with a smaller drop actually coalesce with the bigger drop. If two drops are of similar size and thus settle through the air relatively slowly, their impact speeds may not be large enough to break the surface tension that keeps a drop intact or squeeze out any air that may get trapped between the drops. While we don't know the numbers precisely, the indications are that a large fraction of drops smaller than 100 μm that make impact actually coalesce. Larger raindrops of similar size, on the other hand, have a higher probability of colliding but not coalescing.

Summarizing all these complicated effects, we find that if all the drops in a cloud are very small, say less than 15 μm in radius, the rate of collision and coalescence among those drops is almost negligible. This is because the relative velocities among the drops is small, the cross-sectional areas swept out by the slightly bigger drops relative to the small ones is also small, and the efficiency of collision and coalescence is nearly zero. On the other hand once there exists a few drops greater than 50 or 100 μm in radius in a cloud containing numerous 5–15 μm radius drops (i.e., a cloud containing high liquid water content) those few big drops fall rapidly through the population of small drops and sweep out large cross-sectional areas, and their efficiency of colliding and coalescing with the smaller drops is large. These few larger drops grow rapidly to raindrop size.

The problem is how do those favored drops, which are large enough to grow by collision and coalescence, get to a size of 25 μm and greater? Remember that the rate of change of droplet radius by vapor deposition growth in smooth unmixed updrafts diminishes as droplets become larger. It takes a long time for droplets to grow from 15 μm to 25 μm in radius. We also know that some clouds produce warm rain very rapidly, particularly clouds that reside in a tropical oceanic environment. As noted previously warm cloud precipitation is favored in a tropical marine environment due to lower droplet concentrations and higher liquid water content.

Thus, clouds most favored to create rain by collision and coalescence are clouds that are warm-based and maritime, while those least likely are cold-based and continental.

There are many clouds between these two extremes; however, some produce rain by collision and coalescence and others do not. The reasons why some do and others do not is still being debated by scientists. Some theorize that the particular properties of the turbulence in clouds favors a more rapid growth of a few droplets by vapor deposition to a size that collision and coalescence can operate efficiently. Others theorize that many clouds sweep up very large dust particles called ultra-giant particles that can function as coalescence embryos. It is generally difficult to determine which if either of these processes is operational in any given cloud.

2.4.3 Strategies for enhancing rainfall from warm clouds

The goal of most attempts at enhancing rainfall from warm clouds by cloud seeding has been to introduce particles in the cloud that are large enough to function as embryos for collision and coalescence growth. The approach has been to seed clouds with actual water drops or to seed with hygroscopic particles such as sodium chloride which take on water vapor at rapid rates in a supersaturated cloud to become droplets larger in size than the natural population of droplets. Braham *et al.* (1957) carried out a water drop seeding experiment in maritime tropical trade wind cumulus clouds. They released about 1250 liters of water per kilometer from an aircraft flying at middle levels in cumulus clouds. The initiation of precipitation was determined from radar observations of the clouds. They found that seeded clouds initiated precipitation 6 minutes earlier than unseeded clouds on the average. Furthermore, twice as many of the seeded clouds observed produced radar-detectable rainfall than the unseeded clouds. Their experiments provided convincing evidence that water-spray seeding can enhance the rate of formation of precipitation and in some cases produce rainfall where naturally caused rain would not occur. Direct measurements of rainfall at the surface were not made, so one cannot be certain how effective this technique is in increasing the amount of rain reaching the surface. The primary limitation of this approach is that large amounts of water must be carried aloft by an aircraft. In fact the costs of hauling the water aloft is so great that the technique has not been viewed as being economically feasible.

A more popular approach to seeding warm clouds has been to introduce hygroscopic aerosol particles (salt) ranging in size from 5 to $100\,\mu$m or greater. Once the salt particles experience a supersaturated cloudy environment, they grow rapidly in size by vapor deposition becoming droplets 25–$30\,\mu$m in radius. Mason (1971) calculated that 100 g of salt would be equivalent to seeding with 1 gallon of water (~ 3600 g) in $25\,\mu$m radius drops. Thus hygroscopic seeding has a much greater potential for economic payoff in enhancing rainfall. Furthermore, a number of investigators have performed ground-based hygroscopic seeding experiments thereby eliminating the use of costly aircraft.

Unfortunately most of the hygroscopic seeding experiments have been "black box" experiments in which clouds are randomly seeded from aircraft or the ground and rainfall is measured. These experiments yield little insight into the actual physical responses to seeding and can provide only the answer to the question – does seeding increase rainfall by a measurable amount? Many of those experiments have been inconclusive while a few suggested strong increases in rainfall (Roy *et al.*, 1961; Biswas *et al.*, 1967; Murty and Biswas, 1968) and others decreases in rainfall (Fournier d'Albe and Aleman, 1976). Why some experiments yielded increases while others decreases in rainfall is not known. It is quite possible that the large *natural variability* of rainfall could be a factor. That is the experiments with decreases in rainfall could have been strongly influenced by a few natural, heavy rainfall events in the unseeded population of clouds. Conversely, those experiments with apparent large increases in rainfall could have experienced well below normal rainfall in the unseeded population of clouds.

The salt seeding experiments carried out in northwestern India are another excellent example of a "black box" experiment (Roy *et al.*, 1961; Biswas *et al.*, 1967; Murty and Biswas, 1968). These ground-based seeding experiments suggested that a 41.9% increase in rainfall occurred on seed days in the downwind direction. Many scientists remain skeptical about the results (Mason, 1971; Simpson and Dennis, 1972; Warner, 1973) largely due to the lack of supporting physical evidence that the seeded clouds exhibited a significantly different microstructure from unseeded clouds. This illustrates that even a well-designed statistical experiment will not be accepted by the scientific community as being credible unless that experiment is supported by physical evidence that: (1) the seeding material actually entered the clouds; (2) the seeded clouds exhibited broader droplet spectra than unseeded clouds; (3) raindrops were initiated earlier, lower in the cloud, and with higher drop concentrations; and (4) larger amounts of rainfall actually reached the ground.

Some scientists have hypothesized that hygroscopic seeding will initiate a premature Langmuir chain reaction process (Biswas and Dennis, 1971; Dennis and Koscielski, 1972). If hygroscopic seeding initiates a Langmuir chain reaction, then measurably larger concentrations of raindrops should be observed in seeded clouds. Unfortunately no documentation of such higher raindrop concentrations have been reported.

While the main thrust of most hygroscopic seeding experiments has been to increase rainfall by enhancing the collision and coalescence of liquid drops, there also exists the possibility that hygroscopic seeding may enhance ice phase precipitation in taller, colder clouds. It has been known for some time that the speed of conversion of a cloud from an all-liquid supercooled cloud to an all-ice cloud (what we call a glaciated cloud) is greater if the cloud contains a broad droplet spectrum (Cotton, 1972b; Koenig and Murray, 1976; Scott and Hobbs, 1977). Supercooled raindrops speed up the glaciation process by rapidly collecting ice crystals which causes the raindrops to freeze. The frozen raindrops, in turn, collect supercooled cloud droplets which freeze on impact with the frozen raindrop.

A broad droplet spectrum also aids faster glaciation of a cloud by favoring secondary ice crystal production by the rime-splinter mechanism (Chisnell and Latham, 1976a,b; Koenig, 1977; Lamb *et al.*, 1981). As noted previously, Hallett and Mossop (1974) and Mossop and Hallett (1974) showed in laboratory experiments that when a frozen raindrop or graupel particle collects supercooled raindrops, secondary ice crystals are produced (see Fig. 2.4). The process occurs most vigorously between $-4\,^{\circ}$C and $-8\,^{\circ}$C and when a broad droplet spectrum is present including drops smaller than about $7\,\mu$m and greater than $12.5\,\mu$m radius.

Hygroscopic seeding, therefore, has the potential for creating a broad droplet spectrum and therefore initiating, or enhancing, rime-splinter production of small ice crystals, and enhancing the formation of supercooled raindrops that can accelerate the glaciation of a cloud. Hygroscopic seeding could also create dynamic responses such as postulated with silver iodide seeding described in Section 2.3. No observational confirmation of such responses to hygroscopic seeding has been made, however.

Enthusiasm for the potential of hygroscopic seeding has grown in recent years due to results of experiments on cold convective clouds using hygroscopic flares in South Africa and Mexico and warm convective clouds using larger hygroscopic particles in Thailand.

The South African experiment was motivated by a report by Mather (1991) which suggested that large liquid raindrops at $-10\,^{\circ}$C found in a cumulonimbus were the result of active coalescence processes caused by the effluent from a Kraft paper mill. Earlier, Hobbs *et al.* (1970) found that the effluent from paper mills can be rich in CCN. Moreover, Hindman *et al.* (1977a,b) found paper pulp mill effluent to have high concentrations of large and ultra-giant hygroscopic particles, which is consistent with the idea that the paper pulp mill effectively "seeded" the storm.

Another reason for optimism is that Mather *et al.* (1996) applied a pyrotechnic method of delivering salt, based on a fog dispersal method developed by Hindman (1978). This reduced a number of technical difficulties associated with preparing, handling, and delivery of very corrosive salt particles. Seeding with this system is no more difficult than silver iodide flare seeding. Compared to conventional methods of salt delivery, the flares produce smaller-sized particles in the size range of 0.5–$10\,\mu$m. Thus, not as much mass must be carried to obtain a substantial yield of seeding material. The question of effectiveness of this size range will be discussed below. Seeding trials with this system suggested that the pyrotechniques produced a cloud droplet spectrum that was broader and with fewer numbers, which would be expected to increase the chance for initiation of collision and coalescence processes.

Mather *et al.* (1996) analyzed radar-defined cells over a period of about an hour to identify the seeding signatures for 48 seeded storms compared to 49 unseeded storms. They showed that after 20–30 minutes, the seeded storms developed higher rain masses and maintained those higher rain masses for another

25–30 minutes. Bigg (1997) performed an independent evaluation of the South African exploratory hygroscopic seeding experiments and also found that the seeded storms clearly lasted longer than the unseeded storms. Bigg also suggested that there was a clear dynamic signature of seeding. He argued that hygroscopic seeding initiated precipitation lower in the clouds, which, in turn, was not dispersed horizontally as much as the unseeded clouds by vertical wind shears. As a result, Bigg speculated that low-level downdrafts became more intense, which yielded stronger storm regeneration by the downdraft outflows, and longer-lived precipitation cells. Bigg's (1997) hypothesis is a plausible scenario that should be examined thoroughly with numerical models and coordinated, high-resolution Doppler radars.

Cooper *et al.* (1997) performed simulations of the low-level evolution of droplet spectra in seeded and unseeded cloudy plumes. Following a parcel ascending in the cloud updrafts they calculated the evolution of droplet spectra by vapor deposition and collection. The calculations were designed to emulate the effects of hygroscopic seeding with the South African flares. The calculations showed that introduction of particles in the size range characteristic of the flares resulted in an acceleration of the collision and coalescence process. If the hygroscopic particles were approximately $10\,\mu$m in size, precipitation was initiated faster. But, when more numerous $1\,\mu$m hygroscopic particles were inserted, high concentrations of drizzle formed. For a given amount of condensate mass, if the mass is on more numerous drizzle drops than on fewer but larger raindrops, then evaporation rates are greater in the subcloud layer. This could lead to more intense dynamic responses as proposed by Bigg, suggesting that seeding with smaller hygroscopic particles may have some advantages. Keep in mind, however, that this is a very simple model. More comprehensive model calculations should also be performed.

The Mexican hygroscopic seeding experiment described by Bruintjes *et al.* (1999, 2001) was a randomized experiment on convective storms, based on a floating target design, aimed at replicating the South African hygroscopic seeding experiment using the same flare design. Bruintjes *et al.* (2001) showed that the results from the Mexican experiments were similar to the South African experiment with a statistically significant increase in rainfall from radar-estimated rain mass for "floating" convective clusters. The statistical results which were analyzed for a maximum of 60 minutes following seeding exhibited a seeding response 20–60 minutes following seeding.

The Thailand experiment (Woodley *et al.*, 1999b) differed from the Mexican and South African experiments in that the clouds were purely warm clouds and seeding was not done with pyrotechnic flares. Instead, much larger calcium chloride particles were released in the convective clouds. Like the other two experiments, the analysis was performed for radar-identified floating targets. Analysis of the radar-defined rain volumes showed a statistically significant increase in rain volume in seeded clouds. However, Silverman and Sukarnjanaset (1996) found that the main seeding response was 2–6 hours following seeding in untreated clouds that were potentially spawned by the seeded clouds. Like in the South African experiment, Silverman and Sukarnjanaset (1996) speculated that changes in the timing, location, and/or increased intensity of the rain or alteration in the size spectrum of raindrops may produce an enhanced downdraft and associated gust front which can trigger the successive development of more vigorous convective cells.

In summary, there are some exciting new results of hygroscopic seeding, especially those using flares. Since the analyzed statistical results are for radar-defined floating targets, they still do not prove that rainfall can be increased by hygroscopic seeding on the ground for specific watersheds. Moreover, since seeding may alter the size spectrum of raindrops which alters the radar return, uncertainties exist in the evaluation of actual rain amounts for seeded versus not-seeded floating targets. Finally, since the main response to seeding is delayed in time for as much as 6 hours following the cessation of seeding, we lack a clear understanding of the actual processes that can lead to such a physical response.

2.5 Hail suppression

2.5.1 Introduction

For hundreds if not thousands of years man has sought techniques for suppressing hail. This has ranged from ringing bells, to firing cannons, to the modern era of cloud seeding (Changnon and Ivens, 1981). The motivation for suppressing hail is strong since hail can devastate a wheat or grape crop, and when a major hailstorm occurs over a modern urban area millions of dollars of property damage can result and humans and livestock can be severely injured or even killed (e.g., the Fort Collins, Colorado hailstorm of 1983). Scientific programs aimed at developing a hail suppression effort using cloud seeding strategies began in the 1950s in the Soviet Union and in Alberta, Canada. In the early 1960s reports of major success in hail suppression in the Soviet Union were made by visiting US scientists. Subsequently several delegations of US scientists visited the various hail suppression programs in the Soviet Union (e.g., Battan, 1969; Marwitz, 1972).

Not to be outdone by the Soviet Union during the cold war era, in May 1965 the Interdepartmental Committee on Atmospheric Sciences (ICAS), which serves as a coordinating group for US federal meteorological activities, asked the National Science Foundation to prepare plans for a national program of hail suppression. This led to the formation of a pilot program called Hailswath which was carried out in South Dakota and Colorado in the summer of 1966. This was followed by the formation of the National Hail Research Experiment (NHRE) that operated both a randomized cloud seeding program patterned to some degree on the Soviet hail suppression program and basic physical studies of hailstorms and hail growth mechanisms. NHRE operated field programs during the period 1972 to 1976.

In this section we summarize basic concepts of hailstorms and hail formation processes, review the concepts for suppressing hail by cloud seeding, and examine the status of our ability to suppress hail.

2.5.2 *Basic concepts of hailstorms and hail formation*

In order to understand how hail forms, one must first understand the varying behavior of thunderstorms which are the factories where hailstones are produced. A thunderstorm produces the liquid water content upon which hailstones grow. The updrafts then suspend the embryo hailstones, which ultimately determines how long a growing hailstone will remain in a water-rich environment.

The primary energy driving a thunderstorm is the buoyancy experienced by updrafts as latent heat is released as vapor condenses to form cloud droplets, supercooled drops freeze, and vapor deposits on ice crystals. The buoyancy, which is determined by the temperature excess of an updraft relative to its environment multiplied by the acceleration due to gravity, is a local measure of the acceleration of the updraft. The actual magnitude of the updraft strength at any height in the atmosphere is largely determined by the integrated buoyancy that an updraft experiences as it ascends from cloud base to a given height in the atmosphere. The integrated buoyancy through the atmosphere is called convective available potential energy (CAPE). In general, the greater the CAPE, the stronger the strength of updrafts in a thunderstorm.

The likelihood that hail will form, the size of hailstones, and the extent of the hailswath, is not only determined by how much CAPE there is in the atmosphere at any given time. Other environmental factors also are important in initiating thunderstorms and determining the particular flow structure of the storm system. For example, as an updraft in a thunderstorm rises through the atmosphere, it carries with it the horizontal momentum that is characterized by the winds at the updraft source level. As the updraft ascends it encounters air having differing horizontal momentum (i.e., different wind speeds and direction). The vertical variation in horizontal wind speed and direction is called *wind shear*. The interaction of the updraft with ambient air having different horizontal momentum causes the updraft to tilt from the vertical and creates pressure anomalies in the air that can also accelerate the air. Thus the complicated interactions of updraft and downdraft air with an environment having vertical shear of the horizontal wind can alter the storm structure markedly. For example, *ordinary thunderstorms* develop in an atmosphere containing moderate amounts of CAPE and weak to moderate vertical wind shear.

The ordinary cumulonimbus cloud or thunderstorm is distinguished by a well-defined life cycle as shown in Fig. 2.12, which lasts 45 to 60 minutes. Figure 2.12a illustrates the growth stage of the system. The growth stage is characterized by the development of towering cumulus clouds, generally in a region of low-level convergence of warm moist air. Often during the growth stage, towering cumulus clouds merge to form a larger cloud system or a vigorous cumulus cloud expands to a larger cloud. During the growth stage, updrafts dominate the cloud system

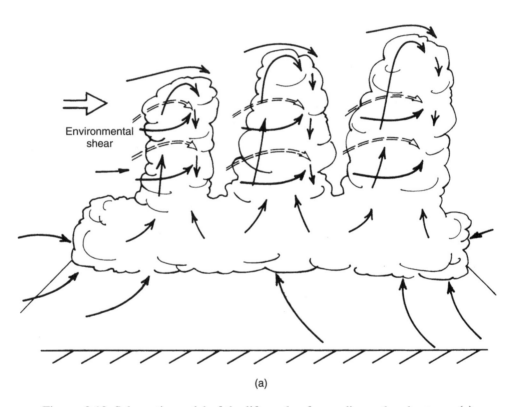

(a)

Figure 2.12 Schematic model of the life cycle of an ordinary thunderstorm. (a) The cumulus stage is characterized by one or more towers fed by low-level convergence of moist air. Air motions are primarily upward with some lateral and cloud top entrainment depicted. (b) The mature stage is characterized by both updrafts and downdrafts and rainfall. Evaporative cooling at low levels forms a cold pool and gust front which advances, lifting warm moist, unstable air. An anvil at upper levels begins to form. (c) The dissipating stage is characterized by downdrafts and diminishing convective rainfall. Stratiform rainfall from the anvil cloud is also common. The gust front advances ahead of the storm preventing air from being lifted at the gust front into the convective storm. From Cotton and Anthes (1989).

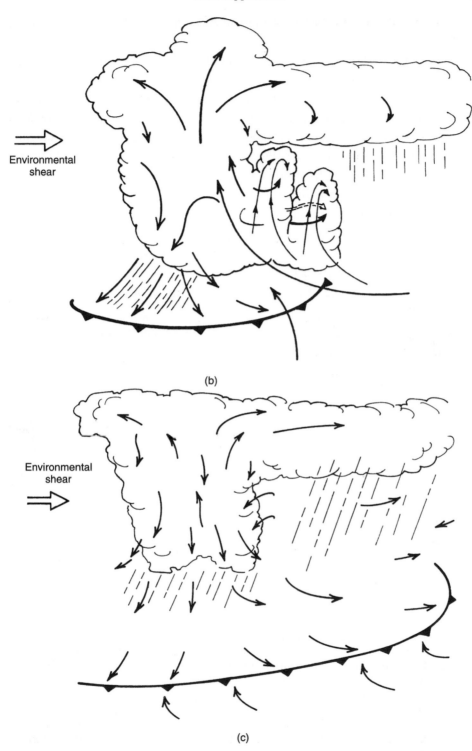

(b)

(c)

Figure 2.12 (cont.)

and precipitation may be just beginning in the upper levels of the convective towers.

The mature stage of a cumulonimbus cloud commences with rain settling in the subcloud layer. Upon encountering the surface, the downdraft air spreads horizontally, where it can lift the warm moist air into the cloud system. At the interface between the cool, dense downdraft air and the warm, moist air, a gust front forms. Surface winds at the gust front are squally; rapidly changing direction and speed. The warm, moist air lifted by the gust front provides the fuel for maintaining vigorous updrafts. Upon encountering the extreme stability at the top of the troposphere, called the tropopause, the updrafts spread laterally, spewing ice crystals and other cloud debris horizontally to form an anvil cloud. In many cases, the updrafts are strong enough that they penetrate into the lower stratosphere creating a cloud dome. Often the stronger updrafts form a thin layered cloud that caps the cloud and is separated from the main body of the cloud which is called *pileus*. The presence of pileus is a visual clue that strong updrafts exist in the storm.

Water loading and the entrainment of dry environmental air into the storm generate downdrafts in the cloud interior, which rapidly transport precipitation particles to the subcloud layer. The precipitation particles transported to the sub-cloud layer partially evaporate, further chilling the subcloud air and strengthening the low-level outflow and gust front. Thus, continued uplift of warm, moist air into the cloud system is sustained during the mature stage. Lowering of pressure at middle levels in the storm as a result of warming by latent heat release and diverging air flow results in an upward-directed, pressure-gradient force which helps draw the warm, moist air lifted at the gust front up to the height of the level of free convection. Thus, the thunderstorm becomes an efficient "machine" during its mature stage, in which warming aloft and cooling at low levels sustains the vigorous, convective cycle.

The intensity of precipitation from the storm reaches a maximum during its mature stage. Therefore, the mature cumulonimbus cloud is characterized by heavy rainfall and gusty winds, particularly at the gust front (see Fig. 2.13). The propagation speed of the gust front increases as the depth of the outflow air increases and the temperature of the outflow air decreases. The optimum storm system is one in which the speed of movement of the gust front is closely matched to the speed of movement of the storm as a whole.

Once the gust front advances too far ahead of the storm system, air lifted at the gust front does not enter the updraft air of the storm, but may only form fair-weather cumulus clouds. This marks the beginning of the dissipating stage of the thunderstorm shown in Fig. 2.12c. During the dissipating stage, updrafts

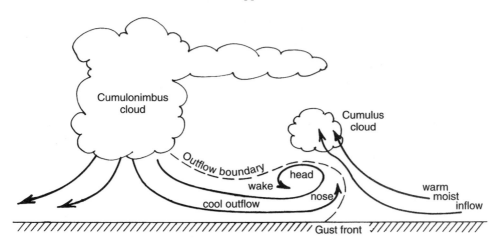

Figure 2.13 Illustration of a gust front formed at the leading edge of the down-draft outflow from a thunderstorm. From Cotton (1990).

weaken and downdrafts become predominant, and rainfall intensity subsides often turning into a period of light steady rainfall.

Many hail-producing thunderstorms are composed of a number of cells, each undergoing a life cycle of 45 to 60 minutes. The thunderstorm system, called a *multicell storm*, may have a lifetime of several hours. Figure 2.14 illustrates a conceptual model of a severe multicell storm as depicted by radar. At 3 minutes, the vertical cross-section through a vigorous cell, C1, shows a weak echo region (WER) at low levels, where there is an absence of precipitation particles large enough to be seen on radar. Updrafts in this region are so strong that there is not enough time for precipitation to form. The region of intense precipitation at middle levels is due to a cell that existed previous to C1. By 9 minutes, cell C1 exhibits a well-defined precipitation maximum at middle levels, while the WER has disappeared. By 15 minutes, precipitation from C1 has reached the surface and a new cell, C2, is evident on the right forward flank of the storm (relative to storm motion). By 21 minutes, the maximum of precipitation associated with C1 has lowered and C2 has grown in horizontal extent. Cell C2 will soon become the dominant cell of the storm, only to be replaced by another cell shortly.

Multicell storms, where updrafts reach 25–$35 \, \mathrm{m \, s^{-1}}$, can produce extensive moderate-sized hailstones (i.e., golfball-sized). Severe multicell storms typically occur in regions where the atmosphere is quite unstable and where the vertical wind shear is moderate in strength.

The grandaddy of all thunderstorms is the *supercell* thunderstorm. It is noted for its persistence, lasting for 2 to 6 hours, in a single cell structure. Figure 2.15

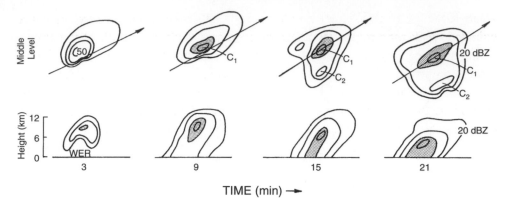

TIME (min) →

Figure 2.14 Conceptual model of horizontal and vertical radar sections for a multicell storm at various stages during its evolution, showing reflectivity contours at 10 dBZ intervals. Horizontal section is at middle levels (~6 km) and the vertical section is along the arrow depicting cell motion. The life cycle of cell C_1 is depicted on the right flank of the storm beginning at 15 minutes. Adapted from Chisholm and Renick (1972).

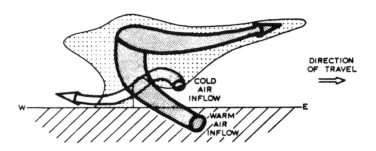

Figure 2.15 Model showing the airflow within a three-dimensional severe right-turning (SR) storm traveling to the right of the tropospheric winds. The extent of precipitation is lightly stippled and the up- and downdraft circulations are shown more heavily stippled. Air is shown entering and leaving the updraft with a component into the plane of the diagram. However, the principal difference of this organization is that cold air inflow, entering from outside the plane of the vertical section, produces a downdraft ahead of the updraft rather than behind it. From Browning (1968). © Royal Meteorological Society.

is a schematic illustration of the dominant airflow branches in a supercell storm. The storm is often characterized by a broad, intense updraft entering its southeast flank, rising vertically and then, in the Northern Hemisphere, turning clockwise in the anvil outflow region. Updrafts in supercell storms may exceed $40\,\mathrm{m\,s^{-1}}$, capable of suspending hailstones as large as grapefruit. Horizontal and vertical cross-sections of a supercell storm as viewed by radar are shown in Fig. 2.16a and b. A distinct feature of the supercell storm is the region that is free of radar echo that can be seen on the southeast flank of the storm at the 4 and

7 km levels and in the vertical cross-section. This persistent feature, called a bounded weak echo region (BWER), or echo-free vault, is a result of the strong updrafts in that region which do not provide enough time for precipitation to form in the rapidly rising air. As noted previously, severe multicell thunderstorms exhibit WERs, but they are neither as persistent as in supercell storms, nor do they maintain the characteristic bounded structure of supercells. Cyclonic rotation (or counter-clockwise in the Northern Hemisphere) of the updrafts may also contribute to the BWER by causing any precipitation elements that do form to be centrifuged laterally out of the rotating updraft region. Supercell storms are also known as severe right-moving storms because, in the Northern Hemisphere, most supercells move to the right of the mean flow due to the interaction of the updraft with environmental shear. The rotational characteristic of supercells, as well as their strong updrafts, results in a storm system which produces the largest hailstones.

Between the severe multicell thunderstorm type and the supercell-type storm exist a continuum of thunderstorm types. Some of these are quite steady, exhibiting a dominant cell for a time and a rotating updraft, but also contain transient cell groups for part of the storm lifetime or at the flanks of the dominant cell and thus do not meet all the criteria of a supercell storm. The more a thunderstorm resembles a supercell thunderstorm, the more likely it is to produce very large hailstones and long, continuous swaths of hail.

Hailstorms generally occur in an environment with large values of CAPE. In such an environment, thunderstorms develop significant positive buoyancy and associated strong updrafts capable of suspending hailstones falling through the air at speeds of 15–25 m s^{-1}. Often, severe thunderstorms producing large hail also produce tornadoes and even flash floods. Storms producing the largest hailstones normally develop in an environment with strong wind shear which favors the formation of supercell thunderstorms. The height of the melting level is also important in determining the size of hailstones that will reach the surface. It has been estimated that as many as 42% of the hailstones falling through the 0 °C level melt before reaching the ground over Alberta, Canada, while it may be as high as 74% over Colorado and 90% in southern Arizona. This is consistent with observations indicating that the frequency of hail is greater at higher latitudes. Any thunderstorm with a radar echo top over 8 km near Alberta, Canada has a significant probability of producing damaging hail!

Hailstone growth is a complicated consequence of the interaction of the airflow in thunderstorms and the growth of precipitation particles. Hailstones grow primarily by collection of supercooled cloud droplets and raindrops. At temperatures colder than 0 °C, many cloud droplets and raindrops do not freeze and can remain unfrozen to temperatures as cold as −40 °C. A few ice particles do freeze,

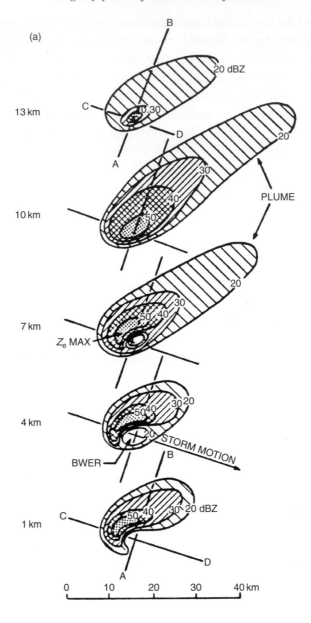

Figure 2.16 (a) Schematic horizontal sections showing the radar structure of a unicellular supercell storm at altitudes of 1, 4, 7, 10, and 13 km AGL. Reflectivity contours are labeled in dBZ. Note the indentation on the right front quadrant of the storm at 1 km, which appears as a weak-echo vault (or BWER, as it is labeled here) at 4 and 7 km. On the left rear side of the vault is a reflectivity maximum extending from the top of the vault to the ground (see Fig. 2.16b.) From Chisholm and Renick (1972).

however, perhaps by collecting an aerosol particle that can serve as a freezing nucleus. If the frozen droplet is small, it will grow first by vapor deposition forming snowflakes such as dendrites, hexagonal plates, needles, or columns. After some time, perhaps 5–10 minutes, the ice crystals become large enough to settle relative to small cloud droplets, which immediately freeze when they impact the surface of the ice particle. If enough cloud droplets are present or the supercooled liquid water content of the cloud is high, the ice particle can collect enough cloud droplets so that the original shape of the vapor-grown crystal becomes obscured and the ice particle becomes a graupel particle of several millimeters in diameter. At first, the density of the graupel particle is low as the collected frozen droplets are loosely compacted on the surface of the graupel particle. As the ice particle becomes larger, it falls faster and sweeps out a larger cross-sectional area, and its growth by collection of supercooled droplets increases proportionally (see Fig. 2.17). As the growth rate increases, the collected droplets may not freeze instantaneously upon impact, and thus flow over the surface of the hailstone filling in the gaps between collected droplets. The density of the ice particle, therefore, increases close to that of pure ice as the dense hailstone falls still faster, growing by collecting supercooled droplets as long as the cloud liquid water content is large. The ultimate size of the hailstone is determined by the amount of

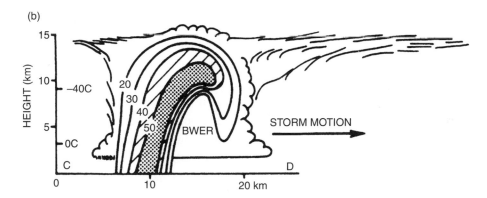

Figure 2.16 (cont.) (b) Schematic vertical section through a unicellular supercell storm in the plane of storm motion (along C–D in Fig. 2.16a). Note the reflectivity maximum, referred to as the hail cascade, which is situated on the (left) rear flank of the vault (or BWER, as it is labeled here). The overhanging region of echo bounding the other side of the vault is referred to as the *embryo curtain*, where it is shown to be due to millimetric-sized particles some of which are recycled across the main updraft to grow into large hailstones. From Chisholm and Renick (1972).

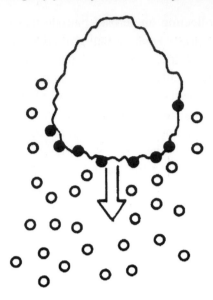

Figure 2.17 Illustration of a hailstone growing by collecting supercooled droplets. From Cotton (1990).

supercooled liquid water in the cloud and the time that the growing hailstone can remain in the high rainwater region. The time that a hailstone can remain in the high liquid water content region, in turn, is dependent on the updraft speed and the fall speed of the ice particle. If the updraft is strong, say 35–$40 \, \text{m s}^{-1}$, and the particle fall speed through the air is only of the order of 1–$2 \, \text{m s}^{-1}$, then the ice particle will be rapidly transported into the anvil of the cloud before it can take full advantage of the high liquid water content region. The ideal circumstance for hailstone growth is that the ice particle reaches a large enough size as it enters the high liquid water content region of the storm so that the ice particle fall speed nearly matches the updraft speed. In such a case, the hailstone will only slowly ascend or descend while it collects cloud droplets at a very high rate. Eventually, the hailstone fall speed will exceed the updraft speed or it will move into a region of weak updraft or downdraft. The size of the hailstone reaching the surface will be greatest if the large airborne hailstone settles into a vigorous downdraft, as the time spent below the $0\,°C$ level will be lessened and the hailstone will not melt very much. Thus, a particular combination of airflow and particle growth history is needed to produce large hailstones. Let us now examine several conceptual models of hailstone growth in the different thunderstorm models we have described previously.

The Soviet hail model

The Soviet hail model is built on the ordinary thunderstorm model illustrated in Fig. 2.18. If the storm develops particularly vigorous updrafts and high liquid water contents during the growth stage, raindrops may form by collision and coalescence with smaller cloud droplets. As the growing raindrops are swept aloft, they continue to grow and eventually ascend into supercooled regions. If the updraft exhibits a vertical profile as shown in the insert of Fig. 2.18, with

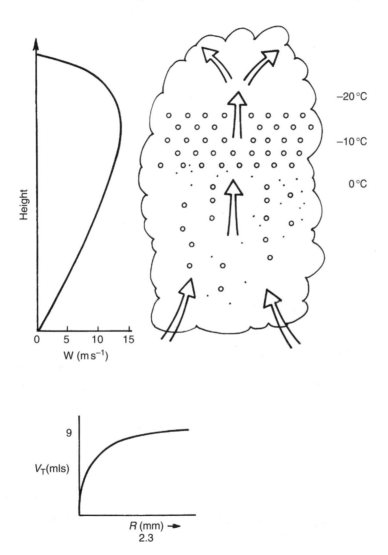

Figure 2.18 The Soviet hailgrowth model. Left panel illustrates a favorable updraft profile. Right panel illustrates the formation of an accumulation zone. Bottom panel illustrates variation in terminal velocity of raindrops with size. From Cotton (1990).

a maximum updraft speed (w) in the layer between $-10\,°C$ and $-20\,°C$, many raindrops may become suspended just above the updraft maximum. This is because the fall speed of raindrops approaches a maximum value slightly greater than $9\,m\ s^{-1}$ for raindrops larger than 2.3 mm in radius (see lower insert Fig. 2.18). The region just above the updraft maximum serves as a trap for large raindrops, and rainwater accumulates in the region as long as an updraft speed greater than $9\,m\ s^{-1}$ persists. Supercooled liquid water contents greater than $17\,g\ m^{-3}$ have been reported in such regions. Normal values of supercooled liquid content rarely exceed $2–4\,g\ m^{-3}$.

If now a few supercooled raindrops freeze in the zone of accumulated liquid water content, they will experience a liquid-water-rich environment and hailstone growth can proceed quite rapidly. The frozen supercooled raindrops serve as very effective hailstone embryos, greatly accelerating the rate of formation of millimeter-sized ice particles. Observations of thunderstorms near Huntsville, Alabama, with multiparameter Doppler radar revealed regions of very high radar reflectivity where ice particles were not detected. Light hailfall was observed at the surface. These observations are consistent with the Soviet hail model.

Conceptual model of hail formation in ordinary multicell thunderstorms

Previously we described a multicell thunderstorm as a storm containing several cells each undergoing a life cycle of 45 to 60 minutes. Over the High Plains of the United States and Canada, multicell thunderstorms are found to be prolific producers of hailstones; if not the very largest in size, at least the most frequent. Large hailstones grow during the mature stage of the cells illustrated in Fig. 2.14 when updrafts may exceed $30\,m\ s^{-1}$. In such strong updrafts, the time available for the growth of hailstones from small ice crystals to lightly rimed ice crystals, to graupel particles or aggregates of snowflakes, to hailstone embryos, is only 5 or 6 minutes! This time is far too short, as it takes some 10–15 minutes for an ice particle to grow large enough to begin collecting supercooled cloud droplets or aggregating with other ice crystals to form an embryonic hailstone. The mature stage of each thunderstorm cell provides the proper updraft speeds and liquid water contents for mature hailstones to grow, but they must be sizeable precipitation particles at the time they enter the strong updrafts in order to take advantage of such a favored environment. Here is where the growth stage of each cell is very important to hailstone growth, as the weaker, transient, updrafts provide sufficient time for the growth of graupel particles and aggregates of snow crystals, which can then serve as hailstone embryos as the cell enters its mature stage. The growth stage of the multicellular thunderstorm thereby preconditions the ice particles and allows them to take full advantage of the high water contents of the mature stage of the storm. Upon entering the mature stage, the millimeter-sized ice particles settle through the air at 8–10 m s^{-1} and therefore rise slowly as the updrafts increase in speed from 10–15 m s^{-1} at low levels to 25–35 m s^{-1} at higher levels.

As mentioned previously, the optimum growth of a hailstone occurs when the updraft speed exceeds the particle fall speed by only a few meters per second. In that case, the hailstone rises slowly only a few kilometers as it collects super-cooled droplets. Fortunately, not all multicellular thunderstorms develop hailstone embryos of the appropriate sizes during the growth stage nor do the embryos enter the updrafts of the mature cell at the right location for optimum hailstone growth.

Conceptual model of hailstone growth in supercell thunderstorms

We have seen that the supercell thunderstorm is a steady thunderstorm system con-sisting of a single updraft cell that may exist for 2 to 6 hours. Updraft speeds are so strong that they are characterized by having a bounded weak echo region (see Fig. 2.16a and b) in which precipitation particles of a radar-detectable size do not form. Nonetheless, the supercell thunderstorm produces the largest hailstones, some-times over very long swaths. Consider for example, the Fleming hailstorm that occurred on June 21, 1972. Figure 2.19 illustrates that this storm first reached super-cell proportions in northeast Colorado and produced a nearly continuous swath of damaging hail 300 km long over eastern Colorado and western Kansas. How can a storm system consisting of a single, steady updraft with speeds in excess of 30 m s^{-1} develop hailstones before the ice particles are thrust into the anvil of the storm?

Browning and Foote (1976) visualized hail growth in a supercell thunderstorm as a three-stage process illustrated in Fig. 2.20a and b. During stage 1, hail embryos form in a relatively narrow region on the edge of the updraft, where speeds are on the order of 10 m s^{-1}, allowing time for the growth of millimeter-sized hail embryos. Those particles forming on the western edge of the main updraft have a good chance of sweeping around the main updraft and entering the region called the *embryo curtain* on the right-forward flank of the storm. These particles will follow the trajectory labeled 1 in Fig. 2.20. By contrast, particles that enter the main updraft directly follow the trajectory labeled 0 and do not have sufficient time to grow to hailstone size. They are thrust out into the storm anvil.

During stage 2, the embryos formed on the western edge of the main updraft are carried along the southern flank of the storm by the diverging flow field. Some of the larger embryos settle into the region of weak updrafts that characterizes the embryo curtain. The particles following the trajectory labeled 2 experience further growth as they descend in the embryo curtain region. Some of the particles settle out of the lower tip of the embryo curtain and re-enter the base of the main updraft, commencing stage 3.

Stage 3 represents the mature and final stage of hail growth, in which the hailstones experience very high liquid water concentrations and grow by collecting

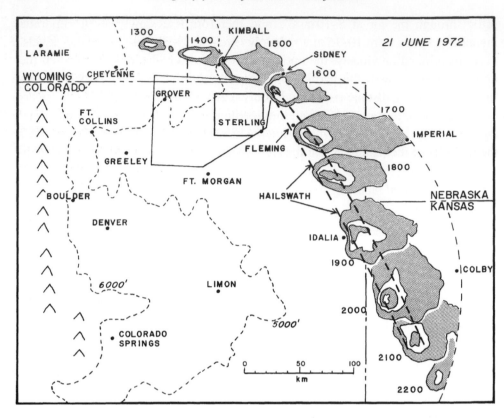

Figure 2.19 Hourly positions of the Fleming hailstorm as determined by the NWS Limon radar (CHILL radar data used 1300-1500 MDT). The approximate limits of the hailswath are indicated by the bold, dashed line. Continuity of the swath is not well established but total extent is. Special rawinsonde sites were located near the towns of Grover, Ft. Morgan, Sterling, and Kimball. Contour intervals are roughly 12 dBZ above 20 dBZ. From Browning and Foote (1976). © Royal Meteorological Society.

numerous cloud droplets during their ascent in the main updraft. The growth of hailstones from embryos is viewed as a single up-and-down cycle. Those embryos which enter the main updraft at lowest levels, where the updraft is weakest, are likely to have their fall speed nearly balanced by the updraft speed. As a result of their slow rise rate, they will have plenty of time to collect the abundant supercooled liquid water. Eventually their fall velocities will become large enough to overcome the large updraft speeds and/or they will move into the downdraft region and descend to the surface on the northern flank of the storm.

Some researchers argue that hail embryos are not formed on the flanks of a single, main updraft, where updraft speeds may be weaker, but instead the embryos

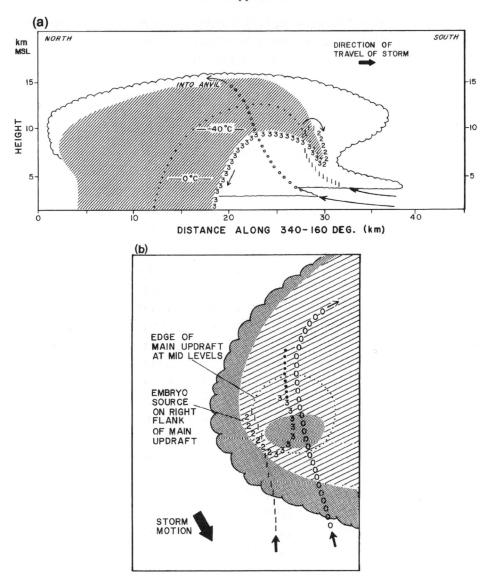

Figure 2.20 Schematic model of hailstone trajectories within a supercell storm based upon the airflow model inferred by Browning and Foote (1976). (a) Hail trajectories in a vertical section along the direction of travel in the storm. (b) These same trajectories in plan view. Trajectories 1, 2, and 3 represent the three stages in the growth of large hailstones discussed in the text. The transition from stage 2 to stage 3 corresponds to the re-entry of a hailstone embryo into the main updraft prior to a final up-and-down trajectory during which the hailstone may grow large, especially if it grows close to the boundary of the vault. Other, slightly less favored, hailstones will grow a little farther away from the edge of the vault and will follow trajectories resembling the dotted trajectory. Cloud particles growing "from scratch" within the updraft core are carried rapidly up and out into the anvil along trajectory 0 before they can attain precipitation size. From Browning and Foote (1976). © Royal Meteorological Society.

form in towering cumulus clouds that are flanking the main updraft of the storm. They argue that the towering cumulus clouds are obscured by the precipitation debris falling out of the main updraft. Embryos can form in the transient, weaker towering convective elements and "feed" the main updraft with millimeter-sized embryos. In some thunderstorms, such transient, flanking cumulus towers are visually separated from the dominant parent cell. Such thunderstorms are a hybrid between the single supercell storm and the ordinary multicell storm and are called *organized multicell thunderstorms* or *weakly evolving thunderstorms*. Such thunderstorms are characterized by a single, dominant cell which maintains a steady flow structure similar to a supercell thunderstorm, including a BWER and a rotating updraft. They also contain distinct flanking towering cumulus clouds that can serve as the manufacturing plant for hailstone embryos that can settle into the main updraft of the steady cell. The so-called *feeder* cells have to be relatively close to the parent cell in order to be effective suppliers of hailstone embryos. Whether or not a supercell actually contains such embedded feeder clouds or is just a single cell is not known at the present time.

In summary, three factors are necessary to produce hail: (1) large amounts of supercooled liquid water, (2) millimeter-sized hydrometeors called hail embryos which are usually supercooled raindrops, graupel particles, or aggregates, and (3) strong enough updrafts to suspend the growing hailstones. If any of the three is absent, hail does not develop. We will see that hail suppression concepts are based on altering at least one of the three factors.

2.5.3 Hail suppression concepts

The approaches to suppressing hail vary considerably depending on the conceptual model one considers the most appropriate for the storms in a region of interest. Differences in conceptual storm models result from differing environmental conditions and from different perceptions of storm structure. Perceptions of storm structure, in turn, vary from one scientist to another and are dependent on the storm observing systems and numerical models available to a researcher. For example, one's perception of a storm structure changes as one moves from observing a storm with a radar providing only reflectivity values, to observing a storm with multiple Doppler radar capable of defining storm air motions, to observing a storm with multiple Doppler radar and a multiparameter radar capable of also identifying ice versus liquid precipitation elements.

The Soviet hail suppression scheme

Hail suppression techniques in the Soviet Union were largely based on the Soviet conceptual model of a hailstorm described above. In that conceptual model hailstones grow from embryos of frozen supercooled raindrops. In a region of high liquid water content resulting from an accumulation of supercooled raindrops, a few drops freeze naturally and grow rapidly to hailstones in the water-rich environment. The Soviets, therefore, developed several techniques for direct injection of silver iodide seeding material in the region of high radar reflectivity and presumably large amounts of supercooled water. The techniques involve the use of several types of rockets and cannons which carried seeding material to heights as much as 8 km where the material is explosively dispersed over the target region (Bibilashvili *et al.*, 1974). The dispersed silver iodide is then hypothesized to be swept up by the supercooled raindrops (or promote freezing of small ice crystals which are swept up by the supercooled raindrops) and promote their freezing. The numerous frozen raindrops then compete "beneficially" for the available supercooled liquid water, thus inhibiting the formation of hailstones. As described by Battan (1969) the projects required the coordinated use of radars to detect regions of high rain water content and ground-based rocket launchers or cannons.

The results reported from a number of seeding operations in the Soviet Union suggested 50% to 100% reduction in hail damage (Battan, 1969; Burtsev, 1974; Sulakvelidze *et al.*, 1974). Attempts to replicate the Soviet hail suppression scheme in Switzerland and the United States have been unsuccessful (Atlas, 1977; Federer *et al.*, 1986). The failure in the United States, however, may have been due to the inability of the scientists to successfully deploy rockets. Most of the seeding was done by burning flares on aircraft flying in updrafts. It is also possibly a result of the fact that the storms in the NHRE experimental area were quite different from the hail-producing storms over the Soviet Union. Firstly, there is clear evidence that the embryos for hailstones in northeastern Colorado are primarily graupel particles rather than frozen drops (Knight and Squires, 1982). Therefore, it is unlikely that the Soviet concept that an accumulation zone of supercooled raindrops serves as the main source region of hailstone growth in the High Plains of the United States. Secondly, the most severe hail-producing storms over northeast Colorado are supercell storms whereas supercells appear to be rather rare in the Soviet hail regions.

The reasons for the failure of the Swiss hail experiment is less obvious since the characteristics of hailstorms in that region more closely resemble the Soviet storms and the Swiss more faithfully modeled their experiment on the Soviet hail suppression scheme.

The glaciation concept

The aim of hail suppression by glaciation is to introduce so many ice crystals via seeding that the ice crystals consume all the available supercooled liquid water as they grow by vapor deposition and riming of cloud droplets. To be effective this technique requires the insertion of very large amounts of seeding materials in the storm updrafts. Modeling studies (Weickmann, 1964; Dennis and Musil, 1973; English, 1973; Young, 1977) have suggested that unless very large amounts of seeding material are used, the strongest updrafts remain all liquid and hail growth is not substantially affected. Therefore, the glaciation concept is generally thought not to be a feasible approach to hail suppression. The glaciation concept is also not popular because many scientists think that it may result in a reduction in rainfall along with hail. Since most hail-prone areas are semi-arid, the loss of rainfall can have a greater adverse impact on agriculture than economic gains from hail suppression.

The embryo competition concept

The competing embryo concept, first introduced by Iribarne and de Pena (1962), involves the introduction of modest concentrations of hailstone embryos (on the order of 10 per cubic meter) in the regions of major hailstone growth. The idea is that millimeter-sized ice particles will then compete beneficially for the available supercooled water and result in numerous small hailstones or graupel particles rather than a few large, damaging hailstones. Because it is not economically feasible to introduce hailstone embryos directly in the cloud, one must use a seeding strategy which utilizes the storm's natural hailstone embryo manufacturing process. For example, the Soviet concept of hail suppression can be considered an embryo competition strategy. In their case the hypothesized hailstone embryos are frozen supercooled raindrops. When seeding material is dispersed into a region containing supercooled raindrops, the raindrops readily freeze and immediately become millimeter-sized hailstone embryos. The numerous hailstone embryos then beneficially compete for the available supercooled water resulting in the formation of numerous small hailstones, many of which will melt before reaching the ground.

Now consider a cloud in which supercooled raindrops are not present. In such clouds millimeter-sized ice particles first must form by vapor deposition until ice crystals of the size of a few hundred micrometers (0.1 mm) form. This takes a significant amount of time, on the order of 5 to 10 minutes. The larger vapor-grown ice crystals can then settle through a population of cloud droplets and grow rapidly by riming those droplets to form graupel particles. The larger ice crystals can also collide with each other to form clusters of ice crystals called *aggregates*. Both the graupel particles and aggregates can serve as hailstone embryos since they have significant fall velocities and cross-sectional areas to enable them to grow rapidly by accreting supercooled cloud droplets to form hailstones.

As a result of the significant amount of time for hailstone embryos to form, seeding intense updrafts, such as exist in supercell storms and the mature cell of severe multicell storms with WERs, is unlikely to have any significant effect on hail growth. The ice crystals formed from seeding would probably be swept aloft into the anvil before becoming large enough to serve as embryos of hailstones. In the case of multicell storms, the recommended approach is to seed in the flanking towering cumulus clouds where updrafts are weaker and transient. If the cell is a daughter cell or a cell that eventually becomes a mature cell, it may be laden with numerous artificially produced hailstone embryos. Likewise, if the flanking cell is in the right location to serve as a feeder cell, then the natural and artificially produced hailstone embryos will be entrained into the mature cell to beneficially compete for the supercooled liquid water and reduce the size of hailstones.

The problem is how can one implement an embryo competition strategy in supercell storms? Remember that supercells have steady updraft speeds of $15–40\,\text{m s}^{-1}$ and that there do not appear to be any flanking towering cumuli associated with them. Modeling studies (e.g., Young, 1977) have suggested that significant embryo growth is only possible in regions with cloud base updraft speeds are less than $3\,\text{m s}^{-1}$. A conclusion drawn from the NHRE is that it is probably not feasible to suppress hail growth in supercell storms (Atlas, 1977).

Early rainout and/or trajectory lowering liquid water depletion by salt seeding

The idea behind early rainout is to initiate ice phase precipitation lower into the feeder or daughter cell clouds where temperatures are in the range $-5\,°\text{C}$ to $-15\,°\text{C}$. If the prematurely initiated precipitation settles below an otherwise rain-free base, it could fall into the inflow into the storm and impede the flow of moisture into the storm, which, in turn, would reduce supercooled liquid water contents deeper in the storm. In addition, initiation of ice lower in the smaller turrets has the potential of reducing supercooled liquid water available for hail growth in the larger turrets, where updrafts are stronger and more conducive to the growth of larger hailstones.

Essentially the same seeding strategy can be used to achieve early rainout and trajectory lowering since if the precipitation particles initiated lower in the cloud do not precipitate from the rain-free cloud base, the number of hail embryos is increased causing beneficial competition.

Another proposed approach to hail reduction is to seed the base of clouds with salt particles or some other hygroscopic material and thereby initiate a warm rain process in the lower levels of the cloud. The concept, also called the *trajectory-lowering technique*, is that the precipitation settling out of the lower part of the cloud will deplete the liquid water in the cloud and therefore limit hailstone growth. It is based on the hygroscopic seeding strategies discussed above in which salt particles or some other hygroscopic material are introduced into the base of flanking clouds, thereby initiating a more vigorous warm rain process in the lower levels of the cloud. Some cloud modeling studies (Young, 1977) suggest that this may be a feasible approach in regions such as the High Plains of the United States or Canada where cloud base temperatures are cold and cloud droplet concentrations are large.

It has also been proposed to use a combination of salt seeding and ice phase seeding to both deplete supercooled liquid water and to promote beneficial embryo competition (Dennis and Musil, 1973). Only limited exploratory field studies have been performed to examine the feasibility of this approach.

These concepts are summarized in Fig. 2.21. Developing flanking line cells with weaker updrafts are shown on the left of the figure and the mature cell with strong updrafts on the right. One can also interpret the figure as being a time–height cross-section of a storm with zero time on the left and the time of the dissipating cells on the far right. We emphasize here that it is the weaker updraft regions of developing cumulus congestus, rather than the main cumulonimbus cell that is the preferred region for seeding.

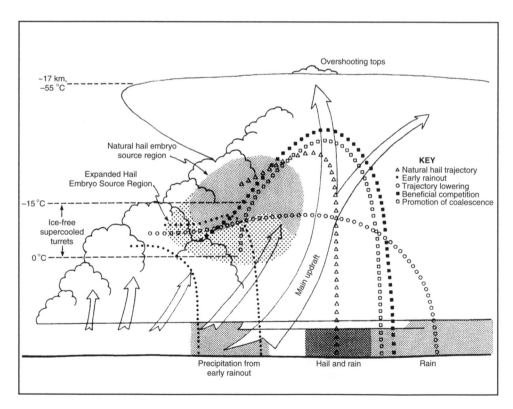

Figure 2.21 Hail suppression concepts. From World Meteorological Organization (1996).

It should be noted that hailstorms are often severe storms that can produce severe winds, tornadoes, and flash floods in addition to hail. Thus seeding hailstorms has the potential for undesirable consequences such as enhancing flood potential or severe wind damage. For example, in their modeling study of the impact of hail size on simulated supercell storms, van den Heever and Cotton (2004) found that decreasing hail size led to stronger low-level downdrafts and colder cold pools over larger areas beneath the storm. This is by virtue of the fact that melting and evaporation are surface phenomena on the hailstones and since, for a given hail water content, net surface area is increased when the hailstones are smaller (the surface-to-volume ratio is greater for smaller particles), a reduction of hail size leads to greater low-level cooling. The greater divergence associated with the more intense cold pools results in stronger divergence of flow from the forward flank downdraft into the region of the updraft and stronger convergence of air into the updraft. The enhanced low-level convergence favors stronger low-level tornado-scale circulations much like an ice-skater spinning faster as they bring in their arms. Thus those simulations imply that seeding thunderstorms to reduce hail size has the potential for increasing the chance of inducing tornadoes. Clearly we need a more holistic understanding of thunderstorm dynamics and microphysics before hailstorm modification is done routinely.

2.5.4 Field confirmation of hail suppression techniques

As noted previously, the reports of major success in reducing crop hail damage in the Soviet Union spawned a number of operational and scientific research programs on hail suppression in the United States, Canada, Switzerland, South Africa, and elsewhere. The Swiss and US programs were attempts to replicate the Soviet strategy. Unfortunately, the US/NHRE did not replicate the Soviet strategy for a variety of reasons. The meteorology over the High Plains of the United States is quite different than in the Soviet hail regions with the storms being cold-based and continental so that a warm rain process is not very active. An accumulation zone of supercooled raindrops is, therefore, not likely in that region. Furthermore, there appears to be a greater preponderance of supercell storms in the NHRE area than in the Soviet hail regions (Battan, 1969; Marwitz, 1972). The storms in NHRE were thus less amenable to hail suppression. Finally, scientists in NHRE were never able to successfully implement a rocket-based seeding strategy. Instead, most of the seeding was conducted as broadcast seeding from aircraft flying beneath the base of the strongest updrafts; a procedure that is least likely to produce competing embryos. The NHRE seeding experiment was therefore curtailed prematurely without definitive results (Atlas, 1977).

In contrast, the Swiss experiment (Federer *et al.*, 1986) had everything going for it. The meteorology in Switzerland is relatively similar to the Soviet hail regions. The experimenters successfully implemented a rocket-based seeding strategy and carried out a well-designed and well-implemented field program. Again, this experiment did not yield positive results of hail suppression. A possible contributor to the inability of finding a significant decrease in hailfall is that the 5-year experimental period was too short. The original design of the experiment called for a 5-year period to discern a seeding signal based on the optimistic expectation of a 60% reduction in kinetic energy of falling hail (Federer *et al.*, 1986). Another factor is that the seeding material only entered a small fraction of the storms thought to be seeded. Using chemical tracers, Linkletter and Warburton (1977) could find "seeded" silver in only 50% of the storms in 1973 and 70% in 1974 during NHRE. Based on those findings they predicted that less than 10% of the storms had enough silver to represent a significant seeding effect. Similar results were found by chemical analyses in the Swiss hail project (Lacaux *et al.*, 1985) where they found on one day that two cells exhibited only 7% and 25% coverage of seeding silver.

Similarly hail suppression experiments in Canada and South Africa have been inconclusive. Only long-term statistical analyses of non-randomized, operational programs have provided more convincing evidence suggesting that seeding can significantly reduce hail frequency. Mesinger and Mesinger (1992), for example, examined 40 years of operational hail suppression data in eastern Yugoslavia. After attempting to remove the effects of climatic fluctuations during the period, they estimated that the hail suppression projects reduced the frequency of hail between 15% and 20%. One can not rule out the possibility that natural climatic variations in hail frequency lead to the "apparent" reductions in hail frequency in such studies. Likewise, Smith *et al.* (1997) found a 45% decrease in crop damage due to hail suppression in a non-randomized operational project in North Dakota. Other evidence exists indicating decreased hail damage during operational hail suppression efforts in Greece (Rudolph *et al.*, 1994) and France (Dessens, 1998). An advantage of evaluating an operational program is that often one can work with long-period records such as 40 years in former Yugoslavia whereas randomized research programs typically cannot get funding for more than 5 years or so. The disadvantage is that one cannot totally eliminate concerns about *natural variability* in the climate.

Another concern about hail suppression is its impact on rainfall. Because hail-storms often occur in semi-arid regions where rainfall is limited, Changnon (1977) estimated that in general, the destructive effects of hail damage are often out-weighed by the positive benefits of rainfall from those storms. This is, of course, not true for certain high-risk crops such as tobacco, grapes, or certain vegetables.

Modeling studies like those of Nelson (1979) and Farley *et al.* (1977) suggested that rainfall and hailfall are positively correlated so that reductions of hailfall coincide with reductions in rainfall. But an evaluation of rainfall from an operational hail suppression program in Alberta, Canada by Krauss and Santos (2004) suggested that seeding to reduce hail damage also resulted in an increase in rain volume by a factor of 2.2.

Overall the scientific basis of hail suppression remains not fully resolved. Moreover, we urge scientists and operators to take a more holistic view of hail suppression activities by considering other potential responses of the storms to seeding including its affects on rainfall, cloud-to-ground lightning, severe winds, and tornadoes.

2.6 Modification of tropical cyclones

As mentioned in Chapter 1, the first attempt at modifying a tropical cyclone or hurricane occurred during Project Cirrus in 1947. Because of the extensive damage produced by hurricanes, interest in developing an economical technique to modify hurricanes developed to a high level. For example Gentry (1974) stated: "For scientists concerned with weather modification, hurricanes are the largest and wildest game in the atmospheric preserve. Moreover, there are urgent reasons for 'hunting' and taming them." Therefore in 1962 a joint project, between the US Weather Bureau and the US Navy, called Project STORMFURY, was created. Before we examine the results of that study, let us review the basic concepts of tropical cyclones.

2.6.1 Basic conceptual model of hurricanes

The tropical cyclone or hurricane is a large cyclonically circulating (counter-clockwise rotation in the Northern Hemisphere) weather system that is composed of bands of deep convective clouds (see Fig. 2.22) (Pielke and Pielke, 1997). Inside a radius of about 400 km, the low-level flow is convergent and associated lifting of warm, moist air produces extensive cumulonimbus clouds and precipitation. The storm is composed of an eyewall (a generally circular ring of intense cumulonimbus convection surrounding the often cloud-free eye), a region of stratiform cloud and precipitation outside the eyewall, and, beyond the eyewall, spiral bands of convective clouds and rainfall. Like a giant flywheel, the region within 100 km of the storm center is inertially stable and is not affected strongly by outside weather systems. Only through enhancements or reductions in the divergence pattern associated with jet streaks is the strength of the storm affected by larger-scale weather systems. Likewise it does not respond rapidly to

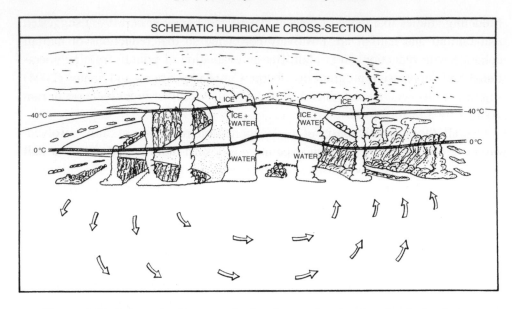

Figure 2.22 Schematic diagram of a hurricane, showing low-level circulation and cloud types. The highest clouds, composed of cirrus and cirrostratus, occur at the tropopause, which is about 16 km. From STORMFURY (1970).

small changes in heat release but only to sustained, large-amplitude changes in heating.

The primary energy driving the tropical cyclone comes from the sea. Air flowing over a warm ocean surface receives energy primarily in the form of sensible and latent heat. Small-scale turbulent eddies near the ocean surface transfer the heat and moisture upward to levels where it becomes saturated and cumulus clouds form. By condensing water to form cloud droplets and consequently releasing latent heat, the moisture and latent heat transferred from the ocean surface warm the cloudy air at a rate which is roughly proportional to the precipitation rate in the clouds. Thus cumulus clouds and subcloud eddies transfer the sensible and latent heat from the ocean surface to the middle and upper troposphere. As the air moves laterally outward from the central region of the storm at upper levels in stratiform anvil clouds, much of the energy gained at the ocean surface is radiated to space by infrared radiation.

These features of a tropical cyclone motivated Emanuel (1988) to suggest that a tropical cyclone is like an idealized heat engine called a Carnot engine. In a Carnot engine, heat is input at a single high temperature and all the heat output is ejected at a single low temperature. The amount of work produced by the Carnot engine is proportional to the difference between the input and output temperatures, and is the maximum amount of energy that can be extracted from a heat source. In the case of a tropical storm, the amount of work or the maximum strength of the winds in a storm is proportional to the difference in temperature between the heat input level (i.e., at the ocean surface), and the heat output level, or the tops of the stratiform–anvil clouds. However, this is not the total story. If the sea surface temperature and temperature at the tropopause were the only determinant factors on hurricane formation and strength, there would be more than ten times as many hurricanes as normally occur. Other factors which determine the extent to which larger scales of motion and convective scales interact in an optimum way to utilize the energy flowing from the ocean surface must also be important as discussed in Pielke (1990).

2.6.2 *The STORMFURY modification hypothesis*

The original STORMFURY hypothesis was first advanced by Simpson *et al.* (1963) and Simpson and Malkus (1964). They proposed that the additional latent heat released by seeding the supercooled water present in the eyewall cloud would produce a hydrostatic pressure drop that would modestly reduce the surface pressure gradient and as a consequence the maximum wind speed.

The original hypothesis was subsequently modified following a series of numerical hurricane simulations (Gentry, 1974). Those numerical experiments suggested that application of the individual cloud dynamic seeding hypothesis to towering cumuli immediately outward of the eyewall would cause enhanced vertical development of the towering cumuli and removal of low-level moisture from the boundary layer immediately beneath them. The loss of moisture outward from the eyewall would starve the clouds of moisture in the eyewall region, causing a shift in the eyewall convection outward to greater radii (as illustrated in Fig. 2.23). Like ice-skaters extending their arms, the storm should rotate slower and the winds diminish appreciably.

2.6.3 *STORMFURY field experiments*

For nearly a decade, STORMFURY performed seeding experiments in an attempt to reduce the intensity of hurricanes. Only a few storms were actually seeded, however, due to the fact the hurricanes are a relatively infrequent phenomenon and the experiment was constrained to operate in a limited region of the North Atlantic well away from land. Some encouraging results were obtained from seeding Hurricane Debbie in 1969 with 30% and 15% reductions in wind speeds following seeding on two days separated by a no-seed day. This led to greatly expanded field programs for a few years but the program was eventually curtailed in the late 1970s with no definitive results. STORMFURY succumbed to the very large *natural variability* of hurricanes including a period of very low hurricane

BEFORE

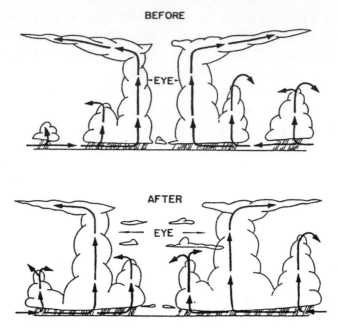

Figure 2.23 Hypothesized vertical cross-sections through a hurricane eyewall and rain bands before and after seeding. Dynamic growth of seeded clouds in the inner rain bands provides new conduits for conducting mass to the outflow layer and causes decay of the old eyewall. From Simpson *et al.* (1978).

frequency from the middle 1960s to the middle 1980s. The conclusions of the project were summarized in Sheets (1981) in which 10–15% decreases in the maximum wind speed with associated damage reductions of 20–60% should occur if the STORMFURY hypothesis (Gentry, 1974) were implemented. No significant changes in storm motion or storm averaged rainfall at any specific location would be expected.

3

The fall of the science of weather modification by cloud seeding

For nearly two decades vigorous research in weather modification was carried out in the United States and elsewhere. As shown in Fig. 3.1, federal funding in the United States for weather modification research peaked in the middle 1970s at nearly $19 million per year. Even at its peak, funding for weather modification research was only 6% of the total federal spending in atmospheric research (Changnon and Lambright, 1987) and this amount included considerable support for basic research on the physics of clouds and of tropical cyclones. Nonetheless, research funding in cloud physics, cloud dynamics, and mesoscale meteorology was largely justified based on its application to development of the technology of weather modification. Research on the basic microphysics of clouds particularly benefited from the political and social support for weather modification.

By 1980, the funding levels in weather modification research began to fall appreciably and by 1985 they had fallen to the level of $12 million. After 1985, funding in weather modification research became so small and fragmented that no federal agency kept track of it. Currently the Bureau of Reclamation has only about $0.25 million per year that can be identified as weather modification. They have operated a program in Thailand that was supported by the Agency for International Development. Basic research in the National Science Foundation that can be linked to weather modification is on the order of $1 million. Likewise the Department of Commerce has no budgeted weather modification program, but has supported a cooperative state/federal program at about the $3.5 million level. This on again–off again "pork barrel" program is supported by congressional write-ins rather than a line item in the National Oceanic and Atmospheric Administration (NOAA) budget. In FY-2003, the Bureau of Reclamation administered this program, but no such funds were earmarked for either the Bureau of Reclamation or NOAA in FY-2004 or FY-2005. In this program, states having strong political lobbying support for weather modification are earmarked for support in this program. Overall the total federal program for weather modification in the United States

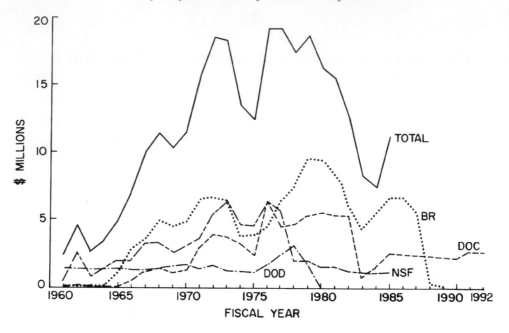

Figure 3.1 Estimates of federal spending levels in the United States for weather modification research. The agencies are BR (Bureau of Reclamation), DOC (Department of Commerce), DOD (Department of Defense), and NSF (National Science Foundation). Data provided courtesy of the National Science Foundation.

is on the order of 10% of its peak level in the middle 1970s. What caused this virtual crash in weather modification research?

Changnon and Lambright (1987) listed the following reasons for this reduction in funding:

- poor experimental designs;
- widespread use of uncertain modification techniques;
- inadequate management of projects;
- unsubstantiated claims of success;
- inadequate project funding; and
- wasteful expenditures.

Changnon and Lambright concluded that the primary cause of the rapid decline in weather modification funding was the lack of a coordinated federal research program in weather modification. However, there are other factors that also must be considered:

- weather modification was oversold to the public and legislatures;
- demands for water resource enhancement declined due to an abnormal wet period;

- impact of "Reaganomics";
- change in public attitude toward the environment; and
- the limited period of government and public interest in specific environmental problems and the resultant diversion of public attention to other weather and climate issues.

Changnon and Lambright argued that Congress' decision to terminate the National Science Foundation's (NSF's) lead agency role in weather modification research in 1968 was a major blunder. Instead, support for weather modification research in the United States became fragmented between the US Bureau of Reclamation (BR) of the Department of the Interior, the Department of Defense (DOD), the NOAA, the NSF, and the Department of Agriculture (DOA). With the exception of NSF, these agencies were mission oriented and as such the primary support was not for basic research, but development of a technology to be applied to water resource enhancement, severe weather abatement, agricultural uses, and military applications. This led to a great deal of interagency rivalry and to research programs having short-term goals of establishing a weather modification technology. As a result, most of the research programs were of a "black box" nature in which clouds were seeded in a randomized statistical design where the only measurements were the amount of rain on the ground or the amount of hail damage, etc. There were few attempts to design studies in which the entire sequence of hypothesized responses to seeding were measured. Did seeding produce more numerous ice crystals? Is the precipitation particle size spectra different in seeded clouds? What are the sequence of dynamic responses of a cloud to seeding? It was only in very recent years, about the time of the crash in weather modification funding, that major attempts to answer these fundamental questions were performed in weather modification programs. Clearly the lack of a coordinated federal research program had a major adverse impact on weather modification research.

Another factor affecting the decline in weather modification research was that from the early days of Project Cirrus onward, weather modification was oversold to the funding agencies and to the public. The cry was weather modification is good! It can enhance rainfall, suppress hail, weaken winds in hurricanes, inhibit lightning, and put out forest fires. Many scientists and program managers argued that only a few years of research were needed to put cloud seeding on a sound scientific basis and for it to be ready for routine applications. Such overselling was often undoubtedly performed by a few unscrupulous scientists, program managers, and commercial seeders. However, more importantly, this behavior reflected the rather naive perceptions of many in the scientific community of the difficult problems faced by the weather modification community. Not only are the physical problems faced by weather modification scientists extremely complex, but scientists generally underestimated the impact of the *natural variability* of precipitation and weather on discerning a seeding signal from the natural background. Thus this overselling has led to a lack of scientific credibility since after more than 50 years, scientists are still disagreeing about the outcome of cloud seeding projects and the scientific status of the field. This loss of scientific credibility came to a head in a meeting of atmospheric scientists organized by the United States National Academy of Sciences to assess research themes of the 1980s (National Academy of Sciences, 1980). The consensus of the attendees was that weather modification should not be given a high priority for federal research funding in the 1980s. Many were vigorously opposed to support for weather modification research. They argued that climate, atmospheric chemistry, and mesoscale meteorology should be given the highest priority.

Major support for weather modification research has traditionally come from the semi-arid western states of the United States where the demand for water often exceeds the supply. However, in the middle 1970s to the 1980s, precipitation was often above normal so that demand for cloud seeding projects and weather modification research dwindled. During the same period, tropical cyclones striking the east coast of the United States also reached a low level (Gray, 1990, 1991). These factors plus the Reagan administration's attempts at reducing federal expenditures combined to make it easy for Congress to cut federal funding in weather modification research.

Finally, enthusiasm for weather modification developed during a time when the prevailing attitude was to restructure the environment to suit the needs of mankind by building dams, cutting trees, building sea walls, and other means of altering the environment. This philosophy has been replaced, by and large, with an environmental awareness ethic. No longer is it acceptable to build nuclear power plants, dams, or highways without a major assessment of the total environmental impact. Weather modification, with its primary aim to change the weather, no longer fits the environmental ethic that prevails in many developed nations.

As a consequence of the above factors, weather modification research experienced a sharp decline in funding in the middle 1980s in the United States. A similar decline in funding was also experienced in many developed countries during the same period, largely due to the loss of scientific credibility of the field. Some countries such as China, however, have maintained vigorous weather modification research programs. More than 100 operational cloud seeding programs currently exist throughout the world including the United States. All that remains of weather modification research in the United States are "pork barrel" congressional write-in programs that are on again and off again every budgeting cycle. Moreover, the support for the research is constrained to those lobbying states and not generally available to the scientific community. Such funding of research does not lend itself to quality, long-term scientific investigations.

There is also some evidence of a rebirth of weather modification research. In Australia, for example, several operational and research programs for precipitation enhancement have been instigated. During the drought of 1988 and in the extended drought in the early 2000s in the southwest United States, operational cloud seeding projects for snowpact enhancement have flourished. Perhaps motivated by this activity a National Research Panel was established to examine the status of weather modification. The panel (National Research Council, 2003a) recommended "that a coordinated national program be developed to conduct a sustained research effort in the areas of cloud and precipitation microphysics, cloud dynamics, cloud modeling, and cloud seeding; it should be implemented using a balanced approach of modeling, laboratory studies, and field measurements designed to reduce the key uncertainties." With the droughts of 1988 and 2002 (Pielke *et al.*, 2005a), and renewed enhanced hurricane activity along the east coast of the United States such as in 2004 and 2005, will demands for weather modification research be increasing? Certainly several operational programs were begun in the drought of 1988 in the United States.

Clearly there is a great need to establish a more credible, stably funded scientific program in weather modification research, one that emphasizes the need to establish the physical basis of cloud seeding rather than just a "black box" assessment of whether or not seeding increased precipitation. We need to establish the complete hypothesized physical chain of responses to seeding by observational experiments and numerical simulations. We also need to assess the total physical, biological, and social impacts of cloud seeding, or what we call taking a holistic approach to examining the impacts of cloud seeding. E. K. Bigg, for example (Bigg, 1988, 1990b; Bigg and Turton, 1988) suggested that silver iodide seeding can trigger biogenic production of secondary ice nuclei. His research suggests that fields sprayed with silver iodide release secondary ice nuclei particles at 10-day intervals and that such releases could account for inferred increases in precipitation 1–3 weeks following seeding in several seeding projects (e.g., Bigg and Turton, 1988). If Bigg's hypothesis is verified, an implication of biogenic production of secondary ice nuclei is that many seeding experiments have thus been contaminated such that the statistical results of seeding are degraded. This effect would be worst in randomized cross-over designs and in experiments in which one target area is used and seed days and non-seed days are selected over the same area on a randomized basis. Thus, not only is the weather modification community faced with very difficult physical problems and large natural variability of the meteorology, but they also are faced with the possibility of responses to seeding through biological processes.

We shall see later that scientists dealing with human impacts on global change are also faced with very difficult physical problems, large *natural variability* of climate, and the possibility of complicated feedbacks through the biosphere. There is also a great deal of overselling of what models can deliver in terms of "prediction" of human impacts over timescales of decades or centuries.

Part II

Inadvertent human impacts on regional weather and climate

In Part I, we discussed man's purposeful attempts at altering weather and climate by cloud seeding. We saw that there is strong evidence indicating that clouds and precipitation processes can be altered through cloud seeding. In general, however, our knowledge about clouds is still not sufficient to enable cloud seeders to alter precipitation processes and severe weather in anything but the simplest weather systems (e.g., orographic clouds, supercooled fogs and stratus, and some cumuli). In Part II we examine the mechanisms and evidence indicating that there have been changes in regional weather and climate through anthropogenic emissions of aerosols and gases, and through alterations in landscape. Regional scale refers to horizontal scales of less than a few thousand kilometers. On these scales we examine the possible changes in rainfall, severe weather, temperatures, and cloud cover caused by anthropogenic activity.

4

Anthropogenic emissions of aerosols and gases

A variety of human activities result in the release of substantial quantities of aerosol particles and gases which may influence cloud microstructure, precipitation processes, and other weather phenomena. In this section we examine evidence that these particulate and gaseous releases are influencing regional weather and climate. In this chapter, however, we will not focus on urban emissions of particulates and gases; that discussion will be reserved for Chapter 5.

4.1 Cloud condensation nuclei and precipitation

In our discussion of purposeful modification of clouds in Part I, we described attempts to enhance precipitation from warm clouds by seeding them with hygroscopic materials. The hygroscopic particles which are called cloud condensation nuclei (CCN) can alter the microstructure of a cloud by changing the concentrations of cloud droplets and the size spectrum of cloud droplets (Matsui *et al.*, 2004, 2006). There are numerous examples of anthropogenic sources of CCN, including automobile emissions (Squires, 1966), certain urban industrial combustion products, and the burning of vegetative matter, especially sugar cane.

Warner and Twomey (1967) observed substantial increases in CCN concentrations beneath the base of cumulus clouds and increases in cloud droplet concentrations above their bases downwind of areas in which the burning of sugar cane fields was taking place. The practice of burning sugar cane fields to remove leaf and trash before harvesting is quite common in most areas where sugar cane is grown. Warner and Twomey hypothesized that the larger numbers of CCN and cloud droplets would slow down collision and coalescence growth of precipitation by virtue of their smaller size and, consequently, small collection efficiencies and small collection kernels. Slower coalescence growth should therefore lead to less rainfall, at least from smaller clouds which do not contain large amounts of liquid water. Warner (1968) performed an analysis of precipitation records downwind of sugar cane burning areas near Bunderburg, Queensland, Australia and those in upwind "control" areas. The analysis was performed only during the 3-month burning periods and over a 60-year record. He found substantial reductions in rainfall amounting to a decrease of approximately 25% downwind of the burning areas. Attempts to confirm these findings in other areas of Australia (Warner, 1971) and over the Hawaiian Islands (Woodcock and Jones, 1970) have been unsuccessful. Nonetheless, Warner's findings strongly suggest that enhanced con- centrations of CCN particles can lead to reductions in rainfall, at least in some clouds and in some regions. Other studies by Hobbs and Radke (1969) and Kaufman and Fraser (1997) suggest that smoke from biomass burning can be prolific sources of small CCN. Rosenfeld (1999) used satellite-based radar on the Tropical Rainfall Measuring Mission (TRMM) satellite to estimate rainfall as well as the Visible and Infrared Sensor (VIRS) to infer cloud droplet sizes. He presented evidence that widespread smoke from biomass burning in Indonesia effectively shuts off precipitation in the affected regions.

There is also evidence suggesting that changes in the size spectrum of CCN by anthropogenic emissions may be responsible for enhancing precipitation (Hobbs *et al.*, 1970; Mather, 1991). Hobbs *et al.* found that pulp and paper mills and certain other industries are prolific sources of CCN. Observations of some of the clouds downwind of the source factories suggest that they actually produce precipitation more efficiently than other clouds in the region. The analysis of precipitation and streamflow records for a period before construction of the mills to after the mills were in operation suggested that rainfall downwind of the pulp and paper mills was 30% greater in the later period than before the mills were put in operation (Hobbs *et al.*, 1970). Subsequent studies (Hindman *et al.*, 1977a) indicated that large ($>0.2 \mu m$) and giant ($>2 \mu m$) CCN (GCCN) were increased in concentration downwind of the plants, while small CCN ($<0.2 \mu m$) were not significantly altered in concentration. It was concluded by Hobbs *et al.* (1970) and Hindman (1977a,b) that the large and giant nuclei emitted by the mills increased the efficiency of the collision and coalescence process thereby enhancing precipitation.

Attempts to simulate the response of cumulus clouds and stratocumulus clouds to injections of large and ultra-giant CCN having similar concentrations to those emitted by paper mills (Hindman *et al.*, 1977b) revealed little change in simulated precipitation. They speculated that heat and moisture emitted by the mills, in combination with the CCN, may have been responsible for increased rainfall. This conclusion is consistent with the finding by Hindman *et al.* (1977a) that liquid water contents in clouds downwind of the paper mills were 1.3 to 1.5 times greater than surrounding clouds, though the results were not statistically significant.

Modeling studies of marine stratocumulus clouds by Feingold *et al.* (1999) showed that increased concentrations of giant CCN (GCCN) enhanced precipitation in clouds with moderately high CCN concentrations. However, if CCN concentrations were quite low, the natural precipitation process was so efficient that GCCN had little influence on precipitation.

Another example of a possible link between anthropogenic emissions of CCN and precipitation is evident from ship track trails viewed on satellite imagery (Coakley *et al.*, 1987; Scorer, 1987; Porch *et al.*, 1990). The ship track trails appear as a line of enhanced brightness in satellite imagery particularly at $3.7\,\mu m$ wavelength. Figure 4.1 shows clear evidence of the much brighter ship trails. The tracks are often as long as 300 km or more and about 9 km wide (Durkee *et al.*, 2000). They typically form in relatively shallow boundary layers between 300 m and 750 m deep. They do not form in boundary layers deeper than 800 m (Durkee *et al.*, 2000). The implications of those brighter clouds to global climate

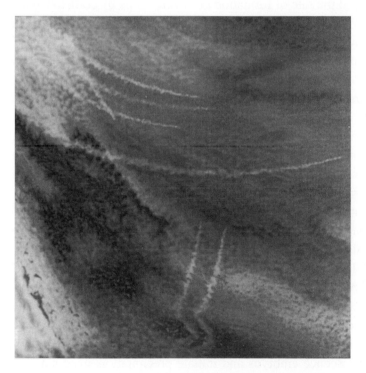

Figure 4.1 Images constructed from 1 km by NOAA-9 AVHRR data $3.7\,\mu m$ radiance for a $500 \times 500\,km$ region of ocean off the coast of California. The data were taken at 2246 UTC on 3 April 1985. The ship tracks evident at $3.7\,\mu m$ may be due to a shift toward smaller droplets for the contaminated clouds. The shift causes an increase in reflected sunlight at $3.7\,\mu m$. From Coakley *et al.* (1987).

will be discussed in Part III. Here, we will concentrate mainly on evidence of the relationship of ship CCN emissions on cloud structure.

The prevailing hypothesis is that the ship's trails appear brighter on satellite imagery because the effluent from the ships is rich in CCN particles. The more numerous CCN particles create larger concentrations of cloud droplets which then reflect more solar energy than the surrounding clouds. Aircraft observations in the ship track clouds as well as surrounding clouds (Radke *et al.*, 1989) reveal that the ship track clouds exhibit higher droplet concentrations, smaller droplet sizes, and higher liquid water content than surrounding clouds. In some cases, deeper clouds in the ship tracks are also observed (Ackerman *et al.*, 2000b). The higher droplet concentrations and smaller droplet sizes are consistent with the hypothesis that the higher cloud brightness is due to a higher concentration of CCN in the ship effluent. The greater liquid water content and deeper clouds in the ship trails, however, is at first a somewhat surprising result.

Albrecht (1989) hypothesized that the higher droplet concentration in ship track clouds reduced the rate of formation of drizzle drops by collision and coalescence similar to Warner and Twomey's (1967) hypothesis for cloud microphysical structure downwind of sugar cane fields. The reduced rate of drizzle formation then resulted in higher liquid water contents and higher droplet concentration in ship track clouds compared to surrounding clouds. Radke *et al.* (1989) found that the concentration of drizzle drops (droplets of diameter $\geq 200\,\mu m$) in the ship track was only 10% of that in surrounding clouds. This supported Albrecht's hypothesis that the liquid water content in the tracks was higher because the higher droplet concentration and smaller droplet sizes limited collision and coalescence.

Ackerman *et al.* (1995) performed modeling studies which suggest that if the marine boundary layer is sufficiently pristine, natural clouds readily drizzle out to the extent that they become so optically thin that cloud top radiative cooling is insufficient to destabilize the marine boundary. Boundary layer clouds thereby cannot persist. Ship effluent, by enriching the boundary layer with CCN, will inhibit the drizzle process and thereby produce boundary layer clouds sustained by cloud top radiative cooling.

We have also seen that effluent from paper pulp mills, though high in CCN concentrations, appear to result in enhanced precipitation downwind. It has been hypothesized that the large and ultra-giant aerosol particles emitted by paper mills serve as coalescence embryos and initiate precipitation-sized particles. Thus the susceptibility of the drizzle process in marine stratocumulus clouds to anthropogenic emissions of CCN may depend on the presence or absence of large and ultra-giant aerosol particles in the subcloud layer (see Feingold *et al.*, 1999). Over the open sea, the dominant large and ultra-giant aerosol particles are sea salt particles. While these particles contribute only 10% or less to the total CCN population, they represent major contributors to the large end of the aerosol size spectrum. Their concentration in the marine boundary layer varies with wind speed, being in greater numbers with stronger winds. We hypothesize that the susceptibility of the drizzle process in marine stratocumuli to anthropogenic sources of CCN such as ships will depend on wind speed, being less susceptible at higher wind speeds than lower.

To understand the importance of drizzle to the cloudy marine boundary layer, consider the LES/bin-microphysics[1] simulations performed by Stevens *et al.* (1998). In the absence of drizzle they simulated a stable stratocumulus cloud. When drizzle was present, however, evaporation of drizzle destabilized the subcloud layer. As a result, the cloud structure changed to a cumulus-under-stratus regime, which has very different optical properties than a solid stratocumulus field. Thus, as found in the simulations in Jiang *et al.* (2002), higher CCN concentrations suppressed drizzle which resulted in weaker penetrating cumulus and an overall reduction in the water content of the clouds. However, the destabilizing effect of drizzle only occurs when drizzle does not reach the surface. When drizzle settles to the surface, such as in heavier drizzle rate situations or when the drops are larger, the entire boundary layer is cooled and is stabilized (Paluch and Lenschow, 1991; Jiang *et al.*, 2002).

Another hypothesis is that the ship track clouds exhibit higher liquid water content because the heat and moisture emissions from the ships invigorate the air motions in the clouds, thereby creating deeper and wetter (hence brighter) clouds. Porch *et al.* (1990) examined this hypothesis and showed that ship tracks are characterized not only by greater brightness but also by clear bands on the edges of the cloud tracks (see Fig. 4.2). They speculated and provided some modeling evidence that the heat and moisture fluxes from the ship effluent excited a dynamic mode of instability which, in some marine stratocumulus environments, led to enhanced upward and downward motion associated with the cloud circulations. Measurements during the Monterey Area Ship Track Experiment (MAST), however, suggest that the heat and moisture emissions from ships is much too small to have any impact on the clouds (Hobbs *et al.*, 2000). An alternative explanation for the formation of deeper and wetter clouds in the ship tracks and perhaps clearing to the sides of the ship tracks is that when drizzle is suppressed, cloud top radiative cooling is enhanced. The enhanced cloud top cooling destabilizes the cloud layer which would result in strong ascending motions in the clouds, transporting more moisture aloft making the clouds wetter and deeper.

---

[1] Large eddy simulation (LES) refers to simulation that explicitly represents large eddies in the boundary layer and/or clouds. Bin microphysics refers to explicit representation of droplets and/or ice particles in discrete bins.

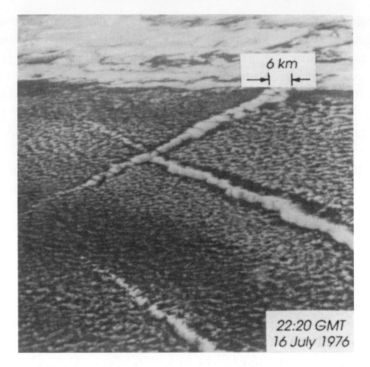

Figure 4.2 Ship trail photography from Apollo-Soyuz on 16 July 1976 at 2220 GMT. From Porch *et al.* (1990). © Pergamon Press PLC.

In addition, in compensation for the enhanced upward motions, sinking motions surrounding the regions of enhanced ascent could cause clearing outside of the ship tracks.

Other evidence of aerosol influences on precipitation are what Rosenfeld (2000) calls "pollution tracks" as viewed by Advanced Very High Resolution Radiometer (AVHRR) satellite imagery. Figure 4.3 illustrates pollution tracks in the Middle East, Canada, and South Australia. It is inferred that the "pollution tracks" are clouds composed of numerous small droplets that suppress precipitation. It is interesting to note that this effect is not limited to warm clouds. Some of the clouds exhibiting "pollution tracks" are in Canada where ice precipitation processes are prevalent. Borys *et al.* (2000, 2003) also provide evidence that pollution can suppress precipitation in wintertime orographic clouds. Their analysis suggests that pollution enhances the concentration of CCN and as a consequence cloud droplets are smaller, which reduces the efficiency of ice crystals collecting supercooled droplets or riming. Givati and Rosenfeld (2004) analyzed orographic precipitation records downwind of major urban centers in Israel and California and inferred that precipitation is suppressed by 15–25% downwind of those urban areas. Perhaps most significant in Rosenfeld's analysis is the conspicuous absence of pollution tracks over the

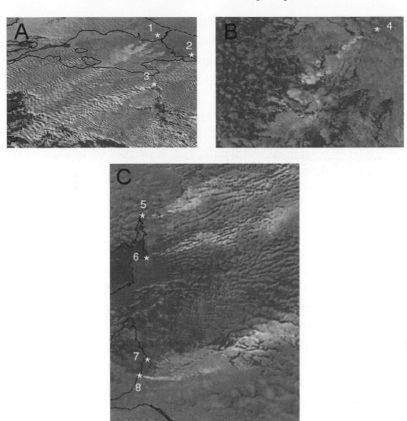

Figure 4.3 Satellite visualization of NOAA AVHRR images, showing the microstructure of clouds for three cases over three different continents with streaks of visibly smaller drops due to ingestion of pollution originating from known pollution sources that are marked by white numbered asterisks. (A) A 300×200 km cloudy area containing yellow streaks originating from the urban air pollution of Istanbul (*1), Izmit (*2), and Bursa (*3) on 25 December 1998 at 12:43 UT. (B) A 150×100 km cloudy area containing yellow streaks showing the impact of the effluents from the Hudson Bay Mining and Smelting compound at Flin-Flon (*4) in Manitoba, Canada (54° 46′ N 102° 06′ W), on 4 June 1998 at 20:15 UT. (C) An area of about 350×450 km containing pollution tracks over South Australia on 12 August 1997 at 05:25 UT originating from the Port Augusta power plant (*5), the Port Pirie lead smelter (*6), Adelaide port (*7), and the oil refineries (*8). All images are oriented with north at the top. The images are color composites, where the red is modulated by the visible channel; blue is modulated by the thermal infrared; and green is modulated by the solar reflectance component of the $3.7\,\mu$m channel, where larger (greener) reflectance indicates smaller droplets. The composition of the channels determines the color of the clouds, where red represents cloud with large drops and yellow represents clouds with small drops. The blue background represents the ground surface below the clouds. From Rosenfeld (2000). Reprinted with permission from D. Rosenfeld, © 2000 American Association for the Advancement of Science. See also color plate.

United States and western Europe. The implication is that these regions are so heavily polluted that local sources cannot be distinguished from the widespread pollution-induced narrow droplet spectra in those regions. In Part III, we examine the evidence that such widespread pollution can impact global radiation budgets and the hydrological cycle.

In summary, the evidence is compelling that anthropogenic emissions of CCN can result in decreases or increases of precipitation, depending on their size distribution. Research suggests that the impacts of aerosol on precipitation can be quite significant in some regions.

4.2 Aircraft contrails

It has become quite common to observe jet contrails covering extensive parts of the sky in regions of heavy jet traffic. Often several independent contrails merge to form an almost solid overcast of thin clouds, much like natural cirrus. The primary source of contrail formation is the water vapor emitted by the jet aircraft engine during combustion. Murcray (1970) noted that a typical medium-sized commercial jet such as a Boeing 727 burns 3100 kg of fuel per hour while cruising and produces over 1.2 times as much water (more than a kilogram per second) as a result of the combustion in the presence of atmospheric oxygen. At very cold temperatures such as exist in the upper troposphere, the saturation vapor pressures with respect to water and ice are very small in magnitude and since there is little difference between those saturation values and typical water vapor mixing ratios, small additions of water vapor to the air can lead to water saturation and also supersaturation with respect to ice. Thus the water emitted by jets is the primary cause of contrail formation.

When the moisture-laden jet exhaust enters the very cold atmosphere, it is rapidly cooled forming a cloud of small droplets. Evidence suggests that the liquid phase is very short lived (on the order of 1 s) and the droplets freeze to form a cloud composed of nearly spherical ice crystals a few micrometers in diameter (Murcray, 1970). As a consequence of their small size, and hence their small settling velocities, and because the atmosphere may be slightly supersaturated or only weakly subsaturated with respect to ice, these crystals can survive for periods of several hours or more.

We now examine what impact such clouds of ice crystals in the upper troposphere can have on regional weather and climate. The most obvious impact is that they can alter the radiation budget that affects surface temperatures. That is, the contrails reflect incoming solar radiation back to space and absorb and reradiate upwelling terrestrial or longwave radiation back toward the ground (for a more detailed discussion of atmospheric radiation the reader is referred to Part III.)

Figure 4.4 illustrates that a portion of the incoming solar energy (R_{sd}) incident on the Earth's surface is reflected back to the atmosphere (R_{su}), a part is conducted into the ground (G_0), some is transferred into the atmosphere by sensible heat convection (S_0) and by latent heat transfer (L_0), and a portion is emitted by terrestrial radiation (R_{lu}). Part of the upwelling terrestrial radiation is transmitted through the atmospheric window (see Part III) while the remainder of the terrestrial radiation is absorbed in the atmosphere by water vapor, carbon dioxide and other trace gases, and by clouds. A percentage of this terrestrial radiation absorbed in the atmosphere is reradiated back towards the ground (R_{ld}) and is absorbed at the Earth's surface. The surface temperature at any time is controlled by a balance between these contributions to the energy transfer. If incoming solar radiation is reduced (and all other contributions remain the same), then surface temperatures will be cooler. If R_{ld} is increased and all other contributions remain the same, surface temperatures will rise.

Kuhn (1970) observed the radiation budget of contrails from an aircraft. He found a 500-m thick contrail depleted incoming solar radiation (R_{sd}) by 15% and increased downward terrestrial radiation (R_{ld}) by 21%. The enhanced terrestrial radiation, however, could not make up for the loss of solar radiation, so that surface temperatures are reduced by the presence of contrails during the daytime. At nighttime, when there is no incident solar radiation, the enhanced downward flux of terrestrial radiation leads to a net warming. Thus the presence of contrails reduces afternoon maximum temperatures, and raises nighttime minimums, causing a moderation of local climate. Kuhn (1970) calculated that if the contrails were

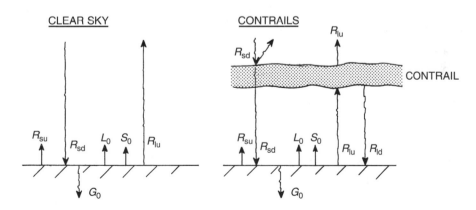

Figure 4.4 Energy budget at Earth's surface, where R_{sd} is incoming solar radiation, R_{su} is reflected solar radiation $\left(\text{albedo} = \frac{R_{su}}{R_{sd}}\right)$, G_0 is heat diffusion into the ground, L_0 is latent heat transfer to the atmosphere, S_0 is sensible heat transfer to the atmosphere, R_{lu} is upwelling terrestrial radiation, and R_{ld} is downward terrestrial radiation.

persistent over a 24-hour period then the net effect would be a 5–6 °C cooling of surface temperatures. This result, however, depends on the latitude and length of day, and the optical thickness of the cloud. Warming is the net effect for shorter days and higher latitudes and optically thin contrails. But contrails typically last only 2–6 hours and moreover, most of the jet aircraft traffic is during the day. So the major impact of contrail cirrus would be a cooling response. Moreover, as discussed by Sassen (1997) the sign of the climatic impact of contrails is a function of the particle size, with clouds containing relatively small particles increasing reflectance and those with larger particles trapping terrestrial radiation. As noted by Khvorostyanov and Sassen (1998) contrails differ in structure from natural cirrus clouds since they are composed mainly of very small ice crystals, especially early in their lifetimes. Thus modeling studies of their potential impacts treating them as being similar to ordinary cirrus and concluding that they lead to surface warming (Liou *et al.*, 1991; Schumann, 1994) may be in error.

Changnon (1981a) performed a climatological analysis of cloudiness, sunshine, and surface temperatures over the midwestern United States during the period 1901–1977. He found that in the period since 1960, there was more cloud cover and a decrease in sunshine, and that surface temperature extremes were moderated (less extreme minimum and maximum monthly averages), especially in the area of the midwest where jet traffic is greatest. In a study of sky cover over the United States, Seaver and Lee (1987) found a decrease in cloudless days over large regions of the United States since 1936. The results are consistent with the hypothesis that increased contrail formation led to the changes, but *natural fluctuations* in climate trends cannot be ruled out.

It has also been proposed that contrails can seed lower-level clouds with ice crystals and thereby enhance surface precipitation (Murcray, 1970). Natural cirrus ice crystals are known to survive long fall distances and potentially seed lower-level clouds (Braham and Spyers-Duran, 1967). Whether or not contrail crystals, which are smaller than natural cirrus crystals, can survive falling through great depths of the troposphere is not known. If, indeed, high concentrations of contrail crystals can survive descent into low-level, water-rich clouds such as orographic clouds and frontal clouds, they could seed those clouds with ice crystals and enhance surface precipitation in a "seeder–feeder" type process (e.g., Bergeron, 1965; Browning *et al.*, 1975; Hobbs *et al.*, 1980; see Cotton, 1990 for a review of the "seeder–feeder" process). This hypothesis has not been supported by any further studies, however.

In summary, contrails have the potential of altering regional and possibly even global climate if they become extensive enough. Their biggest potential impact is on moderating surface temperature extremes (i.e., maximum and minimum temperatures). It is also possible that they could impact daily average surface temperatures, but whether they will cause a net increase or decrease in surface temperatures depends on their thickness, persistence, latitude, time of the year, and aircraft operations.

4.3 Ice nuclei and precipitation

In Part I we examined cloud seeding strategies in which artificial ice nuclei (IN) were purposely inserted in clouds to enhance precipitation. We have seen that it is no simple matter to relate enhanced concentrations of ice nuclei to increase in surface precipitation. Although cloud seeding clearly increases ice crystal concentrations, the impact of those higher ice crystal concentrations on surface precipitation has only been identified in a limited number of cloud types and locations. The same can be said for inadvertent emissions of IN.

Schaefer (1966) showed that automobile exhaust emissions from lead-burning gasoline were prolific sources of IN, especially when those exhaust products react with iodine vapor. Many industries, especially those associated with steel production and refining of metal ores, are well-known sources of IN. Schaefer (1969) described measurements of IN downwind of several eastern United States cities in which IN concentrations were substantially enhanced above those found in pollution-free regions. In some cases, Schaefer measured concentrations in excess of 1000 per liter, which would likely create a stable cloud of very small ice crystals leading to reduced precipitation; a phenomenon called *overseeding*. Schaefer also described examples of plumes of ice crystal clouds extending downwind from major industrial effluents while no such ice crystal clouds could be seen in the surrounding countryside. One author (Cotton) has also observed such localized plumes of ice crystal clouds downwind of Buffalo, New York state during research flights in that region in the 1960s. Whether those enhanced ice crystal concentrations have any significant impact on surface precipitation, however, is unknown. Schaefer (1969) also suggested that dust from plowed fields can serve as ice nuclei, again causing enhanced ice crystal emissions by anthropogenic means. We will discuss the regional impacts of dust in Section 4.5 and the potential role of dust on climate more fully in Part III. There we will show that both Saharan and Asian dust are not only sources of IN but also sources of CCN, and GCCN or ultra-giant particles.

Bigg (1990a) has suggested that there has been a systematic decrease in IN concentrations at several sites in the Southern Hemisphere as well as Hawaii over the last 25 years. It is not known at this time if such observations are a direct result of the increased use of lead-free gasolines or contamination of natural ice nuclei by pollutants, or just a measurement anomaly. Nonetheless, we agree with Bigg

(1990b) that persistent, baseline measurements of IN along with CCN should be routinely made at a number of locations throughout the world.

4.4 Other pollution effects

Besides being sources of CCN and IN, air pollution can be high in concentrations of total aerosol particles. These particles are sufficiently numerous in urban areas that they can deplete direct solar radiation in cities by about 15%, sometimes more in winter and less in summer (Landsberg, 1970). Welch and Zdunkowski (1976) used a model to show that solar radiative heating of a polluted boundary layer can cause warming of $4\,°C\,h^{-1}$ in the pollution layer with a zenith angle of 45°. Hänel *et al.* (1972) reported observed heating rates from absorption of solar radiation by aerosols as large as about $0.5\,°C\,h^{-1}$ during the middle of the day under clear sky conditions over Frankfurt, Germany. Welch *et al.* (1978) applied a two-dimensional model to a polluted urban area for stagnant synoptic conditions and found temperatures at the ground to be reduced by $2\,°C$ because of low-level pollution sources and up to $7\,°C$ when the pollution was situated higher above the surface. During the Yellowstone National Park fires of 1988, as the plume spread over Golden, Colorado, the daily total global horizontal irradiance was 91% of the solar flux measured on a clear day (Hulstrom and Stoffel, 1990). While this reduction in solar radiation would lead to cooling of the ground, it is usually dominated by the urban heat island effect which we will discuss in Chapter 5.

In many parts of the world, widespread haze spreads from highly populated continental areas over the open oceans. A major field program called the Indian Ocean Experiment (INDOEX) described by Ramanathan *et al.* (2001) was carried out to study the haze layers that spread over the North Indian Ocean and South and Southeast Asia. The haze layer, which extends to a height of 3 km, is mainly of anthropogenic origin and is composed of inorganic and carbonaceous particles, including carbon black clusters, fly ash, and mineral dust. Many of these aerosol particles are highly absorbing and thereby contribute to substantial warming of the dust layer and cooling of the underlying surface. Modeling studies (Kiehl *et al.*, 1999; Ackerman *et al.*, 2000a) suggest that the heating in the haze layer can be large enough to totally desiccate boundary layer clouds including trade wind cumuli.

Many industries are also major sources of moisture. Power plant steam plumes and cooling ponds (Murray and Koenig, 1979; Orville *et al.*, 1980, 1981) release sufficient amounts of moisture into the atmosphere to cause cloud and fog formation. Especially in the cold winter months when saturation mixing ratios are small in magnitude, these moisture sources can lead to the persistence of cloud or fog. Highway departments often place fog caution signs along roads near power plants to warn motorists of the increased likelihood of fog in the area. Occasionally, these persistent plumes have been observed to produce snowfall and streaks of snow-covered ground downwind of the moisture sources when air temperatures are less than 0 °C (Kramer *et al.*, 1976). These supercooled cloud plume ice crystals are probably nucleated on natural IN, although some coal-burning power plant steam plumes could be strong sources of IN as well as moisture.

Moisture emissions due to anthropogenic activity in winter months at high latitudes can also lead to persistent ice fog. Cities such as Fairbanks, Alaska (Ohtake and Huffman, 1969) and many industrialized cities in Siberia are prolific sources of ice fog, which affects aircraft operations and automobile travel. The physics of the formation of ice fog is similar to contrail formation. At very cold temperatures, small additions of moisture to the air by burning fossil fuels, automobile exhaust emissions, and even human respiration can lead to persistent clouds of small ice crystals.

4.5 Dust

We discuss dust, and specifically desert dust, as an aerosol produced by anthropogenic activities. Dust from deserts and semi-arid regions is a natural phenomenon. But this can be misleading because many human activities such as overcultivation of poor soils, poor irrigation practices, exposing soil surfaces to wind erosion, overgrazing, deforestation, and urban activities associated with road and building construction and road maintenance, and off-paved road use can lead to dust formation. Tegen and Fung (1995) estimate that human activities contribute 30–50% of the total atmospheric dust loading.

Dust can affect regional and global climate directly by absorbing and reflecting solar and terrestrial radiation, and indirectly by serving as CCN, GCCN, and IN.

4.5.1 Direct radiative forcing

Dust absorbs and scatters solar radiation and in addition absorbs terrestrial radiation. Over a diurnal cycle, the net effect of dust is to heat the layer of air in which it embeds (Quijano *et al.*, 2000). At the same time less solar energy reaches the surface so that dust cools the surface. The heating of the dust layer produces a temperature inversion which means the layer is stabilized. Thus the Saharan dust layer, for example, increases the strength of the trade wind inversion, and generates a dry well-mixed layer which can extend to the middle troposphere

(Prospero and Carlson, 1972). Karyampudi and Carlson (1988) showed that radiative heating by Saharan dust can strengthen the mid-level easterly jet, and reduce convection within the equatorial zone. Dunion and Velden (2004) provided convincing evidence that direct solar heating by Saharan dust including stabilization of the trade wind boundary layer, generating a dry middle troposphere, and developing a stronger easterly jet, can have a substantial impact on Atlantic hurricane activity. They found that Saharan dust suppresses the intensity of hurricanes that it engulfs whereas hurricanes that emerge from its influence can rapidly intensify into strong hurricanes.

4.5.2 Indirect effects of dust

Dust can also serve as CCN, GCCN, and IN, and as such can have conflicting consequences on cloud radiative properties and rainfall. Rosenfeld *et al.* (2001), for example, present evidence that dust suppresses rainfall by virtue of serving as CCN and producing numerous small droplets. On the other hand, Levin *et al.* (1996) show through modeling studies that dust can become coated with sulfates and thereby act not only as CCN but GCCN as well. They suggest that the increased concentrations of GCCN will enhance precipitation. This is supported by the satellite observations of Rosenfeld *et al.* (2001) that show clouds in polluted air are non-precipitating over land but rapidly transform into precipitating clouds once the airmass advects over the sea and sea salt particles (GCCN) become entrained into the clouds.

It has been known for some time from laboratory studies that dust can serve as efficient IN (Schaefer, 1949, 1954; Isono *et al.*, 1959; Roberts and Hallett, 1968; Zuberi *et al.*, 2002; Hung *et al.*, 2003). IN measurements by Gagin (1965) and Levi and Rosenfeld (1996) show further that desert dust is an effective IN. Rosenfeld and Nirel (1996) suggested that desert dust serving as GCCN and IN can enhance precipitation and thereby influence the interpretation of the effectiveness of cloud seeding experiments. During the CRYSTAL-FACE (Cirrus Regional Study of Tropical Anvils and Cirrus Layers – Florida Area Cirrus Experiment) field program conducted by the National Aeronautics and Space Administration (NASA) (Jensen *et al.*, 2004) over the Florida peninsula, Sassen *et al.* (2003) showed, using aircraft and lidar data, that a mildly supercooled altocumulus cloud over southern Florida was glaciated in the presence of Saharan dust, which formed effective IN. During the same field program, DeMott *et al.* (2003) found extremely high concentrations of IN (in excess of $1\,\mathrm{cm}^{-3}$ for particles less than $1\,\mu\mathrm{m}$ in size) within dust layers over Florida, confirming the efficiency of dust aerosols in the nucleation of ice. It is clear that dust can substantially alter the microstructure of clouds in a variety of ways.

But in simulations of entrainment of Saharan dust into Florida thunderstorms, van den Heever and Cotton (2004) found that dust impacts not only the cloud microphysical characteristics but also the dynamical characteristics of convective storms as well. The variations in cloud microstructure and storm dynamics by dust, in turn, alter the accumulated surface precipitation and the radiative properties of anvils. These results suggest that the whole dynamic structure of the storms is influenced by varying dust concentrations. In particular, the updrafts are consistently stronger and more numerous when Saharan dust is present compared with a clean airmass. This suggests that the clouds respond to dust in a similar way to the dynamic seeding concepts discussed in Chapter 2. That is, the dust results in enhanced glaciation of convective clouds which leads to dynamical invigoration of the clouds, larger amounts of processed water, and thereby enhanced rainfall at the ground (Simpson *et al.*, 1967; Rosenfeld and Woodley, 1989, 1993). However, van den Heever *et al.*'s simulations resulted in reduced rainfall on the ground at the end of the day. During the first few hours dust enhanced precipitation but by the end of the day, the clean aerosol simulations produced the largest surface rain volume.

Overall, the evidence is compelling that anthropogenic emissions of aerosols and gases are having an impact on the stability of cloud layers, cloud coverage, cloud microstructure, cloud dynamics, precipitation processes, and cloud/fog occurrence.

5

Urban-induced changes in precipitation and weather

5.1 Introduction

In Chapter 4, we examined the possible effects of particulate and gaseous emissions on precipitation and weather on the regional scale in a general sense rather than specific urban-induced changes. In this chapter, we examine the evidence suggesting that pollutants as well as other urban effects are causing changes in the weather and climate in and immediately surrounding urban areas.

There is considerable evidence which suggests that major urban areas are causing changes in surface rainfall, increased occurrences of severe weather, especially hailfalls, and alterations to surface temperatures (Ashworth, 1929; Kratzer, 1956; Landsberg, 1956, 1970; Changnon, 1968, 1981a; Changnon and Huff, 1977, 1986; Hjelmfelt, 1980; Oke, 1987). Some of the hypothesized causes of those changes include:

- urban increases in CCN concentrations and spectra, and IN concentrations;
- changes in surface roughness and low-level convergence;
- changes in the atmospheric boundary layer and low-level convergence caused by urban heating and land-use changes; and
- addition of moisture from industrial sources.

A major cooperative experiment was carried out in the St. Louis, Missouri area in the 1970s to identify urban-induced changes in weather and climate and to identify the primary causes of those changes. A comprehensive review and summary of the experiment and its results are described in the monograph *METROMEX: A Review and Summary* (Changnon, 1981b). In this section we draw heavily on those findings to discuss the potential mechanisms causing urban-induced changes in weather and climate.

First of all METROMEX and related studies showed that St. Louis exhibits a major summertime precipitation anomaly relative to the surrounding rural area. The area-average urban-related increase is about 25%. Much of the enhanced rainfall occurs during the afternoon (1500 to 2100 local daylight time (LDT)), over the city and the close-in area east and northeast. The clouds producing those changes are deep convective clouds and thunderstorms. In fact the frequency of thunderstorms is enhanced in that region by 45% and hailstorms increased by 31%. Not only is the hailstorm frequency higher, but hailstones are larger and of greater number. The rainfall observations also indicated a maximum around midnight extending from approximately 2100 to 0330 LDT located northeast of the city. Changnon and Huff (1986) estimated that the area experienced a 58% increase in nocturnal rainfall relative to the surrounding countryside. The storms responsible for the nocturnal maxima were well-organized storms such as squall line thunderstorms that swept across the urban area and moved across the affected region.

How does an urban area cause those changes? Let us examine each of the hypothesized mechanisms and see how well each fits the METROMEX observations.

5.2 Urban increases in CCN and IN concentrations and spectra

Not surprisingly, anthropogenic activity in the St. Louis urban area caused major increases in CCN concentrations; as much as 94%. Droplet size distributions as a result were found to be narrower with larger concentrations of droplets in the clouds downwind of the city compared to upwind. Large numbers of large, wettable particles, having radii greater than $10\,\mu$m with many as large as $30\,\mu$m were found over the city. These "ultra-giant" particles can serve as embryos for initiation of collision and coalescence. This is consistent with the finding that clouds over the city did have a greater number of larger droplets. The METROMEX scientists cautioned, however, that they had less confidence in those observations compared to the observed higher concentrations of small cloud droplets.

Similar to the study of paper pulp mills, the METROMEX modeling studies revealed that the time required to initiate precipitation in upwind and downwind clouds was only different by a few minutes. It was therefore concluded that the anthropogenic CCN do not play a major role in the creation of the urban rainfall anomaly.

It was also found that the concentrations of IN were not greatly altered over and downwind of the urban area. If anything, it was found in the winter months that the IN concentrations were actually less over the urban region. This suggested that the coagulation of the few IN with the more numerous anthropogenic aerosol actually deactivated or "poisoned" the IN.

In summary it does not appear that the anthropogenic emissions of aerosols can by themselves cause the observed increases in rainfall. It is possible that changes in the cloud and raindrop spectra can have an impact on the rate of glaciation of a cloud and thereby the subsequent cloud behavior. We will examine this hypothesis next as the *glaciation* mechanism.

5.3 The glaciation mechanism

As noted in Part I, it is generally accepted that cumuli containing supercooled raindrops glaciate more readily than more continental, cold-based cumuli that do not contain supercooled raindrops. There are several reasons for this. First of all, larger drops freeze more readily than smaller drops by immersion freezing. More importantly, as noted in Part I, the coexistence of large, supercooled drops and small ice crystals, nucleated by some mechanism of primary nucleation, favors the rapid conversion of a cloud from a liquid cloud to an ice cloud (i.e., glaciation) (Cotton, 1972a,b; Koenig and Murray, 1976; Scott and Hobbs, 1977). Thus the ultra-giant particles observed over St. Louis could produce more supercooled raindrops which would accelerate the glaciation process. This process does not require any change in IN concentrations.

A second factor potentially affecting the rapid glaciation of urban clouds is that the altered drop-size spectra could initiate secondary production of ice crystals. Laboratory studies have indicated that copious quantities of ice splinters are produced when an ice particle collects supercooled cloud droplets when cloud temperatures are within the range of -3 to $-8\,°C$, and when the cloud is composed of a mixture of large drops (greater than $12.5\,\mu m$ radius) and small droplets (less than $7\,\mu m$). All these criteria were met in the clouds observed over St. Louis during METROMEX.

As noted by Keller and Sax (1981), however, in broad, sustained fast-rising updrafts, even when all the criteria for secondary ice production are met, the secondary ice particles and graupel particles will be swept upwards out of the limited temperature range favorable for secondary ice crystal production. Until the updrafts weaken and graupel particles settle into the secondary production zone, the positive feedback mechanism of secondary production is broken. Therefore the opportunities for rapid and complete glaciation of a cloud are greatest if the cloud has a relatively weak, steady updraft or the updraft is a pulsating convective tower. We will show that the clouds over the St. Louis urban area had less buoyant energy or CAPE (as evidenced by lower values of θ_e)[1] than rural clouds. Thus the

---

[1] θ_e, called equivalent potential temperature, is a conservative variable for wet adiabatic processes. See Cotton and Anthes (1989) and Pielke (1984) for a mathematical definition of θ_e.

clouds over the urban area would be expected to have weaker updraft strengths as they enter freezing levels than rural clouds, further enhancing the potential for rapid and complete glaciation.

The hypothesis then builds on the dynamic seeding hypothesis described in Part I. That is, the rapidly glaciated, urban clouds would explosively deepen after they penetrate into subfreezing temperatures, process more moisture through their greater depths, live longer, and rain more. Evidence supporting the glaciation hypothesis is as follows. First of all, it was observed during METROMEX that cumuli over the adjacent rural areas exhibited a distribution in cloud top heights that was bimodal, with many clouds terminating at a height of about 6 km and many others rising to 12 km, but with few clouds penetrating just to heights of around 9 km. In contrast, urban cloud top heights had a more continuous distribution with cloud top heights at all levels between 5 and 13 km. One interpretation of these measurements is that enhanced glaciation of the urban clouds allowed more clouds to penetrate upward through an arresting level, such as an inversion in temperature or a dry layer, and thereby rise to greater heights.

The fact that there is a downwind maximum in thunderstorm activity and hail is consistent with more vigorous glaciation of the clouds as well. Finally the finding that merger of clouds was more frequent over the urban area is consistent with the dynamic seeding hypothesis (Simpson, 1980).

Unfortunately enhanced glaciation of urban clouds was never directly observed during METROMEX. This is because the airborne sensors used were not capable of discriminating between glaciated and unglaciated clouds. Thus this mechanism remains an unproven hypothesis.

5.4 Impact of urban land use on precipitation and weather

Except for the Ohio River valley in the immediate area of the city, St. Louis is located in a vast farmland on a relatively flat plain. The presence of a major urban area changes the surface properties markedly. Firstly, the presence of buildings, particularly tall downtown structures, alters surface roughness from the relatively smooth cropland and occasional forest to a very rough surface. This rough surface creates surface drag which slows the winds near the ground. As shown in Fig. 5.1, air approaching the city would slow down and tend to divert around the city something like flow around an isolated rock in a stream. On the downwind side of the city, air streaming around the city would tend to converge, causing upward motion in that region. There are documented cases where changes in surface roughness have led to a slowing down of cold fronts upwind of New York City, and acceleration of the front downwind (Loose and Bornstein, 1977). Bornstein and

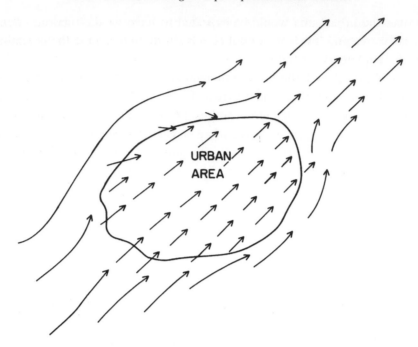

Figure 5.1 Schematic illustration of low-level airflow over and around a major urban area due to changes in surface roughness.

Leroy (1990) found that moving thunderstorms split upon experiencing a barrier-induced divergence around the New York City complex, resulting in enhanced precipitation along the lateral edges of the city and downwind of the city. Analysis of winds and precipitation over Atlanta, Georgia by Bornstein and Lin (2000) suggests that a similar barrier-induced effect is present there as well.

Even more important are the changes in the heat and water budget at the surface caused by the presence of a city. In the countryside, the Earth's surface consists of fallow and plowed fields, grasslands, and small forested areas. The soils are, relative to much of the urban area, rather moist and contain vegetation which can transfer significant amounts of moisture to the atmosphere. By contrast, the surface in the city is a rather impermeable layer, consisting of a mixture of concrete, asphalt, and buildings with a relatively small area of undisturbed soils and vegetation. A greater fraction of rainfall therefore runs off in urban areas than in the countryside.

These changes in surface properties alter the surface energy budget in two ways. First of all, in an urban area such as found in the central United States, a greater fraction of the incoming solar radiation is reflected over the cities as the concrete and buildings are more reflective than plowed fields and cropland. This greater reflectance, or what we call *albedo*, has a cooling effect over the urban areas. Secondly, the more moist land surfaces over the countryside cause a greater fraction of the solar energy absorbed at the surface to be converted into latent heat release rather than sensible heat transfer. In other words, much of the absorbed energy goes into evaporating water from the soil and transpiration from the vegetative canopy. This causes a cooling effect in rural areas relative to the drier, less vegetated urban areas. Because the impact of creating a drier, less vegetation-covered soil in the urban area is much greater than the cooling effect of increased albedo over urban areas, the urban areas in a humid climate such as St. Louis warm more quickly than the surrounding countryside. In addition, the heat stored in concrete and asphalt leads to the urban area remaining warmer later into the evening than the surrounding countryside. Other factors such as heat and moisture emissions by industry, automobiles, and buildings contribute to heating of the urban area relative to the countryside. All heating leads to what is called an *urban heat island.*

During METROMEX, St. Louis was shown to have a well-defined heat island centered over the downwind commercial district, northeast of the core of the urban area. Its maximum size and intensity occurred between midnight and 0600 LDT. It was also found that the air immediately above the urban area was usually drier than over nearby rural areas. Let us consider the hypothetical diurnal variation of the boundary layer of the urban area.

At sunrise, air temperatures begin to rise over both the rural and urban areas. Owing to a shallower nocturnal inversion over the country than the city, air temperatures rise more quickly over the countryside at first. As the ground is heated in both the urban and rural areas, however, a mixed layer forms which deepens more rapidly over the city than the rural areas. This is because the low-level nocturnal inversion strength is weaker over the city. By midday, heating proceeds more rapidly over the city because more of the absorbed energy goes into sensible heat rather than latent heat. The boundary layer thus becomes increasingly deeper and drier over the city. On typical afternoons, the urban boundary layer was found to be 100 to 400 m deeper over St. Louis than the rural areas.

Figure 5.2a illustrates a late afternoon vertical cross-section over the city showing a deeper urban boundary layer, that is warmer and drier than the countryside. Associated with this warmer and drier pool of air over the city is rising motion which produces a sea-breeze-like circulation between the city and the countryside. As seen in Fig. 5.2b this rising motion over the city draws low-level air into the city causing low-level convergence. Such low-level convergence has been found to be favorable for producing deep, precipitating cumulus clouds and also increases the likelihood that those clouds will merge in this low-level convergence zone to produce bigger, heavier raining clouds (Pielke, 1974; Ulanski and Garstang, 1978a,b; Chen and Orville, 1980; Simpson *et al.*, 1980; Tripoli and

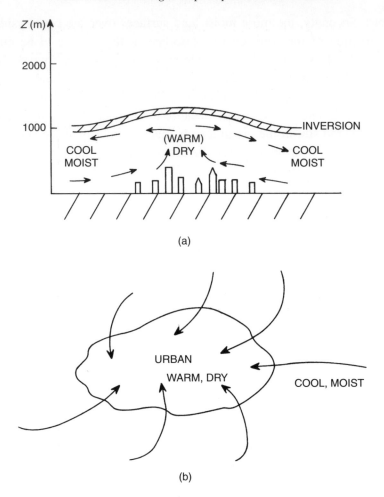

Figure 5.2 (a) Schematic vertical cross-section over a major urban area during the late afternoon in a humid climate region illustrating the effects of the urban heat island. (b) Similar to (a) except a horizontal map of the altered low-level winds by the heat island in the absence of large-scale prevailing flow.

Cotton, 1980). The maximum convergence would occur somewhat downwind of the urban area as the heated boundary layer is advected in that direction (Mahrer and Pielke, 1976; Hjelmfelt, 1980).

During the evening hours, heat conduction from the ground in the urban area limits the rate of cooling compared to the countryside. The surface remains warmer and the low-level air is less stable than in the rural areas. Thus the heat island remains stronger over the city throughout the night.

In the next subsections, the observed behavior of clouds and precipitation over and downwind of St. Louis are discussed to see if they are consistent with changes in the urban boundary layer.

5.4.1 Observed cloud morphology and frequency

Clouds over the St. Louis urban area were found to have bases 600 to 700 m higher than rural clouds. This is consistent with the observation that the air over the city is warmer and drier. The exception was clouds downwind of refineries, where it is believed that moisture injections by the refineries caused lower cloud bases. Air motion into the bases of the clouds were stronger which is consistent with the expected more vigorous thermals due to the heat island.

Cloudiness (defined as the percent coverage of clouds over an area) was found to be greater over the urban area in the later afternoon (1600 LDT) consistent with the observed convergence and upward motion due to the heat island.

5.4.2 Clouds and precipitation deduced from radar studies

The first detectable radar echoes is a measure of the initiation of precipitation. Echoes were found to be more frequent over the urban area during the late morning, about 1400 LDT, and after 1930 LDT. This suggests that the heat-island-induced convergence field played a major role in creating precipitating cumuli. Moreover, individual cumulus cells over the urban area were found to grow deeper and have slightly longer durations than over the rural areas. Again, this is consistent with stronger convergence over the urban heat island favoring deeper, longer-lasting precipitating cumuli.

Clouds over the urban area were also found to merge more frequently with cells over the city, grew taller, and lasted longer than did merged cells over rural areas. As noted previously, this is consistent with observations and modeling studies which suggest that cloud merger is enhanced by low-level convergence, such as that caused by the urban heat island effect. Because it is generally found that taller and longer-lasting cells create more rain and a greater likelihood for hail, these findings are consistent with the hypothesis that the urban heat island enhances convective rain systems over and downwind of the city.

Analysis of cells that contributed to the nocturnal rainfall maxima downwind of St. Louis (Changnon and Huff, 1986) suggested that this urban-related anomaly was associated with the enlargement of rain areas from well-organized storms that existed upwind of St. Louis and then moved over and downwind of the city, as well as the development of new cells over the urban area. Changnon and Huff (1986) and Braham (1981) speculated that this behavior of the storms may have been due to the injection of drier air into the storms as they passed over the urban area, causing them to weaken prematurely and release stored water downwind of the city. This interpretation, however, is inconsistent with the observation that organized, nocturnal storms normally draw on air that has its origin over a large area 50 or more kilometers away from the storm. This warm, moist air typically glides over the nocturnal low-level inversion so that the nocturnal storms do not readily ingest much surface air. Even if the weaker nocturnal inversion over the city allows the storms to tap the drier urban surface air, it seems that the volume of urban air ingested would be a small fraction of the total volume of air ingested into the storm. It is our opinion that enhanced mesoscale ascent associated with the urban heat island could have intensified the nocturnal storms. The fact that nocturnal storms are typically less severe than afternoon convection means that they could strengthen without exhibiting any increase in severe weather. Further studies are needed (probably with multiple Doppler radar) to determine if the storms contributing to the nocturnal urban rainfall anomaly were actually weakening or strengthening, on average.

In summary, there is considerable evidence indicating that the St. Louis urban area enhances rainfall and possibly the occurrence of severe weather. The actual physical processes responsible for those effects, however, have not been fully identified. Both the glaciation mechanism and urban heat-island-induced mesoscale changes are leading contenders. Further observational and modeling studies are required to identify the actual causal mechanisms.

One may ask: is it really necessary to identify the actual mechanisms responsible for an urban precipitation anomaly? Can't we be satisfied that the rainfall analysis shows a strong rainfall anomaly downwind of the urban area? *The answer is clearly no!* For one thing we cannot be sure that the statistical analysis of the rainfall records did not produce an urban "signal" purely by chance. Another reason for establishing a cause and effect is that St. Louis, like many major urban areas, is situated in a river valley. Could local physiographic features such as the higher terrain of the valley sidewalls and moisture sources from the low, relatively wet river bottomland or channeling of the moist, low-level jet through the river valley be the primary causal factors in creating a rainfall anomaly? Attempts to isolate contributions to the rainfall anomaly were made during METROMEX using mesoscale models (Vukovich *et al.*, 1976; Hjelmfelt, 1980). These models revealed that there may be important interactions between the local topography and the downwind thermal plume of the heat island. It was concluded by the METROMEX team that these effects were small, at least in the afternoon hours. They could not dismiss the possibility that natural physiographic effects could have contributed to the nighttime maximums, however. It should be noted that models used at that time could not simultaneously simulate both the mesoscale responses to the physiography and urban heat island, and the response of deep precipitating convection to those forcings.

Recently, Rozoff *et al.* (2003) performed three-dimensional simulations of an actual case study day over St. Louis, in which both the urban heat island and deep convection were explicitly represented. They simulated the urban heat and moisture sources including a generalized urban canyon scheme for dense urban land use. They found that the urban heat island plays a central part in initiating storms that are enhanced by the urban surface. This is consistent with the two-dimensional modeling results of Thielen *et al.* (2000) which indicated that sensible heat flux variations, due to the urban surface, provide the largest impact upon convection. Observational case studies of Bornstein and Lin (2000) are also consistent with the results in this study, but their pre-convective surface convergence values were much smaller (with maximum convergence on the order of $1 \times 10^{-4}\,\text{s}^{-1}$). The position of the urban heat-island-induced boundary layer updraft cell and resulting thunderstorm qualitatively agrees with Baik and Kim (2001), in that the basic state wind places convection downwind of the heating source.

Rozoff *et al.* (2003) found that storms are not initiated over St. Louis when the urban heat island is turned off and only momentum flux representative of the city is considered. That is not to say that modifications in momentum flux by the city are not important. They found that feedbacks between the momentum flux and sensible/latent heat fluxes cause substantial modification in urban circulations. Enhanced momentum flux, created by the urban area, slowed environmental wind and wind driven by the heat island circulation, decreasing the strength of the heat island. As a result, downwind convergence was dampened. This phenomenon led to later downwind convective development. Thus the influence of urban-induced changes in surface momentum fluxes and that due to heat/moisture changes over the city on mesoscale circulations and precipitation are essentially inseparable. Likewise, Rozoff *et al.* (2003) performed simulations in which local topographic effects were removed from the model. They found that the Ozark foothills had some impact on precipitation but the bluffs along the Mississippi River and bottomlands north of the city had little effect. Overall, the urban heat-island-induced convergence downwind of the city dominated over the topographic influences.

A third reason for isolating the causes of the urban precipitation anomaly is that it may become necessary to reduce the rain anomaly and enhanced severe weather occurrences. Without a clear identification of the causal factors, one cannot decide if reductions in emission of gases and aerosols contributing to CCN, or alterations in land-use patterning, is required to reduce the anomaly. At this time it has not been determined how strong the influence of gases and aerosol emissions are on precipitation relative to land-use changes.

The results of METROMEX apply to urban environments in midlatitude humid areas in which the natural vegetation is a deciduous forest which has been replaced by agriculture. Avissar and Pielke (1989) modeled such an environment where the urban area was assumed to contain 20% built-up areas, 10% bodies of water, 40% agricultural crops, and 30% forests. Shown in Fig. 5.3a, for 1400 local standard time (LST), a substantial modification of the boundary layer over the urban

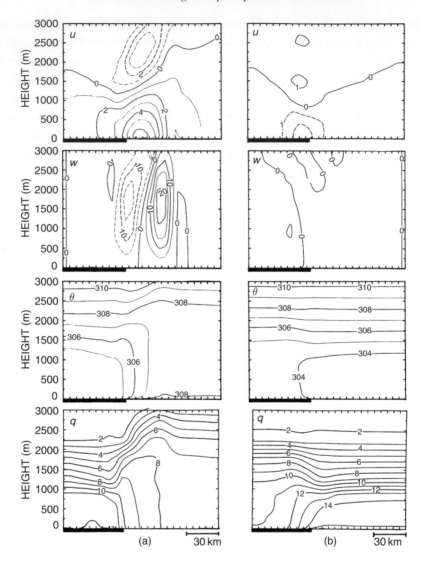

Figure 5.3 Vertical cross-section of the simulated region at 1400 LST for: (i) the horizontal wind component parallel to the domain (u) in m s^{-1}, positive from left to right; (ii) the vertical wind component (w) in cm s^{-1}, positive upward; (iii) the potential temperature (θ) in K; and (iv) the specific humidity (q) in g kg^{-1}, resulting from the contrast of a 60 km wide, heterogeneous land surface region (indicated by the dark underbar) which consists of 20% built-up areas and wastelands, 10% bodies of water, 40% agricultural crops, and 30% forests. The adjacent region is (a) bare and dry, and (b) completely covered by unstressed vegetation. From Avissar and Pielke (1989).

area and the development of low-level convergence into the city were simulated, which is consistent with the interpretation of the METROMEX data. In contrast, when this same heavily irrigated urban area is inserted into an arid or semi-arid environment (such as Denver, Colorado), the impact on the local environment is even more pronounced as shown in Fig. 5.3b with the urban area acting as an oasis during the day, rather than a daytime urban heat island. Large differences in boundary layer structure between the urban area and the surrounding desert terrain result in a well-defined local wind circulation.

In desert environments, the urban island effect has also been documented. For example, the nighttime summer temperature at Sky Harbor Airport in Phoenix, Arizona increased an average of 1.1 °C every decade from 1948 to 1984 (Balling and Brazel, 1987) apparently as a result of the reduction in the urban area that contains irrigated vegetation and an increase in coverage by buildings, concrete, and asphalt. The demand for air conditioning resulted in an increase of peak electricity demands of 1% to 2% per degree Celsius (Akbari *et al.*, 1989). McPherson and Woodard (1990) suggest that the ratio of water and energy costs determine the optimal landscape type which should be used in this environment to minimize their costs. The types of landscaping include *zeroscape*, which is primarily rock-covered ground, *xeriscape*, which utilizes drought-tolerant vegetation such as mesquite, palo verde, and heritage oak, and *mesiscape*, that includes moderate or high water users such as magnolias and ashes. The optimal planting is one that permits shading and cooling by transpiration to minimize air conditioning needs, yet water loss is constrained as much as possible. McPherson and Woodard (1989) estimate that in Tucson, Arizona, the projected annualized cost of a mature tree is $7.76, while its benefits are $26.18 with $19.20 of this amount resulting from cooling due to transpiration.

The influence of land use on climate and weather is discussed in more detail in the next chapter. It is clear, however, that the urban effect will vary depending on its geographic location and we need to explore a range of urban environments in even more detail than was achieved during METROMEX.

Finally, Karl and Jones (1989) compared urban and rural temperature records to show that the growth of cities during this century has resulted in a 0.4 °C bias in the United States climate record. However, since as we have discussed, the urban effect varies geographically, the correction of an urban bias when constructing regional and global analyses could still introduce systematic biases (Zhou *et al.*, 2004). An underrepresented urban bias would suggest the global climate is warming more rapidly than it actually is.

Additional reading

Oke, T.R., 1987. *Boundary Layer Climates*, 2nd ed. New York: Routledge.

6

Other land-use/land-cover changes

6.1 Landscape effects

Land-use/land-cover changes continue to accelerate at the beginning of the twenty-first century (Lepers *et al.*, 2005). Figure 6.1, from the Australia Conservation Foundation (2001), illustrates the global extent of these changes between 1990 and 2000. The role of this conversion on the regional and global climate system as a first-order climate forcing was recognized in the National Research Council (2005) report. The reasons for its importance are reviewed in this chapter.

6.1.1 Surface effects

Concepts

The surface energy and moisture budgets for bare and vegetated soils are schematically illustrated in Figs. 6.2 and 6.3. These surface budgets can be written as

$$R_N = Q_G + H + L(E + TR) \tag{6.1}$$

$$P = E + TR + RO + I \tag{6.2}$$

where R_N represents the net radiative fluxes $= Q_s(1 - A) + Q_{LW}^{\downarrow} - Q_{LW}^{\uparrow}$; P is the precipitation; E is evaporation (this term represents the conversion of liquid water into water vapor by non-biophysical processes, such as from the soil surface and from the surfaces of leaves and branches); TR is transpiration (represents the phase conversion to water vapor, by biological processes, through stoma on plants); Q_G is the soil heat flux; H is the turbulent sensible heat flux; $L(E + TR)$ is the turbulent latent heat flux; L is the latent heat of vaporization; RO is runoff; I is infiltration; Q_s is insolation; A is albedo; $Q_{LW}^{\downarrow}$ is downwelling longwave radiation; $Q_{LW}^{\uparrow}$ is upwelling longwave radiation $= (1 - \epsilon)Q_{LW}^{\downarrow} + \epsilon\sigma T_s^4$; ϵ is the surface emissivity; and T_s is the surface temperature.

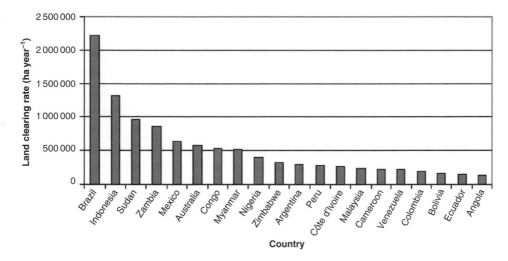

Figure 6.1 International annual land clearing rates for 1990–2000. From the Australia Conservation Foundation (2001); reproduced by permission of the Australian Conservation Foundation.

Detailed discussion of these terms is given in Pielke (1984; 2002a, ch. 11). Equations (6.1) and (6.2) are not independent of each other. A reduction in evaporation and transpiration in Eq. (6.2), for example, increases Q_G and/or H in Eq. (6.1) when R_N does not change. This reduction can occur, for example, if runoff is increased (such as through clear-cutting a forest). The precipitation rate (and type) also influence how water is distributed between runoff, infiltration, and the interception of water on plant surfaces.

The relative amounts of turbulent sensible (H) and latent heat fluxes [$L(E + TR)$] are used to define the quantity called the Bowen ratio (B), and evaporative fraction, e_f:

$$B = \frac{H}{L(E + TR)}; \quad e_f = L(E + TR)/R_N. \tag{6.3}$$

The denominator $L(E + TR)$ has been called "evapotranspiration," although since evaporation and transpiration involve two distinct pathways for liquid water to convert to water vapor, the use of the term "evapotranspiration" should be discouraged. It is preferable to refer to E as "physical evaporation" and TR as "transpiration." The relation of R_N to H and $L(E + TR)$, following Segal *et al.* (1988), can be written as

$$H \cong \frac{R_N - Q_G}{(1/B) + 1}. \tag{6.4}$$

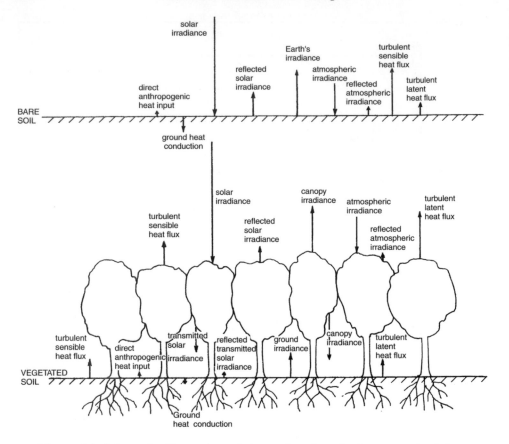

Figure 6.2 Schematic illustration of the surface heat budget over (top) bare soil, and (bottom) vegetated land. The roughness of the surfaces (and for the vegetation, its displacement height), will influence the magnitude of the heat flux. Dew and frost formation and removal will also influence the heat budget. Adapted from Pielke and Avissar (1990) with kind permission from Kluwer Academic Publishers.

With $Q_G \ll |H|$ and $Q_G \ll |E + TR|$, as discussed in Segal *et al.* (1988),

$$H \cong \left(\frac{1+B}{B}\right) R_N \qquad (6.5)$$

Segal *et al.* (1995) showed that with the same value of R_N, with a smaller Bowen ratio, the thermodynamic potential for deep cumulus convection increases.

   Therefore, any land-use/land-cover change that alters one or more of the variables in Eqs. (6.1) and (6.2) will directly affect the atmosphere. For instance, a decrease in A (i.e., a darkening of the surface) would increase R_N; thus making more heat energy available for $Q_G, H, E,$ and TR. The heat that goes into H increases the potential temperature, $\theta(\theta = T[1000 \text{ mb}/p \text{ (mb)}]$, because

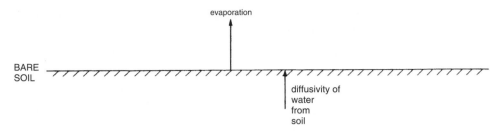

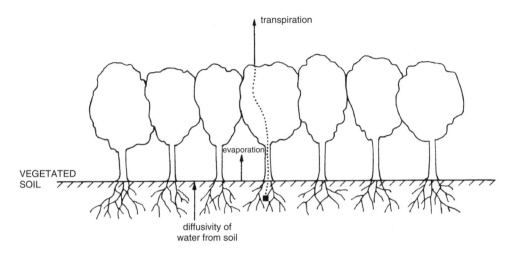

Figure 6.3 Schematic illustration of the surface moisture budget over (top) bare soil, and (bottom) vegetated land. The roughness of the surface (and for the vegetation, its displacement height), will influence the magnitude of the moisture flux. Dew and frost formation and removal will also influence the moisture budget. Adapted from Pielke and Avissar (1990) with kind permission from Kluwer Academic Publishers.

temperature increases. The heat that goes into E or TR goes into the equivalent potential temperature, θ_E, because w (mass of water vapor per unit mass of air) increases $\left(\theta_e = \theta \exp\left(\frac{Lw}{C_p T}\right)\right)$. If the surface were dry and bare, all of the heat energy would necessarily go into Q_G and H as shown by Pielke (1984, p. 381) for the Empty Quarter in Saudi Arabia.

Lyons *et al.* (1996), for example, found a reduction of H in southwestern Australia as a result of the conversion of land to agriculture. Bryant *et al.* (1990) found higher sensible heat fluxes in the Sonoran Desert of Mexico due to overgrazing. Fitzjarrald *et al.* (2001), and Schwartz (1994) found that the leafing out of vegetation in the spring has a dramatic effect on a reduction in H. Schreiber *et al.* (1996) and Rabin *et al.* (1990) discuss how cumulus cloud base height is directly related to surface heat and moisture fluxes, as modulated by the characteristics of the underlying heterogeneous surface. As discussed in Pielke (2001a), changes in θ and θ_E change convective available potential energy (CAPE) and the other cumulus convective indices.

When snow and/or frozen soils are present, figures similar to Figs. 6.2 and 6.3 can be presented. With snow cover, for example, the albedo will be greater than for bare soil so a greater fraction of incident solar radiation is reflected upwards. When vegetation is present which is not covered by snow, the protruding vegetation will result in a lower albedo than would occur when it is completely covered. Strack *et al.* (2004) has shown that large differences in the turbulent fluxes result when snow is covering the ground and is influenced by protruding vegetation.

Surface air moist enthalpy

Although climate change and variability involves all aspects of the climate system (Pielke, 1998), the assessment of anthropogenically forced climate change has focused on surface temperature as the primary metric (Mann and Jones, 2003; Soon *et al.*, 2004). The term "global warming" has been used to describe the observed surface air temperature increase in the twentieth century. However, this concept of global warming requires assessments of units of heat (that is, joules). Temperature, by itself, is an incomplete characterization of surface air heat content.

Pielke (2003) used the concept of heat changes in the ocean, for example, to diagnose the radiative imbalance of the Earth's climate system. The oceans, of course, are the component of the climate system in which the vast majority of actual global warming or cooling occurs. Here we use the more limited application of the term global warming to refer to surface air changes.

The heat content of surface air (i.e., z right above ground level, so that $z = 0$ can be assumed) can be expressed as:

$$M = C_p T + Lq \tag{6.6}$$

where C_p is the specific heat of air at constant pressure, T is the air temperature, L is the latent heat of vaporization, and q is the specific humidity (Haltiner and Williams, 1980). The quantity, M, is called moist enthalpy and can be expressed in units of J kg^{-1}. The surface dry enthalpy can be written as

$$S = C_p T. \tag{6.7}$$

Surface air temperature trends that have been reported monitor only S. The monitoring of H, however, is the more appropriate metric to assess surface air warming.

To investigate the effect of monitoring variations of M in time, both M and S were calculated for the year 2002 in Fort Collins, Colorado, and at the Central Plains Experimental Range (CPER) of the US Department of Agriculture's Agricultural Research Service located 60 km northeast of the city (Fig. 6.4). Both locations offer high-quality temperature and humidity observations. The Fort Collins site is on a university campus with nearby buildings, parking lots, and irrigated grass, while the CPER site is an ungrazed natural grass area. To facilitate the comparison with temperature, an effective temperature as $T_E = M/C_p$ is calculated.

As shown by Pielke (2001a), in terms of heat content, at 1000 mb, an increase of 1 °C in the dewpoint temperature produces the same change in M as a 2.5 °C increase in temperature. This means, for example, that a decrease of 1 °C of the dewpoint temperature, but a 1 °C increase in the temperature, actually is a reduction of heat content in terms of J kg^{-1} of the air!

The plots of T and T_E (with the corresponding values of M and S on the right axis) for 2002 illustrate that when the absolute humidity is low (such as on cold winter days), T and T_E are nearly equal. However, there are large differences in

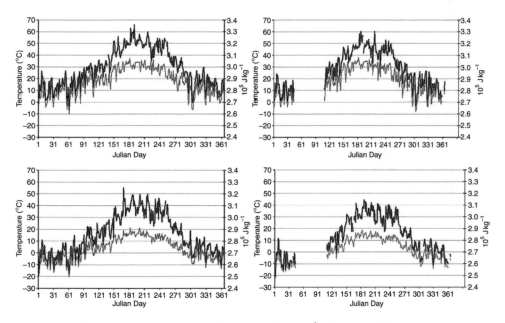

Figure 6.4 T and T_E in °C (S and M, in 10^5 J kg^{-1}) for Fort Collins, Colorado (left panels) and the CPER ungrazed site (right panels) are shown for 2002. The top two panels are for maximum daily temperature while the bottom two panels are for minimum daily temperature. The gray lines represent T (and S) while the black lines represent T_E (and M). From Pielke *et al.* (2004a). © 2004 American Geophysical Union. Reproduced by permission of the American Geophysical Union.

these values in the growing season when the absolute humidity is higher. Dry days, however, have less heat content than more humid days with the same air temperature.

The average differences of the annual averaged maximum and minimum temperature for the two sites for dates where data were available from both locations (the value at Fort Collins minus the value at the CPER site) are $0.25\,°C$ and $1.86\,°C$, respectively. The differences in T_E, however, are larger ($2.69\,°C$ and $4.20\,°C$).

For the growing season part of 2002, the differences in the average maximum and minimum value of T were $0.91\,°C$ and $1.82\,°C$, while the corresponding difference values of T_E were $3.48\,°C$ and $4.96\,°C$. The value of T_E provides a more accurate characterization of surface heat content and is the more appropriate metric for assessing surface air warming.

The different variation of M, as contrasted with S, as a function of land use could help explain the results reported in Kalnay and Cai (2003), in which they concluded that land-use change could explain at least part of the observed temperature changes in the eastern United States in recent decades. Davey *et al.* (2006) has independently confirmed the robustness of the Kalnay and Cai conclusions. The difference in temporal trends in surface and tropospheric temperatures (National Research Council, 2000), which has not yet been explained, could be due to the incomplete analysis of the surface and troposphere for temperature, and not the more appropriate metric of heat content. Recent analyses of satellite data have reduced the differences, but have not eliminated the disagreement (Christy *et al.*, 2003; Mears *et al.*, 2003; Mears and Wentz, 2005; Christy *et al.*, 2006).

This analysis shows that surface air temperature alone does not capture the real changes in surface air heat content. Even using the limited definition of the term global warming in order to refer to surface air warming, the moisture content of the surface air must be included. Future assessments should include trends and variability of surface heat content in addition to temperature.

A conclusion of Subsection 6.1.1 is that changes in the Earth's surface can result in significant changes in the surface energy and moisture budgets. These changes will influence the heat and moisture fluxes and the surface air heat content.

6.1.2 Boundary-layer effects

Once the surface energy budget is altered, fluxes of heat, moisture, and momentum within the planetary boundary layer are directly affected (Segal *et al.*, 1989). As an example, Fig. 6.5 illustrates an idealization of the vertical structure of the convective boundary layer, where the surface heat flux, H, depth of the layer z_i, and the temperature stratification just above z_i determine the vertical profile of

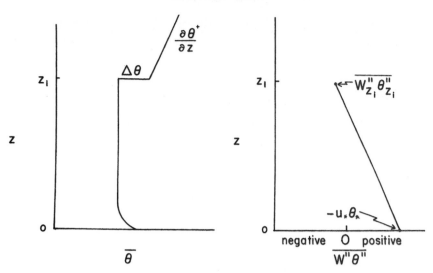

Figure 6.5 The potential temperature and heat flux profiles assumed in the "jump" model. Reprinted from Pielke (1984) with permission from Academic Press.

temperature and heat flux. Deardorff (1974) suggested a growth rate equation for z_i, in the absence of large-scale wind flow, which is proportional to

$$\frac{\partial z_i}{\partial t} \sim H^{\frac{2}{3}} z_i^{-\frac{4}{3}}. \tag{6.8}$$

The entrainment of air from above z_i to heights below z_i is given by

$$H_{z_i} = -\alpha H \tag{6.9}$$

where α is the entrainment coefficient ($\alpha \cong 0.2$, although there are suggestions it is different from this value; Betts *et al.*, 1992). McNider and Kopp (1990) discuss how the size of thermals generated from surface heating are a function of z_i, H, and height within the boundary layer. The rate of growth of the boundary layer during the day, and the ingestion of free atmospheric air into the boundary layer are, therefore, both dependent on the surface heat flux, H.

A simplified form of the prognostic equation for θ can be used to illustrate how temperature change is related to the surface heat flux, H_s,

$$\frac{\partial \theta}{\partial t} = \frac{\partial}{\partial z}\left(\frac{H}{\rho C_p}\right) \tag{6.10}$$

where ρ is the air density, and C_p is the specific heat at constant pressure. Integrating from the surface to z_i and using the mean value theorem of calculus yields

$$\frac{\partial \bar{\theta}}{\partial t} = \frac{1}{z_i \rho C_p} \left[H_s - H_{z_i} \right] = \frac{1.2}{\rho C_p z_i} H_s \tag{6.11}$$

where Eq. (6.9) with $\alpha = 1.2$ has been used. Using this equation, a heating rate of a 1 km deep boundary layer of $2\,^{\circ}\mathrm{C}$ over 6 h is produced by a surface heat flux of 100 W m^{-2}.

Figures 6.6 and 6.7 show how H, and therefore other characteristics of the boundary layer, including z_i, as based on actual observations, are altered as a result of different land-surface characteristics. Segal *et al.* (1989) discuss how wet soils and canopy temperatures affect the growth of the boundary layer. Amiro *et al.* (1999) measured elevations of surface radiometric temperatures by up to $6\,^{\circ}\mathrm{C}$, which remained elevated even for 15 years after forest fires in the Canadian boreal forest.

At night when longwave radiative fluxes become important, the boundary-layer structure is quite different than in the daytime (e.g., see Gopalakrishnan *et al.*, 1998). The temperature usually increases with height above the surface.

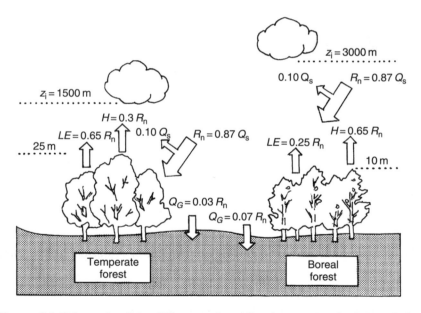

Figure 6.6 Schematic of the differences in surface heat energy budget and planetary boundary layer over a temperate forest and or boreal forest. The symbols used refer to Eq. (6.1). Horizontal fluxes of heat and heat storage by vegetation are left out of the figure. From Pielke (2001a). © 2001 American Geophysical Union. Reproduced by permission of the American Geophysical Union.

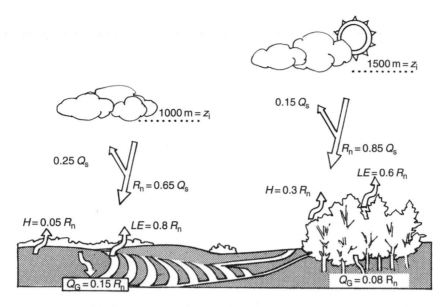

Figure 6.7 Same as Fig. 6.6 except between a forest and cropland. From Pielke (2001a). © 2001 American Geophysical Union. Reproduced by permission of the American Geophysical Union.

Moisture and temperature profiles left over from the daylight period dominate the evolution of the nighttime boundary layer in the absence of strong horizontal wind advection. Under lighter wind conditions, changes in the radiative heat fluxes due to land-use/land-cover change will result in altered vertical temperature and moisture profiles (Pielke and Matsui, 2005).

The conclusion from the analyses in this section is that the *boundary-layer structure, including its depth, is directly influenced by the surface heat and moisture fluxes*. If the surface fluxes change, so will the boundary-layer structure.

6.1.3 Local wind circulations

Local wind circulations can subsequently result from horizontal variations in H and z_i (Segal and Arritt, 1992). Such wind circulations are referred to as *solenoidal circulations* and are the reason sea and land breezes occur (Simpson, 1994; Pielke, 1984, ch. 13). The reason that these local wind circulations can develop is described in Appendix A of Pielke (2001a) and in Pielke and Segal (1986).

Mesoscale circulations produce focused regions particularly favorable for deep cumulus convection (Pielke *et al.*, 1991b). In these areas, CAPE and other measures of the potential for deep cumulus convection are increased in response to boundary wind convergence associated with local wind circulations (Pielke *et al.*,

1991a). Convective inhibition is reduced in these areas. These wind convergence zones can also provide specific vertical motion "triggers" with which to initiate deep cumulus convection. Therefore, *the spatial structure of the surface heating, as influenced by landscape, can produce focused regions for deep cumulonimbus convection, as well as other mesoscale systems.*

6.1.4 Vertical perspective

As overviewed in Subsection 6.1.1 to 6.1.3, land-surface characteristics influence the heating and moistening of the atmospheric boundary layer. Therefore, vertical radiosonde soundings over adjacent locations that have different surface conditions offer opportunities to observationally assess the effect of landscape variations, while models and observations can be used to evaluate the importance of spatially varying boundary-layer structure in generating mesoscale circulations. For example, Segal *et al.* (1991; see Fig. 4 in that paper) present measured differences in boundary-layer structure between adjacent areas with and without snow cover. The influence of landscape conditions on cumulus cloud and thunderstorm development have been evaluated using models and observations, for example, in Garrett (1982), Clark and Arritt (1995), Cutrim *et al.* (1995), Hong *et al.* (1995), and Crook (1996).

6.1.5 Mesoscale and regional horizontal perspective

Since land–water contrasts permit the development of sea breezes which focus thunderstorm development over islands and coastal regions in the humid tropics, and in humid middle and high latitudes during the summer (e.g., see Fig. 12-13 from Pielke, 1984; also see Pielke, 1974; Pielke *et al.*, 1991b; Marshall *et al.*, 2004a), it would be expected that similar variations in surface heating associated with landscape patterns and patchiness would also produce mesoscale circulations of a similar magnitude.

Avissar and Schmidt (1998) explored how landscape patchiness influences cumulus development using a large eddy simulation. They reported preferential locations within the heterogeneous landscape where pockets of relatively high moisture concentrations occurred. As shown in Figs. 6.8 and 6.9, the shape of the heterogeneity strongly influences the ability of mesoscale flows to concentrate CAPE within local regions so as to permit a greater likelihood of stronger thunderstorms. The large square-shaped area, for example, is able to focus the lower tropospheric winds so as to optimize the accumulation of CAPE. This focusing of CAPE is analogous to what occurs with round islands (Neumann and Mahrer, 1974).

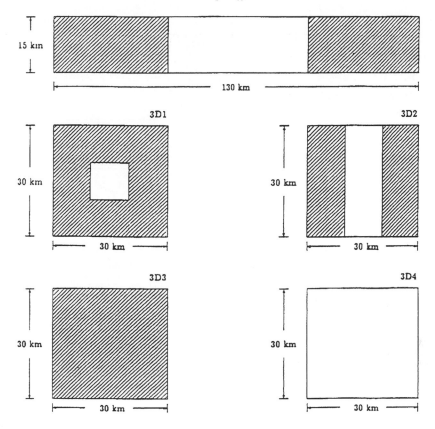

Figure 6.8 Schematic representation of the simulated three-dimensional domains. From Avissar and Liu (1996). © 1996 American Geophysical Union. Reproduced by permission of the American Geophysical Union.

Dalu *et al.* (1996) used a linear model to conclude that the Rossby radius (λ_R) is the optimal spatial scale for landscape heterogeneities to produce mesoscale flows (which is on the order of tens of kilometers). Landscape heterogeneities smaller than λ_R simply excite gravity waves which rapidly disperse throughout the atmosphere whereas landscape heterogeneities larger than λ_R can produce persistent circulations.

Avissar and Pielke (1989), Hadfield *et al.* (1991), Shen and Leclerc (1995), Zeng and Pielke (1995a,b), Wang *et al.* (1997), and Avissar and Schmidt (1998) also explored the issue of the size of landscape patchiness that is needed before the boundary-layer structure is significantly affected and a mesoscale circulation produced. Segal *et al.* (1997) found that cumulus clouds are a minimum downwind of mesoscale-sized lakes during the warm season as a result of mesoscale-induced subsidence over the lake and the resultant suppression of z_i.

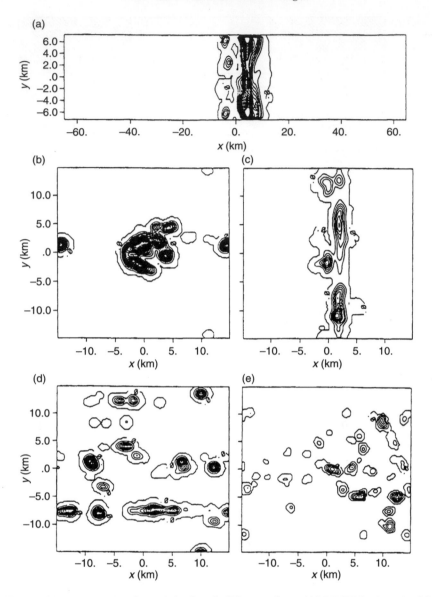

Figure 6.9 Accumulated precipitation (millimeters), at 1800 LST in domain (a) 3D0, (b) 3D1, (c) 3D2, (d) 3D3, and (e) 3D4. Contour intervals are 2 mm in 3D0, 1 mm in 3D1, 3D2, and 3D3, and 0.05 mm in 3D4. From Avissar and Liu (1996). © 1996 American Geophysical Union. Reproduced by permission of the American Geophysical Union.

Other studies that have explored the influence of landscape heterogeneity on cumulus convection include Segal *et al.* (1989), Rabin *et al.* (1990), Chang and Wetzel (1991), Fast and McCorcle (1991), Segal and Arritt (1992), Chen and Avissar (1994a,b), Li and Avissar (1994), Clark and Arritt (1995), Cutrim *et al.*

(1995), Lynn *et al.* (1995a,b, 1998), Rabin and Martin (1996), and Wang *et al.* (2000).

Pielke *et al.* (1997) present a sensitivity experiment to evaluate the importance of land-surface conditions on thunderstorm development. Using identical lateral boundary and initial values, two model simulations for 15 May 1991 were performed for the Oklahoma–Texas Panhandle region. One experiment used the current landscape (which includes irrigated crops and shrubs, as well as the natural shortgrass prairie), while the second experiment used the natural landscape in this region (the shortgrass prairie). Figure 6.10 provides the results at 1500 LST for both experiments. The simulation with the current landscape (Fig. 6.10, top) produced a thunderstorm system along the dryline, while only a shallow line of cumulus clouds were produced using the natural landscape (Fig. 6.10, bottom). A thunderstorm was observed in this region on 15 May 1991 with the other meteorological quantities also realistically simulated (Grasso, 1996; Shaw *et al.*, 1997; Ziegler *et al.*, 1997). The thunderstorm developed when the current landscape was used since the enhanced vegetation coverage (higher leaf area) permitted more transpiration of water vapor into the air than would have occurred with the natural landscape. The result was higher CAPE with the current landscape.

Lyons *et al.* (1993, 1996) and Huang *et al.* (1995a), in a contrasting result, found that the replacement of native vegetation with agriculture reduced sensible heat flux with a resultant decrease in rainfall. Wetzel *et al.* (1996), in a study in the Oklahoma area, found that cumulus clouds form first over hotter, more sparsely vegetated areas. Over areas covered with deciduous forest, clouds were observed to form 1 to 2 hours later due to the suppression of vertical mixing. Rabin *et al.* (1990) also found from satellite images that cumulus clouds form earliest over regions of large sensible heat fluxes and are suppressed over regions with large latent heat flux during relatively dry atmospheric conditions.

Clark and Arritt (1995), however, found while the cumulus cloud precipitation was delayed when the soil moisture was higher, the total accumulation of precipitation was greater. The largest rainfall was generally predicted to occur for moist, fully vegetated surfaces. De Ridder and Gallée (1998) reported significant increases in convective rainfall in southern Israel associated with irrigation and intensification of agricultural practices, while De Ridder (1998) found that dense vegetation produces a positive feedback to precipitation. Baker *et al.* (2001) explored the influence of soil moisture and other effects on sea-breeze-initiated precipitation in Florida.

Emori (1998) showed, using idealized simulations, how cumulus rainfall and soil moisture gradients interact so as to maintain a heterogeneous distribution of soil moisture. Taylor *et al.* (1997) concluded that such a feedback occurs in the Sahel of Africa which acts to organize cumulus rainfall on scales of about

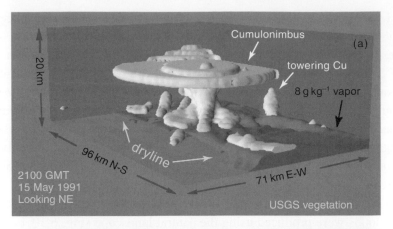

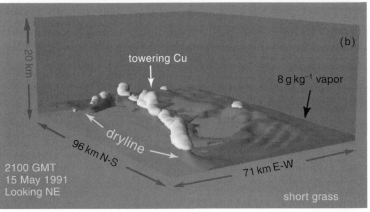

(C. Ziegler, NSSL)

Figure 6.10 (a) and (b) Model output cloud and water vapor mixing ratio fields on the third nested grid (grid 4) at 21:00 UT on 15 May 1991. The clouds are depicted by white surfaces with $q_c = 0.01$ g kg^{-1}, with the Sun illuminating the clouds from the west. The vapor mixing ratio in the planetary boundary layer is depicted by the shaded surface with $q_v = 8$ g kg^{-1}. The flat surface is the ground. Areas formed by the intersection of clouds or the vapor field with lateral boundaries are flat surfaces, and visible ground implies $q_v < 8$ g kg^{-1}. The vertical axis is height, and the backplanes are the north and east sides of the grid domain. Reproduced from Pielke *et al.* (1997) with permission from *Ecological Applications* and the Ecological Society of America. See also color plate.

10 km. Simpson *et al.* (1980, 1993) demonstrated that cumulus clouds that merge together into a larger scale produce much more rainfall.

Chen and Avissar (1994b) used a model to show that land-surface moisture significantly affects the timing of onset of cumulus clouds, and the intensity and distribution of precipitation. Eltahir and Pal (1996) also explored the relation between surface conditions and subsequent cumulus convective rainfall. Mölders

(1999b) found that natural flooding and anthropogenic land-surface changes such as the drainage of marshes influence the water vapor supply to the atmosphere, clouds, and precipitation. Grasso (2000) has shown that dryline formation in the central Great Plains of the United States is critically dependent on the spatial pattern of soil moisture, while Pan *et al.* (1996) concluded that increases in soil moisture enhanced local rainfall when the lower atmosphere was thermodynamically unstable and relatively dry, but decreased rainfall when the atmosphere was humid and lacked sufficient thermal forcing to initiate deep cumulus convection.

These results illustrate that the effect of landscape evaporation and transpiration on deep cumulonimbus convection is complex and quite nonlinear. Opposing effects help explain the apparent contradiction between the results reported in Lyons *et al.* (1996) and Pielke *et al.* (1997), for example, that are discussed earlier in this section. While increased moisture flux into the atmosphere can increase CAPE, the triggering of these deep cumulus clouds may be more difficult since the sensible heat flux may be reduced. The depth of the planetary boundary layer, for example, will be shallower, if the sensible heat flux is less. Other examples of studies that explore how vegetation variations organize cumulus convection include Anthes (1984), Vidale *et al.* (1997), Liu *et al.* (1999), Souza *et al.* (2000), and Weaver *et al.* (2000).

There are also studies of the *regional* importance of spatial and temporal variations in soil moisture and vegetation coverage (e.g., Fennessy and Shukla, 1999; Pielke *et al.*, 1999a). Using a model simulation covering Europe and the North Atlantic, for example, Schär *et al.* (1999) determined that the regional climate is very dependent on soil moisture content. They concluded that wet soils increase the efficiency of convective precipitation processes, including an increase in convective instability. Delworth and Manabe (1989) discuss how soil wetness influences the atmosphere by altering the partitioning of energy flux into sensible and latent heat components. They found that a soil moisture anomaly persists for seasonal and interannual timescales so that anomalous fluxes of sensible and latent heat also persist for long time periods. A similar conclusion was reported in Pielke *et al.* (1999a). Wei and Fu (1998) found that the conversion of grassland into a desert in northern China would reduce precipitation as a result of the reduction in evaporation. Jones *et al.* (1998) discussed how surface heating rates over regional areas are dependent on surface soil wetness. Viterbo and Betts (1999) demonstrated significant improvement in large-scale numerical weather prediction when improved soil moisture analyses were used. Betts *et al.* (1996) reviewed these types of land–atmosphere interactions, as related to global modeling. Nicholson (2000) reviewed land-surface processes and the climate of the Sahel. Other recent regional-scale studies of the role of landscape processes in cumulus convection and other aspects of weather include Lyons *et al.* (1993), Carleton *et al.* (1994), Copeland *et al.* (1996), Huang *et al.* (1996), Bonan (1997),

Sun *et al.* (1997), Bosilovich and Sun (1999), Liu and Avissar (1999a,b), Adegoke (2000), and Li *et al.* (2000).

Segal *et al.* (1998) concluded that average rainfall in North America is increased as a result of irrigation, which is consistent with the influence of irrigation on CAPE as shown by Pielke and Zeng (1989). Pan *et al.* (1995) concluded that soil moisture significantly affects summer rainfall in both drought and flood years in the midwest of the United States. Kanae *et al.* (1994) concluded that deforestation in southeastern Asia since 1951 has resulted in decreases in rainfall in September in this region, when the large-scale monsoon flow weakens. Kiang and Eltahir (1999), Wang and Eltahir (1999, 2000a,b), Eastman *et al.* (2001a), Lu *et al.* (2001), and Niyogi and Xue (2006) have used coupled regional atmospheric–vegetation dynamics models to demonstrate the importance of two-way interaction between the atmosphere and vegetation response. Hoffman and Jackson (2000), for example, propose that as a result of atmospheric–vegetation interactions in tropical savanna regions, anthropogenic impacts can exacerbate declines in precipitation. Shinoda and Gamo (2000) used observations to demonstrate a correlation between vegetation and convective boundary-layer temperature over the African Sahel.

A clear conclusion from these studies is that *both mesoscale and regional landscape patterning and average landscape conditions exert major controls on weather and climate.* In the following sections, we will focus on specific examples of the effect of the deliberate modification of the landscape by humans on weather and climate.

6.2 Influence of irrigation

6.2.1 Colorado

Two examples of observed surface temperature contrasts between agricultural irrigated areas in juxtaposition to a grazed shortgrass prairie region in Colorado are presented in Fig. 6.11 as originally reported in Segal *et al.* (1988). The examples provide the Earth surface blackbody Infrared (IR) temperature derived from the GOES geostationary satellite (with a pixel footprint of 4×8 km) which were taken during the summer of 1986. Using the Denver radiosonde data, a correction for atmospheric water vapor absorption was included. Each image is a composite of the first 15 days of August at 1300 LST (when the surface temperature is around its daily peak value). Only areas with clear sky conditions at that hour were included in the construction of the composites. The vegetated areas (indicated by shading on the figures) were identified based on the Landsat image of Colorado for the end of June – beginning of July 1976 as well as from the US Department of Agriculture vegetation cover maps of Colorado.

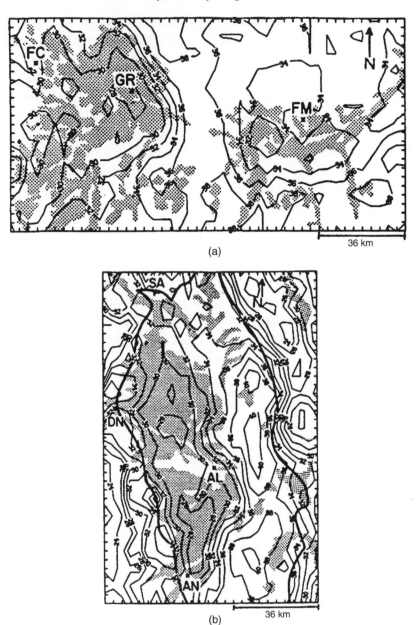

(a)

36 km

(b)

36 km

Figure 6.11 Composite of GOES-derived surface temperature at 1300 LST for the period 1 August 1986 to 15 August 1986 for (a) northeast Colorado (FC, Fort Collins; FM, Fort Morgan; GR, Greeley), and (b) the San Luis Valley in Colorado (AL, Alamosa; AN, Antonito; DN, Del Norte; SA, Saguache). The lower valley is outlined by a dark line separating it from significant elevated terrain. Irrigated areas are shaded. Reproduced from Segal *et al.* (1988) with permission from the American Meteorological Society.

The first example (Fig. 6.11a) is for northeast Colorado which includes the agricultural areas along the Front Range and the South Platte River. The topographical variations in the area presented are in the range of 200 m.

The second example is for the San Luis Valley (Fig. 6.11b) where there is intensive irrigated summer agricultural activity in the central domain. The valley is nearly flat, and it is surrounded by steep mountains to the southwest and northeast as indicated in the figure.

In both composites, there is a significant correspondence between the cooler areas and vegetative cover. The highest temperature occur in the uncultivated areas. Maximum IR surface temperature gradients of 10 °C and 12 °C over distances of 10–20 km were observed in northeast Colorado and the San Luis Valley, respectively. Since the emittance of the Earth surface is less than 1, the actual surface temperatures involved with these cases are somewhat higher than those presented. The emissivity of the irrigated area, however, is likely to be somewhat higher than that of its dryland surroundings (e.g., Lee, 1978). Thus the actual surface temperature gradients are suggested to be larger than those illustrated. Therefore noticeable circulations should be expected to occur with such temperature gradients.

Figure 6.12 illustrates two soundings made over two locations in northeast Colorado at 1213 LST on 28 July 1987 (Pielke and Zeng, 1989; Segal *et al.*,

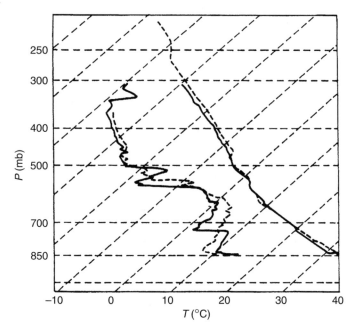

Figure 6.12 Radiosonde measurements of temperature (right side) and dewpoint temperature (left side) for a dryland area (dashed line) and an irrigated area (solid line) in northeast Colorado at 1213 LST on 28 July 1987. Reproduced from Pielke and Zeng (1989) with permission from the National Weather Association.

1989). The soundings were made prior to significant cloud development. The radiosonde sounding over an irrigated location had a slightly cooler, but moister lower troposphere than the sounding over the natural, shortgrass prairie location. Aircraft flights at several levels between these two locations on 28 July 1987 (Figs. 6.13 and 6.14) demonstrate that the moistening and cooling occurred over the entire region of irrigation. Using a convective index to 500 mb, the lifted index, assuming surface parcel ascent, was −2 over the irrigated land but zero over the shortgrass prairie (Pielke and Zeng, 1989). For this case, the moistening of the lower atmosphere over the irrigated area was more important in increasing CAPE than was the slight cooling in decreasing CAPE.

6.2.2 Nebraska

As discussed in Adegoke *et al.* (2003), over the last five decades, the total acreage under irrigation in the US High Plains increased from less than 1.2 million hectares to over 8 million hectares (Kuzelka, 1993). The rapid development of

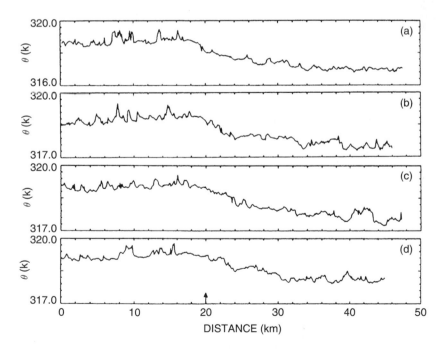

Figure 6.13 Measured potential temperature from Briggsdale to Windsor at the altitude of (a) 140 m, (b) 240 m, (c) 345 m, and (d) 440 m above the ground. The observed crop–dryland boundary is indicated by an arrow, with cropland to its right. Adapted from Segal *et al.* (1989) with permission from the American Meteorological Society.

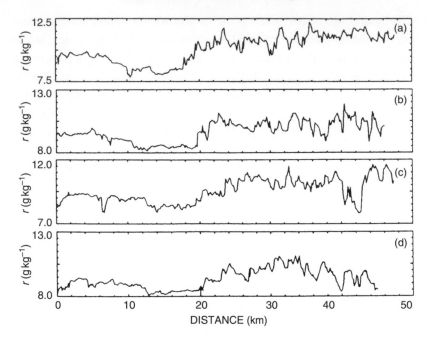

Figure 6.14 Same as Fig. 6.13 except for moisture mixing ratio. Adapted from Segal *et al.* (1989) with permission from the American Meteorological Society.

irrigation in this area, which stretches from Nebraska through western Kansas to the Texas Panhandle, enabled the transformation of the area into one of the major agricultural areas of the United States. In Nebraska, as in much of the High Plains, corn is the dominant crop cultivated during the warm season months (Williams and Murfield, 1977). Irrigated corn, which represented about 10% of total corn-producing areas during the early 1950s, now composes nearly 60% of the total corn-producing areas in Nebraska (Fig. 6.15). Figure 6.16 illustrates how this irrigated landscape appears.

Data from the National Agricultural Statistics Services (NASS) of the United States Department of Agriculture (USDA) for York County in east-central Nebraska further underscore these changes. Between 1950 and 1998 the irrigated corn area in York County increased from 3500 ha to 80 000 ha (a 2300% increase) while the rain-fed corn area declined rapidly during the same period (National Agricultural Statistics Service, 1998). This rapid land-use change was achieved largely by converting rain-fed corn areas to irrigation. Land-use conversion of this magnitude could affect energy and moisture exchanges between the land surface and the lower atmosphere by altering transpiration and evaporation thus generating complex changes in the surface energy budget.

Regional Atmospheric Modeling System (RAMS) (Pielke *et al.*, 1992; Cotton *et al.*, 2003) simulations were performed which consisted of four land-use

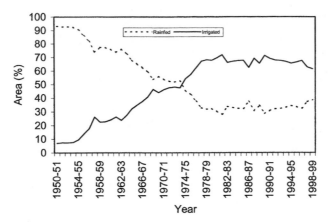

Figure 6.15 Area (%) of rainfed and irrigated corn farming in Nebraska (1950–98). Reproduced from Adegoke *et al.* (2003) with permission from the American Meteorological Society.

Figure 6.16 South of the South Platte River, south and west of North Platte, Nebraska, looking north on 1 June 2004 at approximately 1300 LST. A number of pivot irrigators are not watering. Photo courtesy of Kelly Redmond. See also color plate.

scenarios covering the 15-day period from 1 to 15 July 1997. The first scenario (control run) represented current farmland acreage under irrigation in Nebraska as estimated from 1997 Landsat satellite and ancillary data (Fig. 6.17a). The second and third scenarios (OGE wet and dry runs) represented the land-use conditions from the Olson Global Ecosystem (OGE) vegetation dataset (Fig. 6.17b), and the fourth scenario (natural vegetation run) represented the potential

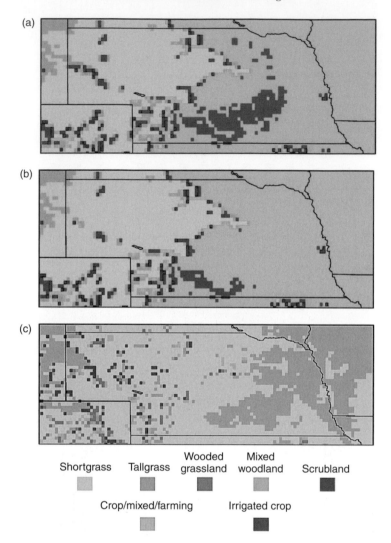

Figure 6.17 Land-cover datasets used for RAMS simulations for (a) 1997 Land-sat and ancillary data irrigation, (b) OGE, and (c) Küchler potential vegetation. From Adegoke *et al.* (2003) reproduced with permission from the American Meteorological Society. See also color plate.

(i.e., pre-European settlement) land cover from the Küchler vegetation dataset (Fig. 6.17c). In the control and OGE wet run simulations, the topsoil of the areas under irrigation, up to a depth of 0.2 m, was saturated at 0000 UTC each day for the duration of the experiment (1–15 July 1997). In both the OGE dry and natural runs, the soil was allowed to dry out, except when replenished naturally by rainfall.

The "soil wetting" procedure for the control and OGE wet runs was constructed to imitate the center-pivot irrigation scheduling under dry synoptic atmospheric conditions such as observed in Nebraska during the first half of July 2000 (i.e.,

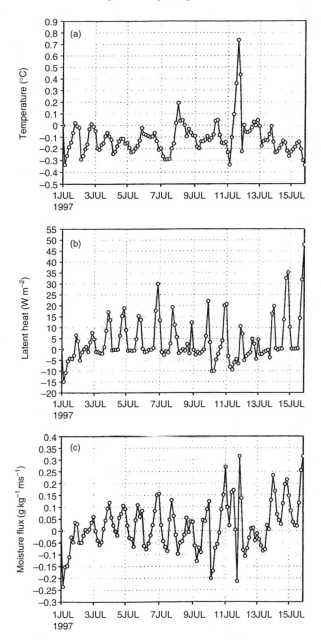

Figure 6.18 Control minus OGE wet run inner-domain area-averaged (a) 2 m temperature, (b) surface latent heat, and (c) moisture flux at 500 m. Reproduced from Adegoke *et al.* (2003) with permission from the American Meteorological Society.

Table 6.1 *Inner-domain area-averaged model parameters for 7–15 July 1997 including scenario comparisons (% change)*

	Control	OGE (wet)	OGE (dry)	Natural vegetation
Temperature (°C)	24.1	24.9 (0.8%)	25.5 (1.4%)	27.6 (3.5%)
Surface sensible heat (W m^{-2})	76.2	79.8 (4.7%)	86.9 (15%)	98.4 (29%)
Latent heat (W m^{-2})	102.4	98.2 (4%)	74.5 (35%)	72.0 (42%)
Vapor flux at 500 m (g kg^{-1} m s^{-1})	11.1	10.4 (7%)	9.1 (22%)	8.2 (34%)

Source: From Adegoke *et al.* (2003).

little or no rainfall recorded throughout the state). The observed atmospheric conditions from the National Centers for Environmental Prediction (NCEP) reanalysis data (Kalnay *et al.*, 1996) were used to create identical lateral boundary conditions in the four cases. A two-grid nested model domain configuration was adopted with a 10-km grid centered over Nebraska nested inside a larger 40-km grid centered over Nebraska nested inside a larger 40-km grid, which extends over most of the central United States.

Results for the inner-domain area-averaged model parameters between the control run and OGE wet run (Fig. 6.18) showed very moderate differences (see Table 6.1 for a summary of scenario comparisons). This reflects the rather small change (less than 10%) in the irrigated portion of the OGE vegetation data (Fig. 6.17b) compared to the more recent Landsat satellite-based land-cover estimates (Fig. 6.17a). In both simulations, the soil-wetting procedure was implemented for the irrigated areas (i.e., land-use class 16 in LEAF-2 (Land-Ecosystem Atmospheric Feedback Model Version 2)). Larger changes were observed when the control run was compared to the OGE dry run; midsummer 2 m temperature over Nebraska might be cooler by as much as 3.4 °C under current conditions (Fig. 6.19a). The average difference between the control and OGE dry runs computed for the 6–15 July 2000 period was 1.2 °C. The irrigation-induced surface cooling was accompanied by a 36% increase in the surface in the surface latent heat flux (Fig. 6.19b) and a significant increase (28%) in water vapor flux at 500 m above the ground (Fig. 6.19c). A corresponding reduction in surface sensible heat (15%) and a 2.6 °C elevation in dewpoint temperature were also observed (not shown).

The cooling effect and the surface energy budget differences identified above intensified in magnitude when the control run results were compared to the potential natural vegetation scenario. For example, the near-ground average temperature for 6–15 July 2000 was 3.3 °C cooler, the surface latent heat flux was 42% higher, and the water vapor flux (at 500 m) 38% greater in the control run compared to the natural landscape run (Fig. 6.20). The first 5 days of the simulation were

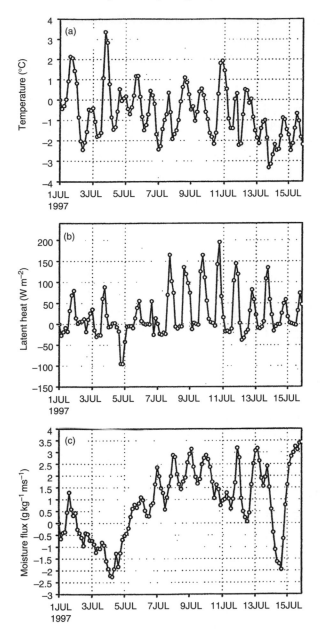

Figure 6.19 Control minus OGE dry run inner-domain area-averaged (a) 2 m temperature, (b) surface latent heat, and (c) moisture flux at 500 m. Reproduced from Adegoke *et al.* (2003) with permission from the American Meteorological Society.

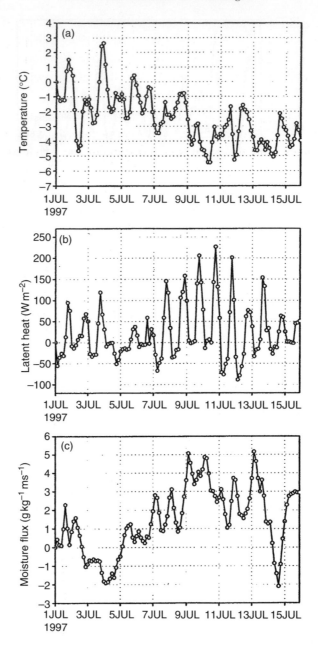

Figure 6.20 Control minus natural vegetation run inner-domain area-averaged (a) 2 m temperature, (b) surface latent heat, and (c) moisture flux at 500 m. Reproduced from Adegoke *et al.* (2003) with permission from the American Meteorological Society.

excluded in computing this temporal average because a fairly strong mid-level low stalled over the study area during that period. This synoptic environment was responsible for the unusual depression in both the minimum and maximum temperatures recorded in the early part of July 2000 in Nebraska (Figs. 6.21 and 6.22).

Important physical changes between the natural shortgrass prairie of this region and the current land-use patterns include alterations in the surface albedo, roughness length changes, and increased soil moisture in the irrigated areas. These changes are capable of generating complex changes in the lower atmosphere (planetary boundary layer, PBL) energy budget. For example, the simulated increase in the portion of the total available energy being partitioned into latent heat

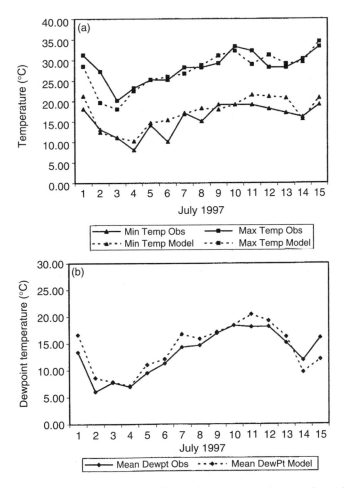

Figure 6.21 Observed minimum and maximum temperatures for Ainsworth municipal compared to model 2 m temperature. Reproduced from Adegoke *et al.* (2003) with permission from the American Meteorological Society.

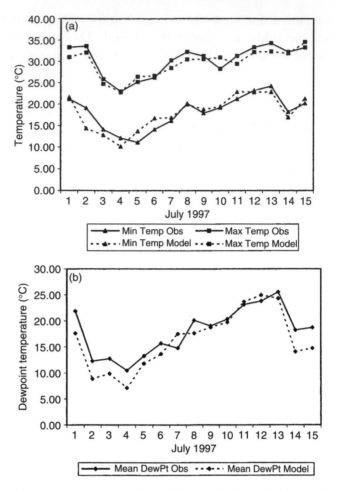

Figure 6.22 Same as 6.21 except from Omaha/Eppley Field. Reproduced from Adegoke *et al.* (2003) with permission from the American Meteorological Society.

rather than sensible heat resulted directly from the enhanced transpiration and soil evaporation in the control run.

The analyses of domain-averaged surface temperature and surface fluxes for this Nebraska study clearly indicate that key components of the planetary boundary layer, particularly the partitioning of the available surface energy in this region, are very sensitive to changes resulting from increased irrigation. RAMS simulations forced, in part, by four different land-cover scenarios indicate a cooling in the near-ground temperature, and significant increases in vapor and latent heat fluxes into the atmosphere in response to this human disturbance of the landscape. These changes are shown to be related to the conversion of the natural prairie vegetation in this region to irrigated and dry cropland. Corroborating evidence from analyses of long-term surface climate data indicates a steady decreasing trend in mean and maximum air temperature at locations located within heavily irrigated areas thus supporting the irrigation-induced cooling effect suggested by the RAMS simulations.

6.3 Dryland agriculture: Oklahoma

McPherson *et al.* (2004) analyzed a number of related observations and demonstrated that Oklahoma's winter wheat belt had a significant impact on the near-surface temperature and moisture fields during the period when winter wheat was growing and during the period after harvest. As they reported, as vegetation grew across the wheat belt, maximum daily temperatures were cooler than those measured over adjacent regions of dormant grasslands. Monthly averaged values of maximum temperature from crop year 2000 displayed a cool anomaly over the growing wheat from November 1999 through April 2000. Using data from a network of surface stations from 1994 through 2001, the cooler temperatures over the wheat belt were shown to be statistically significant at the 95% confidence level for November, December, January, February, and April. As green-up of grasslands occurred during May, the cool anomaly over the wheat belt disappeared. After the wheat was harvested, maximum and minimum daily temperature data revealed a warm anomaly across the wheat belt during June, July, and August. The warmer temperatures also were shown to be statistically significant for all three months.

Cooley *et al.* (2005) also found a substantial effect on regional climate due to dryland farming. In the southern Great Plains of the United States, where winter wheat accounts for 20% of the land area, within the harvested region simulated 2 m air temperatures were 1.3 °C warmer at midday averaged over the 2 weeks following the harvest.

6.4 Desertification

6.4.1 Historical overview

Desertification has an opposite effect on climate to that of irrigation. Desertification was originally defined by Aubreville (1949) as cited in Odingo (1990), as land degradation in semi-arid and subhumid regions such that arid conditions develop and persist. Vegetation loss due to human activities can, therefore, create deserts. The Rajasthan Desert in northwest India is an example of desertification. In that area, in the millennium before the Christian era, a well-developed civilization existed. Currently the region is a tropical desert with only archeological ruins remaining. It has been suggested that overpopulation in the area denuded the vegetation, as the populace used firewood and cleared land for agriculture. While the change in weather could have been part of the natural evolution of climate, major landscape changes by humans on the scale of this desert region have occurred elsewhere.

In the Middle East, Neumann and Parpola (1987), for example, have documented that the political, military, and economic decline of Assyria and Babylonia in the twelfth through tenth centuries BC coincide with a notable period of warming and drying in the region which started around 1200 BC and lasted until about 900 BC. While natural large-scale fluctuations in climate could have caused this aridity, as implied by their paper, desertification provides another explanation. The return of weather to wetter conditions could have occurred in response to a rejuvenation of vegetation as human stress on the landscape was reduced, and the larger-scale weather patterns which influence the region were not more permanently displaced as apparently has occurred in northwest India.

6.4.2 North Africa

Charney (1975) proposed a mechanism of desertification over northern Africa in which the removal of vegetation increased the albedo of the land. Since more solar radiation was reflected back into space, the result was increased subsidence in the lower atmosphere in order to compensate for the loss of heat energy. In a later study, Charney *et al.* (1977) used the Goddard Institute for Space Studies (GISS) model to conclude that local evaporation rates are as important as albedo in influencing rainfall patterns in semi-arid regions, with their relative effects dependent on location. This study found a more complicated interaction between the land surface and the atmosphere than was originally hypothesized in the Charney (1975) paper, with significant research work supporting this conclusion repeated in Claussen *et al.* (1999). They found that rapid natural desertification in the mid-Holocene could only be reproduced in a model if atmospheric–vegetation feedback dynamics were included.

The process of desertification continues today in the Sahel regions of Africa as illustrated in Fig. 6.23 in which a portion of Chad is shown. The darker region in the figure corresponds to an area in which the government controlled grazing while the adjacent areas were extensively utilized by cattle and goats. The significantly higher albedo observed in the overgrazed area is evident in the figure.

6.4.3 Western Australia

In western Australia, as reported by Lyons *et al.* (1993), the use of fencing to prevent the movement of rabbits into an agricultural area is evident from satellite

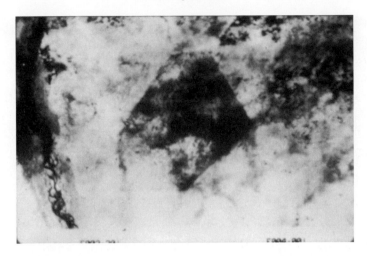

Figure 6.23 Landsat imagery of Chad.

imagery. In this region, 130 000 km^2 of native perennial vegetation has been transformed into winter-growing annual species. Lyons reported a 30% decline of winter rainfall in this region which he attributes to the change of vegetation types. Presumably, the winter crop is harvested such that subsequent transpiration of water into the atmosphere is lost, whereas the perennial plants would be reaching their peak growth (and transpiration) in the latter part of the winter and in the spring.

6.4.4 Middle East

Otterman (1974) and Otterman and Tucker (1985) also used an area of contrasting land use in the Negev Desert–Sinai Peninsula region to demonstrate how desertification works. Figure 6.24a presents satellite imagery of this region, in which the political boundary between Egypt and Israel is clearly shown. In Israel grazing was controlled such that dark plant debris and limited clumps of living vegetation reduced the albedo of the surface. In the Egyptian area, in contrast, overgrazing by goats permitted few plants to survive so that the soil became more mobile and covered the vegetation debris. The result was a higher albedo from the light bare soil. Figure 6.24b documents using a numerical model simulation that surface ground temperature differences over 5 °C can result with the desertified area being cooler. Otterman suggests that these cooler temperatures result in less annual precipitation than otherwise would occur due to the reduced surface supplied buoyant energy for showers. Thus it would be less likely for the vegetation to regenerate even if the grazing pressure were removed.

(a)

(b)

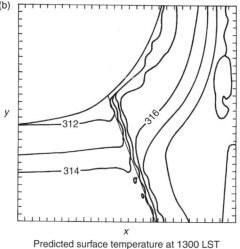

Predicted surface temperature at 1300 LST

Figure 6.24 (a) This MODIS view shows the denuded high albedo regions of the Sinai and Gaza Strip, in contrast to the darker western Negev. Sensor: Terra/MODIS; Datastart: 2000-09-10; Visible Earth v1 ID: 5606; Visualization date: 2000-10-12. Courtesy of NASA Visible Earth and Jacques Descloitres, MODIS Land Science Team. See also color plate. (b) Predicted surface temperature at 1300 LST for typical fall meteorological conditions over the same area. From Mahrer and Pielke (1978).

Mahrer and Pielke (1978) performed a model simulation of this region to show that the gradient of surface temperatures due to the albedo differences will cause preferential ascent (and thus more likely showers when the atmosphere is favorable for cumulus convection over the Negev side of the boundary). Otterman *et al.* (1990) suggest that natural vegetation in a semi-arid or arid region could have a more positive impact on rainfall than the tarring of high albedo sandy desert areas as proposed by Black and Tarmey (1963). Otterman *et al.* concluded that land management has a significant impact on climate.

6.5 Deforestation

6.5.1 Historical perspective

Deforestation is another type of landscape change. Deforestation occurs when lumbering and fires remove extensive areas of trees from a region. Such deforestation occurred, for example, in the eastern United States in the nineteenth century and resulted in an almost total removal of the original uneven aged, climax forest. Only in the Great Smoky Mountain Forest, where about 40% of the virgin trees remain, can one appreciate the large diameter deciduous and evergreen trees and sparse understory vegetation that originally covered the area prior to European settlement. During the last 50 to 75 years, as small farms became less economically viable, large areas of the forest has returned as the nutrient-rich organic soils in the midlatitude region permitted regrowth from root sprouts, natural seedings, and forestry planting of seedlings. Thus while the deforestation was extensive, a new forest, albeit of a different species composition and age, has evolved in non-agricultural and non-urban areas in the eastern United States.

6.5.2 Amazon

More recent deforestation has occurred in the Amazon region and the influence of this removal of trees on global climate has been the subject of considerable interest (Dickinson, 1987; Lean and Warrilow, 1989). There are regional effects of this deforestation as well. Currently, it is estimated that the water vapor for about 50% of rainfall within the Amazon Basin is from local evaporation and transpiration (Salati *et al.*, 1978). Figure 6.25 shows that the removal of trees is not completed as one vast clear-cutting but is accomplished through a series of strips of several kilometers in width and in large clear-cut areas. Local wind circulations develop in response to these landscape alterations

Figure 6.25 Examples of clear-cutting of the tropical forest in two areas of the Amazon. Photos provided by Carlos Nobre of the Center for Weather Prediction and Climate Studies – CPTEC, INPE, Brazil. See also color plate.

that result in an enhancement of local rainfall and cloudiness (Avissar and Liu, 1996).

A major difference between the midlatitude forests and forests in the humid tropics such as the Amazon is the absence of substantial organic material and trace metals in the soils in the tropical environment. The heavy rains in the region leach nutrients to below the root layers with the result that much of the nutrients required for vegetation development and growth are within the living and recently dead plants. A rapid recycling of nutrients occurs as new vegetation generates from the rapidly decomposing plants. In the absence of this rapid recycling, rains would deplete the root zones of the needed plant food. Thus, the removal of trees by lumbering short-circuits the recycling. In the absence of these nutrients, even an increase of rainfall due to the patchiness of the timber cutting may not benefit the regeneration of the forest.

Several field studies have been undertaken in the Amazon in order to better understand the influence of this natural vast forest area on climate and weather. In 1985 and 1987, the Amazon Boundary Layer Experiment (ABLE) was completed (Garstang *et al.*, 1990). Among the results it was confirmed that local wind circulations apparently develop in response to large rivers in the area being adjacent to the forests (Miller *et al.*, 1988). A similar atmospheric response would be expected when large areas are cleared of trees. The Anglo-Brazilian Amazonian Climate Observation Study (ABRACOS) was initiated at the end of 1990 to investigate the influence of clear-cutting in the region. Results from the field study (see Fig. 6.26) demonstrate that the forest captures more radiative energy than an area which is pasture. During the daytime, the albedo, on average, is about 2.5% larger over the cleared area with the resultant wind speed, temperature, and moisture profiles shown in Figs. 6.27, 6.28, and 6.29.

These measurements show that the temperature change is twice as great over the cleared area with even a greater variation in wind speed. Early morning fog or mist is observed in the clearings, while such moisture saturation is almost unknown in the forest.

6.5.3 Africa

The influence of deforestation in another tropical area, southern Nigeria, is discussed in Ghuman and Lal (1986). Figure 6.30, from that study, is consistent with the results from the Amazon in that the diurnal range of temperature is substantially greater over the cleared ground. This figure also shows that the influence of forest results in somewhat cooler temperatures deeper in the soil (i.e., 50 cm).

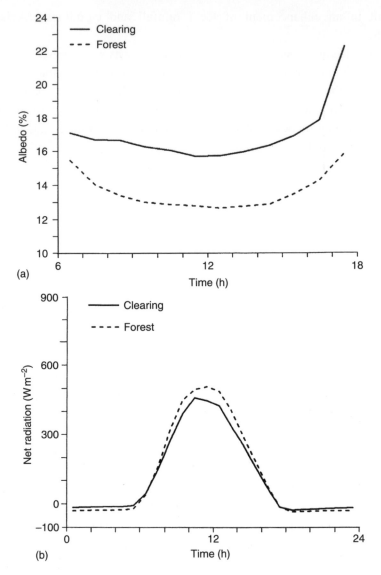

Figure 6.26 (a) Mean hourly albedo at the clearing and forest sites in the Amazon for the study period from 12 October to 10 December 1990. (b) Mean hourly net all-wave radiation at the clearing and forest sites for the same study period. From Bastable *et al.* (1993).

6.6 Regional vegetation feedbacks

Eastman *et al.* (2001a,b) have shown that land-use change, grazing, and the biogeochemical effect of increased carbon dioxide can significantly alter the regional climate system in the central Great Plains of the United States. Figure 6.31

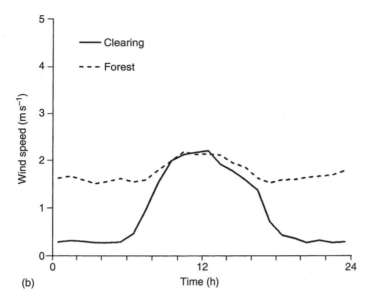

Figure 6.27 Mean hourly wind speed at clearing and forest sites in the Amazon for the period 12 October to 10 December 1990. From Bastable *et al.* (1993).

shows these effects on maximum and minimum temperature, rainfall, and above-ground biomass growth, during a growing season averaged over the central Great Plains. For example, the effect of enhanced atmospheric concentrations of carbon dioxide on plant growth on a seasonal timescale is shown to be much more important than the radiative effect of enhanced atmospheric carbon dioxide for this region.

The nonlinear effect of vegetation–atmospheric feedback on this scale also results in a complex spatial and temporal pattern of response. Not only is there a teleconnection of atmospheric conditions to locations distant from where the land feedback occurs, but the landscape at distant locations itself is influenced by the altered weather. In manipulative vegetation experiments where carbon dioxide concentrations are arbitrarily increased, for example, this nonlinear feedback between the atmosphere and land surface is missed since there is no feedback to the regional weather (with greater vegetation cover resulting in greater summer rainfall and cooler maximum temperatures).

The biogeochemical effect of enhanced carbon dioxide and trace gas concentration, and of aerosol deposition (such as nitrogen; Raddatz, 2003; Galloway *et al.*, 2004; Holland, 2005) on landscape dynamics has also not been adequately considered. For example, Jenkinson *et al.* (1991) demonstrated a significant positive carbon dioxide radiative feedback where soils released carbon to the atmosphere under warming conditions. Lenton (2000), using a simple box model, and

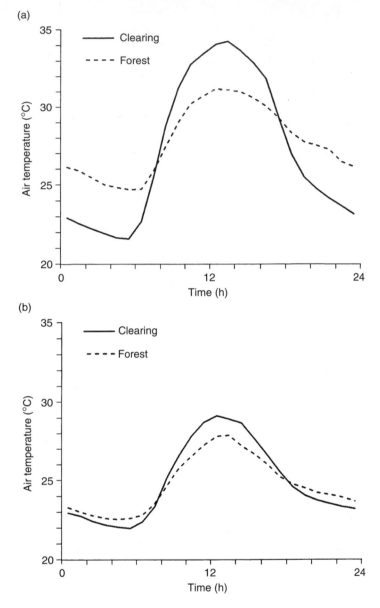

Figure 6.28 Mean hourly air temperature at screen height at the clearing and forest sites in the Amazon for (a) the dry period from 12 to 21 October 1990 and (b) the wet period from 1 to 10 December 1990. From Bastable *et al.* (1993).

Cox *et al.* (2000) using general circulation model (GCM) sensitivity experiments, showed that biogeochemical feedbacks in conjunction with an increased carbon dioxide radiative warming produced an amplified regional and global temperature increase. These results are in contrast to those of Eastman *et al.* (2001a) where a cooler daytime and warmer nighttime in the central Great Plains was simulated in

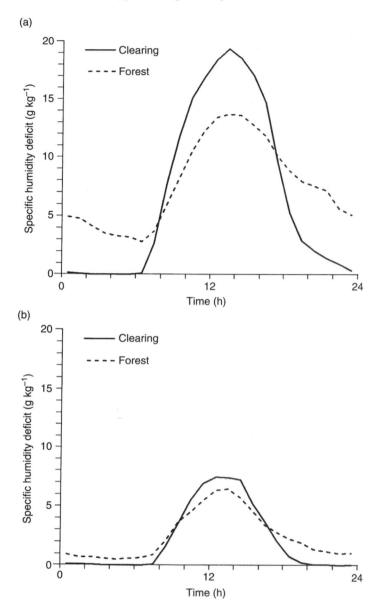

Figure 6.29 Mean hourly specific humidity deficit at screen height at the clearing and forest sites in the Amazon for (a) the dry period from 12 to 21 October 1990, and (b) the wet period from 1 to 10 December 1990. From Bastable *et al.* (1993).

response to greater plant growth in a doubled carbon dioxide atmosphere. Zeng *et al.* (2004) showed that abrupt transitions can occur in semi-arid regions associated with the dynamical interactions of soil water and vegetation which further illustrates the complexity of the nonlinear interactions between vegetation and the atmosphere.

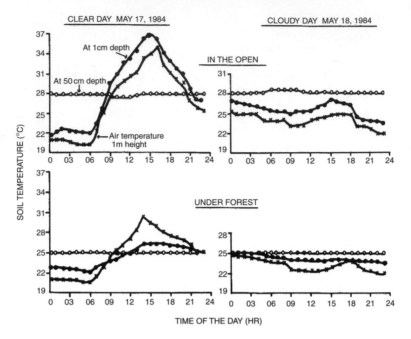

Figure 6.30 Diurnal fluctuations in soil and air temperature under forest and on cleared land in southern Nigeria. From Ghuman and Lal (1986).

Niyogi and Xue (2006) used a coupled process-based model to show that the carbon assimilation potential for each of the GCM land-use categories (comprising both C3 and C4 photosynthesis pathways) is sensitive to the soil moisture availability. Niyogi *et al.* (2004) and Gu *et al.* (2003) have shown that atmospheric aerosols, through their increase of the fraction of diffuse sunlight, enhance photosynthesis. Nemani *et al.* (2002) demonstrated how increased precipitation and humidity over the United States, particularly during the period from 1950 to 1993 increased the growth of natural vegetation more than increased carbon dioxide and warmer temperatures did. Ozone pollution in China has been shown to affect crop yields which will necessarily affect transpiration in the region (Chameides *et al.*, 1999).

The presence of drought and hydrological feedbacks associated with land-use change locally or through teleconnections, therefore, have a direct impact on the source/sink capabilities of the terrestrial ecosystem. These studies illustrate the significant role that biogeochemistry has within the climate system. This feedback, along with other climate forcings and feedbacks (Pielke, 2001b), make climate prediction on timescales of years and longer a particularly difficult problem.

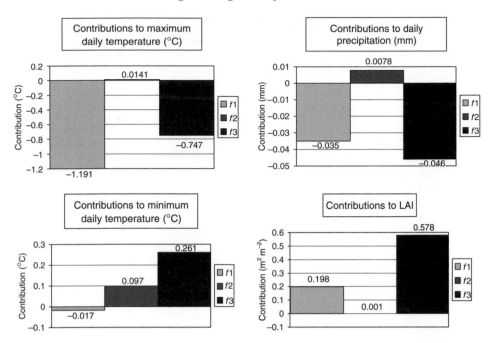

Figure 6.31 RAMS/GEMTM nonlinear coupled model results – the seasonal domain-averaged (central Great Plains) for 210 days during the growing season, contributions to maximum daily temperature, minimum daily temperature, precipitation, and leaf area index (LAI) due to: $f1$ = natural vegetation, $f2$ = $2 \times CO_2$ radiation, and $f3 = 2 \times CO_2$ biology. From Rial *et al.* (2004) with kind permission of Springer Science and Business Media.

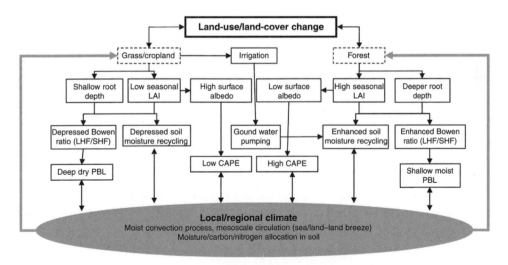

Figure 6.32 Hypotheses of the influence of land-use/land-cover change on regional climate. From Pielke *et al.* (2006a).

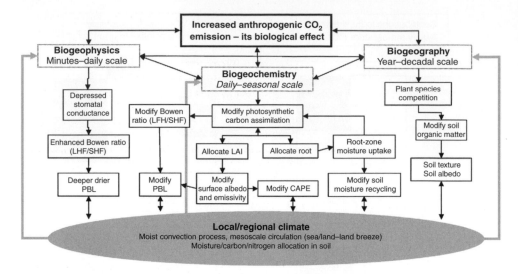

Figure 6.33 Increased carbon dioxide on biological activity as it influences weather. From Pielke *et al.* (2006a).

6.7 Conclusion

Figures 6.32 and 6.33 illustrates the pathways that land-use/land-cover change and vegetation dynamics in response to increased concentration of carbon dioxide act as climate forcings and feedbacks. Table 6.2 provides a summary list of additional examples of papers which document the role of these climate forcings and feedbacks in the climate system.

Table 6.2 *Regional land-use/land-cover assessments: effect on weather and climate*

Geographic region	Reference
Africa	Brovkin *et al.* (1998, 2004), Claussen (1997), Claussen and Gayler (1997), Claussen *et al.* (1999, 2002, 2003), Eltahir (1996), Eltahir and Gong (1996), Gong and Eltahir (1996), Irizarry-Ortiz *et al.* (2003), Kiang and Eltahir (1999), Kim and Eltahir (2004), Maynard and Royer (2004), Pinty *et al.* (2000), Polcher and Laval (1994a,b), Snyder *et al.* (2004a), Wang and Eltahir (2000a,b,c,d,e, 2002), Wang *et al.* (2004), Xue *et al.* (1990, 2004a), Zhang *et al.* (2001), Zheng and Eltahir (1997, 1998), Zheng *et al.* (1999)
Alaska, USA	Mölders and Olson (2004)
Amazon	Baidya Roy and Avissar (2002)[a], Chagnon and Bras (2005), Chagnon *et al.* (2004), Chen *et al.* (2001), Costa and Foley (2000),

	Dickinson and Kennedy (1992), Dirmeyer and Shukla (1994), Dolman *et al.* (1999), Durieux *et al.* (2003)[a], Eltahir and Bras (1993a, 1994a,b), Eltahir *et al.* (2004), Gash and Nobre (1997)[a], Gedney and Valdes (2000), Hahmann and Dickinson (1997), Henderson-Sellers and Gornitz (1984), Henderson-Sellers *et al.* (1993), Kleidon and Heimann (1999, 2000), Kleidon and Lorenz (2001), Lean and Rowntree (1997), Lean and Warrilow (1989), Mylne and Rowntree (1992), Negri *et al.* (2004)[a], Nobre *et al.* (1991), Oyama and Nobre (2003, 2004), Polcher and Laval (1994a,b), Shukla *et al.* (1990), Silva Dias *et al.* (2002), Wang *et al.* (2000), Werth and Avissar (2002), Zeng (1998), Zeng *et al.* (1996), Zhang *et al.* (1996a, 2001)
Atlanta, Georgia, USA	Shepherd *et al.* (2004)
Australia	Eberbach (2003)[a], Gero *et al.* (2006), Gordon *et al.* (2003)[a], Huang *et al.* (1995a,b), Lyons (2002), Lyons *et al.* (1996), Miller *et al.* (2005), Narisma and Pitman (2003, 2004, 2006), Narisma *et al.* (2003), Peel *et al.* (2005), Pitman and Narisma (2005), Pitman *et al.* (2004), Ray *et al.* (2003), Timbal *et al.* (2002)[a], Timbal and Arblaster (2006)
Boreal Forest	Betts (2000), Bonan *et al.* (1992), Brovkin *et al.* (2003), Steyaert *et al.* (1997), Vidale *et al.* (1997)
Brazil	Oyama and Nobre (2004)
Canadian Prairie	Hanesiak *et al.* (2004), Raddatz (1998, 1999, 2000[a], 2003, 2005), Raddatz and Cummine (2003)[a]
Central Africa	Eltahir *et al.* (2004), Mylne and Rowntree (1992), Werth and Avissar (2005)
Central America	Ray *et al.* (2006)
China	Gao *et al.* (2003), Ming and Xinmin (2002)[a], Sen *et al.* (2004a, b), Simmonds *et al.* (1999)[a], Suh and Lee (2004)[a], Wang *et al.* (2003), Wang *et al.* (2004), Wei and Fu (1998), Xue (1996), Xuejie *et al.* (2003)[a], Zhao and Pitman (2005), Zheng *et al.* (2002)[a]
Colorado, USA	Borys *et al.* (2003), Chase *et al.* (1999)[a], Hanamean *et al.* (2003)[a], Pielke and Zeng (1989)[a], Segal *et al.* (1988, 1989)[a], Stohlgren *et al.* (1998)
Costa Rica	Lawton *et al.* (2001), Nair *et al.* (2003)
East Asia	Suh and Lee (2004)[a]
East of 100°, USA	Bonan (1999)
Eurasian Boreal Forest	Betts (2000), Douville and Royer (1997)
Europe	Dolman *et al.* (2004), Gaertner *et al.* (2001), Gates and Liess (2001), Heck *et al.* (1999, 2001)
Florida, USA	Marshall *et al.* (2003, 2004a,b), Pielke *et al.* (1999b)
Germany	Mölders (1998, 1999a, 2000a,b), Mölders and Rühaak (2002)
Great Plains, USA	Adegoke *et al.* (2006) Cheng and Cotton (2004), Eastman *et al.* (2001a,b), Grossman *et al.* (2005), Holt *et al.* (2006) Mahmood and Hubbard (2002, 2003)[a], Mahmood *et al.* (2004[a], 2006) Muhs and Holliday (1995), Niyogi *et al.* (2006), Segal *et al.* (1995)[a], Strack *et al.* (2004)

Table 6.2 *(cont.)*

Geographic region	Reference
Houston, Texas, USA	Jin *et al.* (2005), Shepherd and Burian (2003)
India	Bosilovich and Schubert (2002)[a], Douglas *et al.* (2006), Niyogi *et al.* (2003)
Indonesia	Polcher and Laval (1994a)
Israel (southern)	De Ridder and Gallée (1998)[a], Otterman *et al.* (1990)[a]
Japan	Ichinose (2003), Suh and Lee (2004)[a]
Kansas, USA	Clark and Arritt (1995)[a]
Korea	Suh and Lee (2004)[a]
Mexico	Asner and Heidebrecht (2005)
Middle Europe	Millán *et al.* (2005), Mölders (1999b)
Midlatitudes	Mahfouf *et al.* (1987)[a], Segal *et al.* (1988)[a], Seth and Giorgi (1996)[a]
Midwest, USA	Bonan (2001), Carleton *et al.* (2001)[a], Changnon *et al.* (2003)[a], Diffenbaugh (2005), Georgescu *et al.* (2003), Rozoff *et al.* (2003), Weaver and Avissar (2001, 2002)
Mississippi River Basin, USA	Twine *et al.* (2004)
Nebraska, USA	Adegoke *et al.* (2003)[a]
New York, USA	Jin *et al.* (2005)
North Africa	Cathcart and Badescu (2004), Xue (1997)
North America	Bosilovich and Schubert (2002)[a], Segal *et al.* (1998)[a], Skinner and Majorowicz (1999)
Oklahoma, USA	Basara and Crawford (2002)[a], McPherson *et al.* (2004)[a]
Oklahoma and Kansas, USA	Weaver and Avissar (2001)[a]
Phoenix, Arizona, USA	Diem and Brown (2003)
Puerto Rico	van der Molen *et al.* (2005)
Regional paleo-studies	Rivers and Lynch (2004)
Sahel	Charney (1975), Clark *et al.* (2001, 2004), Claussen and Gayler (1997), De Ridder (1998), Dolman *et al.* (2004), Hutjes *et al.* (2003), Nicholson (2000), Nicholson *et al.* (1998), Taylor *et al.* (2002), Wang and Eltahir (2000a), Wang *et al.* (2004), Xue (1997), Xue and Shukla (1993), Xue *et al.* (2004b), Zeng *et al.* (1999)
Sonoran Desert, Mexico	Balling (1988)[a]
Southeast Asia	Henderson-Sellers *et al.* (1993), Kanae *et al.* (2001), Sen *et al.* (2004a), Zhang *et al.* (2001)
South America	Beltran, (2005), Cook and Vizy (2006), Eltahir and Bras (1993a,b, 1994a,b), Eltahir and Humphries (1998); Eltahir and Pal (1996), Oyama and Nobre (2003), Vizy and Cook (2005), Wang *et al.* (1998, 2000)
Spain	Hasler *et al.* (2005), Miao *et al.* (2003)[a]
Tennessee, USA	Banta and White (2003)[a]
Texas, USA	Barnston and Schickedanz (1984)[a], Beebe (1974)[a], Moore and Rojstaczer (2002), Shepherd *et al.* (2002)
Tropical Savannas	Hoffmann *et al.* (2002)[a]
United States	Alig *et al.* (2004), Baidya Roy *et al* (2003)[a], Bonan (1997, 1999), Christy *et al.* (2006), Civerolo *et al.* (2000)[a], Copeland *et al.* (1996), Givati and

	Rosenfeld (2004), Hale *et al.* (2006), Kalnay and Cai (2003), Kalnay *et al.* (2006), Lim *et al.* (2005), Pan *et al.* (2004), Shepherd *et al.* (2002), Xue *et al.* (1996, 2001)
West Africa	Cathcart and Badescu (2004), Mohr *et al.* (2003)[a], Xue and Shukla (1993), Zeng *et al.* (2002), Zheng and Eltahir (1997, 1998[a])
Other studies of mesoscale effects	Avissar and Chen (1993), Avissar and Liu (1996), Chase *et al.* (1999), Chen and Avissar (1994a), Goutorbe *et al.* (1994), Lynn *et al.* (1995a), Mahfouf *et al.* (1987), Mahrt *et al.* (1994), Ookouchi *et al.* (1984), Pielke *et al.* (1991b), Segal and Arritt (1992), Segal *et al.* (1988), Stohlgren *et al.* (1998), Taylor *et al.* (1998), Wang *et al.* (1997, 1998), Weaver and Avissar (2001)

[a]Information was provided by Dr. Rick Raddatz.

Additional reading

There are several useful texts that are available to provide additional in-depth information on the material in this chapter. These include:

Arora, V., 2002. Modeling vegetation as a dynamic component in soil–vegetation–atmosphere transfer schemes and hydrologic models. *Revs. Geophys.*, **40**, 1006, doi:10.1029/2001RG 000103.

Bonan, G. B., 2002. *Ecological Climatology: Concepts and Applications.* New York: Cambridge University Press.

Dolman, A. J., A. Verhagen, and C. A. Rovers, 2003. *Global Environmental Change and Land Use.* Dordrecht, The Netherlands: Kluwer.

iLEAPS, 2004. Integrated Land Ecosystem-Atmosphere Processes Study (iLEAPS): The IGBP – Land-Atmosphere Project. Science Plan and Implementation Strategy. Available online at: www.atm.helsinki.fi/ILEAPS/index.php?page=draft

Lambin, E. F., H. J. Geist, and E. Lepers, 2003. Dynamics of land-use and land-cover change in tropical regions. *Ann. Rev. Environ. Resources*, **28**, 205–241.

Lee, T. J., R. A. Pielke, and P. W. Mielke Jr., 1995: Modeling the clear-sky surface energy budget during FIFE87. *J. Geophys. Res.*, **100**, 25585–25593.

Lepers, E., E. F. Lambin, A. C. Jenetos, *et al.*, 2005. A synthesis of information on rapid land-cover change for the period 1981–2000. *BioScience*, **55**, 115–124.

McCumber, M. C., and R. A. Pielke, 1981. Simulation of the effects of surface fluxes of heat and moisture in a mesoscale numerical model. I. Soil layer. *J. Geophys. Res.*, **86**, 9929–9938.

Northern Eurasia Earth Science Partnership Initiative (NEESPI), 2004. *NEESPI Science Plan*, P. Y. Groisman and S. A. Bartalev, eds. Available online at NEESPI.gsfc. nasa.gov/

Orme, A. R. (ed.), 2002. *Physical Geography of North America.* Oxford, UK: Oxford University Press.

Pitman, A. J., 2003. Review: The evolution of, and revolution in, land surface schemes designed for climate models. *Int. J. Climatol.*, **23**, 479–510.

Waring, R. H., and S. W. Running, 1998: *Forest Ecosystems: Analysis at Multiple Scales*, 2nd edn. San Diego, CA: Academic Press.

7

Concluding remarks regarding deliberate and inadvertent human impacts on regional weather and climate

We have seen that there is considerable evidence suggesting that anthropogenic activity, either in the form of constructing major urban areas, changing natural landscape to agricultural and grazing areas, or emission of particles and gases, has contributed to changes in weather and climate on the regional scale. There is a diversity of views on the demonstration of this evidence, however, with a distinction between deliberate and inadvertent weather and climate changes. It is curious, for example, that the scientific community has accepted the rainfall change results obtained in studies such as METROMEX as being valid, yet has questioned the validity of cloud-seeding-induced changes in rainfall inferred from well-designed, randomized cloud seeding experiments.

The answer to this paradox lies in human psychology. As an example, Dr. Stan Changnon described a conversation he had with Dr. John Tukey, one of the world's leading statisticians, following a meeting of the Weather Modification Advisory Board in the 1990s. Stan asked John why the statisticians had been very critical of attempts to prove planned weather modification of clouds and rainfall was successful, yet were not so critical of inadvertent weather modification (i.e., the cities are not randomized). John looked up at Stan and said, "Well, Stan, in the end it is just a lot more believable that a big city can cause clouds, rain, and hail than it is that a small amount of seeding material can." In other words, no matter how objective we attempt to be, a certain amount of subjectivity is involved in accepting the results of any scientific study.

In summary, while there is considerable evidence supporting the hypothesis that human activity is inadvertently modifying weather and climate on the regional scale, much more research is required to pinpoint the causes of inferred human-related weather anomalies and to strengthen the statistical inferences and physical understanding.

Unfortunately, United States federal funding of inadvertent weather modification fell sharply in the 1980s along with that of deliberate weather modification.

For some incomprehensible reason, funding of inadvertent weather modification has apparently been tied to funding of advertent weather modification. To some extent, funding for inadvertent weather modification fell through the cracks of the funding agencies. It does not fit into the mission-oriented agencies that have supported planned weather modification, so little support came from those agencies. It does fit within the mission of the Environmental Protection Agency (EPA), but the research program there, with respect to inadvertent local and regional weather and climate changes, was seriously weakened under the Reagan administration. Because inadvertent modification and planned weather modification were placed within the same program in the National Science Foundation, cuts in the planned weather modification program also cut the already meager program in inadvertent weather modification. Again, the lack of a lead agency or a funded, coordinated program in weather modification appears to be limiting progress in furthering our quantitative understanding of deliberate and inadvertent human impacts on weather and climate on the local and regional scale.

Part III
Human impacts on global climate

8

Overview of global climate forcings and feedbacks

8.1 Overview

The Earth's global *climate system* consists of the atmosphere, oceans, land, and continental glaciers as illustrated in Fig. 8.1. In this framework, variables such as salinity, soil moisture, and flora are integral to the functioning of this dynamic system. All of these variables are climate variables. Several of these climate forcings were discussed in a regional context in Chapters 5 through 7. Here we present a global perspective. This definition of climate is broader than the definition of climate as long-term weather statistics (Pielke, 1998).

A *climate forcing* is defined as "an energy imbalance imposed on the climate system either externally or by human activities" (National Research Council, 2005). Climate forcing can be separated into radiative and non-radiative forcing following the definitions provided in National Research Council (2005). A *radiative forcing* is reported in the climate change scientific literature as a change in energy flux at the tropopause, calculated in units of watts per square meter. A *non-radiative forcing* is a climate forcing that creates an energy imbalance that does not immediately involve radiation. An example is the increasing latent heat flux resulting from agricultural irrigation. A *direct forcing* is a climate forcing that directly affects the radiative budget of the Earth's climate system. For example, this perturbation may be due to a change in concentration of the radiatively active gases, a change in solar radiation reaching the Earth, or changes in surface albedo. A *climate feedback* is an amplification or dampening of the climate response to a specific forcing due to changes in the atmosphere, oceans, land, or continental glaciers. These definitions are summarized in Appendix C as given in National Research Council (2005).

Figure 8.2 schematically illustrates the relation of climate forcings to the climate response. Also included in this figure are identified forcings, which are discussed in this chapter.

153

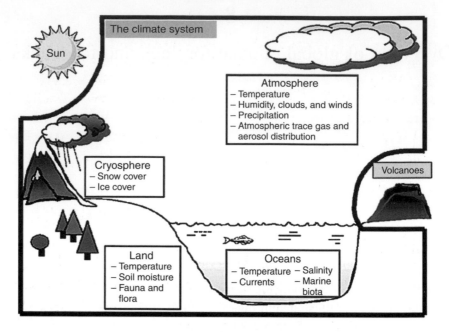

Figure 8.1 The climate system, consisting of the atmosphere, oceans, and land. Critical state variables for each sphere of the climate system are listed in the boxes. For the purposes of this report, the Sun, volcanic emissions, and human-caused emissions of greenhouse gases or changes to the land surface are considered external to the climate system. From National Research Council (2005).

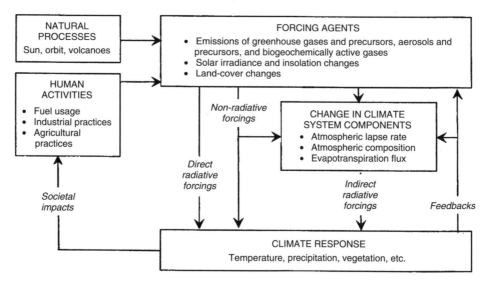

Figure 8.2 Conceptual framework for the climate forcings that impact the physical components of the climate system under present-day conditions. From National Research Council (2005).

8.2 Atmospheric radiation

The principal source of energy driving the atmospheric/ocean system comes from the Sun. In order to understand the potential for human impact on global climate, we must have at least a general understanding of how the Sun's energy is distributed through the atmosphere–Earth–ocean system. The Sun emits most of its radiant energy over wavelengths ranging from 0.2 to 1.8 μm, with a peak in intensity at 0.470 μm. As shown in Fig. 8.3, the spectrum of energy emitted by the Sun closely approximates the energy spectrum emitted by a perfect absorber–emitter substance (or what we call a blackbody) having a temperature of 5900 K. On the other hand the spectrum of energy emitted by the Earth emits radiant energy with a peak in spectral density at wavelengths between 10 and 15 μm. Thus the combination of radiation energy emitted by the Sun as received by the Earth at its orbital distance and emitted by the Earth span from less than 0.2 μm to greater than 50 μm. The spectrum of energy emitted by the Sun and the Earth, however, exhibit very little overlap. We therefore refer to the energy emitted by the Sun as *shortwave radiation* and the energy emitted by the Earth and its atmosphere as *longwave radiation*.

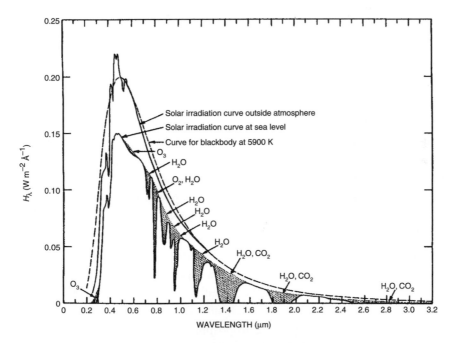

Figure 8.3 Spectral distribution curves related to the Sun; shaded areas indicate absorption at sea level, due to the atmospheric constituents shown. From Gast *et al.* (1965). © McGraw-Hill; Reprinted by permission.

As atmospheric radiation passes through the atmosphere it interacts with the gases and particulates. Some of the radiant energy is absorbed, some scattered, and some unaffected. The amount of energy absorbed and scattered varies with the wavelength of the radiant energy, the type of gases, and the size and chemical composition of the particles. The absorbed energy is converted to heat in the atmosphere and that energy is reradiated at a wavelength and intensity that corresponds to the temperature and composition of the absorbing gases or particulates. The amount of energy that is scattered, however, is simply redirected in space (or reflected) with no change in intensity. The amount of energy scattered and the directions in which it is scattered depends on the wavelength of the radiation as well as the types of gases and size and composition of particles.

8.2.1 Absorption and scattering by gases

As shown in Fig. 8.3, the energy spectrum incident at the top of the atmosphere is significantly attenuated by the time it reaches the Earth's surface at sea level. It is not attenuated uniformly across the spectrum, however, but instead major energy losses occur over rather narrow bands, called *absorption bands*. In a cloud-free clean atmosphere the primary absorbers of shortwave radiation are ozone and water vapor. Airborne particles, or aerosols, in typical non-polluted air contribute less to absorption. At wavelengths less than $0.3\,\mu m$, oxygen and nitrogen absorb nearly all the incoming solar radiation in the upper atmosphere. Between 0.3 and $0.8\,\mu m$, however, little gaseous absorption occurs. Only weak absorption by ozone takes place over this spectral range. At wavelengths less than $0.8\,\mu m$ Rayleigh scattering by oxygen and nitrogen molecules also depletes the radiant energy that reaches the surface. At longer wavelengths absorption in various water vapor bands is quite pronounced while weak absorption by carbon dioxide and ozone also occurs.

Absorption of longwave radiation occurs in a series of bands. The principal natural absorbers of longwave radiation are water vapor, carbon dioxide, and ozone. Figure 8.4 illustrates the infrared spectrum obtained by a scanning interferometer looking downward from a satellite over a desert. Strong absorption by carbon dioxide at a wavelength band centered at $14.7\,\mu m$ is shown by the emission of radiance at a temperature of 220 K which is representative of stratospheric temperatures. The stratosphere contributes mainly to the peak of this band. Water vapor absorption bands at 1.4, 1.9, 2.7, and $6.3\,\mu m$, and greater than $20\,\mu m$, cause emissions corresponding to midtropospheric temperatures. Little absorption is evident in the region called the *atmospheric window* between 8 and $14\,\mu m$. Here the satellite observed radiance corresponds to the surface temperature of the desert except for a slight depression in magnitude due to departures of the

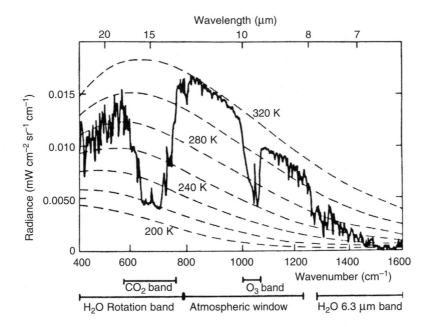

Figure 8.4 Atmospheric spectrum obtained with a scanning interferometer on board the Nimbus 4 satellite. The interferometer viewed the Earth vertically as the satellite was passing over the North African desert. After Hänel *et al.* (1972). © [1972] American Geophysical Union. Reproduced with permission from the American Geophysical Union.

emittance of sand from its block body value of unity. A distinct ozone absorption band is evident at about 9.6 μm in the middle of the window. Although not very evident in the figure, weak continuous absorption also occurs across the window believed to be due to the presence of clusters or dimers of water vapor molecules.

In summary, the principal natural gaseous absorbers of terrestrial radiation are carbon dioxide and water vapor. These are the so-called principal "greenhouse" gases. The basic greenhouse concept was probably first proposed by Fourier (1827). Later, Arrhenius (1896) calculated the effects of varying carbon dioxide concentrations on surface temperatures and even included the impact of water vapor absorption in his calculations. He estimated that a 2.5- to 3-fold increase in carbon dioxide would increase temperatures in Arctic regions by 8–9 °C. His intent was to examine the impact of greenhouse gases on ice ages and glaciation and was not particularly concerned about human sources. As early as 1938, Callendar (1938) became concerned about the impacts of anthropogenic emissions of carbon dioxide on global temperatures. Therefore, the basic concept that anthropogenic sources of carbon dioxide and other greenhouse gases can warm the atmosphere is not new, but reports suggesting that global average surface temperatures are rising has caused a great deal of concern about potential human impact on climate, and resulted in major research activities in global climate change.

In the absence of so-called greenhouse gases, the average surface temperature of the Earth would be over 30 °C cooler than it is today. This is because the greenhouse gases absorb upwelling terrestrial radiation and radiate the absorbed gases, both upward and downward, causing a net gain in energy (reduced loss) at the Earth's surface. Without those greenhouse gases, a larger fraction of the upwelling terrestrial radiation would escape through the atmospheric window to space. The major greenhouse gas is water vapor which varies naturally in space and time due the Earth's hydrological cycle (e.g., see Randall and Tjemkes, 1991). The second most important greenhouse gas is carbon dioxide. In contrast to water vapor, carbon dioxide is rather uniformly distributed throughout the troposphere, although the radiative forcing associated with it is more heterogeneous as a result of spatial (e.g., latitudinal) and temporal variations in tropospheric temperature and water vapor concentrations, and in surface emissions and absorption.

We discuss the greenhouse effect further in Subsection 8.2.4. Other strong absorbers of terrestrial radiation are methane, nitrous oxide, and chlorofluorocarbons (CFCs). Currently the concentrations of these gases are so small that their contributions to infrared absorption cannot be easily detected by satellite. The concentrations of CFCs, for example, are six orders of magnitude (a million times) smaller than carbon dioxide. The concern about CFCs is that per molecule they are 20 000 times more effective at absorbing infrared radiation than carbon dioxide.

8.2.2 Absorption and scattering by aerosols

The radiative properties of aerosol particles is a complicated function of their chemistry, shape, and size spectra. Moreover, if the aerosol particles are hygroscopic, their radiative properties change with the relative humidity of the air. At relative humidities greater than 70%, the hygroscopic particles (called haze) take on water vapor molecules and swell in size, thus changing their radiative properties not only because of size effects but also because of changes in their complex indices of refraction as the water-solution–particle mixture changes in relative amounts.

Dry aerosol particles in high concentrations such as over major polluted urban areas and over deserts can cause substantial absorption and scattering of solar radiation. In some polluted boundary layers, aerosol absorption has been estimated to result in heating rates on the order of a few tenths to several degrees per hour (Braslau and Dave, 1975; Welch and Zdunkowski, 1976). In the Saharan dust layer, aerosol absorption has been calculated to produce a heating of 1–2 °C per day. As discussed more fully in Chapter 9, modeling studies suggest that radiative heating of haze and dust can alter the atmospheric general circulation (Chung and Ramanathan, 2003) and even desiccate clouds (Hansen *et al.*, 1997a; Ackerman *et al.*, 2000a).

The effects of aerosols on infrared radiation transfer is less, being mainly limited to the atmospheric window where gaseous absorption is weak (Ackerman *et al.*, 1976; Welch and Zdunkowski, 1976; Carlson and Benjamin, 1980). Some evidence of aerosol effects on longwave radiation heating/cooling have been detected over very deep Saharan dust layers and very polluted airmasses (Welch and Zdunkowski, 1976; Carlson and Benjamin, 1980; Saito, 1981). There is also some evidence that the swelling of haze particles at high relative humidities enhances their effectiveness in absorbing infrared radiation.

8.2.3 Absorption and scattering by clouds

Like aerosol particles, cloud droplets, raindrops, and ice particles (hydrometeors) interact with radiation in complex ways. The amount of absorption and scattering depends on the wavelength of the incident radiation and on the size of the hydrometeors, and the hydrometeor phase and shape. Numerous small, liquid cloud droplets are strong reflectors of solar radiation, while raindrops, though much fewer in number, are strong absorbers of solar radiation, but small contributors to scattering. Likewise, numerous small ice crystals can be strong reflectors of solar radiation while being weak absorbers. The amount of energy reflected by ice crystals depends on the ice crystal habit (i.e., dendrites vs. needles or columns) and on the preferred fall orientation of the ice crystals relative to the direction of incident radiation.

Important to the radiative properties of clouds is their *liquid water (or frozen water) path*. The liquid water path is the cumulative condensed water (or frozen water) along the direction of the incident radiation in a cloud. If this radiative energy becomes sufficiently scattered such that it is isotropic, the liquid water path becomes equal to the remaining depth of the cloud. A shallow cloud, such as some stratocumulus clouds, may have a relatively small liquid water content through a depth of a kilometer or so. As a result much of the Sun's energy can pass through the cloud without being appreciably attenuated. It will still appear relatively bright beneath the clouds and the disk of the Sun may still be visible. The same is true for high thin cirrus clouds which often cause only weak attenuation of solar radiation. In contrast deep cumulonimbus clouds contain appreciable amounts of condensed liquid water through their depths. As a result these clouds often appear very dark, even black, because solar radiation is nearly completely attenuated.

Except for very deep, wet clouds, absorption of solar radiation by cloud droplets and ice crystals is small. By contrast, absorption of longwave radiation by clouds is quite large, while they reflect longwave radiation rather poorly. As much as 90% of incident longwave radiation can be absorbed in less than 50 meters in moderately wet clouds. Clouds such as deep convective clouds and thick stratus are such effective absorbers of longwave radiation that they are often viewed as blackbodies, or perfect absorbers and emitters of terrestrial radiation. A cumulonimbus cloud, for example, behaves like a blackbody after longwave radiation has penetrated a distance of only 12 meters. By contrast, thin cirrus clouds must be greater than several kilometers in depth before they behave as blackbodies, which is generally greater than their depths.

We noted previously that in a cloud-free atmosphere, little gaseous absorption takes place between 8 and 14 μm, or the atmospheric window. In a cloudy atmosphere, on the other hand, there are no spectral regions where gaseous absorption of longwave radiation is small. Clouds effectively slam the atmospheric window shut, thereby limiting the amount of radiation emitted to space and increasing the amount of longwave radiation re-emitted downward towards the ground. See Chapter 9 for a more in-depth examination of the radiative impacts of clouds.

8.2.4 Global energy balance and the greenhouse effect

To illustrate schematically the role of radiation on global climate change, consider the globally averaged energy budget as shown in Fig. 8.5. Of the 100 units of solar energy (this corresponds to 380 W m^{-2}) entering the atmosphere, 19 units are absorbed by water vapor, dust, and ozone, 4 units are absorbed by clouds, 8 units are scattered by air molecules, 17 units are reflected by clouds, 6 units are reflected by the Earth's surface back into space, and 46 units are absorbed at the Earth's surface. The Earth's albedo as viewed from space is 0.31, which means that 31% of the solar radiation incident on the Earth is reflected back to space.

At the Earth's surface, 115 units of longwave radiant energy are emitted with 9 units escaping directly to space through the atmospheric window. A total of 106 longwave radiation units are absorbed by water vapor, carbon dioxide, ozone, and clouds. Of that amount, 100 units are re-emitted and absorbed at ground, while 40 units are re-emitted to space by water vapor, carbon dioxide, and ozone, and 20 units are emitted to space at the tops of clouds. In addition, 7 units of heat energy are input into the atmosphere from the surface as sensible heat and 24 units are input as latent heat.

The climate of Earth will remain constant as long as the apportionment of energy contributions to the global budget remain the same. Any systematic change in one

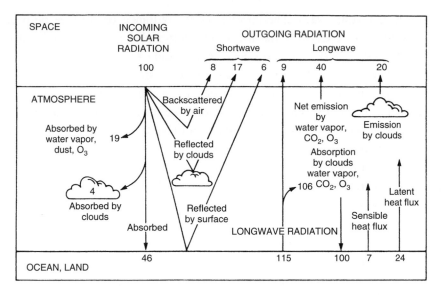

Figure 8.5 Schematic diagram of the global average components of the Earth's energy balance to space with the values expressed as percentages where the total solar radiation of 380 W m^{-2} corresponds to 100 units in the figure. Adapted from MacCracken (1985) © US Government.

or more components can lead to a radiative imbalance in the global budget and lead to warming or cooling of globally averaged heat content (Pielke, 2003). For example, if all existing greenhouse gases were removed from the atmosphere, the amount of longwave energy emitted to space would be greatly enhanced. This would result in an average surface temperature of the Earth that is 30 °C cooler than it is today. That is why the warming caused by absorption of longwave radiation by water vapor, carbon dioxide, etc. is called the "greenhouse effect."[1]

Let us consider some of the possible changes in the global energy budget that can occur naturally.

8.2.5 Changes in solar luminosity and orbital parameters

The total energy output from the Sun cannot be viewed as being a constant (Lean, 2005). There is evidence suggesting that early in Earth history the Sun's output was 25% less than it is today. In recent history solar luminosity has been observed to decline from 1980 to 1986 and to increase since 1986 (Willson *et al.*, 1986;

---

[1] The term "greenhouse" is somewhat of a misnomer as discussed by Bohren (1989). Actual greenhouses work primarily through their influence in preventing the turbulent removal of heat from the building as a result of the glass panes. The emissivity of the glass panes to infrared radiation is generally insignificant. The glass panes permit sunlight to enter and heat the interior of the greenhouse.

Willson and Hudson, 1988). The current solar irradiance on a plane perpendicular to the incident energy at the top of the atmosphere is about $1365\,\mathrm{W\,m^{-2}}$ (Barkström *et al.*, 1990). Attempts have been made to relate changes in solar luminosity to observed solar parameters such as sunspot activity, solar diameter, and umbral–penumbral ratio (Wigley *et al.*, 1986; Pecker and Runcorn, 1990; Friis-Christensen and Lassen, 1991). The data suggest that solar luminosity is positively correlated to sunspot number with an 11-year cycle having an amplitude of $0.1\,\mathrm{W\,m^{-2}}$ at the top of the atmosphere (Willson and Hudson, 1988). Estimates of global mean surface temperature changes due to observed changes in sunspot number are generally quite small, being on the order of less than $0.1\,°\mathrm{C}$ and as a result undetectable (Wigley, 1988). Nonetheless, convincing statistical studies suggest a correlation between sunspot number and meteorological parameters such as temperatures at the 30 mb level in the Arctic (Labitzke, 1987). The implication of those findings to surface temperatures remains unknown, however. There is evidence (Friis-Christensen and Lassen, 1991; Kerr, 1991) that the *length* of the solar cycle correlates very closely to the global average surface temperature. It still remains a mystery, however, as to how such small changes in solar irradiance caused by variations in sunspot number and the length of the solar cycle could have a significant climatic impact. Some scientists speculate that some unknown indirect amplification mechanism must exist.

There have been a number of attempts to use indirect measures of solar activity to identify possible climatic impacts. Wigley (1988), for example, attempted to infer changes in solar luminosity associated with variations in $^{14}\mathrm{C}$ concentrations in tree rings. Because the production rate of $^{14}\mathrm{C}$ in the atmosphere is related to the output of energetic particles from the Sun or solar wind, variations in $^{14}\mathrm{C}$ concentrations may be indicative of variations in solar irradiance. Historical records, for example, show a correlation between positive $^{14}\mathrm{C}$ anomaly and sunspot minima. According to Wigley and Kelly (1990) the $^{14}\mathrm{C}$ concentration peaked during the Little Ice Age in the seventeenth century and during earlier intervals of glacial advance. Assuming that the $^{14}\mathrm{C}$–climate link is real, Wigley (1988) estimated the change in solar irradiance associated with observed changes in $^{14}\mathrm{C}$ concentrations and inferred changes in surface temperatures over the last several hundred years. He estimated that changes in solar irradiance associated with $^{14}\mathrm{C}$ anomalies would translate into changes in global mean temperature of a magnitude of 0.1–$0.3\,°\mathrm{C}$.

Variations in solar activity modulates both cosmic ray fluxes and solar irradiance. It has been speculated that because enhanced solar activity enhances cosmic ray fluxes which, in turn, will generate larger concentrations of ions in the atmosphere, these ions will serve as cloud condensation nuclei (Carslaw *et al.*, 2002). Motivated by studies by Svensmark and Friis-Christensen (1997) that there is a statistical correlation between solar-modulated galactic cosmic fluxes and cloud cover, Carslaw *et al.* argue that variations in galactic cosmic ray fluxes will enhance CCN concentrations and thereby contribute to enhanced cloud cover. The basis of this hypothesis was first discussed by Dickinson (1975). However, a long chain of physical processes is required to go from enhanced galactic cosmic ray fluxes to enhanced cloud cover, much like many of the cloud seeding hypotheses discussed in Part I. First of all, ions by themselves are very poor CCN as they require supersaturations for activation that are much higher than occur in the atmosphere. The ions must first coagulate with other ions and aerosols and while doing so either absorb gases like sulfur dioxide or coagulate with small hygroscopic particles to become soluble particles of sizes $0.1\,\mu m$ or greater that can serve as CCN at typical cloud supersaturations. Typically only 1% to 10% of the total aerosol population meet these criteria. The hypothesis then builds on the Twomey hypothesis described in Chapter 9 including Albrecht's (1989) hypothesis that enhanced CCN concentrations will suppress drizzle formation and lead to enhanced cloud cover. We show in Chapter 9 that due to the strong non-linearity of cloud systems once drizzle is involved, the response of clouds to reduced drizzle may not always lead to enhanced cloud cover. Certainly, variations in CCN concentrations can at best be considered to be a second-order effect in determining cloud cover. Thus other factors may be involved in producing the correlations between cloud cover and cosmic ray fluxes as suggested by Udelhofen and Cess (2001) or Dickinson (1975), or the statistical correlations are just that, and do not represent any real physical linkage.

The net output of energy from the Sun is not only important to climate variability, but the distribution of the Sun's energy on the Earth–atmospheric system is also important in controlling global temperatures and whether or not Earth may be moving into an ice age. Variations in the pattern of energy reaching the Earth's surface, in turn, are caused by slow changes in the geometry of the Earth's orbit around the Sun and in the Earth's axis of rotation. The fundamental theory predicting the onset of ice ages in response to changes in orbital parameters is attributed to M. Milankovitch (see Imbrie and Imbrie, 1979; Berger, 1982). As shown in Fig. 8.6 the important orbital parameters are (a) the eccentricity of the orbit, (b) the axial tilt which affects the distribution of sunlight, and (c) the precession of the equinoxes. The amount of radiation reaching polar latitudes in summer is important to the onset of ice ages, since it is the amount of summer melting of glaciers which largely determines whether glaciers are growing or receding. Thus a decrease in axial tilt causes a decrease in summer radiation. Likewise an increase in Earth–Sun distance in any season causes a decrease in radiation during that season and the strength of those effects varies systematically with latitude. The effect on the net solar radiation received at the surface in response to the

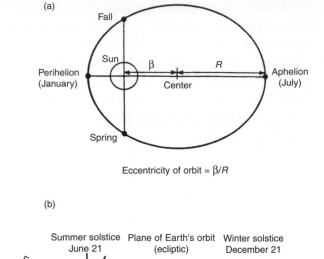

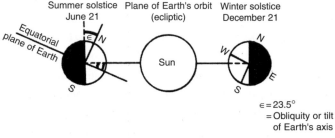

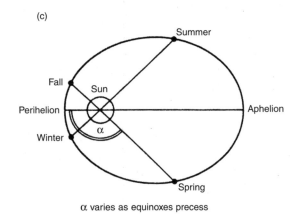

Figure 8.6 Important components of Earth–Sun geometry. Important orbital parameters: (a) eccentricity of orbit, (b) axial tilt, and (c) precession of the equinoxes. From Griffiths and Driscoll (1982).

41 000-year oscillation in the Earth's tilt is large at the poles and small at the equator. On the other hand, the influence of the 22 000-year precession cycle is small at the poles and large near the equator.

The Milankovitch theory predicts that Earth will gradually be moving into an ice age over the next 5000 years. It remains to be seen if the impact of anthropogenic activities on climate can alter the effect of the orbital forcing on climate trends.

8.2.6 Natural variations in aerosols and dust

We have seen that aerosol particles can significantly attenuate solar radiation. A major source of natural dust and aerosol particles in the upper troposphere and stratosphere are volcanoes. A major volcanic eruption such as Tambora in 1815 and Pinatubo in 1991 can spew large quantities of aerosols and gases into the lower stratosphere where they can reside for months to several years. The gases such as sulfur dioxide are converted to sulfate aerosols through photochemical processes. The high-level aerosol particles reflect some of the incoming radiation, thus increasing planetary albedo, and absorb solar radiation in the stratosphere, thus reducing the amount of energy reaching the Earth's surface. Therefore, a single large volcanic eruption can reduce surface temperatures by several tenths of a degree for several years (Hansen *et al.*, 1978, 1988; Robock, 1978, 1979, 1981, 1984).

Because of the thermal inertia associated with the oceans, a 2–3-year period of reduced solar heating can impact average surface temperatures for several decades. Periods of active volcanism can have a substantial counteracting effect to global warming scenarios. In fact, a period of global warming between 1920 and 1940 has been attributed to very low volcanic activity (Robock, 1979). Because large volcanic eruptions occur rather frequently and they cannot be predicted, they represent a major source of uncertainty in predicting climatic trends.

Another potentially major source of dust and aerosol in the atmosphere can result from a collision between the Earth and a large meteor or comet. It is hypothesized that major meteor impacts sent up a cloud of dust into the atmosphere that was so dense it caused such dramatic cooling that major life forms such as dinosaurs were destroyed (see review by Simon, 1981). Support for the hypothesis was provided by the Alverez team who found high concentrations of iridium as well as other rare metals on Earth that are abundant in extraterrestrial material in the geologic strata corresponding to the Cretaceous–Tertiary geological boundary. Thus large meteor collisions provide another source of uncertainty in climate prediction, although the frequency of such events is so low that they are generally thought to be a minor factor in climate change over the next several hundred years.

8.2.7 Surface properties

As shown in Fig. 8.5, the Earth's radiation balance can be altered by variations of surface properties. The net albedo of the surface of the Earth is determined by the percent coverage of ocean versus land, the amount of glacial coverage, and properties of the land surface such as the amount of desert versus forested lands. Dust ejected into the atmosphere by wind from deserts (e.g., Kallos *et al.*, 1998; Nickovic *et al.*, 2001; Rodriguez *et al.*, 2001) also results in a major influence on the radiative balance. The surface properties introduce a major non-linearity into the system, since if summer temperatures are cooler at high latitudes, glaciers advance, which in turn increases net surface albedo contributing to cooler temperatures and so forth. Likewise, as the glaciers advance, less water is available for the oceans and the percent cover of ocean surface is decreased, also altering the net albedo. The ocean albedo itself can change and alter large-scale atmospheric circulations (Shell *et al.*, 2003). More importantly, lower sea levels may block certain ocean currents from transporting warm sea water into high latitudes, enhancing sea ice coverage, which again affects the net surface albedo.

Vegetation can also respond to changes in surface temperature and rainfall causing another complicated feedback through changing global albedo. All these complicated feedbacks must be included in any climate change model. Also the local albedo of the surface can influence cloud cover, thereby modifying the radiation balance as the cloud cover is increased or decreased or becomes thinner or thicker. Moreover, the effect of the surface albedo on the planetary albedo will depend on this cloud coverage since the surface would be shaded when it is cloudy.

8.2.8 Assessment of the relative radiative effect of carbon dioxide and water vapor

Let us now examine some of the basic concepts and philosophy of modeling global change.

To examine the impact of changing carbon dioxide and water vapor concentrations on radiative fluxes and heating rates, single column radiative transfer calculations were performed on standard atmospheric profiles for which carbon dioxide and water vapor concentrations were varied.[2] Three profiles, representative of tropical, subarctic summer, and subarctic winter clear-sky conditions (McClatchey *et al.*, 1972), were used for the calculations. Temperature and water vapor mixing ratio as a function of pressure for each of the profiles are shown in Fig. 8.7, 8.8, and 8.9. The atmosphere was discretized into 19 layers and the BUGSrad model (Stephens *et al.*, 2001) was used.

---

[2] These calculations and the summary text were provided by Norman Wood and Graeme Stephens of Colorado State University.

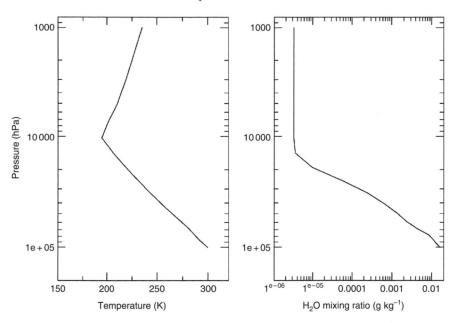

Figure 8.7 Profiles of temperature and water vapor mixing ratio for tropical atmosphere. Figure courtesy of Norman Wood, Colorado State University.

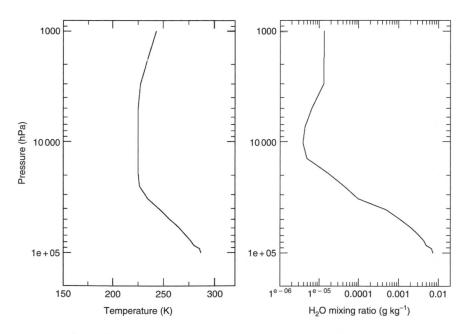

Figure 8.8 Profiles of temperature and water vapor mixing ratio for subarctic summer atmosphere. Figure courtesy of Norman Wood, Colorado State University

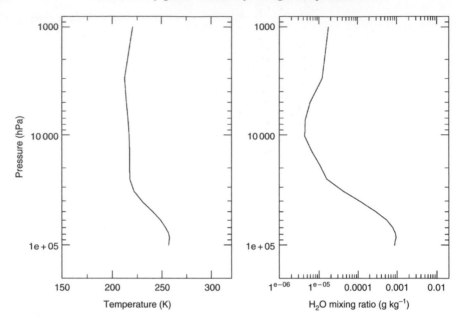

Figure 8.9 Profiles of temperature and water vapor mixing ratio for subarctic winter atmospheres. Figure courtesy of Norman Wood, Colorado State University.

Carbon dioxide was treated as uniformly mixed. Concentrations of 0, 280 (pre-industrial), 360 (current), and 560 (doubling of pre-industrial) ppmv were used for the calculations. For water vapor, the mixing ratios associated with the standard profiles were scaled by factors of 0, 1.0, 1.05, and 1.1. The resulting effects on longwave heating rates in the atmospheric column are shown in Figs. 8.10, 8.11, and 8.12 for carbon dioxide and Figs. 8.13, 8.14, and 8.15 for water vapor. In each figure, the upper panel shows the heating rates for the particular scenario in $K\,day^{-1}$. The lower panel shows the changes in heating rates attributable to the perturbations in carbon dioxide or water vapor mixing ratio. The effects on downwelling longwave fluxes at the surface are shown in Tables 8.1 and 8.2. The results suggest that the radiative changes induced by perturbations to carbon dioxide and water vapor are substantially different. For water vapor, modest increases beyond the base profile mixing ratios have minimal impact on longwave heating rates, but cause significant increases in downwelling longwave fluxes to the surface. For carbon dioxide, increasing the concentration beyond the base profile of 280 ppmv contributes to enhanced heating in the lower troposphere and to significantly enhanced cooling in the stratosphere, but causes minimal increases in downwelling longwave flux, particularly for the tropical

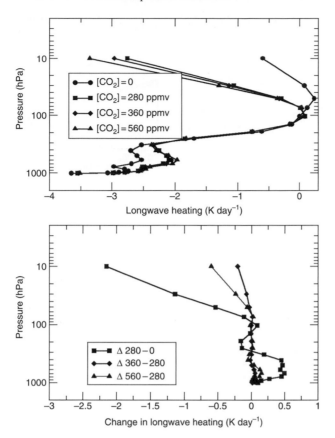

Figure 8.10 Longwave heating rates (upper) and change in longwave heating rates (lower) for tropical atmosphere with perturbed carbon dioxide. Figure courtesy of Norman Wood, Colorado State University.

and subarctic summer profiles. For the subarctic winter profile, this doubling of carbon dioxide concentration produces an increase in downwelling longwave flux similar in magnitude to that for a 10% increase in water vapor mixing ratio.

A number of factors potentially contribute to these differences in the effects of water vapor and carbon dioxide on heating rates and fluxes. Firstly, water vapor is significantly more prevalent in the lower troposphere than in the upper troposphere and stratosphere. Consequently, the scaling approach used here to perturb the trace gas amounts tend to produce stronger perturbations of water vapor mass mixing ratio in the lower troposphere than at higher altitudes. This effect is less significant for carbon dioxide since this gas is more uniformly mixed.

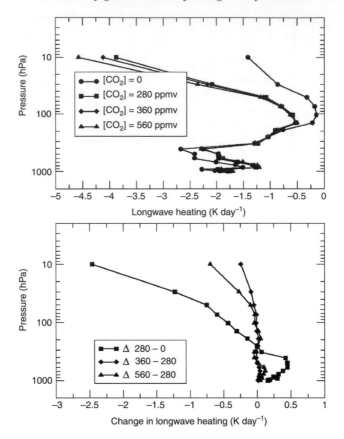

Figure 8.11 Longwave heating rates (upper) and change in longwave heating rates (lower) for subarctic summer atmosphere with perturbed carbon dioxide. Figure courtesy of Norman Wood, Colorado State University.

In terms of radiative factors, the heating rate in a layer of the atmosphere is a function of the spectrally varying absorption–emission characteristics of the layer, the spectral fluxes incident on the layer, and the layer temperature. The absorption by a layer is a function of the abundances of absorbing gases in the layer. In the longwave spectral region in which carbon dioxide is a significant absorber (for wavelengths of about $12.5\,\mu m$ and longer) water vapor is also radiatively active. For a wavelength at which water vapor is already significantly absorbing, the addition of an amount of carbon dioxide to the layer will cause relatively little increase in the flux absorbed by the layer and thus cause relatively little increase in the radiative heating of the layer.

The spectral fluxes incident on the layer are also a function of the temperatures and emission characteristics of the layer's surroundings. For a layer with given absorption characteristics, a stronger incident flux will cause more flux to

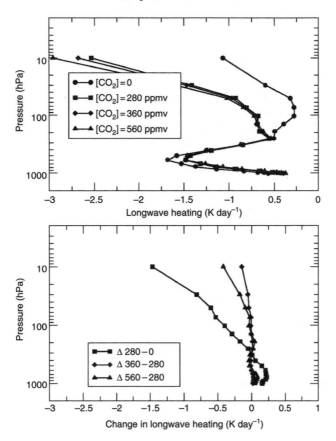

Figure 8.12 Longwave heating rates (upper) and change in longwave heating rates (lower) for subarctic winter atmosphere with perturbed carbon dioxide. Figure courtesy of Norman Wood, Colorado State University.

be absorbed by the layer and contribute to heating. In particular in the lower troposphere, the proximity of the warm surface of the Earth contributes to this effect. A warmer surface temperature (as in, for example, the tropical or subarctic summer profiles used here) will contribute to enhanced heating in the lower troposphere as opposed to a cooler surface (as in the subarctic winter profile).

Finally, the temperature of the layer itself influences the amount of flux emitted by the layer. A warmer layer will emit flux more strongly, and thus have a greater tendency for cooling, than will a cooler layer. Profiles with warmer temperatures in the lower troposphere, such as the tropical profile, have stronger cooling in the lower troposphere than will a profile, such as the subarctic winter, that has cooler temperatures in the lower troposphere.

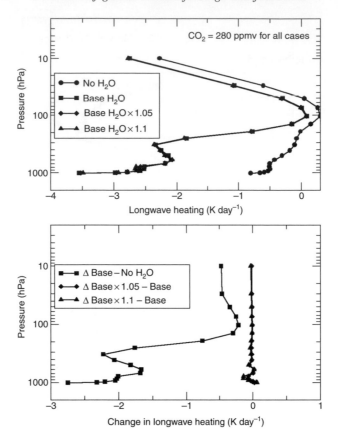

Figure 8.13 Longwave heating rates (upper) and change in longwave heating rates (lower) for tropical atmosphere with perturbed water vapor. Figure courtesy of Norman Wood, Colorado State University.

The downwelling flux at the surface is a function of the emission characteristics of the atmosphere, in particular the profile of the derivative of transmission with respect to height, known as the weighting function, and the temperature profile of the atmosphere. The height at which the weighting function peaks generally indicates the level of the atmosphere from which emission from the atmosphere most effectively reaches the surface, and this peak can be either broad (indicating the flux reaching the surface is strongly blended from different levels of the atmosphere) or narrow (indicating the flux reaching the surface is strongly selected from that particular level of the atmosphere). As trace gas concentrations in the lower troposphere increase, the general tendency is for the weighting function to shift lower in the atmosphere. For temperature profiles which decrease with height, this shift leads to increased emission to the surface.

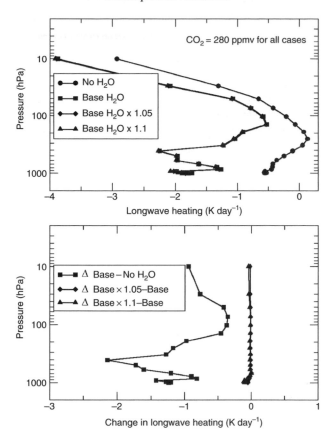

Figure 8.14 Longwave heating rates (upper) and change in longwave heating rates (lower) for subarctic summer atmosphere with perturbed water vapor. Figure courtesy of Norman Wood, Colorado State University.

The downwelling fluxes at the surface for the subarctic profile appear less sensitive to changes in carbon dioxide and water vapor concentrations than do the fluxes for the tropical and subarctic summer profiles. The subarctic winter profile has a relatively weak lapse rate in the lowest part of the troposphere, so changes in the position of the weighting function may have had little effect on the downwelling fluxes. In addition, the water vapor amounts in the subarctic winter profile are considerably smaller than those in the two other profiles. Since a scaling factor was used to perturb water vapor amounts for this study, the change in water vapor mixing ratio in the lower troposphere would be considerably smaller for the subarctic winter profile than for the two other profiles. This approach probably contributed to cause a less significant change in the weighting function for the subarctic winter case than that for the other two cases.

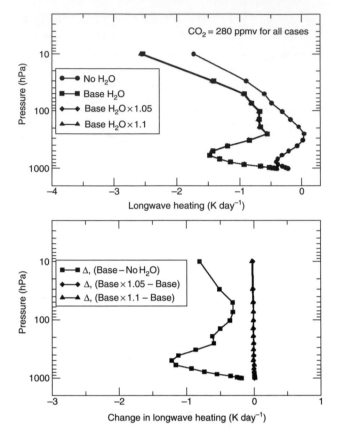

Figure 8.15 Longwave heating rates (upper) and change in longwave heating rates (lower) for subarctic winter atmosphere with perturbed water vapor. Figure courtesy of Norman Wood, Colorado State University.

8.3 Climate feedbacks

8.3.1 Water vapor feedbacks

Water vapor is the principle greenhouse gas, which is quite clearly shown in Tables 8.1 and 8.2. Therefore, any changes in water vapor concentration in response to other greenhouse gases would substantially alter the net greenhouse heating. As the atmosphere and the ocean warm in response to enhanced greenhouse warming, more water vapor evaporates from the ocean and land surfaces. The higher water vapor content of the atmosphere causes further greenhouse warming, which causes more evaporation. There are other possible feedbacks associated with higher moisture contents, however, which complicate this feedback process. Clouds can form, and as we will see in the following discussion, this can lead to both warming and cooling of the atmosphere through their interactions

Table 8.1 *Downwelling longwave flux at the surface for carbon dioxide scenarios*

CO_2 concentration, ppmv	Flux, $W m^{-2}$		
	Tropical	Subarctic summer	Subarctic winter
0	407.84	306.36	161.16
280	408.19	309.05	174.74
360	408.25	309.30	175.59
560	408.34	309.77	176.68

Source: Prepared by Norman Wood, Colorado State University.

Table 8.2 *Downwelling longwave flux at the surface for water vapor scenarios*

H_2O mixing ratio scale factor	Flux, $W m^{-2}$		
	Tropical	Subarctic summer	Subarctic winter
0	104.35	89.47	58.36
1.0	408.19	309.05	174.82
1.05	412.07	311.78	175.52
1.10	415.71	314.48	176.19

Source: Prepared by Norman Wood, Colorado State University.

with radiation. Simulation of the moisture feedback in GCMs requires realistic models of the hydrological budget (rainfall and physical evaporation and transpiration of water vapor from land and ocean surfaces) and of the response of the ocean to heating of the atmosphere. Unfortunately, the responses of the various GCMs to simulated surface energy budgets are diverse (e.g., see Wild, 2005).

The fluxes of moisture at the ocean surface, for example, are a function of the ocean surface temperature as well as the surface wind stresses (Pielke, 1981). Thus, ocean surface temperatures in tropical regions can warm appreciably in response to tropospheric warming, but in regions of weak winds, such as over the intertropical convergence zone, only weak vertical moisture fluxes will be experienced. Moreover, crucial to water vapor feedback is the vertical distribution of moisture in the middle and upper troposphere. This, in turn, is related not only to local convective transports, but also to the strength of regional circulations such as the Hadley and Walker circulations (Philander, 1990) both of which transport moisture over long distances horizontally and vertically. Thus, while the moisture feedback can amplify globally averaged surface air warming by as much as a factor of three (Ramanathan, 1981), consideration of realistic cloud feedbacks and realistic ocean–atmospheric circulation responses requires more skill in the simulation of climate responses than is possible in current models.

8.3.2 Cloud feedbacks

We have seen that one of the major positive feedbacks to carbon dioxide or other greenhouse gas warming is that as the ocean surface and ground temperatures rise, an increase in flux of water vapor into the atmosphere is expected. In a cloud-free atmosphere, enhanced water vapor content provides a strong enhancement of greenhouse warming. In response to the enhanced water vapor content of the atmosphere, however, one would also expect that cloud coverage would increase and some clouds would become optically thicker. Moreover, clouds alter the stability of the atmosphere in response to their release of latent heat, and their associated rising and sinking motions transfer heat, moisture, momentum, and various particles and trace gases vertically and horizontally in the atmosphere. In this section, we will examine the radiative feedback of clouds as well as the feedback associated with changes of atmospheric stability and transport.

The global average radiative effects of high clouds warm the atmosphere while low clouds cool the atmosphere and middle-level clouds are in near balance between cooling by reflection of solar radiation and warming by absorption of longwave radiation. Averaged over the entire Earth, the enhanced albedo of clouds is slightly greater than their greenhouse warming, so clouds radiatively cool the atmosphere compared to a cloud-free Earth (Ramanathan *et al.*, 1989a,b; Randall *et al.*, 1989). In the case of an atmosphere–ocean system warmed by added greenhouse gases, what is the feedback of clouds to that response? Unfortunately, clouds are forced by relatively small-scale atmospheric motions, thus climate models must parameterize clouds in rather crude ways, with the cloud cover being specified based on climatology or parameterized as a function of relative humidity and as a by-product of convective parameterization schemes. Even in the more sophisticated GCMs which have a prognostic equation for cloud liquid water or ice water amounts, the horizontal and vertical resolution of those models is not sufficient to simulate the major ascending motions leading to cloud formation. As a result, the simulated feedbacks of clouds on the greenhouse theory vary from model to model, depending on the nature of the cloud/radiative scheme and resolutions of the models.

Randall *et al.* (1989) suggest that cumulus anvil clouds exert a very powerful influence on tropical convection by radiatively destabilizing the upper troposphere, and by trapping terrestrial radiation. They argue that the skill in predicting the effects of those clouds in GCMs, while being qualitatively correct, are not sufficiently accurate to be relied upon quantitatively in considering their strong influence on simulated climates. Their conclusion is particularly valid when one considers the important role that organized mesoscale systems play in producing long-lasting upper tropospheric anvil clouds in both the tropics and middle latitudes. These systems produce large areas of optically thick, stratiform–anvil clouds in response to detrainment of rising moist air from deep convective towers as well as cloud formation in slowly ascending mesoscale motion. Current cloud parameterization schemes in GCMs do not generally consider the organized mesoscale motions in their estimates of anvil cloud cover or optical thicknesses. The proposal to embed cloud-resolving models within GCMs (Randall *et al.*, 2003) is computationally very expensive, and moreover, up to the present, this method used two-dimensional (not more realistic three-dimensional) cloud field models.[3]

Some cloud systems such as mesoscale convective complexes and tropical cloud clusters exhibit a well-defined nocturnal maximum in their frequency of occurrence (see Cotton and Anthes, 1989). Thus, if a warmed ocean creates more tropical cloud clusters, then their net impact would be to cause a positive feedback since they would have little effect on solar radiation. It is important, therefore, to consider the diurnal variations of cloudiness when examining the impacts of clouds on climate.

Lindzen (1990) argues that deep convective clouds should produce a net drying of the upper troposphere due to the drying influence of compensating subsidence. He bases his argument on the behavior of simple cumulus parameterization schemes that he has developed. While drying effects of compensating subsidence do indeed take place in the middle troposphere, it appears that the moistening effects of detrainment of water substance from cumulus towers and stratiform–anvil clouds of mesoscale convective systems override the drying effects. As a result, a general moistening in the upper troposphere associated with deep convection is observed (Rind *et al.*, 1991). This issue of an atmospheric "iris" effect is still being debated in the literature.

Another interesting feedback from clouds has been hypothesized by Mitchell *et al.* (1989). They found a strong negative feedback of clouds related to a reduction in the depletion of cloud water content in a carbon dioxide warmed atmosphere due to a reduction in ice phase precipitation formation in middle latitudes. As a result, cloud cover increased and the stronger albedo caused a reduced rate of greenhouse-induced warming than in the version of the model using relative humidity as the primary parameter determining cloud amounts.

---

[3] An alternate approach where parameterizations are replaced by much more computationally efficient look-up tables has been proposed in Pielke *et al.* (2006b).

Several researchers have attempted to evaluate cloud feedbacks to a warming atmospheric–ocean system by examining the observed relationship between sea surface temperature (SST) anomalies and changes in cloudiness and net radiative budgets. Ramanathan and Collins (1991), for example, diagnosed changes in cloudiness and net radiation associated with warm SST anomalies during the 1987 El Niño. They inferred that in warm ocean regions where SSTs were less than 300 K, the net effect of enhanced water vapor and cloudiness resulted in a positive greenhouse effect. However, when SSTs exceeded ~300 K, they found that cirrus clouds formed by the flux of water substance into the upper troposphere by cumulonimbus clouds, became optically thick and, as a result, they reflect more solar radiation than is absorbed and reradiated downward by terrestrial radiation. They argue that the highly reflective, optically thick cirrus clouds acts as a thermostat, which prevents further warming of the oceans. They suggest that the implications to greenhouse warming is that "it would take more than an order-of-magnitude increase in atmospheric carbon dioxide to increase the maximum SSTs by a few degrees in spite of a significant warming outside the equatorial regions."

In another study, Peterson (1991) examined the relationship between SST anomalies and anomalies of high, middle, and low-level cloudiness using satellite data. He found that over much of the tropics and the south Pacific convergence zone, high clouds increased with warm SST anomalies, while in subtropical stratocumulus regions, low cloud coverage decreased with positive SST anomalies. Averaged over the entire region he sampled, total cloudiness increased with positive SST anomalies. Because the coverage of optically thick low clouds decreased while optically thin high clouds increased over regions of warm SSTs, he calculated that the average, net radiative flux to space decreased in response to warm SST anomalies. He concluded that his results provide observational evidence that clouds provide a positive feedback loop to global warming scenarios.

There are other effects of clouds that can create positive or negative feedbacks to greenhouse-gas-induced warming. Some of these are associated with changes in the stability of the troposphere including the height of the tropopause. In response to a warming climate, for example, deep convective overturning will raise the height of the tropopause in the tropics, yielding a colder tropopause. The temperature of the tropical tropopause has a strong influence on the water vapor content of the stratosphere. The colder the tropopause, the lower the water vapor content of the stratosphere, since the tropopause acts as an effective trap of moisture as more water will be condensed and precipitated out of the air. As a result, with a colder tropopause, there will be little absorption of terrestrial radiation in the stratosphere causing a cooling effect on surface temperatures. Overall, simulation of the feedback of clouds on climate change can be considered to be one of the major "Achilles heels" of climate models since those feedbacks can be quite large and cause a major moderating influence on greenhouse-gas-induced warming. Unfortunately, most of the cloud processes occur on scales considerably below the resolution of current GCMs.

8.3.3 Surface albedo feedbacks

Changes in surface snow and ice coverage are a well-studied feedback to greenhouse warming, while other surface albedo feedbacks are related to changes in surface vegetation coverage. The so-called "ice albedo" feedback relates melting of sea ice and snow cover to greenhouse warming. Snow and ice reflect more solar radiation than open water, bare soil, or soil covered by vegetation. As a result, a warming high latitude troposphere is expected to reduce sea ice cover appreciably, and to cause earlier seasonal melting of snow and a retreat in glaciers. It is estimated that these effects would positively amplify greenhouse warming by 10% to 20% globally (Lian and Cess, 1977; Hansen *et al.*, 1986), but it can have a 2- to 4-fold amplification in polar oceans and near the sea ice margins. Major uncertainties in modeling the ice albedo feedback are related to distinguishing between the albedo of old snow and fresh snow, identifying albedo effects of thick versus thin sea ice, and simulating the strong influence of ocean circulations on snow and sea ice cover. In addition, cloud cover can mask albedo changes associated with increases or decreases of snow and sea ice, and increases in precipitation at higher latitudes can increase snow cover even if warming occurs. There are indications, for example, that changes in cloud distribution associated with snow cover changes can produce a reversal of the sign of this feedback (Cess *et al.*, 1991). Likewise, snow albedo changes over forested regions can be partially masked.

Observations up through 2005 show that sea ice and snow cover trends do not match what were predicted by the GCMs to have occurred over the last few decades. Pielke *et al.* (2004b), for example, document a slower areal decrease of Arctic sea ice areal coverage with no reduction in Antarctic areal sea ice coverage. Northern Hemisphere snow cover has not decreased in its coverage. Indeed, while snow cover did decrease in total annual coverage in the 1980s, more recently it has even increased slightly. The ice albedo and snow albedo feedbacks are clearly more complicated than simulated by the models.

An even more complicated feedback in surface albedo and surface fluxes is associated with changes in vegetation coverage. If the tundra–boreal forest boundary should shift poleward in response to a warming planet then this would cause a decrease in albedo and represent a positive feedback. If, on the other hand, deserts increase in semi-tropical areas due to reductions in rainfall, this could create a negative effect. Likewise, it is possible that enhanced carbon dioxide concentrations will serve as a fertilizer (King *et al.*, 1985; Houghton, 1987, 1988; Idso, 1988) and result in increased biomass coverage causing a decrease in albedo and a positive feedback. The increase of biomass, however, would provide an enhanced atmospheric sink of carbon, at least until new biomass decays.

8.3.4 Ocean feedbacks

Ocean feedbacks affect almost every aspect of climate ranging from global responses to regional responses. As mentioned previously, the amplitude of the moisture and ice albedo feedbacks is strongly influenced by the response of ocean circulations to a warming troposphere. One of the most important effects of the ocean is the large amount of heat it is capable of storing (Levitus *et al.*, 2005). As a result, the time that it takes the atmosphere–ocean system to respond to greenhouse warming is largely controlled by the ocean response. Pielke (2003) has shown that the Earth's radiative imbalance was around $0.3 \, \mathrm{W \, m^{-2}}$ for the period of the mid-1950s to the mid-1990s using ocean heat storage changes to diagnostically measure this imbalance. Willis *et al.* (2004) analyzed the ocean heat storage data for more recent years and diagnosed heat changes corresponding to $0.62 \, \mathrm{W \, m^{-2}}$ to a depth of 750 m (Pielke and Christy, 2005). If exchanges of heat with the deep ocean are small, then the upper mixed layer of the ocean, which is commonly about 70 to 100 m deep, will respond to a warming atmosphere on the timescale of decades. If, on the other hand, the rate of exchange of heat with deep ocean layers is greater, it may require timescales on the order of a century or more for the upper levels of the ocean to respond appreciably to greenhouse warming. This will substantially delay the timescale that greenhouse warming will be detectable in the Earth–atmospheric system. Levitus *et al.* (2000, 2001) found that half of the ocean heat changes were below 300 m.

Not only is the ocean a large reservoir of heat, it is also a large reservoir of carbon dioxide. The ocean contains as much as 50 times the amount of carbon as resides in the atmosphere. The simplest feedback is that as the ocean warms, the solubility of carbon dioxide decreases and more carbon dioxide is released from the ocean to the atmosphere creating a positive feedback. However, cold, upwelling regions of the ocean (which require high spatial resolution models to simulate) may provide a region for carbon dioxide sinks from the atmosphere even if the ocean surface were to warm overall (Pielke, 1991).

Coupled GCMs of the atmosphere and the ocean provide insight on these inter-actions. These models permit significant changes in ocean circulations in response to climate forcings. The ocean circulations are driven by wind stress which, in turn, is driven by changes in wind speed and direction and in thermodynamic stability of the atmosphere immediately above the air–water interface. Moreover, the ocean circulations are also driven by thermohaline circulations or circulations induced by spatial gradients in the density of the water. The density of water is a function of both temperature and of salinity. Thus, warm ocean water is less dense. Likewise ocean water low in salt content is less dense.

For example, water flowing out of the warm, salty Mediterranean Sea is dense and undercuts warmer, less salty water in the Atlantic. In our current climate, the warm Gulf Stream flows northward towards the British Isles, where it moderates the climate over northern Europe. This current flows northward to replace cold, dense deep water that flows southward from Arctic regions while spreading out across the ocean bottom. Some coupled ocean models (e.g., Washington and Meehl, 1984; Stouffer *et al.*, 1989; Houghton *et al.*, 1990) respond to enhanced greenhouse warming by producing increased rainfall in the summer and fall months at high latitudes, and thereby forming a layer of fresher, less dense water on the surface of the North Atlantic. The strength of the thermohaline circulation thus weakens and the westerly winds also weaken. This reduces the flow of warmer water across the Atlantic, causing cooler surface temperatures than in the current climate. This response, while physically plausible, is dependent upon many details of the models (i.e., precipitation parameterization schemes, melting of sea ice, vertical and horizontal resolution of both the atmospheric and ocean models) and thus should be viewed as one of many complicated feedback scenarios that could occur in a coupled atmosphere–ocean system.

8.4 Views of the Intergovernmental Panel on Climate Change and the National Research Council of climate forcings

The Intergovernmental Panel on Climate Change (IPCC) summarized their perspective of climate forcings (see Houghton *et al.*, 2001) which are shown here in Fig. 8.16. This figure lists 12 climate forcings that focus exclusively on radiative forcings. The level of scientific understanding for each forcing is shown with only one listed as having a high level, and two with a medium level of understanding. One is listed as having a low level of understanding, while the remaining have a very low level of scientific understanding. It is important to recognize that despite listing as low or very low level of scientific understanding, specific uncertainty brackets are placed around each forcing. Also, the vertical axis indicate a "warming" or "cooling" radiative forcing for each, yet as presented in the same figure in the "Statement of Policymakers" of Houghton *et al.* (2001, p. 8), these values do not represent a current forcing, but "The global mean radiative forcing of the climate system for the year 2000, relative to 1750." The IPCC report also does not recognize any climate forcings except for radiative forcings, in contrast to

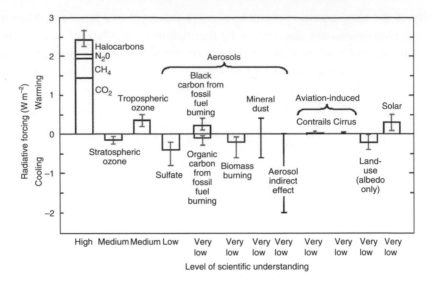

Figure 8.16 Estimated radiative forcing since pre-industrial times for the Earth and troposphere system (TOA radiative forcing with adjusted stratospheric temperatures). The height of the rectangular bar denotes a central or best estimate of the forcing, while each vertical line is an estimate of the uncertainty range associated with the forcing, guided by the spread in the published record and physical understanding, and with no statistical connotation. Each forcing agent is associated with a level of scientific understanding that is based on an assessment of the nature of assumptions involved, the uncertainties prevailing about the processes that govern the forcing, and the resulting confidence in the numerical values of the estimate. On the vertical axis, the direction of expected surface temperature change due to each radiative forcing is indicated by the labels "warming" and "cooling." From Intergovernmental Panel on Climate Change (2001).

National Research Council (2005) which expands the identification of climate forcing.

What climate forcings are missing from their figure? The National Research Council recently completed a study to identify the important climate forcings, and defined several of the IPCC recognized ones in more detail and added several. The diverse types of indirect aerosol climate forcings as given in National Research Council (2005) are summarized in Table 8.3 while other climate forcings that influence global averaged radiative forcing are listed in Table 8.4. The radiative forcing effects can be either positive or negative depending on the effect. For some of the forcings, the sign is not even known. Thus even using the IPCC estimate of climate forcings, the real-world climate forcings are not completely understood. With such limited understanding of the forcings, can we expect to predict the future climate?

Table 8.3 *Overview of different aerosol indirect climate forcings*

Effect	Cloud type	Description	Sign of global averaged radiative forcing
First indirect aerosol effect (cloud albedo or Twomey effect)	All clouds	For the same cloud water or ice content more but smaller cloud particles reflect more solar radiation	Negative
Second indirect aerosol effect (cloud lifetime or Albrecht effect)	Warm clouds	Smaller cloud droplets decrease the precipitation efficiency thereby prolonging cloud lifetime	Negative
Semi-direct effect	Warm clouds	Absorption of solar radiation by soot leads to an evaporation of cloud droplets	Positive
Glaciation indirect effect	Mixed-phase clouds	An increase in ice nuclei increases the precipitation efficiency	Positive
Thermodynamic effect	Mixed-phase clouds	Smaller cloud droplets inhibit freezing causing supercooled droplets to extend to colder temperatures	Unknown
Surface energy budget effect	All clouds	The aerosol induced increase in cloud optical thickness decreases the amount of solar radiation reaching the surface, changing the surface energy budget	Negative

Source: From National Research Council (2005).

Climate feedbacks add even more complexity. Houghton *et al.* (2001, Ch. 7) has an extensive discussion of climate processes and feedbacks. Unfortunately, they do not provide a summary figure on the level of uncertainty, or their relative importance. They do use the term "physical climate" which as illustrated in Fig. 8.1 should more appropriately be referred to as "physical components of the climate." The National Research Council (2003b) report summarizes climate feedbacks as follows:

(1) Feedbacks that primarily affect the magnitude of climate change

- Cloud, water vapor, and lapse rate feedbacks
- Ice albedo feedback
- Biogeochemical feedbacks and the carbon cycle
- Atmospheric chemical feedbacks

Table 8.4 *Overview of other different climate forcings that influence global averaged radiative forcing, which are not included in Fig. 8.16*

Effect	Description	Sign of global averaged radiative forcing
Land-use/land-cover change (other than albedo) (Chase *et al.*, 2000; Zhao *et al.*, 2001a,b)	Different vegetation types, fractional coverage, and phenology affect the surface heat and moisture fluxes	Unknown; regions of positive and negative forcing expected
Biogeochemical forcing due to increased CO_2 (Betts *et al.*, 2000; Friedlingstein *et al.*, 2001)	Alters stomatal conductance and plant growth with resultant influences on surface heat and moisture fluxes	Unknown
Biogeochemical forcing due to N deposition (Holland *et al.*, 2005)	Alters plant growth which changes surface heat and moisture fluxes	Unknown
Aerosol effect on the ratio of direct to diffuse solar insolation (e.g., Niyogi *et al.*, 2004)	Alters stomatal conductance and plant growth	Unknown

(2) Feedbacks that primarily affect the transient response of climate

- Ocean heat uptake and circulation feedbacks

(3) Feedbacks that primarily influence the pattern of climate change

- Land hydrology and vegetation feedbacks
- Natural modes of climate system variability
- Circulation feedbacks, i.e., changes in the Gulf Stream.

The National Research Council (2003b) report concluded that the two most important areas for near-term study are (a) cloud, water vapor, and lapse rate feedback, and (b) ice albedo feedback.

An important question is whether we can use our knowledge of climate forcings and feedbacks for skillful prediction of future climate. Pielke (2002b) provided classes of prediction as shown in Fig. 8.17. The levels for prediction are guessing (not shown), sensitivity studies, projections, and perfect foresight. A *sensitivity* study involves imposing a subset of important climate forcings and feedbacks in the prediction model. A *scenario* includes each of the important forcings and feedbacks, but only runs one realization of a prediction with a specific set of initial and surface boundary conditions. A *projection* runs an ensemble of realizations that includes the expected spectrum of surface and surface boundary conditions.

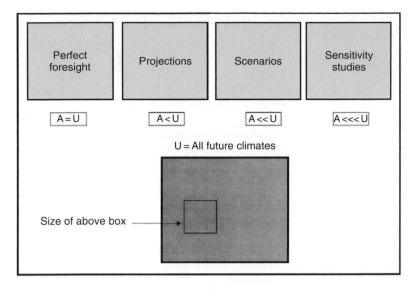

Figure 8.17 Schematic of different classes of prediction. The size of the box labeled "U" represents the range of future climate, while the box labeled "A" indicates the relative subset of possible future climate estimated using the different classes of prediction. From Pielke (2002b). With kind permission of Springer Science and Business Media.

A projection, using this definition, is a forecast. *With these definitions, the IPCC and US National Assessment model simulations were sensitivity studies since they only included a subset of forcings and feedbacks. They are not forecasts.* It is also important to recognize that the box A, representing the space of the prediction, can also lie outside of U.

There are different uses of these terms, however, which obscures this distinction. MacCracken (2002), in response to Pielke (2002b), uses the term "projection" to actually represent what Fig. 8.17 shows as a sensitivity study. The term "scenario" is also used in this context. Since prediction means different things to different communities, however, it is important to clearly define what is actually meant, and Fig. 8.17 provides such a framework. In the context of climate change and the definitions given in Fig. 8.17, there have not yet been forecasts of climate years into the future.

Additional reading

Betts, A. K., 2004. Understanding hydrometeorology using global models. *Bull. Am. Meteor. Soc.*, **11**, 1673–1688.

Castro, C. L., R. A. Pielke Sr., and G. Leoncini, 2005. Dynamical downscaling: Assessment of value restored and added using the Regional Atmospheric Modeling

System (RAMS). *J. Geophys. Res. Atmospheres*, **110, D5**, 5108, doi:10.1029/2004JD004721.

Foley, J. A., R. DeFries, G. P. Asner, *et al.*, 2005. Global consequences of land use. *Science*, **309**, 570–574.

Houghton, J. T., Y. Ding, D. J. Griggs, *et al.* (eds.), 2001. *Climate Change 2001: The Scientific Basis*, Contribution of Working Group I to the Third Assessment Report of the Intergovernmental Panel on Climate Change. Cambridge, UK: Cambridge University Press.

iLEAPS, 2004. Integrated Land Ecosystem-Atmosphere Processes Study (iLEAPS): The IGBP – Land-Atmosphere Project. Science Plan and Implementation Strategy. Available online at www.atm.helsinki.fi/ILEAPS/index.php?page=draft

Kabat, P., Claussen, M., Dirmeyer, P. A., *et al.* (eds.), 2004. *Vegetation, Water, Humans and the Climate: A New Perspective on an Interactive System*. New York: Springer-Verlag.

Kleidon, A., 2004. Beyond Gaia: Thermodynamics of life and Earth system functioning. *Climat. Change*, **66**, 271–319.

Kondrat'yev, K. Ya., 2005. *Atmospheric Aerosol Formation Processes, Properties, and Climatic Impacts*. Saint Petersburg, Russia: Research Center for Ecological Safety of the Russian Academy of Sciences.

Lohmann, U., and J. Feichter, 2005. Global indirect aerosol effects: A review. *Atmos. Chem. Phys.*, **5**, 715–737.

Marland, G., R. A. Pielke Sr., M. Apps, *et al.*, 2003: The climatic impacts of land surface change and carbon management, and the implications for climate-change mitigation policy. *Climate Policy*, **3**, 149–157.

National Research Council, 2003b. *Understanding Climate Change Feedbacks*, Panel on Climate Change Feedbacks, Climate Research Committee, Board on Atmospheric Sciences and Climate, Division on Earth and Life Studies. Washington, DC: National Academy Press.

Pielke, R. A. Sr., G. Marland, R. A. Betts, *et al.*, 2002. The influence of land-use change and landscape dynamics on the climate system- relevance to climate change policy beyond the radiative effect of greenhouse gases. *Phil. Trans. Roy. Soc. A*, **360**, 1705–1719.

Tsonis, A. A., 2001. The impact of nonlinear dynamics in the atmospheric sciences. *Int. J. Bifurcation and Chaos*, **11**, 881–902.

Wang, G., and D. Schimel, 2003. Climate change, climate modes, and climate impacts. *Ann. Rev. Environ. Resources*, **28**, 1-28, doi:10.1145/annurev.energy.28.050302. 105444.

9

Climatic effects of anthropogenic aerosols

9.1 Introduction

In Chapter 4, we examined evidence suggesting that human production of aerosol particles has local and regional impacts on clouds, precipitation, and atmospheric temperature. In this chapter we examine the evidence indicating potential impacts of anthropogenic aerosol on global climate.

Estimating the effects of aerosol on climate is particularly challenging. We noted in Chapter 8 that the radiative response to aerosol particles varies with size and chemical composition of the particles relative to the wavelength of the incident radiation. Moreover, because most aerosol particles are heterogeneous in structure, some components of the particles are very absorbing while others are reflecting. Table 2-2 from National Research Council (2005) (reproduced as Table 8.3 in Chapter 8) summarizes the major recognized climate forcings of aerosols. As a consequence only gross estimates of the radiative properties can be made. Aerosol particles also have limited lifetimes in the atmosphere. Particles greater than a few micrometers may survive for only a few days, while particles on the order of $0.1\,\mu m$ and less may reside in the lower troposphere for several weeks. This results in pronounced regional and hemispheric variations in aerosol concentrations. Unfortunately, there have been few systematic long-term observations of aerosols, their size spectra, and chemical composition. It is, therefore, usually necessary to resort to a variety of less-than-direct measurements, regional field campaigns, and global models to infer changes in aerosol concentrations and their radiative effects.

As discussed in Chapter 4, aerosol particles can have direct effects on atmospheric radiation as well as indirect effects through their impact on cloud microstructure. Their direct and indirect effects will be examined separately below.

9.2 Direct aerosol effects

It is well known that aerosol particles in polluted urban areas deplete direct solar radiation by about 15%, sometimes more in winter and less in summer (Landsberg, 1970). Less well known, however, is how far the pollution-caused aerosol extends from the urban areas. Schwartz (1989) summarized measurements of the concentrations of sulfate (or sulfur) aerosol at remote locations in both the Northern and Southern Hemispheres. These particles form from sulfur dioxide emitted naturally largely by decay of plant and animal matter, by wildland fires, by volcanoes, and by anthropogenic activity. The particles form primarily from the in situ oxidation of sulfur dioxide either as a primary gas or as an intermediate stage of oxidation. Two mechanisms for aerosol formation from sulfur dioxide are: (1) dissolving in cloud droplets to create sulfurous acid, which oxidizes further to form sulfuric acid aerosol particles, and (2) photochemical oxidation to form sulfate particles. Although sulfate particles are not the only human-caused aerosol, they are certainly prolific, so identification of their distribution is important to understanding the role of human-caused aerosol on climate.

Anthropogenic sulfur dioxide emissions have increased to their present level almost entirely within the last 100 years (Cullis and Hirschler, 1980) and, furthermore, as summarized by Schwartz (1989), the bulk of those emissions are in the Northern Hemisphere. Aerosol sulfate is quite common at remote sites in both the Northern and Southern Hemispheres, with Northern Hemispheric concentrations substantially exceeding those in the Southern Hemisphere. At several remote sites the observed high aerosol sulfate concentrations have been attributed to transport from regions of industrial activity over 1000 km away (Prahm *et al.*, 1976; Wolff, 1986). There is also evidence that sulfate concentrations have increased substantially over the last century in polar ice at Northern Hemispheric sites (Barrie *et al.*, 1985; Neftel *et al.*, 1985; Mayewski *et al.*, 1986) but there is no such increase in Antarctica (Delmas and Boutron, 1980; Herron, 1982). While it is difficult to make accurate estimates of the concentration of sulfate aerosol, and its spatial and temporal variability, there is little doubt that it is increasing, particularly within a few thousand kilometers of industrial regions. As an example, Fig. 9.1 shows the drastic reduction in visibility that has occurred in the eastern United States since 1948. Much of this visibility degradation is attributable to sulfate particles. Black carbon has also been identified as a major aerosol forcing of climate (e.g., see Menon *et al.*, 2002a; Hansen *et al.*, 2005).

An indirect measure of aerosol concentrations is the electrical conductivity of the air. The electrical conductivity, in turn, is controlled by the concentration and mobility of ions and small, charged aerosol particles. In general, an increase in the concentration of aerosol particles decreases conductivity because the more highly concentrated aerosol particles collect small ions and charged small aerosol

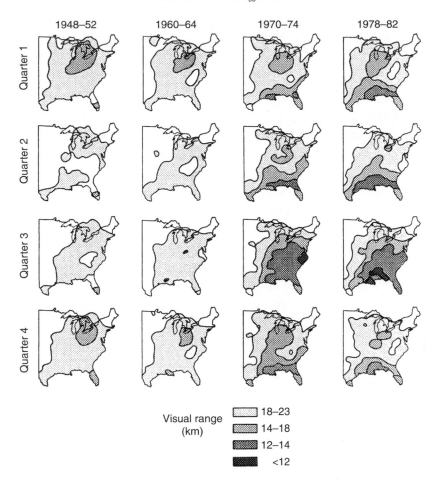

Figure 9.1 Change in visibility over the eastern United States from 1948 to 1982. Reproduced from Malm (1989) with permission of Kluwer Academic Publishers.

particles, thus immobilizing the charge. An exception is radioactive contamination which makes the atmosphere more conductive. Extensive conductivity measurements were made during the first half of the last century, but unfortunately the practice has not been continued in the latter part of the century. Cobb and Wells (1970) summarized the results of a few such measurements that were collected from 1907 to 1970. In general, they suggested a 20% decrease in conductivity took place in the North Atlantic over this period. This corresponds to roughly a doubling of small aerosol particle concentrations over the area. Limited data in the southern Pacific show no such trends.

In general, there is compelling evidence that anthropogenic activity is increasing the concentration of tropospheric aerosol, particularly in the Northern Hemisphere.

Let us now examine the potential impacts of those increased aerosol concentrations on climate. Modeling studies indicate that naturally occurring aerosol particles can affect climate and impact the global circulation of the atmosphere (Hansen *et al.*, 1980; Randall *et al.*, 1984; Tanre *et al.*, 1984; Coakley and Cess, 1985; Hansen *et al.*, 1988; Ramaswamy, 1988). The responses of those models to naturally occurring aerosols vary with the concentration and type of aerosol, the characteristics of the underlying surface (i.e., the surface albedo), the solar zenith angle, cloud cover, and the way the models treat ocean responses. The latter effect is particularly important since, if the ocean temperatures are fixed, the oceans serve as an infinite heat reservoir, and no long-term global temperature responses can be expected (Coakley *et al.*, 1987). Furthermore, unless SSTs can vary, important feedbacks such as variations of the flux of moisture from the ocean cannot occur (e.g., Ramanathan, 1981).

In general, absorption of solar radiation by aerosols reduces solar heating at the surface while it heats the layer of air in which the aerosols reside. The impact of aerosols on the surface, however, varies with the albedo of the underlying surface. If a surface has a relatively low albedo such as over the ocean, a given aerosol may increase the surface albedo, while over the higher albedo deserts, it may decrease it. In general, the impact of aerosol on albedo dominates over absorption at most latitudes, but in higher latitudes over snow- or ice-covered surfaces, aerosol absorption can dominate (Charlson *et al.*, 1992). Recent work, however, has suggested that the darkening of snow and ice by aerosols (particularly black carbon) results in a more important aerosol effect (Hansen and Nazarenko, 2004). The atmospheric response to aerosol heating also varies widely depending on the height and depth of the aerosol layer, and on the basic stability of the layer.

While the major forcing of aerosols is regional in nature, there are indications from model experiments that their influence may impact global circulations. Chung and Ramanathan (2003), for example, examined the influence of the South Asian haze on the general circulation. The South Asian haze extends over an area about the size of the United States and covers the South Asian continent to the Arabian Sea and from the Bay of Bengal to the Indian Ocean Intertropical Convergence Zone. Chung and Ramanathan (2003) estimate that roughly 75% of the particles in the haze are emitted by human activities. In their simulations a radiative heating profile was imposed in the lower atmosphere in the National Center for Atmospheric Research (NCAR) three-dimensional global model (CCM3) that corresponds to that estimated for the South Asian haze layer. The haze heating was imposed only during the dry season of November to April. In the region of haze heating, weak upward motion develops with corresponding divergence flow in the upper troposphere. In response to that regional circulation, rainfall is enhanced over the Indian peninsula and suppressed in southwest Asia. Suppressed rainfall over southwest Asia and the western equatorial Pacific is a result of compensating subsidence in the regions outside the regions of dust-enhanced upward motions. Note that the suppressed convection in the tropical western Pacific produces a weaker zonal gradient in latent heating by deep convection which weakens trade winds, and thereby deepens the ocean thermocline in the eastern Pacific basin. This leads to a weaker zonal gradient of SST which further weakens the trade winds. Thus the El Niño–Southern Oscillation (ENSO) circulation field is modulated. Their simulations also suggest that the regional heating by haze perturb the so-called Arctic Oscillation (Thompson and Wallace, 2000; Thompson *et al.*, 2000) which has been shown to impact Northern Hemisphere climate. While these simulations are quite simple, they do indicate that aerosol-induced regional heating perturbations have the potential of altering circulations and climate over a much larger area.

Heating of the air in which the aerosols reside can result in stabilization of a moist moderately stable layer and shut down deep convection and precipitation which could have important climatic implications through the hydrological cycle. In other regions where there is not sufficient moisture or instability to support deep convection, aerosol impacts would be less. This effect of aerosols has been termed the "semi-direct" effect (Hansen *et al.*, 1997a,b; see also Table 8.3). The reduction in cloud cover associated with this effect can alter the surface energy budget significantly. If the aerosols comprise a large fraction of soot, such as the South Asian haze, then warming in the aerosol layer can nearly totally desiccate stratocumulus cloud layers and alter the properties of the trade wind cumulus layer (Ackerman *et al.*, 2000a). General circulation model simulations by Menon *et al.* (2002b) suggest that black carbon emissions over China may be producing changes in the general circulation which contributes to observed increases in summer flooding in south China and drought in north China. Thus in spite of the fact that anthropogenic aerosol there is regionally concentrated, the potential for global impacts is great.

Another facet of aerosol direct or semi-direct effects is on the nucleation of cloud droplets and thus the concentration of droplets in clouds. Conant *et al.* (2002) computed the effects of carbon black aerosols on cloud droplet nucleation. Figure 9.2 illustrates the equilibrium supersaturation over solution drops containing carbon black aerosols. The peak in the curves, called Kohler curves, represents the supersaturation that must be attained in a cloud in order to form growing cloud droplets. If the cooling rate in clouds which is normally proportional to updraft velocity is not large enough to exceed the peak values in the curves, then the aerosol particles of the size indicated cannot form a cloud droplet. Conant *et al.*'s calculations showed that since carbonaceous particles strongly absorb solar radiation, warming of those aerosol particles elevates

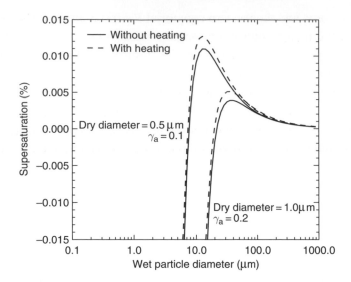

Figure 9.2 Effect of droplet heating on the Kohler curves of particles of 0.5 μm and 1.0 μm dry diameter. Heating parameters of 0.1 and 0.2 are chosen for the two droplet sizes, respectively. The solid curve represents the no-heating case, the dashed curve represents the heating case. Particles are assumed to have the hygroscopic properties of sulfate. From Conant *et al.* (2002). © [2002] American Geophysical Union. Reproduced by permission from the American Geophysical Union.

peak supersaturation. Thus, for example, if many aerosol particles are 0.1 μm in diameter, and peak supersaturations are less 0.01%, those particles will not be activated to form cloud droplets owing to absorption of solar radiation. This effect is most pronounced in low supersaturation clouds such as fogs. Moreover, this effect is greatest for larger aerosol particles since the absorption cross-section is proportional to the area of the particle. Its main impact is on giant CCN (see Nenes *et al.*, 2002) which we have seen in Chapter 4 can initiate or speed up the warm rain collision and coalescence process. Thus if the carboneous particles absorb solar radiation, they may not become big enough to initiate collision and coalescence which, in turn, can act to suppress warm rain processes.

9.3 Aerosol impacts on clouds: the Twomey effect

Clouds, we have seen, are good reflectors of solar radiation and therefore contribute significantly to the net albedo of the Earth system. We thus ask, how might aerosol particles originating through anthropogenic activity influence the radiative properties of clouds and thereby affect climate?

First of all, there are indications that in urban areas aerosols make clouds "dirty" and thereby decrease the albedo of the cloud aerosol layer and increase the absorptance of the clouds (Kondrat'yev *et al.*, 1981). This effect appears to be quite localized, being restricted to over and immediately downwind of major urban areas, particularly cities emitting large quantities of black soot particles. Kondrat'yev *et al.* noted that the water samples collected from the clouds they sampled were actually dark in color.

A potentially more important impact of aerosol on clouds and climate is that they can serve as a source of CCN and thereby alter the concentration of cloud droplets. Twomey (1974) first pointed out that increasing pollution results in greater CCN concentrations and greater numbers of cloud droplets, which, in turn, increase the reflectance of clouds. Subsequently, Twomey (1977) showed that this effect was most influential for optically thin clouds, clouds having shallow depths or little column-integrated liquid water content. Optically thicker clouds, he argued, are already very bright, and are therefore susceptible to increased absorption by the presence of dirty aerosol. In Twomey's words: "an increase in global pollution could, at the same time, make thin clouds brighter and thick clouds darker, the crossover in behavior occurring at a cloud thickness which depends on the ratio of absorption to the cube root of drop (nucleus) concentration. The sign of the net global effect, warming or cooling, therefore involves both the distribution of cloud thickness and the relative magnitude of the rate of increase of cloud-nucleating particles vis-a-vis particulate absorption." Subsequently, Twomey *et al.* (1984) presented observational and theoretical evidence indicating that the absorption effect of aerosols is small and the enhanced albedo effect plays a dominant role on global climate. They argued that the enhanced cloud albedo has a magnitude comparable to that of greenhouse warming (see Chapter 11) and acts to cool the atmosphere. Kaufman *et al.* (1991) concluded that although coal and oil emit 120 times as many carbon dioxide molecules as sulfur dioxide molecules, each of the latter is 50-1100 times as effective in cooling the atmosphere than each carbon dioxide molecule is in warming it. This is by virtue of the sulfur dioxide molecules' contribution to CCN production and enhanced cloud albedo.

Twomey suggests that if the CCN concentration in the cleaner parts of the atmosphere, such as the oceanic regions, were raised to continental atmospheric values, about 10% more energy would be reflected to space by relatively thin cloud layers. He also points out that an increase in cloud reflectivity by 10% is of greater consequence than a similar increase in global cloudiness. This is because while an increase in cloudiness reduces the incoming solar radiation, it also reduces the outgoing infrared radiation. Therefore both cooling and heating effects occur when global cloudiness increases. In contrast, an increase in cloud reflectance due to enhanced CCN concentration does not appreciably affect infrared radiation but does reflect more incoming solar radiation which results in a net cooling effect.

Moreover, we have seen in Chapter 4 that increases in CCN concentration can reduce drizzle and rain production. Because cloud reflectance is increased both by increased droplet concentrations and by increased column-integrated liquid water, a suppression of drizzle by enhanced CCN concentrations would contribute to enhanced cloud albedo. Albrecht (1989) hypothesized that suppression of drizzle formation would lead to longer-lived clouds which by increasing cloud cover would further enhance the albedo of those clouds. This is not necessarily true for all boundary layer clouds. In a marine stratocumulus layer with nearly 100% cloud cover, the amount of condensate in the form of drizzle is only about 10% of the total amount of cloud condensate. Thus if drizzle is suppressed, it is not likely to impact cloud cover. Drizzle could have complicated feedbacks onto the cloudy boundary layer (see Chapter 4). In Jiang *et al.*'s (2002) simulations, for example, they found that higher CCN concentrations suppressed drizzle which resulted in weaker penetrating cumulus and an overall reduction in the water content of the clouds. Drizzle, however, destabilizes the boundary layer only when it does not reach the surface. When drizzle settles to the surface, such as in heavier drizzle rate situations or when the drops are larger, the entire boundary layer is cooled and is stabilized (Paluch and Lenschow, 1991; Jiang *et al.*, 2002). Thus suppressing drizzle formation can either lead to enhanced boundary layer convection and enhanced cloud albedo, or weaker boundary layer convection and thereby lower albedo clouds.

In addition, drizzle formation is not controlled totally by concentrations of CCN. Numerous studies (Houghton, 1938; Johnson, 1982; Tzivion *et al.*, 1994; Levin *et al.*, 1996; Cooper *et al.*, 1997; Feingold *et al.*, 1999) have indicated that giant CCN (GCCN) or ultra-giant particles can serve as precipitation embryos and initiate collision and coalescence. These aerosol particles, which are greater than $5 \mu m$ in radius, can form cloud droplets large enough to initiate collision and coalescence regardless of whether they are considered activated drops according to Kohler theory (Johnson, 1982). For example, modeling studies of marine stratocumulus clouds by Feingold *et al.* (1999) showed that increased concentrations of GCCN enhanced precipitation in clouds with moderately high CCN concentrations. However, if CCN concentrations were quite low, the natural precipitation process was so efficient that GCCN had little influence on precipitation.

Thus the susceptibility of the drizzle process in marine stratocumulus clouds to anthropogenic emissions of CCN depends on the presence or absence of large and ultra-giant aerosol particles in the subcloud layer. Over the open sea, the dominant large and ultra-giant aerosol particles are sea salt particles. While these particles contribute only 10% or less to the total CCN population, they represent major contributors to the large end of the aerosol size spectrum. Their concentration in the marine boundary layer varies with wind speed, being in greater numbers with stronger winds. *We therefore hypothesize* that the susceptibility of the drizzle process in marine stratocumuli to anthropogenic CCN will depend on wind speed, being less susceptible at higher wind speeds than lower. Another potential source of GCCN is desert dust which has both natural and anthropogenic origins (see Chapter 4). But desert dust can also suppress clouds by radiatively heating the cloud layer, and serve as enhanced CCN which will suppress precipitation. It is not clear how desert dust effects the albedo of marine stratocumulus clouds in a climatological sense because of its potentially complicated impacts.

Let us now examine the evidence supporting or refuting the Twomey hypothesis. The major focus of these studies is on the world's oceans where shallow stratocumulus clouds reside. These shallow clouds, which are believed to be most susceptible to the Twomey effect, cover over 34% of the world's oceans. Therefore, any consistent trend towards increasing CCN concentration over the oceans has the potential for increased atmospheric albedo and global cooling. Unfortunately, there have not been long-term systematic measurements of CCN concentration anywhere on Earth.

We must therefore resort to indirect methods of assessing whether global pollution is affecting climate. In general, oceanic CCN concentrations are low; being on the order of 50 to $100 \, \mathrm{cm}^{-3}$. Over continental areas CCN concentrations range from 500 to $1000 \, \mathrm{cm}^{-3}$ with some heavily polluted regions reaching several thousand per cubic centimeter. The main source of natural CCN over the oceans is believed to be dimethylsulfide (DMS) which is excreted by plankton and then liberated into the atmosphere where it is oxidized, probably photochemically, to form non-sea-salt-sulfate aerosol (Bigg *et al.*, 1984; Charlson *et al.*, 1987; Krcidenweis *et al.*, 1991).

In Chapter 4 we saw that ship tracks (Coakley *et al.*, 1987; Scorer, 1987; Radke *et al.*, 1989; Porch *et al.*, 1990) are viewed as a "Rosetta stone" which illustrates the interaction of CCN and cloud albedo. The evidence is quite clear that exhaust from some fossil-fuel-burning ships is producing a plume of aerosol which acts as CCN and thereby enhances droplet concentrations. The enhanced droplet concentrations, in turn, suppress drizzle thus making the clouds wetter. Both the enhanced droplet concentrations and wetter clouds produces a plume or "ship track" that is brighter than surrounding clouds. Another example is what Rosenfeld (2000) calls "pollution tracks" as viewed by AVHRR satellite imagery. These pollution tracks are associated with pollution plumes from specific industrial pollution sources. It is inferred that the pollution tracks are clouds composed of numerous small droplets that suppress precipitation. We noted in Chapter 4 that perhaps the most significant aspect of Rosenfeld's analysis is the conspicuous absence of pollution tracks over the United States and western Europe. The implication is that these regions are so heavily polluted that local sources cannot be distinguished from the widespread pollution-induced narrow droplet spectra in those regions.

One way to examine the influence of widespread sources of CCN-generating aerosols on global climate is to use simulations with general circulation models. Firstly it must be recognized that GCMs have grid spacings of 150–250 km and that clouds often comprise a small fraction of the grid-cell area and the average grid-cell vertical velocities are very small, approximately $0.01\,\mathrm{m\,s^{-1}}$ whereas actual cloud-scale vertical velocities are more like $1\,\mathrm{m\,s^{-1}}$ and often greater. The number of cloud droplets activated in clouds is not only a function of the number of CCN available but also on the peak supersaturations in clouds which is related to cloud-scale vertical velocity. It is, therefore, important to estimate cloud-scale vertical velocities to predict the concentrations of cloud drops. Some modelers have assumed an empirical relationship between predicted sulfate mass concentrations and droplet concentrations (Martin *et al.*, 1994; Boucher and Lohmann, 1995; Kiehl *et al.*, 2000), but that is equivalent to assuming there is only a single value of cloud updraft velocity for all clouds in the model. Others have estimated vertical velocity from predicted turbulent kinetic energy from their boundary-layer models (Ghan *et al.*, 1997; Lohmann *et al.*, 1999). This is a step in the right direction, but does not take account of the fact that cloudy updrafts are at the tail of the probability density function (PDF) of vertical velocity. Chung *et al.* (1997) assume a normal distribution of vertical velocity with a mean given by the GCM gridpoint mean. They then determine the velocity-weighted mean droplet concentration which takes into account the tails of their assumed PDF of vertical velocity. But observed PDFs of vertical velocity in the cloudy boundary are found to be multimodal and better fit by double-Gaussian PDFs (Larson *et al.*, 2001) with a mean that is a function of the root mean square (RMS) vertical velocity not a GCM gridpoint mean. Moreover, the complications of precipitation or drizzle processes on cloud lifetime, cloud water contents, and cloud radiative properties discussed above cannot be well simulated in GCM cloud parameterization schemes. For example, we have seen that precipitation processes are non-linear functions of total condensate water contents. As a result GCM model grid-box mean liquid water content is essentially meaningless for representation of precipitation production (Stevens *et al.*, 1998; Pincus and Klein, 2000). As pointed out by Pincus and Klein, a PDF approach to subgrid modeling may be the optimum approach to resolving these deficiencies (Larson *et al.*, 2005). Another option, though considerably more computationally expensive, is to use what have been called "super-parameterizations" in which cloud-resolving models are activated at model gridpoints (Grabowski, 2001; Randall *et al.*, 2003).[1] These

---

[1] A much more computationally efficient method, which retains the physics of the super-parameterizations, has been proposed (Pielke *et al.*, 2006b) where look-up tables, or functional fits, are used within the parent model while the super-parameterization is run off-line.

models have the capability of predicting cloud-scale vertical velocities and liquid water contents and thus explicitly representing precipitation processes.

Keeping these caveats in mind, let us examine some of the results of GCM simulations of the indirect effects of aerosols on climate. Chung *et al.* (1997) used a coupled aerosol chemistry GCM to examine the influence of anthropogenic aerosols on climate. The model explicitly calculates the conversion of sulfur dioxide gases into sulfate particles. It includes an inventory of natural and anthropogenic emissions of these gases and then predicts the global distribution of the aerosols. Both direct and indirect radiative effects of aerosols are considered. But they do not consider the complexities of drizzle formation and its potential radiative influences. They estimate an indirect radiative forcing ranging from -0.4 to $1.6\,W\,m^{-2}$. They find that the maximum in indirect forcing is over the Atlantic Ocean near the coastline of North America.

This is in contrast with Boucher and Lohmann (1995) who estimated a stronger magnitude of indirect forcing and especially over polluted land regions. Because aerosol pollution is largely concentrated in the Northern Hemisphere (Schwartz, 1989) surface cooling is concentrated over the North Atlantic and North Pacific Oceans due to the higher albedo of contaminated clouds. Several GCMs have thus simulated an alteration in the general circulation which then affects precipitation in areas well beyond those regions (Rotstayn *et al.*, 2000; Williams *et al.*, 2001; Rotstayn and Lohmann, 2002). These models were coupled to an ocean mixed-layer model so that enhanced cloud albedo produced cooler ocean surface temperatures in the Northern Hemisphere. In addition, suppressed rainfall resulted in more extensive cloud cover which also cooled ocean surfaces. The models responded by shifting the Intertropical Convergence Zone (ITCZ) southward which enhanced precipitation in the Southern Hemisphere tropical regions (Rotstayn *et al.*, 2000) and drying in the Sahel zone in Africa (Rotstayn and Lohmann, 2002).

The latter response is consistent with observed reduction in rainfall in the Sahel zone during the twentieth century. Williams *et al.* (2001) also found a similar response to both direct and indirect effects of pollutant aerosols, and in addition they found a reduction in the Indian monsoon precipitation during June, July, and August. The cooling in their model also resulted in expanded sea ice coverage in the Arctic Ocean in summer. This is in response to the southward displacement of storm tracks associated with the shift of the ITCZ southward. Thus the greatest impacts of the enhanced aerosol concentration were over the north Polar regions and secondarily around $40°\,N$ latitude.

We would like to note that one should not interpret the results of those simulations as being quantitative forecasts of the effects of aerosols on patterns and amounts of regional precipitation. As noted previously, there are too many uncertainties in the distribution and concentrations of aerosols in the past and even in the present. In addition, we have seen there are many simplifications in the models that limit their ability to realistically simulate indirect effects of aerosols. Instead, these simulations demonstrate the potential that direct and indirect aerosol forcing, even though being regional in nature, can have wide area responses well beyond the regions directly influenced by aerosol changes in radiation.

We have seen that greenhouse warming as a result of enhanced carbon dioxide concentrations is only significant when the global hydrological cycle is enhanced and greater amounts of water vapor are evaporated into the air principally over the oceans but also over land, since water vapor is the dominant greenhouse gas (as discussed in Chapter 8). The increased amounts of water vapor in the air, in turn, result in a strong positive feedback to CO_2 warming. Recent GCM simulations of both greenhouse warming, and direct and indirect aerosol effects (Liepert *et al.*, 2004) suggest that aerosol indirect and direct cooling reduces surface latent and sensible heat transfer and as a consequence acts to spin down the hydrological cycle and thereby substantially weaken greenhouse gas warming. This is important since most investigators compare top of the atmosphere radiative differences for greenhouse gas warming and aerosol direct and indirect effects separately. But since greenhouse warming depends on a spin-up of the hydrological cycle and aerosol direct and indirect cooling counters that, the potential influence of aerosols on climate could be far more significant than previously thought.

9.4 Aerosols in mixed-phase clouds and climate

We have seen in Chapter 4 that supercooled clouds can exhibit what are called pollution tracks (Rosenfeld, 2000) which are believed to be clouds composed of numerous small droplets that suppress precipitation. Perhaps the most significant aspect of Rosenfeld's analysis to global precipitation and climate is the conspicuous absence of pollution tracks over the United States and western Europe. The implication is that these regions are so heavily polluted that local sources cannot be distinguished from the widespread pollution-induced narrow droplet spectra in those regions. We have also seen that pollution can suppress precipitation in wintertime orographic clouds (Borys *et al.*, 2000, 2003) by enhancing the concentration of CCN and as a consequence cloud droplets are smaller, which reduces the efficiency of ice crystals collecting supercooled droplets or riming. Further evidence of this effect was suggested in the analysis by Givati and Rosenfeld (2004) of orographic precipitation records downwind of major urban centers in Israel and California. They inferred that precipitation is suppressed by 15% to 25% downwind of those urban areas.

We have seen that some pollution sources, particularly those associated with mining and heavy metal industries and leaded gasoline, are high in IN concentrations (Schaefer, 1969). There are indications, however, that there has been a systematic decrease in IN concentrations at several sites in the Southern Hemisphere as well as Hawaii over the last 25 years (Bigg, 1990b). It is not known at this time if such observations are a direct result of the increased use of lead-free gasolines or contamination of natural ice nuclei by sulfate pollutants (often called IN poisoning). There are observations, however, of both enhanced CCN and IN concentrations in the boundary layer. For example during the FIRE/SHEBA field experiment in the Arctic Basin, Yum and Hudson (2001) found CCN concentrations of 100 and $250 \, \text{cm}^{-3}$ (active at 1% supersaturation) below and above the boundary layer inversion, respectively. At the same time Rogers *et al.* (2001) measured IN concentrations ranging from approximately $3 \, l^{-1}$ below the inversion to $85 \, l^{-1}$ above the inversion. The air over the ice surface in the Arctic boundary layer is often very clean. But it has long been known that there are major intrusions of polluted air into the Arctic basin and that they often contain high concentrations of CCN (Borys and Rahn, 1981; Patterson *et al.*, 1982). Whether the pollution sources are also rich in IN is less well known.

Cloud-resolving simulations of Arctic boundary layer clouds during FIRE/SHEBA, first for a particular day (Carrió *et al.*, 2005a) and then for the entire spring season of the field campaign (Carrió *et al.*, 2005b) were carried out. The model was initialized with either the clean subcloud aerosol concentrations throughout the boundary layer or with the observed polluted aerosol concentrations above the inversion and clean below. During the spring season simulations were performed with the model coupled to a sea ice model. The multi-month simulations were performed using two to three daily SHEBA soundings nudged into the cloud-resolving model to represent daily variations in the synoptic atmosphere. Mixed-phase clouds prevailed during the first 2 months of simulation, while predominately liquid clouds were simulated during the last month. The effects of IN entrainment when mixed-phase clouds were present decreased liquid water paths while ice water paths increased. Even though the above inversion IN concentrations are much lower than estimates from midlatitudes using the Meyers *et al.* (1992) formula, the clouds were essentially overseeded. As a result the crystal fall speeds were reduced and so were the precipitation rates. This resulted in longer residence times of the ice particles and increased total condensate paths. This produced enhanced downward longwave radiation. Enhanced albedo associated with enhanced CCN concentrations (hence droplet concentrations) also occurred but this was much smaller in magnitude so that net surface radiation was increased. As a consequence sea ice melting rates were greater. Overall, the model suggests that the entrainment of a polluted air layer overriding the inversion enhances sea

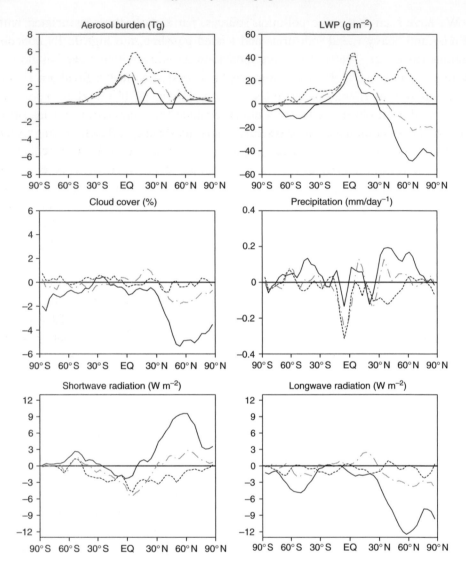

Figure 9.3 Zonal annual mean changes between present-day and pre-industrial times for the experiments where 10% of the hydrophilic black carbon act as ice nuclei (solid line), 1% (dash–dot line), and 0% (dotted line). Adapted from Lohmann (2002). © [2002] American Geophysical Union. Reproduced by permission of the American Geophysical Union.

ice melting rates. Melting rates were approximately 4% higher when the air above the boundary layer was polluted than when the entire layer is composed of clean subcloud air.

Only recently have GCMs been able to simulate the effects of enhanced IN concentrations on climate. Lohmann (2002) assumed that a fraction of hydrophilic soot aerosol particles act as contact ice nuclei at temperatures between $0\,°C$ and $-35\,°C$ based on laboratory studies by Gorbunov *et al.* (2001). She found that increases in aerosol concentration from pre-industrial times to present day pose a new indirect effect, a so-called "glaciation indirect effect," on clouds (see Table 8.3). She showed that increases in contact IN in the present-day climate result in more frequent glaciation of clouds and increase the amount of precipitation via the ice phase. This effect can at least partly offset the solar indirect aerosol effect on water clouds and oppose suppression of drizzle by enhanced CCN concentrations. As a result she concluded that precipitation is enhanced by anthropogenic aerosols (Fig. 9.3). It should be noted that whether enhanced IN concentrations increase or decrease precipitation is very cloud specific. It is also the most controversial aspect of cloud seeding. In clouds with low liquid water content, complete glaciation of clouds can overseed them and result in high concentrations of small ice crystals as reported in Carrió *et al.* (2005a,b). On the other hand, modest increases of ice crystal concentrations in some clouds can convert them from a non-precipitating or weakly precipitating cloud to one that is precipitating. Unfortunately, neither cloud-scale models or GCMs have demonstrated the ability to predict which regime prevails under a variety of cloud regimes.

9.5 Aerosols, deep convection, and climate

Most large-scale numerical prediction models such as GCMs use convective parameterization schemes such as Arakawa–Schubert (Arakawa and Schubert, 1974), Betts–Miller (Betts and Miller, 1986), or Kuo (1974). These schemes provide heating and moistening by deep convection and have a simple precipitation parameterization that is usually linked to a bulk precipitation efficiency. As such, these models do not have sufficient sophistication in cloud microphysics to include aerosol influences on precipitation or mass fluxes, or heating profiles. An exception would be those models that use a so-called super-parameterization approach (Randall *et al.*, 1984; Grabowski, 2001). Since this approach uses cloud-resolving models, it is relatively straightforward to extend them to include explicit aerosol effects.

Using a more classical approach to parameterizing convection, Nober *et al.* (2003) decreased precipitation efficiency in convective clouds having temperatures warmer than 263 K depending on the cloud droplet number concentration. The instantaneous effects on precipitation formation were very large, reducing precipitation formation by as much as 100%. However, because these changes are confined to small areas, the large-scale mean precipitation anomalies were not significantly affected.

Khain *et al.* (2005) postulate that smaller cloud droplets, such as originating from anthropogenic activity, would reduce the production of drizzle drops. When these droplets freeze the associated latent heat release results in more vigorous convection. In a clean cloud, on the other hand, drizzle would deplete the cloud liquid water so that less latent heat is released when the cloud glaciates resulting in less vigorous convection. Thus, they found that a squall line did not form under clean conditions, whereas the squall line developed under continental aerosol conditions which produced more precipitation after 2 hours. Zhang *et al.* (2004) came to similar conclusions for different 3-week periods over the Atmospheric Radiation Measurement (ARM) site in Oklahoma.

In simulations of entrainment of Saharan dust into Florida thunderstorms, van den Heever *et al.* (2006) also found dust not only impacts the cloud microphysical characteristics but also the dynamical characteristics of convective storms as well. We have seen in Chapter 4 that dust can enhance CCN, GCCN, and IN concentrations. Van den Heever *et al.* considered the influence of dust on deep convection where dust served as just CCN, then as GCCN, then only as IN, and then as the entire combination. In each simulation, dust altered the dynamics of convection by producing greater amounts of supercooled water. In response to freezing of the greater amounts of supercooled water, the strengthened updrafts thrust more water into anvil levels and produced less accumulated rainfall on the ground by the end of the day. Earlier in the afternoon precipitation was enhanced by dust, but later most of the dust had been washed out by precipitation. The variations in cloud microstructure and storm dynamics by dust, in turn, alters the accumulated surface precipitation and the radiative properties of anvils. This is in contrast to the dynamic seeding concept in which seeding enhances glaciation of convective clouds which leads to dynamical invigoration of the clouds, larger amounts of processed water, and thereby enhanced rainfall at the ground (Simpson *et al.*, 1967; Rosenfeld and Woodley, 1989, 1993). From a radiative perspective, it may be more important that the anvil properties of the clouds are modified by dust than surface rainfall. That is, if the direct effects of dust are not so strong that convection is totally inhibited, those storms that do form are likely to produce anvils which are optically thicker and thereby reflect incoming solar radiation and radiate more longwave radiation to the ground, with the latter response being dominant. Therefore dust may act to enhance greenhouse warming.

Overall, aerosols have the potential for having substantial impacts on clouds, precipitation, and the radiative properties of clouds. However, the interaction between aerosols and clouds is sufficiently complex that even cloud-resolving models have difficulty in accurately simulating their physics and dynamics. GCMs have not included all of these complexities, thus we cannot yet skillfully simulate the impact of aerosols on global climate.

10

Nuclear winter

10.1 Introduction

The nuclear winter hypothesis, in simplest terms, contends that a large-scale nuclear war would generate large amounts of smoke and dust in the atmosphere which would attenuate solar radiation and cause so much cooling of land areas that winter-like conditions would prevail in the summer months and major crop failures would occur. The rudiments of the hypothesis were first presented by Crutzen and Birks (1982) who calculated the amounts of smoke that would be lofted into the atmosphere by fires following large-scale nuclear warfare resulting in obscuration of solar radiation of the Northern Hemisphere for several weeks or more. They then speculated on the possible climatic responses to the resultant reduced surface temperatures. This paper was followed by Turco *et al.* (1983) or the so-called TTAPS paper in which a one-dimensional, globally and annually averaged, radiative–convective model was used to calculate surface temperatures following massive injections of smoke into the atmosphere. They calculated that the smoke and dust emitted by fires following a massive nuclear exchange would cool land surface temperatures in midsummer below freezing in the Northern Hemisphere, with surface temperatures falling as much as ~35 °C. The term "nuclear winter" was established to describe this modeled cooling.

These papers stimulated intense interest in the topic both in the scientific community and in the political arena. As a result a series of national and international committees were established to assess the potential environmental consequences of large-scale nuclear warfare (e.g., Harwell and Hutchinson, 1985; National Research Council 1985; Pittock *et al.*, 1986). National and cooperative international research programs were also established.

In the United States, research programs on nuclear winter were established by the Department of Energy, the Defense Nuclear Agency, and to a lesser degree the National Science Foundation and the National Oceanic and Atmospheric Administration. Funding for nuclear winter research was mandated by Congress but no specific funds were allocated. As a result funding was achieved largely by internal reprogramming within the various agencies. In some cases, this meant that researchers in or supported by the agencies were polled to see if they were doing research relevant to nuclear winter and if so, their programs were totaled to identify monies in the agencies going to the nuclear winter program. In other words, the scientists did what they had always done, with the exception, perhaps, of attending a few meetings. In other cases, the agencies redirected scientists to work on nuclear winter research, sometimes with less than enthusiastic participation. In other cases, funding was reprogrammed from other areas making the program very vulnerable to outside attack once the political support for nuclear winter research waned. Still, a few scientists and agencies jumped into nuclear winter research with enthusiasm.

Some scientists, such as Carl Sagan, used the nuclear winter hypothesis as a vehicle for proclaiming anti-nuclear war and anti-war policies. As noted by Dyson (1988), this has put professional scientists in an awkward position. Singer (1992) has also briefly discussed the political aspects of the nuclear winter issue. On the one hand, it is the responsibility of scientists to critically examine a new and exciting theory, and in many cases, prove it wrong. This is how science works. Every new theory has to be defended by its proponents against intense and often bitter criticism and scrutiny. This is what keeps science honest! The rare theory that withstands the onslaught of criticism is strengthened and improved by it.

On the other hand, nuclear winter as a political statement makes us want to believe it, because few scientists favor the destructive powers of nuclear warfare, especially if it means the worldwide destruction of society. This is an example of a genuine conflict between the demands of science and the demands of humanity.

Coincident with the "Peristroika" policy and eventual break-up of the Soviet Union, interest in the science of nuclear winter plummeted so that by 1990 no national research program in nuclear winter remained. Nuclear winter research experienced a "rise and fall" in a time-span of only 8 years. In 1991, Carl Sagan attempted to revive interest in nuclear winter by predicting on "Nightline" that the Kuwaiti oil fires which were a result of the Persian Gulf War would produce a nuclear winter effect, causing a "year without a summer," and endangering crops around the world. Sagan stressed this outcome was so likely that "it should affect the war plans." Scientific measurements were taken locally and at long range in the smoke plumes (Cahill *et al.*, 1992; Lowenthal *et al.*, 1992; Pilewskie and Valero, 1992). These analyses suggested that the meteorological and air quality effects were mainly confined to regional impacts.

In this chapter we set aside our humanitarian concerns and focus on the scientific question: is the nuclear winter hypothesis a viable hypothesis and what is the evidence supporting or refuting it?

10.2 The nuclear winter hypothesis: its scientific basis

The nuclear winter hypothesis is another example of a long, multiple-chain hypothesis. It is composed of the following elements.

- Large-scale nuclear warfare involving a total yield in excess of 100 megatons will occur over much of the Northern Hemisphere.
- In addition to the death and destruction due to the direct consequences of the bombing, area-wide fires will ensue for a period of several days, and the smoke will be injected into the upper troposphere and lower stratosphere.
- Much of the injected smoke will consist of submicron-sized soot or elemental carbon particles which are excellent absorbers of solar radiation and, owing to their small size and low fall velocities, have long residence times in the atmosphere.
- The smoke that escapes early scavenging by clouds and precipitation elements will disperse throughout the Northern Hemisphere creating a nearly uniform pall over much of the hemisphere.
- The widely dispersed smoke will have sufficiently large optical depths to cause significant absorption of solar radiation with little absorption of terrestrial radiation (assumes that the smoke layer is cloud-free and particles are small and dry) so that widespread lowering of surface temperatures occurs.
- The thermal inertia or heat storage in the oceans will cause only local (coastal?) modification of the airmasses thus not damping the widespread cooling effect substantially.
- A threshold amount of smoke-induced cooling will trigger a major climatic shift that will take years or centuries for recovery.
- The smoke-induced cooling will result in major crop losses and it will compound food losses due to disrupted production and distribution systems associated with the direct effects of warfare, leading to widespread famine and death.

Let us now examine each of these subhypotheses, respectively.

10.2.1 The war scenarios

In order to estimate the maximum possible extent of widespread nuclear warfare, one must first estimate the total nuclear arsenal in the United States, former Soviet Union, and other countries and their allies. Much of this information is classified but educated estimates can be made. Of the total arsenals one must next estimate what fraction of those weapons would be functional and how many would be made inoperative in a first strike (before the defender can respond). Turco *et al.* (1990) estimate that the combined United States and Soviet arsenals comprise about 25 000 warheads carrying roughly 10 000 megatons (Mt). Crutzen and Birks (1982) used two scenarios, a 5750 Mt detonation and a 10 000 Mt detonation. The original TTAPS study spanned a range of 100 to 25 000 Mt with a baseline scenario of 5000 Mt. The US National Research Council (1985) report considered a 6500 Mt war as a baseline scenario. This is the level of bombing that is most often used in the model calculations. Obviously the actual amount of bombing that could occur is a matter of conjecture and we hope that none of these scenarios would ever become reality. The nuclear arsenals, of course, have also been reduced in recent years through arms control agreements in conjunction with the collapse of the Soviet Union, but very substantial nuclear warheads, of course, still remain in the United States and Russia.

10.2.2 Smoke production

Estimating the amount of smoke produced in any particular bombing scenario is particularly complicated since one must speculate on the nature of targets (i.e., are they urban centers, or rural areas where missile silos, military camps, etc. are located, or a mix of urban–industrial and rural land), the nature of the fuels in each hypothetical target, the magnitude and type of bomb used at each target, and the local weather conditions affecting fire behavior.

In an urban area, for example, smoke production will vary depending on whether the fuel is open to the air or buried beneath piles of rubble. Likewise the size of soot particles and their optical blackness depends on whether the fire burns quickly and at high temperatures or smolders for a long time. Turco *et al.* (1990) cite references to sources of how fuel loading estimates are made. They conclude that the uncertainty in estimating combustible loading for a particular bombing scenario is probably less than 50%.

Taking factors such as these into account, various scientists and groups of scientists have made rough estimates of the amount of soot that is produced for given bombing scenarios. For example, the National Research Council (1985) estimated that a baseline 6500 Mt war would produce 360 Tg (10^{12} g) of smoke containing 65-70 Tg of strongly absorbing soot particles. Penner (1986) estimated the total amount of smoke production for a large-scale nuclear war could range from 24 to 400 Tg, with soot production being between 10 and 225 Tg. It must be recognized that a large uncertainty exists with these probable values.

Once these bulk estimates of smoke production are made, then one must estimate how this smoke is distributed vertically and horizontally and how much smoke is removed by sedimentation and scavenging.

10.2.3 Vertical distribution of smoke

Estimating the vertical distribution of smoke from fires is important in determining how long the smoke will remain in the atmosphere. If most of the smoke remains below 2-3 km above the ground such as occurred in the Kuwaiti oil fires of 1991, turbulent mixing, cloud, and precipitation particle scavenging and sedimentation of particles will result in the removal of the smoke from the atmosphere in a week or so. On the other hand, if much of the smoke rises to the upper troposphere, where turbulence is less common, clouds are low in water content and less able to scavenge soot particles, and precipitation rates are small, the smoke may remain in the atmosphere for periods of several weeks to a month or so. Finally, if a large amount of smoke makes its way into the very stable, lower stratosphere where clouds are uncommon, the smoke could remain in the atmosphere for periods of many months to as much as a year or more, similar to the residence time associated with volcanic aerosols ejected into the stratosphere.

The vertical distribution of smoke produced by a given fire depends on the fuel loading and resultant burn, and on many meteorological conditions including the likelihood that wet deep convection will participate in the vertical transport of the smoke. Typically wildland fires have modest fuel loadings and as a result most of the smoke from them remains below 3–4 km above ground level (Small *et al.*, 1989). Urban fires can be far more concentrated and if there is sufficient moisture and conditional instability in the environment, numerical simulation of urban fire plumes with two- and three-dimensional models suggests that wet deep convection can be triggered by the heat from the fires (Banta, 1985; Cotton, 1985; Penner, 1986; Pittock *et al.*, 1986; Tripoli, 1986). If the heat output is great enough and the atmosphere has sufficient conditional instability, then the models suggest that vigorous convective storms can be triggered that can transport smoke, debris, and water substance into the upper troposphere and lower stratosphere. Stratospheric injection of smoke from burning urban areas, however, is probably relatively little. Turco *et al.* (1990) estimate as much as 10% of the smoke will reach the stratosphere but this is probably on the high side. First of all, the intense burning period following a bombing is likely to last only a few hours and only a few targets will have sufficiently concentrated fuels and sufficiently unstable atmosphere to support stratospheric penetrating convection.

Even intense firestorms such as occurred over Dresden, Germany during World War II are likely to transport only a small proportion of the smoke produced into the stratosphere. Firestorms are intense fires that produce vigorous whirling winds on the scale of tens of kilometers. The strong winds associated with firestorms ventilate the fires, thereby strengthening them. Tripoli's (1986) simulation of such an urban firestorm revealed, however, that the cyclostrophic reduction in pressure associated with the rapidly rotating storm created vertical pressure gradients that weakened the storm's updrafts, and, as a result, the smoke plume was detrained at lower levels. Moreover, the weaker updrafts increase the time that scavenging processes will operate in the rising updrafts, and increase the efficiency of precipitation processes and wet removal of smoke. We will address the scavenging issue more directly in the next subsection. In summary, it appears that the bulk of the smoke from wide-spread nuclear warfare will be deposited below the middle troposphere, probably in the range of 4 to 6 km (Small *et al.*, 1989).

10.2.4 Scavenging and sedimentation of smoke

Another uncertainty is the amount of smoke that is emitted by fires and then transported aloft by dry or wet convection which is removed by scavenging and sedimentation. In the case of cloud-free convection, the main removal mechanism is by slow settling of the particles. This process is enhanced by coagulation among numerous, small, slowly settling particles to form larger, faster falling particles. Coagulation occurs very rapidly in the highly concentrated regions close to the fires, but as turbulent diffusion lowers the concentration appreciably, coagulation takes place very slowly, requiring many days to weeks to create numerous, faster falling particles that settle to the Earth's surface. Only the very large smoke particles (greater than tens of micrometers) will settle out of the plume in a few hours to days in the absence of clouds.

Clouds greatly enhance the removal of smoke particles. Some of the smoke particles, such as those from wood products, are likely to be hygroscopic and serve as CCN. Others, such as from petroleum products, are less likely to serve as CCN. But measurements in the Kuwait oil fire smoke plumes by Hudson and Clarke (1992) showed that 70% of the particles are active as CCN. Those particles may after a time coagulate with hygroscopic particles and thereby become wetted as the hygroscopic component of the mixed aerosol particle takes on water vapor. Estimates of the amount of smoke that is removed by activation of sooty CCN or what we call *nucleation scavenging* varies widely from less than 10% to more than 90%. In our opinion, for wildland fires and most residential fires it is probably closer to 90%, while in industrial petrochemical fires it may be as low as 10-20%. These are just educated guesses, however.

Just because such particles are embedded in cloud droplets does not mean they will be removed from the atmosphere. Many cloud droplets remain small and will eventually evaporate releasing the embedded particles back into the atmosphere. During their brief residence in the cloud droplets, the aerosol particles may partially dissolve, undergo wet chemical reactions, and coagulate with other embedded particles. Thus while the particles may be released back into the atmosphere when the droplets evaporate, they probably have changed both in size and chemical composition. Some of the droplets containing embedded aerosol will participate in the formation of raindrops. Most of those aerosol particles will rain out to the ground and no longer play a role in electromagnetic radiation attenuation. The fraction of aerosol actually rained out is proportional to the precipitation efficiency of the clouds. Precipitation efficiencies vary widely among cloud types ranging from 10% to more than 90%. Fire-triggered clouds having very intense updrafts such as in some model calculations may exhibit precipitation efficiencies that are very low, similar to supercell thunderstorms (~10%), whereas if cloud updrafts are weaker, the longer time available in the updrafts favors higher precipitation efficiencies, probably closer to 70-80%. Because the bulk of the fire-producing clouds are probably in the weaker updraft categories, precipitation efficiencies are probably on the higher side, greater than 50%.

In addition to nucleation scavenging, some remaining particles are directly scavenged by cloud droplets and raindrops. The small aerosol particles that are most likely to attenuate sunlight appreciably are susceptible to scavenging by Brownian diffusion and phoretic scavenging (see Pruppacher and Klett, 1978; Cotton and Anthes, 1989). These processes are relatively slow and probably account for less than 10% of the total removal of small particles. Again, once embedded in droplets, their ultimate fate is determined by the precipitation efficiency of the clouds. Larger aerosol particles are readily scavenged by hydrodynamic capture (see Pruppacher and Klett, 1978) by raindrops. The very large (greater than a few tens of micrometers) non-hygroscopic smoke particles will be capable of colliding with cloud droplets or raindrops thereby becoming large drops and then being rained out as they participate in the precipitation process. Such large particles probably contribute to the "black rains" seen in major urban fires following nuclear bombardment (MacCracken and Chang, 1975; National Academy of Sciences, 1975; Whitten *et al.*, 1975), and such "black rains" were also seen in Basra, Iraq, downwind of the oil fires in Kuwait according to news reports. TTAPS originally estimated that 25% to 50% of the bulk smoke mass in a large-scale war scenario would be scavenged immediately and subsequently, Turco *et al.* (1990) downgraded this estimate to 10% to 25%. In our opinion, the original

TTAPS estimate is probably an underestimate, with 50% to 60% being more likely.

10.2.5 Water injection and mesoscale responses

As seen in the above discussion, major nuclear-triggered fires would not only produce the vertical transport of smoke but also the vertical transport of water substance. The source of the water substance transported upward over the fires arises directly from the combustion process and from the low-level convergence of air feeding the fires. Of the fuel that is burned, particularly wood fuels, roughly 10% of the burned mass is water vapor, whereas the amount of submicron-sized smoke that can effectively absorb sunlight comprises less than 1% of the smoke mass (e.g., Cotton, 1985). The combustion process, however, is not the major source of water transported aloft by fires. Feeding the fires and the rising hot gases is low-level convergence of ambient air. In the summer months (which we shall show is the crucial period for major climatic consequences), surface mixing ratios of water vapor range from 15-20 g kg^{-1} in the semi-tropics and tropical coastal areas, to as low as 8-10 g kg^{-1} in the continental interiors. Even in the driest desert regions, in summer, surface vapor mixing ratios rarely go below 5 g kg^{-1}. As the air converging in the fires rises in the buoyant updrafts, it mixes with the moist, fire exhaust-product air as well as drier ambient air at higher levels. Moreover, the rising air column cools nearly adiabatically raising the relative humidity to over 100% where a cloud forms. Eventually some moisture is lost by being precipitated out of the cloud and as a consequence of mixing with drier, high-level air. Nonetheless, the air that is detrained out of the rising fire plumes in the middle and upper troposphere is considerably more moist than the surrounding environmental air.

Ambient water vapor mixing ratios at heights of 8–10 km in the atmosphere are typically less than 1 g kg^{-1} and so also are saturation values, so that only small additions of water vapor are needed to produce persistent clouds (see discussion in Section 4.2 on contrails).

Turco *et al.* (1990) conclude that the total mass of water injected over a hemisphere would perturb ambient water vapor concentrations by at most a few percent and therefore have little effect. They argue that the clouds formed by the fire injections would eventually be warmed by smoke injections and thereby be dissipated. We must remember, however, that the smoke and water substance are simultaneously injected into the atmosphere and with the exception of precipitation removal, they will remain colocated in the atmosphere for an extensive period. Moreover, water vapor in the upper atmosphere is a trace gas that requires only small additions to become saturated to form clouds. An example are jet contrails, which as we saw in Chapter 4, can lead to persistent cloud cover. The moisture released by jet contrails are many orders of magnitudes less than the expected bulk moisture amounts injected into the middle and upper troposphere by nuclear bomb produced fire plumes. As a result we anticipate that associated with the widespread smoke release would be persistent cirrus and altostratus cloud decks which will substantially alter the radiation balance. We examine its potential consequences next.

Several investigators began preliminary investigations of the mesoscale responses to smoke and moisture emissions by intense urban fires (Golding *et al.*, 1986; Giorgi and Visconti, 1989). Depending on the amount of moisture emitted with the smoke, middle and upper tropospheric clouds formed in their simulations over broad areas (patches greater than 50 km or more) either in the smoke core or at the peripheries of the smoke plumes.

Cotton *et al.* (personal communication), for example, examined the mesoscale responses to just a dry smoke plume introduced into the upper troposphere. The emitted smoke amounts were based on the amounts of smoke detrained from a single simulated smoke plume using a three-dimensional cloud model (e.g., Tripoli, 1986), but no moisture was added to the atmosphere by the fires. They found that the model responded to solar heating of the smoke plume by causing lofting of the plume and a sea-breeze-like circulation in the upper troposphere at the boundaries of the plume. This sea-breeze-like circulation caused the formation of a cirrus-like ice cloud in the rising air entering the base of the smoke plume whenever the ambient relative humidity exceeded 70% relative to water. Thus, in spite of the fact that the heated air in the smoke plume lowered the relative humidity, the ascent of unheated air by the sea-breeze-like, solenoidal circulation triggered ice and liquid cloud formation. Moreover the ice cloud persisted owing to the lower saturation vapor pressures with respect to ice relative to liquid water.

They calculated that the persistent ice cloud absorbed the upwelling terrestrial radiation, causing a greenhouse warming effect, particularly at night. Thus while a simulation with a dry smoke cloud (i.e., no cloud was allowed to form in the rising air column) produced average surface temperatures 6 K cooler than the no-smoke simulation after 24 hours, the case with an ice cloud was only 1.4 K cooler than the no-smoke control experiment. This illustrates the potential moderating influence of cloud formation associated with ascent of a smoke plume in the middle and upper troposphere. Because the smoke plumes that rise into the middle and upper troposphere would bring with them large quantities of water substance (relative to what naturally resides at those levels, but not necessarily in absolute values relative to surface moisture values), the moisture and mesoscale circulations associated with the smoke plumes would likely produce extensive stratiform cloud cover which could persist for periods of weeks or more in direct association with the smoke plume. The moderating influence of the fire-induced cloud would delay the onset of strong surface cooling during the phase that the smoke cloud is most concentrated. If the fire-induced cloud is sufficiently optically thick, it could reflect so much solar radiation that it would create a positive feedback to the smoke-induced cooling. Eventually much of the liquid and ice cloud will precipitate out of the atmosphere (faster than the sedimentation of the smoke). The question remains will the persistent cloud be present long enough to allow diffusion of the smoke concentrations to such small magnitudes that only small surface cooling will occur?

In the absence of any cloud responses, Haberle *et al.* (1985), for example, predicted that the solar heated smoke would be lofted into the stratosphere where it could reside for periods of months or more. We shall show that GCMs simulated similar responses in the absence of any cloud feedbacks. Such behavior of cloud lofting to higher levels, however, did not occur with the Kuwaiti oil fire plumes, even in clear skies, as a result of the strong thermodynamic stability in the midtroposphere. The presence of water clouds provides an additional interaction with the smoke. Clearly, an assessment of cloud feedbacks is important to understanding the ultimate climatic responses to large-scale nuclear warfare. We shall see that a better understanding of cloud feedbacks is important to evaluating any potential impacts of human behavior on climate.

10.2.6 Other mesoscale responses

Atmospheric responses to smoke and water substance injections as a consequence of nuclear warfare on scales of a few hundred kilometers to a few thousand kilometers (the mesoscale) is important in determining the ultimate fate of the the smoke and soot. Global climate models begin with a uniform layer of smoke introduced into the middle and upper troposphere and lower stratosphere. The actual smoke injections initially would be in discrete plumes originating from the fires over individual targets. Will the plumes remain in relatively narrow corridors and not become widely distributed as is assumed in the larger-scale models? Pielke and Uliasz (1993) have shown that dispersion is substantially enhanced due to spatial variations in surface heating as a result of temporal and spatial variations of turbulence and the generation of coherent mesoscale circulations. A similar response would be expected to occur associated with mesoscale patches of heating within the atmosphere. Even without mesoscale forcing, the results of McNider *et al.* (1988) suggest that vertical shear of the horizontal wind in the free atmosphere will significantly enhance dispersion from what is normally simulated in larger-scale models. If they do merge and diffuse into a relatively uniform pall, will this dispersion take so long that the smoke concentrations have diminished below levels for any significant solar heating responses?

It is possible that the consequences of such large-scale warfare would result in only mesoscale and regional responses with no longer-term global effects. One can speculate on a number of other potential mesoscale responses to smoke and water substance emissions. For example, patchy surface heating related to attenuation of solar energy by the smoke and cloud plumes can generate low-level, sea-breeze-like solenoidal circulations that can trigger deep convective clouds and rainfall. In fact the triggered clouds could not only produce severe weather but penetrate into the elevated plume, scavenging the smoke plumes that have spawned them.

The patchy surface heating can modify natural physiographically-driven circulations (i.e., coastal sea breezes, mountain slope flows, circulations driven by differential surface heating such as described in Part II), strengthening them in some cases, and weakening in others. The patchy surface heating can also alter weak synoptic-scale fronts causing those beneath the smoke/cloud cover to penetrate further equatorward during the daytime and possibly less equatorward at night.

These are just a few examples of potential mesoscale and regional scale responses to smoke and water substance injections by large-scale nuclear warfare.

10.2.7 Global climatic responses

The estimates of the global climatic responses to smoke injections during large-scale nuclear warfare are normally considered in two stages: an *acute stage* lasting a month or so and a *chronic stage* lasting several months or more. The global models used to calculate potential climatic impacts normally begin with the smoke being uniformly distributed through the Northern Hemisphere with a specified vertical distribution, and a concentration that varies with the particular war scenario. The first model calculations were done by TTAPS with a one-dimensional radiative–convective model, followed by the application of longitudinally averaged two-dimensional models (e.g., Cess *et al.*, 1985), and a number of GCMs of varying complexity.

The acute phase

The original TTAPS one-dimensional model calculations suggested that maximum summertime decreases in surface temperatures would be as large as 35 °C. The summer months are believed to experience the greatest cooling because too little radiation is present in the winter months to be appreciably diminished in intensity by the presence of the smoke. In addition, during summer the reduction of photosynthetically active radiation by smoke for a sufficiently long period of time below the threshold required by vegetation to sustain their metabolic processes

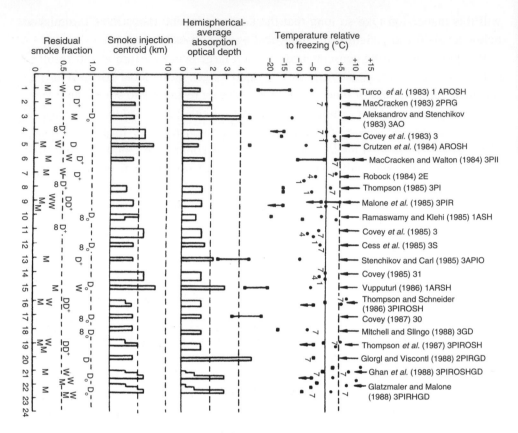

Figure 10.1 Summary of nuclear winter climate model calculations. Data shown for the following: (i) (•) Average land temperatures (coastal plus inland) in regions beneath widespread smoke layers for the coldest 1- to 2-week period in the simulation (some of the references report only temperature changes; the absolute temperatures have been deduced by subtracting the computed average temperature decrease from the temperature offset given below for each season); the month of the simulation is indicated by a numeral; for the one-dimensional radiative and convective models, the average land temperature decreases are taken as one-half of the "all-land" temperature decreases to account for the effect of ocean moderation; annual average solar insolation also applies in these cases. (ii) The box represents minimum land temperatures beneath smoke during the acute phase of nuclear winter simulations (again, where necessary, absolute temperatures were obtained by subtracting decreases from offsets); these temperatures are averaged over at least 1 day. (iii) Hemispheric average absorption optical depth of the smoke injection. (iv) Height centroid of the smoke mass injection. (v) Residual smoke fractions at several times in each simulation. The selected calculations roughly correspond to recommended baseline smoke injection scenarios; less severe and more severe cases have been investigated, but not as frequently as baseline cases. The studies have been ordered from left to right roughly in chronological sequence and are numbered along the bottom of the figure. For a given study, several cases may be illustrated (for example, the Ramaswamy and Kiehl (1985) results are shown for two smoke-injection profiles). The data are organized vertically for each simulation.

can result in the death of the plants. Sagan (1983) speculated that if the smoke concentrations exceeded a critical level, catastrophic changes in climate would occur. This speculation has not been supported by the more sophisticated model calculations. As more realistic physics was added to the models, the magnitude of predicted cooling diminished appreciably from the original TTAPS estimates. Some of the more important improvements included the allowance for smoke to be transported vertically and horizontally, the inclusion of vertical heat transports from ocean surfaces, and better estimates of scavenging and removal. Vertical transport of the heated smoke, for example, created a lofted stable plume of smoke that formed a very stable layer of air aloft much like the natural stratosphere but with its base much lower at ~ 5km (Malone *et al.*, 1986). The strength of that simulated response is only realistic if middle- and high-level clouds do not form in association with the lofted layer.

The inclusion of vertical heat transports from the ocean surface plays a major role in moderating the strength of surface cooling particularly in coastal areas (Schneider and Thompson, 1988). For the National Research Council baseline smoke injection scenario, they calculated maximum summertime, Northern Hemisphere, land surface temperature changes of $5-15\,°C$, or less than half the original TTAPS estimates. Some short-term sporadic cooling events were still evident in their single realization. Because of the moderation in their simulated responses to the introduction of the smoke, they referred to the phenomena as *nuclear fall* rather than nuclear winter. Figure 10.1 illustrates that the amplitude of predicted temperature changes for baseline war scenarios diminished appreciably over the seven short years of nuclear winter research.

Several of the models simulate dramatic changes in precipitation in tropical latitudes during the acute phase (Ghan *et al.*, 1988). Sharp drops in precipitation are associated with weakening of major tropical circulation features such as the Asian monsoon. The actual magnitudes of the reduction in rainfall should not

Caption for Figure 10.1 (cont.). The temperature offset used: $0\,°C$, winter; $13\,°C$ annual, fall, and spring; $25\,°C$, summer; and $35\,°C$ for cases calculated by Lawrence Livermore National Laboratories. The model treatments: 1, 2, and 3 indicate dimensions; A, annual solar insolation; P, patchy smoke injection; I, interactive transport; R, removal by precipitation; O, optical properties evolve; S, scattering included; H, infrared-active smoke; E, energy balance; G, ground heat capacity; D, diurnal variation; and M, mesoscale (48 hours). Smoke removal: D, after 1 day (prompt removal); D_0, arbitrary initial injection; W, after 1 week; M, after 1 month; $^+$, assumption implicit in smoke scenario adopted; and ∞, no smoke removal after injection. From Turco *et al.* (1990), wherein all source papers are referenced. © American Association for the Advancement of Science.

be taken too seriously, however, because quantitative rainfall simulations with GCMs are not particularly reliable.

The chronic phase

The chronic phase or effects of smoke injections by large-scale nuclear warfare on timescales greater than a month are particularly challenging to simulate. Reliable simulations require a coupled ocean model that can simulate changes in SSTs and associated vertical heat fluxes (e.g., Robock, 1984), as well as changes in meridional and zonal heat transports by ocean circulations. Interactions with sea ice formation and melting as well as changes in albedo of snow and ice fields caused by soot fallout should also be considered (e.g., Warren and Wiscombe, 1985; Ledley and Thompson, 1986; Vogelmann *et al.*, 1988). Moreover, cloud feedbacks are also important. Not only are the direct responses of clouds important, such as discussed above, but so are more subtle cloud interactions related to changes in land and sea surface temperatures, and changes in important general circulation patterns (i.e., monsoonal circulations, large-scale ridge/trough patterns). Fundamental to simulating long-term responses is the prediction of the horizontal and vertical distribution of smoke and its concentrations, and the removal of smoke by scavenging and sedimentation.

Long-term survival of smoke in the atmosphere requires that the smoke be injected into the upper troposphere and lower stratosphere where residence times can be on the order of several months to as long as a year or more, respectively. The creation of a very stable, upper troposphere in the heated, lofted smoke layer which Schneider and Thompson (1988) called a *smokeosphere* would greatly extend the survival of smoke into a chronic phase.

Overall, the models used thus far for simulating the longer-term effects of smoke injections by large-scale nuclear warfare are too crude to be considered reliable. We must, therefore, wait for a new generation of GCMs to be implemented, and for a rekindling of interest, to examine those potential consequences quantitatively.

10.2.8 Biological effects

A comprehensive assessment of the ecological and agricultural effects of nuclear war has been done by the SCOPE team (Harwell and Hutchinson, 1985). They concluded that agricultural systems, in particular, are very sensitive to even small changes in temperature, photosynthetically active radiation, and rainfall. They conclude that they are so sensitive that many of the unresolved climatic issues we have discussed above are less relevant, since even lower estimates of many effects (i.e., on temperatures, rainfall, and sunshine) could be devastating to agricultural production and thereby to human populations on regional and global scales.

In our opinion, this conclusion represents a rather naive perception of the level of uncertainty of climatic response estimates to large-scale nuclear war. There is little question that the level of uncertainty is very great indeed with respect to longer-term chronic impacts, which are important to biological effects. At this point one cannot be certain that the potential anomalies in temperature, precipitation, and sunshine triggered by smoke and soot released by large-scale nuclear war, for example, fall outside the envelope of expected *natural variability* of those parameters that agricultural systems must cope with and have coped with in the last century. There is little question that just a naturally poor growing season following a major war would compound the already stressed food distribution system due to disruption of transportation and limited availability of fuels, fertilizers, etc. needed for agricultural productivity. Nuclear fallout effects in such a poor growing season would further compound the problem. On the other hand, if a good growing season followed a devastating war, the possibly minor smoke-induced climatic effects would have little impact (although agriculture and the remainder of the biosphere, including people, would have to cope with the accumulation of a variety of harmful radionuclides on the Earth's surface).

Even during the acute phase, the uncertainty of smoke-induced impacts on weather and climate are so great that SCOPE's conclusion is a bit of an overstatement. In our opinion, the GCM experiments by Schneider and Thompson (1988) represent the most useful simulations for examining potential impacts of smoke-induced changes on agriculture and ecosystems. They emphasized the importance of considering geographical and weather variability rather than just time and spatial averages of temperature and other parameters. They noted, for example, that if temperatures fell below a critical threshold for only a few hours or so, crop production could be severely impacted (e.g., subfreezing temperatures for wheat, or temperatures as cool as 15 °C during the flowering phase of rice). They examined temperature extremes simulated with the NCAR Community Climate Model (CCM) by determining the coldest temperatures reached during a 30-day July control simulation and a smoke-perturbed, baseline war scenario. They found that the regions of subfreezing temperatures that are normally confined to polar latitudes and high mountains, expanded in the smoke-perturbed simulation. They also found considerable regional variability in the simulated responses to smoke. The smoke-perturbed case, for example, exhibited more frequent cooler temperatures over the midwestern United States, but no subfreezing temperatures. They reported some probability of subfreezing temperatures over the Ukraine and some temperatures over China that were low enough to impact rice growing in the smoke-perturbed cases.

Let us ignore, for the moment, the uncertainties in the GCM simulation of the climatic response to smoke perturbations due to the specification of the height and concentration of smoke, the dispersion of the smoke, the impact of clouds on scavenging of smoke and interference with smoke-induced radiative anomalies, and the fact that Schneider and Thompson's simulation did not include a diurnal cycle. Instead let us focus on the *natural variability* question again. It is important to determine where the role of smoke-induced perturbations compared to the *natural variability* that can be expected in July over a given region. Again if the war-produced smoke is to have any climatic or biological impacts of significance, it should produce anomalies greater than the expected *natural variability* over a region. This cannot be addressed from a single GCM realization. To actually define both natural and smoke-perturbed regional variability, a GCM must be run over a number of realizations each of which is initialized by a slightly perturbed initial state. A measure of the validity of the model would be to determine how well the model represents actually observed *natural variability*. Then one could examine the variability of the smoke-perturbed simulations relative to the *natural variability* to determine if the potential smoke-induced changes are substantially different from what could be expected naturally. Such an ensemble approach is now routinely used for climate change simulations.

As it is, these single realization GCM runs represent only a first step sensitivity experiment showing plausible physical responses to smoke. They should not be taken too literally with respect to biological impacts.

10.3 Summary of the status of the nuclear winter hypothesis

The nuclear winter hypothesis has been examined mainly with the use of models of varying complexities. There have been some attempts to examine the hypothesis relative to analogs such as meteor impacts and volcanic emissions, but both of these processes involve the deposition of large amounts of smoke and debris in the stratosphere where the expected residence times are considerably longer than in the troposphere. Measurements of smoke emissions from natural wildland fires, and from industrial fires and even Kuwaiti oil fires have also been made, but these are much smaller than anticipated for nuclear winter with the bulk of the smoke being confined to lower levels in the atmosphere than expected from large urban firestorms. As a result of the almost total reliance on models that have numerous shortcomings, the nuclear winter hypothesis is a long way from being proven scientifically viable. Hopefully, we will never have the opportunity to test the hypothesis experimentally.

One virtue of the hypothesis is that it has triggered many refinements in climate models, particularly in terms of the introduction of aerosol physics and refined radiation physics. It has also brought to the forefront a realization of the many potential ecological consequences of climate change and nuclear war.

It is clear that nuclear winter was largely politically motivated from the beginning, and it is an example of science being subverted to political ends. This is not the way that good science should be conducted.

11

Global effects of land-use/land-cover change and vegetation dynamics

11.1 Land-use/land-cover changes

Estimates of the Earth's landscape which have been disturbed from their natural state vary according to how the disturbance is defined. In terms of global cultivated land, Dudal (1987) indicates that $14.6 \times 10^6 \, \text{km}^2$ of a potential cultivated coverage of $30.31 \times 10^6 \, \text{km}^2$ are presently being utilized. Since the Earth's land surface covers $133.92 \times 10^6 \, \text{km}^2$, this indicates that 10.9% of the landscape is cultivated, with the potential level reaching 22.6% coverage.

This value of land disturbance due to human activities is an underestimate, however. Brasseur *et al.* (2005) report that up to half of the Earth's landscape has been directly altered. Human activities also include domestic grazing of semi-arid regions, urbanization, drainage of wetlands, and alterations in species composition due to the introduction of exotic trees and grasses. In the United States, for example, $426\,000 \, \text{km}^2$ (4.2% of the total land area) have been artificially drained (Richards, 1986).

In China, of the $2 \times 10^6 \, \text{km}^2$ in the temperate arid and semi-arid grassland regions, hundreds of thousands of square kilometers have been degraded due to overgrazing and the overextension of agriculture, often to the extent that desertification has occurred (Committee on Scholarly Communication with the People's Republic of China, 1992).

The influence of vegetation on climate includes its influence on albedo, water-holding capacity of the soil, stomatal resistance to water vapor transfer, aerodynamic roughness of the surface, and effect on snow cover. These effects are seasonally varying and are essential in GCM simulations of the effect of vegetation removal on climate change (Rind, 1984).

The sensitivity of global climate to even small changes in surface properties has been discussed in Pielke (1991). As described in that paper, an increase of the average albedo of the land surface of the Earth as viewed from space of 4% would result in a lowering of the Earth's equilibrium temperature by 2.4 °C. A decrease of this average albedo by 4% would increase the equilibrium temperature by 2.4 °C. This is on the same order as the estimates of a global *surface* equilibrium temperature change in GCM doubled carbon dioxide radiative–convective model calculations of 0.5 °C to around 4 °C.

The simple analysis presented in Pielke (1991) ignores, of course, how such a temperature change would be distributed with height or geographically. Nonetheless even if the change of equilibrium temperature is reduced by one-half as a result of these effects, the importance of the landscape energy budget to global climate is obvious.

There have been almost no global quantitative evaluations of changes in the heat and moisture fluxes at the surface due to human activity. One pioneering work that has been completed in this area is Otterman (1977) who estimated the change in the Earth's surface albedo over the last few thousand years due to overgrazing. He has concluded that overgrazing results in a higher albedo of the trampled, crumbled soil than in the original steppe where there was dark plant debris accumulating on a crusted soil surface (also see Chapter 6 for a discussion of overgrazing). Otterman estimated that the current Earth's surface albedo is 0.154 whereas before anthropogenically caused grazing (6000 years before the present), this albedo value was 0.141. This albedo change could have resulted in an average cooling effect of 0.77 °C in the Earth's equilibrium temperature. Hartmann (1984) suggested that a shift of precipitation between land and ocean areas at low latitude can also lead to planetary albedo changes as a result of alterations in vegetation and through changes in cloud cover distribution resulting from these changes in land surface albedo. He concluded that this albedo change could result in an approximately 5 °C global mean temperature change without any other feedbacks. Gordon *et al.* (2005) have concluded that deforestation and irrigation have significantly altered atmospheric water vapor flows on a global scale.

11.2 Historical land-use change

A documentation of global patterns of land-use change from 1700 to 2000 is presented in Klein Goldewijk (2001). Klein Goldewijk reports on worldwide changes of land to crops of 136 Mha, 412 Mha, and 658 Mha in the periods 1700-1799, 1800-1899, and 1900-1990, respectively. Conversion to pasture was 418 Mha, 1013 Mha, and 1496 Mha in these three time periods. Figure 11.1 illustrates these changes, including an acceleration of tropical deforestation during the twentieth century. O'Brien (2000) also documents land-use change for recent years (Table 11.1).

(a)   (c)

(b)   (d)

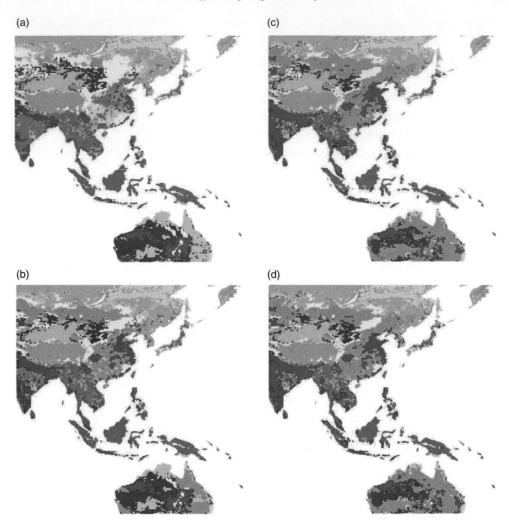

Figure 11.1 Examples of land-use change from (a) 1700, (b) 1900, (c) 1970, and (d) 1990. The human-disturbed landscape includes intensive cropland (red), and marginal cropland used for grazing (pink). Other landscape includes, for example, tropical evergreen and deciduous forest (dark green), savanna (light green), grassland and steppe (yellow), open shrubland (maroon), temperate deciduous forest (blue), temperate needleleaf evergreen forest (light yellow), and hot desert (orange). Of particular importance is the expansion of the cropland and grazed land between 1700 and 1900. Data obtained from the Hyde Database available at www.mnp.nl/hyde/. Reproduced with permission from Kees Klein Goldewijk. See also color plate.

Apart from their role as reservoirs, sinks, and sources of carbon, tropical forests provide numerous additional ecosystem services. Many of these ecosystem services directly or indirectly influence climate. The climate-related ecosystem

Table 11.1 *Tropical forest extent and loss (rain forest and moist deciduous forest ecosystems)*

Country	Rain forest		Moist deciduous forest	
	Extent in 1990 (kha)	Decrease 1981–90 (%)	Extent in 1990 (kha)	Decrease 1981–90 (%)
Brazil	291 597	6	197 082	16
Indonesia	93 827	20	3 366	20
Democratic Republic of the Congo (Zaire)	60 437	12	45 209	12
Columbia	47 455	6	4 101	38
Peru	40 358	6	12 299	6
Papua New Guinea	29 323	36	705	6
Venezuela	19 602	14	15 465	36
Malaysia	16 339	36	0	0
Myanmar	12 094	24	10 427	28
Guyana	11 671	0	5 078	6
Suriname	9 042	0	5 726	4
India	8 246	12	7 042	10
Cameroon	8 021	8	9 892	12
French Guiana	7 993	0	3	0
Congo	7 667	4	12 198	4
Ecuador	7 150	34	1 669	34
People's Democratic Republic of Laos	3 960	18	4 542	18
Philippines	3 728	62	1 413	54
Thailand	3 082	66	5 232	54
Vietnam	2 894	28	3 382	28
Guatemala	2 542	32	731	0
Mexico	2 441	20	11 110	30
Belize	1 741	0	238	0
Cambodia	1 689	20	3 610	20
Gabon	1 155	12	17 080	12
Central African Republic	616	14	28 357	8
Cuba	114	18	1 247	18
Bolivia	0	0	35 582	22

Source: World Resources Institute (1994), adapted from O'Brien (2000); from Pielke *et al.* (2002).

services that tropical forests provide include the maintenance of elevated soil moisture and surface air humidity, reduced sunlight penetration, weaker near-surface winds, and the inhibition of anaerobic soil conditions. Such an environment maintains the productivity of tropical ecosystems (Betts, 1999) and has helped sustain the rich biodiversity of tropical forests.

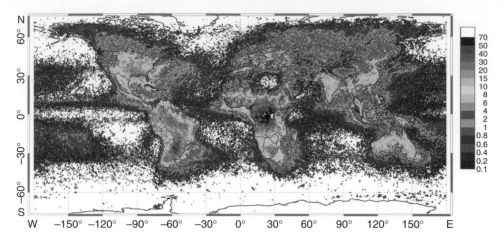

Figure 11.2 Global distribution of lightning from April 1995 through February 2003 from the combined observations of the NASA Optical Transient Detector (OTD) (4/95-3/00) and Lightning Imaging Sensor (LIS) (1/98-2/03) instruments. From http://thunder.nsstc.nasa.gov/images/HRFC_AnnualFlashRate_cap. jpg. See also color plate.

11.3 Global perspective

The effect of well above average ocean temperatures in the eastern and central Pacific Ocean, which is referred to as El Niño, has been shown to have a major effect on weather, thousands of kilometers from this region (Shabbar *et al.*, 1997). The presence of the warm ocean surface conditions permits thunderstorms to occur there that would not have happened with the average colder ocean surface. These thunderstorms export vast amounts of heat, moisture, and kinetic energy to the middle and higher latitudes, particularly in the winter hemisphere. This transfer alters the ridge and trough pattern associated with the polar jet stream (Hou, 1998). This transfer of heat, moisture, and kinetic energy is referred to as "teleconnection" (Namias, 1978; Wallace and Gutzler, 1981; Glantz *et al.*, 1991). Almost two-thirds of the global precipitation occurs associated with mesoscale cumulonimbus and stratiform cloud systems located equatorward of 30° (Keenan *et al.*, 1994). In addition, much of the world's lightning occurs over tropical land masses, with maxima also over the midlatitude land masses in the warm seasons (Lyons, 1999; Rosenfeld, 2000) (Fig. 11.2). These tropical regions are also undergoing rapid landscape change as reported in Section 11.2.

As shown in the pioneering study by Riehl and Malkus (1958), and Riehl and Simpson (1979), 1500 to 5000 thunderstorms (which they refer to as "hot towers") are the conduit to transport this heat, moisture, and wind energy to higher latitudes. Since thunderstorms only occur in a relatively small percentage of the area of the tropics, *a change in their spatial patterns would be expected to have global climate consequences.*

Wu and Newell (1998) concluded that SST variations in the tropical eastern Pacific Ocean have three unique properties that allow this region to influence the atmosphere effectively: large magnitude, long persistence, and spatial coherence. Since land-use change has the same three attributes, a similar teleconnection would be expected with respect to landscape patterns in the tropics. Dirmeyer and Shukla (1996), for example, found that doubling the size of deserts in a GCM model caused alterations in the polar jet stream pattern over northern Europe. Kleidon *et al.* (2000) ran a GCM with a "desert world" and a "green planet" in order to investigate the maximum effect of landscape change. However, these experiments, while useful, do not represent the actual effect of realistic anthropogenic land-use change. Actual documented land-use changes are reported, for example, in Baron *et al.* (1998), Giambelluca *et al.* (1999), Leemans (1999), and O'Brien (1997, 2000). Giambelluca *et al.*, for example, report albedo increases in the dry season of from 0.01 to 0.04 due to deforestation over northern Thailand.

Figure 11.3 illustrates how precipitation patterns in the tropics are altered in Southeast Asia and adjacent regions in a GCM. Two 10-year simulations were performed: one with the current global seasonally varying leaf area index and one with the potential seasonally varying leaf area index, as estimated by Nemani

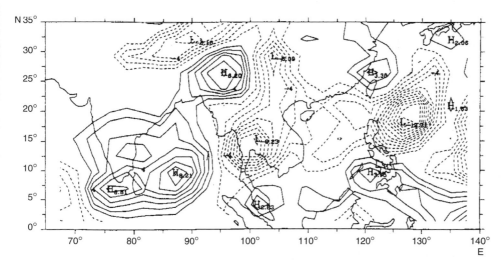

Figure 11.3 Illustration of how precipitation patterns in the tropics are altered in Southeast Asia and adjacent regions in a GCM where two 10-year simulations were performed: one with the current global LAI and one with the potential LAI, as estimated by Nemani *et al.*, (1996). From Chase *et al.* (1996). © 1996 American Geophysical Union. Reproduced by permission of the American Geophysical Union.

et al. (1996). No other landscape attributes were changed. The figure presents the 10-year averaged difference in precipitation for the month of July for the two GCM sensitivity experiments, which illustrates major pattern shifts in precipitation. As with El Niño, this alteration in tropical thunderstorm patterning teleconnects to higher latitudes (Fig. 11.4), where the 10-year averaged 500 mb heights for July are presented; the 10-year averaged 500 mb heights are also shown for January.

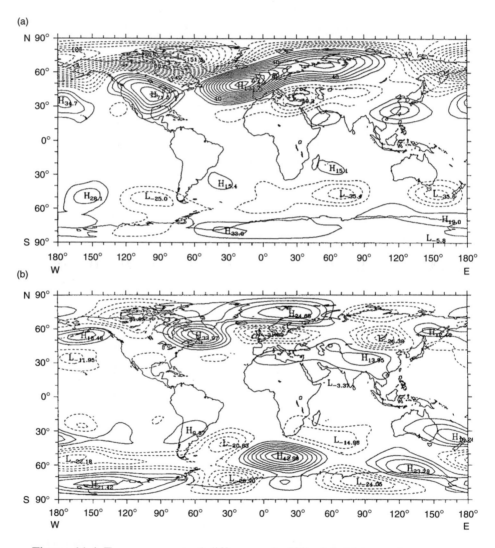

Figure 11.4 Ten-year averaged differences in 500 mb heights zonal wavenumbers 1–6 only (actual LAI minus potential LAI). (a) January, contour 10 m, and (b) July, contour 5 m. From Chase *et al.* (1996). © 1996 American Geophysical Union. Reproduced by permission of the American Geophysical Union.

The GCM produced a major, persistent change in the trough/ridge pattern of the polar jet stream, most pronounced in the winter hemisphere, which is a direct result of the land-use change. Unlike an El Niño, however, where cool ocean temperatures return so that the El Niño effect can be clearly seen in the synoptic weather data, the landscape change is permanent. Figure 11.5 shows how the 10-year averaged surface air temperatures changed globally in this model experiment (Chase *et al.*, 1996).

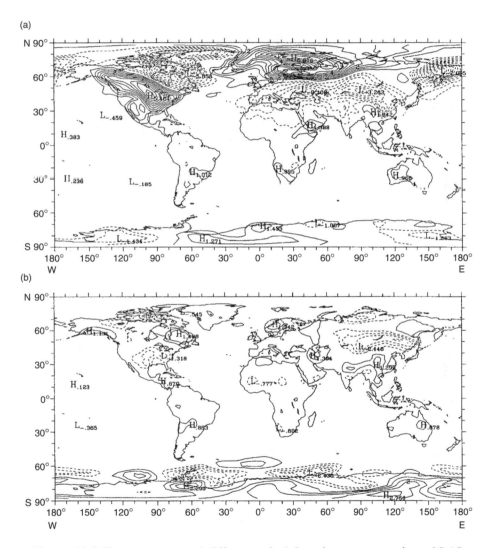

Figure 11.5 Ten-year averaged differences in 1.5 m air temperature (actual LAI minus potential LAI) in K for (a) January, and (b) July. Contour interval is 0.5 K. From Chase *et al.* (1996). ©1996 American Geophysical Union. Reproduced by permission of the American Geophysical Union.

That landscape change in the tropics affects cumulus convection and long-term precipitation averages should not be a surprising result. For example, as reported in Pielke *et al.* (1999b), using identical observed meteorology for lateral boundary conditions, the Regional Atmospheric Modeling System was integrated for July–August 1973 for south Florida. Three experiments were performed: one using the observed 1973 landscape, another the 1993 landscape, and the third the 1900 landscape, when the region was close to its natural state. Over the 2-month period, there was a 9% decrease in rainfall averaged over south Florida with the 1973 landscape and an 11% decrease with the 1993 landscape, as compared with the model results when the 1900 landscape is used. A follow-on study further confirmed these results (Marshall *et al.*, 2004a). The limited available observations of trends in summer rainfall over this region are consistent with these trends.

Chase *et al.* (2000) completed more general landscape change experiments using the CCM3 from the NCAR. In this experiment, two 10-year simulations were performed using current landscape estimates and the potential natural landscape estimate under current climate. In addition to LAI differences, albedo, fractional vegetation coverage, and aerodynamic roughness differences were included. While the amplitude of the effect of land-use change on the atmospheric response was less than when the CCM2 GCM model was used, substantial alterations of the trough/ridge polar jet stream still resulted. Figures 11.6, 11.7, and 11.8 show the January 10-year averaged cumulus convective precipitation, 200 mb height, and near-surface temperature differences between the two experiments. Despite the difference between the experiments with CCM2 and CCM3, both experiments produce a wavenumber three change pattern in the polar jet stream. Pitman and Zhao (2000) and Zhao *et al.* (2001a) have recently performed similar GCM experiments that have provided confirmation of the Chase *et al.* (1996, 2000) results.

Many other studies support the result that there is a significant effect on the large-scale climate due to land-surface processes (see Table 11.2).

Zeng *et al.* (1998) found, for example, that the root distribution influences the latent heat flux over tropical land. Kleidon and Heimann (2000) determined that deep-rooted vegetation must be adequately represented in order to realistically represent the tropical climate system. Dirmeyer and Zeng (1999) concluded that evaporation from the soil surface accounts for a majority of water vapor fluxes from the surface for all but the most heavily forested areas, where transpiration dominates. Recycled water vapor from evaporation and transpiration is also a major component of the continental precipitation. Brubaker *et al.* (1993) found that local contributions of water vapor to precipitation generally lie between 10% to 30%, but can be as high as 40%. Eltahir and Bras (1994a) concluded that there is 25–35% recycling of precipitation water in the Amazon. Dirmeyer (1999)

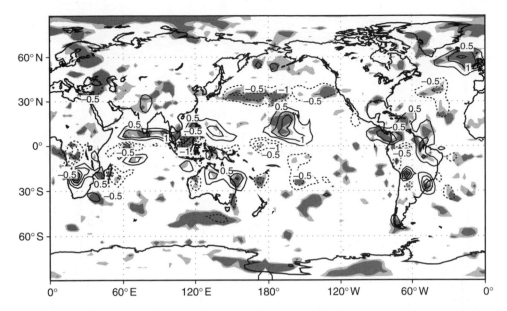

Figure 11.6 Convective precipitation differences (current minus natural, contour by 0.5 mm day^{-1}) using a nine-point spatial filter for easier visibility. Light shading represents the 90% significance level for a one-sided *t*-test. Dark shading represents the 95% significance level. Reproduced from Chase *et al.* (2000) with kind permission of Springer Science and Business Media.

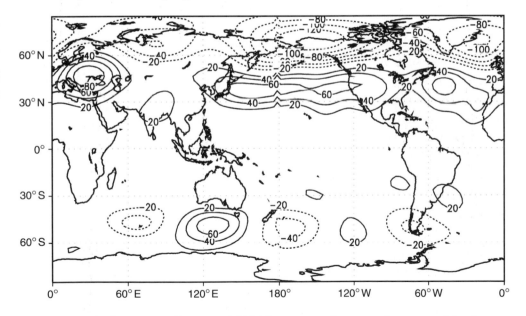

Figure 11.7 Height difference of 200 hPa pressure (current minus natural landscapes). Contour interval is 20 m. Reproduced from Chase *et al.* (2000) with kind permission of Springer Science and Business Media.

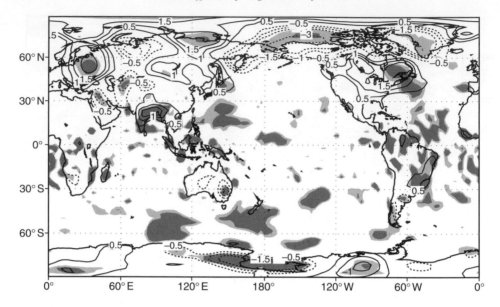

Figure 11.8 Difference in near-surface air temperature (current minus natural landscape) using a nine-point spatial filter for easier visibility. Contour intervals are 0.5, 1.0, 1.5, and 3.0 °C. Light shading represents the 90% significance level for a one-sided *t*-test. Dark shading represents the 95% significance level. Reproduced from Chase *et al.* (2000) with kind permission of Springer Science and Business Media.

concluded that interannual variations of soil wetness are large enough to influence GCM climate simulations.

Deliberate land-use change (afforestation or reforestation) has been accepted as a mechanism to remove carbon dioxide from the atmosphere and sequester carbon in trees and soils. However, as discussed by Betts (2000) and Pielke (2001b), this activity may have unintended consequences with respect to radiative forcing in the atmosphere which again illustrates the complexity of the climate system. For example, in regions subject to significant snow cover, afforestation would result in a lower surface albedo and hence a positive radiative forcing, resulting in a net warming effect despite the removal of carbon dioxide from the atmosphere (Fig. 11.9). Similarly, increases in the surface fluxes of water vapor could result in positive radiative forcing.

The biogeochemical effect of enhanced carbon dioxide and trace gas concentration, and of aerosol deposition (such as nitrogen), on landscape dynamics has also not been adequately considered. For example, Jenkinson *et al.* (1991) demonstrated a significant positive feedback where soils released carbon to the atmosphere under warming conditions. Lenton (2000), using a simple box model, and Cox *et al.* (2000) using GCM sensitivity experiments, showed that biogeochemical

Table 11.2 *A summary list of global land-use/land-cover assessments: effect on weather and climate*

Alpert *et al.* (2005), Avissar and Werth (2005), Baidya Roy and Avissar (2002), Betts (1999, 2001),[a] Betts *et al.* (1997), Bonan *et al.* (1995), Boucher *et al.* (2004), Bounoua *et al.* (2000, 2002),[a] Broström *et al.* (1998), Brovkin *et al.* (1999, 2002, 2004, 2005),[a] Brubaker and Entekhabi (1995), Burke *et al.* (2000), Chase *et al.* (1996, 2000, 2001), Claussen (1995, 1998, 2001a,b), Claussen *et al.* (1998, 1999, 2001, 2002, 2004, 2005), Collatz *et al.* (1991), Costa and Foley (2000), Couzin (1999), Cox *et al.* (2000), DeFries *et al.* (2002), deNoblet-Ducoudre *et al.* (2000), Dirmeyer (1994, 1998), Dirmeyer and Shukla (1996), Eltahir (1996), Entekhabi and Brubaker (1995), Entekhabi *et al.* (1992), Feddema *et al.* (2005a,b), Ferranti and Molteni (1999), Foley *et al.* (1994, 1998, 2003, 2005), Fraedrich *et al.* (1999), Franchito and Rao (1992),[a] Friedlingstein *et al.* (2001), Ganopolski *et al.* (1998), Gedney and Valdes (2000), Govindasamy *et al.* (2001), Guillevic *et al.* (2002), Hoffmann and Jackson (2000), Idso *et al.* (1975), Keith *et al.* (2004), Kleidon *et al.* (2000), Koster and Suarez (2004),[a] Koster *et al.* (2000, 2004),[a] Kubatzki and Claussen (1998), Lamptey *et al.* (2005), Levis *et al.* (1999, 2000), Liu *et al.* (2006), Lofgren (1995a,b), Loveland *et al.* (2000), Lyons *et al.* (1996), Matthews *et al.* (2004), McGuffie *et al.* (1995), Milly and Dunne (1994), Myhre and Myhre (2003), Pielke Sr. (2005), Pielke *et al.* (2002), Pitman and Zhao (2000), Pitman *et al.* (1999), Polcher and Laval (1994a,b), Porporato *et al.* (2000), Ramankutty and Foley (1999), Ramankutty *et al.* (2006), Ramírez and Senareth, (2000), Rodriguez-Iturbe *et al.* (1991a,b), Ruddiman (2003), Snyder *et al.* (2004b), Sud *et al.* (1993, 1995, 1996), Texier *et al.* (2000), Turner *et al.* (1990), van den Hurk *et al.* (2003),[a] Vitousek *et al.* (1997), Werth and Avissar (2002, 2005), Williams (2003), Xue (1996, 1997), Xue and Fennessy (2002), Xue and Shukla (1993, 1996), Xue *et al.* (1996), Zeng and Neelin (1999), Zeng *et al.* (1999, 2002), Zhang *et al.* (1996a,b, 2001), Zhao and Pitman (2002a,b), Zhao *et al.* (2001a,b)[a]

[a] Information was provided by Dr. Rick Raddatz.

feedbacks in conjunction with an increased carbon dioxide radiative warming produced an amplified regional and global surface temperature response. As discussed in Chapter 6, Eastman *et al.* (2001a,b) used a regional climate model in a sensitivity study and suggested a cooler daytime and warmer nighttime in the central Great Plains in response to greater plant growth in a doubled carbon dioxide atmosphere.

The presence of drought and hydrological feedbacks associated with land-use change locally or through teleconnections, therefore, has a direct impact on the source–sink capabilities of the terrestrial ecosystem and therefore on climate.

An important conclusion from these studies is that land-use change directly alters local and regional heat and moisture fluxes in two ways. Firstly, the local and regional CAPE is changed since the Bowen ratio is changed as the surface heat and moisture budgets are altered. Secondly, larger-scale heat and moisture convergence, and associated large-scale wind circulations can be changed as a result of changes in the large-scale atmospheric pressure field due to the landscape change.

As a general conclusion, *these regional and global model studies indicate that the spatial patterning of land-use/land-cover change result in changes over time in regional tropospheric diabatic heating patterns. These changes in the patterns result in alterations in global circulation patterns. Indeed, it appears that such heterogeneous diabatic heating changes on the regional scale have more of an influence on the global climate scale than the more spatially homogeneous climate forcing of the radiative effect of increasing CO_2.*

11.4 Quantifying land-use/land-cover forcing of climate

Carbon has become the currency used to assess the human intervention in the Earth's climate system. Impacts on climate are compared in terms of radiative forcing, which can be considered as perturbations to the Earth's radiation budget prior to feedbacks from the rest of the climate system. The concept of global warming potential (GWP) where

$$\text{GWP} = \frac{\int_0^{TH} a_x x(t) \mathrm{d}t}{\int_0^{TH} a_r r(t) \mathrm{d}t},\qquad(11.1)$$

has been adopted to convert other atmospheric constituents into their equivalent in terms of carbon dioxide atmospheric radiative forcing (Houghton *et al.*, 2001). Here, *TH* is the time period over which the calculation is considered, a_x is the radiative efficiency due to a unit change in atmospheric abundance of the substance x (i.e., $\text{W m}^{-2} \text{kg}^{-1}$), and $x(t)$ is the time-dependent decay of the abundance from an instantaneous release of the substance. The denominator is the same expression but for the reference substance r, defined to be carbon dioxide.

The effects of land-surface albedo change can be quantified in terms of radiative forcing (Hansen *et al.*, 1997a; Betts, 2001), and this has been used in attempts to compare the global significance of historical land-use change with that of other drivers of climate change (Houghton *et al.*, 2001). Betts (2000) suggested that radiative forcing calculations could be used to translate albedo changes into equivalent carbon emissions (Fig. 11.9); this could be useful for quantifying land-use changes in regions where the main impact is on surface albedo, such as areas subject to significant snow cover. However, in other regions, changes in other land-surface properties may not exert a radiative forcing but still significantly influence climate. For example, the partitioning of available energy into latent and sensible heat fluxes exerts a direct impact on near-surface air temperature, so a change in this partitioning should be considered a climate forcing. Radiative forcing, and hence GWP, are therefore not able to represent the full impact

(a) Radiative forcing due to carbon sequestration (nW m^{-2} ha^{-1})

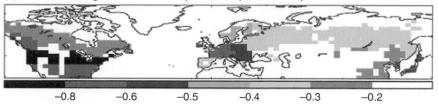

$\quad\quad -0.8 \quad\quad -0.6 \quad\quad -0.5 \quad\quad -0.4 \quad\quad -0.3 \quad\quad -0.2$

(b) Radiative forcing due to albedo change (nW m^{-2} ha^{-1})

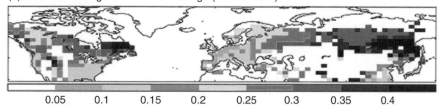

$\quad\quad 0.05 \quad\quad 0.1 \quad\quad 0.15 \quad\quad 0.2 \quad\quad 0.25 \quad\quad 0.3 \quad\quad 0.35 \quad\quad 0.4$

(c) Carbon emissions equivalent to albedo change (t C ha^{-1})

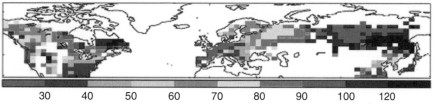

$\quad\quad 30 \quad\quad 40 \quad\quad 50 \quad\quad 60 \quad\quad 70 \quad\quad 80 \quad\quad 90 \quad\quad 100 \quad\quad 120$

(d) Net radiative forcing by afforestation (nW m^{-2} ha^{-1})

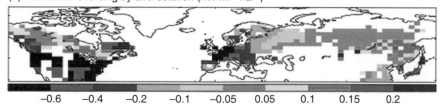

$\quad -0.6 \quad -0.4 \quad -0.2 \quad -0.1 \quad -0.05 \quad 0.05 \quad 0.1 \quad 0.15 \quad 0.2$

Figure 11.9 Radiative forcing of climate by afforestation, considering illustrative 1-ha plantations in the temperate and boreal forest zones. Calculations apply to the time at the end of one forestry rotation period, relative to the start of the rotation period with plantation areas unforested. (a) Global mean longwave radiative forcing due to carbon dioxide removal through sequestration (n W m^{-2} ha^{-1}). (b) Global mean shortwave radiative forcing due to albedo reduction (n W m^{-2} ha^{-1}). (c) Carbon emissions that would give the same magnitude of radiative forcing as the albedo reduction (t C ha^{-1}). (d) Net radiative forcing due to afforestation, found by summing (a) and (b) (n W m^{-2} ha^{-1}). Positive forcing implies a warming influence; where (d) shows positive values, afforestation would warm climate rather than cooling it as would be expected by considering carbon sequestration alone. After Betts (2000); from Pielke *et al.* (2002). See also color plate.

of land-cover change in all regions. Some new means of quantifying land-use forcing is therefore required. Separation of the components of the surface energy budget could provide a possible starting point, with the surface heat energy being separated into

$$Q_{\mathrm{N}} + Q_{\mathrm{H}} + Q_{\mathrm{LE}} + Q_{\mathrm{G}} = 0 \tag{11.2}$$

where

$$Q_{\mathrm{N}} = Q_{\mathrm{S}}(1 - A) + Q_{\mathrm{LW}}^{\downarrow} Q_{\mathrm{LW}}^{\uparrow}. \tag{11.3}$$

Here, Q_{N} is the net radiative flux, Q_{H} is the turbulent sensible heat flux, Q_{LE} is the turbulent latent heat flux (evaporation/transpiration), Q_{G} is the heat flux into the Earth's surface, Q_{S} is the solar irradiance, A is the surface albedo, $Q_{\mathrm{LW}}^{\downarrow}$ is the downward atmospheric irradiance, and $Q_{\mathrm{LW}}^{\uparrow}$ is the upward surface irradiance. The magnitude of these fluxes can be expressed in units of watts per meter squared or joules per unit of time (for example, a globally averaged value of $1\,\mathrm{W\,m^{-2}}$ is equal to $1.61\,10^{23}\,\mathrm{J}$ per decade; $1\,\mathrm{W} = 1\,\mathrm{J\,s^{-1}}$). One measure of land-cover change forcing of climate could be the perturbation to one of the components of the surface energy balance equation (Eq. 11.2) prior to feedbacks from the rest of the climate system.

In the past, climate change metrics have been concerned with globally averaged responses as exemplified in the GWP. However, as discussed previously in this chapter, global-scale climate changes can also occur due to regional land-use changes but where global averaged values are unchanged. This occurred in the situation described by Chase *et al.* (1996, 2000) where there was no significant global averaged change in these quantities. Figure 11.10 illustrates the anthropogenically caused change in LAI (see Nemani *et al.*, 1996), which were used in the Chase *et al.* studies. It was the spatial redistribution of the land surface sensible and latent heat flux pattern that resulted in the global climate change.

Globally averaged climate change may therefore bear no well-defined relation to the real changes experienced in any region and these regional changes, which can be of any sign, are what impact people and will stimulate mitigation strategies to be applied (Matsui and Pielke, 2006). Therefore, we identify a surface "regional climate change potential" (RCCP) that addresses this deficiency. The RCCP would be defined to quantify where a direct human-caused change (either positive or negative) alters any of the individual terms in Eq. (11.2). These changes could be scaled by the surface area of the Earth to place them in a global context.

To provide a land-use forcing term free of feedbacks, as in the case of radiative forcing, a land-surface scheme could be used to calculate perturbations to the surface energy budget excluding feedbacks from the atmosphere. However, assessing the true global significance of these feedbacks is not trivial, since atmospheric circulation changes could give rise to remote climate changes that do not relate

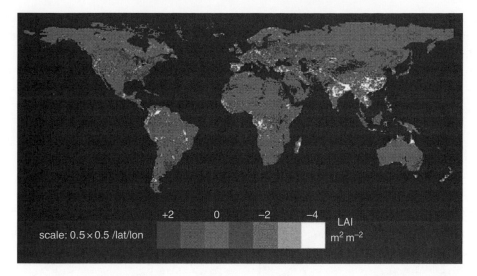

Figure 11.10 Effect of land-use changes on plant canopy density (potential LAI − actual LAI). Scale 0.5 latitude × 0.5 longitude. From Pielke *et al.* (2002), based on Nemani *et al.* (1996). See also color plate.

linearly to the perturbation in the region of land-use change (e.g., Gedney and Valdes, 2000). To illustrate this point, the model results from Chase *et al.* (2000) are used to show the 10-year average absolute-value differences in the surface energy flux for January and July (i.e., Q_{LE} and Q_H from Eq. 11.2) between the climate with the current landscape and the climate with potential landscape that should exist under current atmospheric conditions without human intervention. Figure 11.11 presents the 10-year average value difference in the surface turbulent heat fluxes between the two experiments for just the locations where the human-caused landscape change was prescribed in the model, while Fig. 11.12 presents the 10-year average difference values in the surface turbulent heat fluxes worldwide.

The 10-year average change for the locations where only the land cover was altered were 5.3 W m^{-2} and 4.7 W m^{-2} in January, and 8.8 W m^{-2} and 6.7 W m^{-2} in July for surface sensible heat flux and latent heat flux, respectively. The globally weighted changes in the sum of the absolute values of the surface sensible and latent heat flux changes for Fig. 11.11 were 0.7 W m^{-2} in January and 1.08 W m^{-2} in July. With teleconnections changes also included, however, the globally averaged changes in the surface fluxes were significantly larger with values of 9.47 W m^{-2} in January and 8.90 W m^{-2} in July (Fig. 11.12). These results clearly demonstrate that regional landscape change can result in an amplified perturbation to surface fluxes elsewhere in the world through non-linear feedbacks within the atmosphere's global circulation. If the other terms in the

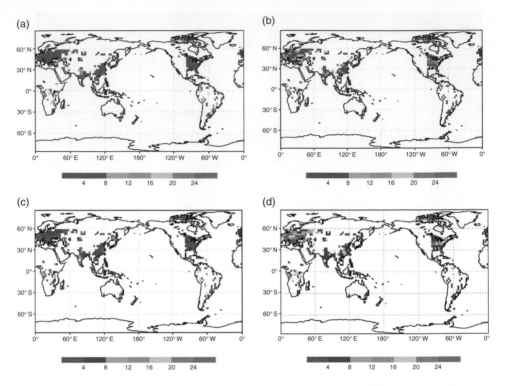

Figure 11.11 The 10-year average absolute value change in (a) January surface latent turbulent heat flux, (b) July surface latent turbulent heat flux, (c) January surface sensible heat flux, and (d) July surface sensible heat flux in W m^{-2} at the locations where land-use change occurred. Based on Chase *et al.* (2000); from Pielke *et al.* (2002). See also color plate.

surface heat budget (Eq. 11.2) were included, the magnitude of the differences (i.e., the RCCP) would presumably be even larger.

A potential based purely on surface flux perturbations would not be analogous to GWP, because the latter is an index requiring an expression in terms of the abundance of atmospheric constituents. Direct comparison of land-cover change effects with greenhouse gas emissions therefore remains a challenge. One possibility might be to dispense with quantification in terms of forcings, and compare anthropogenic influences by modeling whole-system responses to individual causes of climate change.

For example, Claussen *et al.* (2001) compared biogeophysical and biogeochemical effects of land-cover change in terms of the overall climate change (as opposed to the forcing). A number of simulations were performed with an intermediate complexity Earth system model, in each run altering a small part of the land surface and simulating its effects on global mean temperature (through changes in

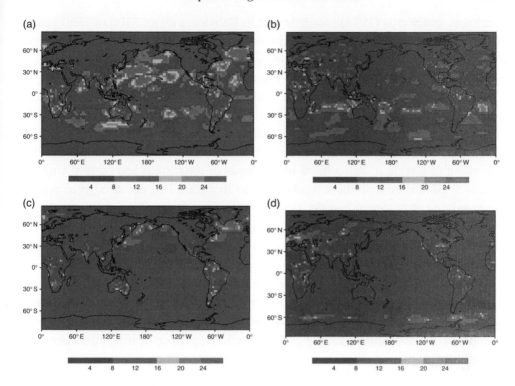

Figure 11.12 The 10-year average absolute value change in surface sensible and latent turbulent heat flux in $W m^{-2}$ worldwide as a result of the land-use changes. (a) January surface latent turbulent heat flux, (b) July surface latent turbulent heat flux, (c) January sensible turbulent heat flux, and (d) July sensible turbulent heat flux. Based on Chase *et al.* (2000); from Pielke *et al.* (2002). See also color plate.

surface properties and carbon exchange). However, as stated above, global mean temperature changes conceal important local and regional changes, so applications of this type of modeling approach again require a global quantification of regional changes.

11.5 Atmosphere–vegetation interactions

The atmosphere and biosphere are coupled, dynamic systems. On the long time-scales, the major vegetation biomes of the Earth propagate in response to climate change. During the Pleistocene, for example, spruce and fir forests covered what is now northern Florida, as a vast continental ice sheet covered much of northeastern North America. In the eastern plains of Colorado, large sand dunes developed about 3000 to 1500 years before the present (BP) at a time of greater aridity than exists today (Muhs, 1985). Human agricultural systems are also affected.

Neumann and Sigrist (1978) conclude that in Babylonia about 3800 to 3650 years BP the barley harvest was 10–20 days earlier than it is now due to a warmer climate, while cooler weather from the years 2600 to 2400 years BP resulted in a harvest 10–20 days later than under the current climate in this region.

On shorter timescales, Schwartz and Karl (1990) report that the seasonal upward progression of temperature in the eastern United States is temporarily halted as vegetation greens in the spring. At least a 3.5°C reduction in surface daily maximum temperature occurred over the subsequent 2-week period after the beginning of leafing in agricultural inland areas of the central and eastern United States, as contrasted with the 2 weeks prior to first leaf. This occurred even when the overlying atmosphere had the same lower tropospheric average temperature. On a hemispheric scale, seasonal pulsations of carbon dioxide concentrations occur as vegetation extracts carbon dioxide in spring and summer in the vast land areas of the Northern Hemisphere.

On daily timescales, vegetation interacts with the atmosphere through its direct influence on the partitioning between latent and sensible heat fluxes, as discussed in Chapter 6. During the day, over regions of transpiring vegetation, huge fluxes of moisture to the atmosphere and a significant reduced surface heating effect occurs. At night the stomata on the leaves close and respiration of oxygen dominates over photosynthesis and the uptake of carbon dioxide by plants.

The importance of biological interactions with the atmosphere and as a component of the climate system is only now becoming recognized. The ability of vegetation to effectively utilize increased carbon dioxide to increase global biomass may result in reduced carbon dioxide warming from what otherwise would occur. We need to determine, however, the impact of deforestation and conversion of vegetated land to asphalt, roads, industrial centers, etc. on this potential storage mechanism for carbon. Also, how are methane and nitrogen oxide inputs to the atmosphere from industrial and agricultural practices influenced by the presence of plants? Does enhanced nitrogen, for example, permit enriched fertilization for plant growth?

Several non-radiative forcings involve the biological components of the climate system. They can be categorized into three types.

(1) *Biophysical forcing* involves changes in the fluxes of trace gases and heat between vegetation, soils, and the atmosphere. For example, a GCM experiment by Sellers *et al.* (1997) showed that, in the presence of increased carbon dioxide, plants open their stomata less and are therefore more water efficient. Thus, increased carbon dioxide impacts the hydrological cycle, in addition to its well-known direct radiative impacts.

(2) *Biogeochemical forcing* involves changes in vegetation biomass and soils. For example, increased nitrogen deposition caused by greater anthropogenic emissions of

NH_3, NO, and NO_2 is a biogeochemical forcing of the climate system (Galloway *et al.*; 2003, Holland *et al.*, 2005). This deposition has altered the functioning of soil, terrestrial vegetation, and aquatic ecosystems worldwide. Galloway *et al.* (2004) document that human activities increasingly dominate the nitrogen budget at the global scale, and that fixed forms of nitrogen are accumulating in most environmental reservoirs. In addition to impacts on ecosystem functioning, which are important in themselves, this forcing modifies physical components of the climate system, such as surface albedo and sensible and latent heat.

(3) *Biogeographic forcing* involves alterations in plant species composition. Such changes can occur slowly in response to changes in the weather over time, or suddenly due to fires or other disturbances. For example, greater shrub growth in the high latitudes of the Northern Hemisphere has been observed (McFadden *et al.*, 2001), which could alter the spatial distribution of drifting snow and subsequent melt pattern and timing (Liston *et al.*, 2002).

Complex interactions among these forcings make it difficult to determine their net climate effects (Eastman *et al.*, 2001a; Narisma *et al.*, 2003; Raddatz, 2003). Eastman *et al.* (2001a), for example, found that with doubled carbon dioxide the grasslands of the central United States were more water efficient on an individual stoma level (biophysical forcing), but grew more biomass (biogeochemical forcing). The net effect was cooler daytime temperatures during the growing season.

There are no widely accepted metrics for quantifying regional non-radiative forcing. Indeed, because non-radiative forcings affect multiple different climate variables, there is no single metric that can be applied to characterize all non-radiative forcings (Marland *et al.*, 2003; Kabat *et al.*, 2004). Non-radiative forcings generally have significant regional variation, making it important that any new metrics be able to characterize the regional structure in forcing and climate response, whether the response occurs in the region, in a distant region through teleconnections, or globally. As is the case for regional radiative forcing, further work is needed to quantify links between regional non-radiative forcing and climate response. Another consideration in devising metrics for non-radiative forcings is enabling direct comparison with radiative forcings, computed in units of watts per meter squared. However, not all non-radiative forcings are easily quantified in these units.

A metric that could prove useful for quantifying impacts on the hydrological cycle is changes in surface sensible and latent turbulent heat fluxes. In Section 11.4, the concept of a surface RCCP was introduced, which is calculated by summing and weighting globally the absolute values of changes in the surface sensible and latent turbulent heat fluxes. Extending this concept to the global water cycle, Pielke and Chase (2003) quantified landscape forcing in terms of precipitation and moisture flux changes. They found globally averaged differences between the current and the natural landscape of 1.2 mm day^{-1} for precipitation and 0.6 mm day^{-1} for moisture flux. However, such metrics do not provide a complete measure of the integrated effect on the climate system due to the regional concentration of changes in diabatic forcing, although it does provide a global average context for the effect of landscape change on the hydrologic cycle.

11.6 The abrupt desertification of the Sahara

Paleoclimatic reconstructions suggest that during the Holocene climate optimum (9000-6000 years ago), North Africa was wetter and the Sahara was much smaller than today (Prentice *et al.*, 2000). Annual grasses and shrubs covered the desert, and the Sahel reached as far as 23° N (Claussen *et al.*, 1999), over 500 km north of its present location. During the Holocene optimum a slightly increased tilt of the Earth's spin axis and perihelion in July led to stronger insolation of the Northern Hemisphere during summer thereby strengthening the North African summer monsoon (Kutzbach and Guetter, 1986). However, the North African climate is sensitive to changes in the land surface's albedo, which can result from vegetation removal. Charney and Stone (1975) recognized that high albedo resulting from vegetation removal can enhance desert expansion by reducing rainfall, which further reduces vegetation, in a strong, desert-expanding positive biogeophysical feedback. This mechanism offers a possible explanation for past climate changes in the Sahara and particularly for increased drought in the Sahel and its southward migration in late Holocene. Actually, when using present-day land cover as initial condition, models based solely on atmospheric processes do not yield an increase in precipitation large enough to lead to a substantial reduction in the Sahara 6000 years ago (Joussaume *et al.*, 1999). However, when feedbacks between atmosphere and vegetation are incorporated, the models simulate a vegetation distribution in good agreement with paleobotanic reconstructions (Claussen and Gayler, 1997; deNoblet-Ducoudre *et al.*, 2000; Doherty *et al.*, 2000). The interpretation is that precessional forcing led to an enhancement of the African monsoon, creating conditions that were then amplified mainly by atmosphere–vegetation feedbacks, and to a lesser extent, by atmosphere–ocean interaction (Ganopolski *et al.*, 1998; Braconnot *et al.*, 1999). These lead to multiple equilibrium states (Claussen, 1997) with the possibility of abrupt changes when thresholds are crossed (Brovkin *et al.*, 1998), as shown in Fig. 11.13 (modified from Claussen *et al.*, 1999 and deMenocal *et al.*, 2000). This figure shows a model simulation of an abrupt decline in precipitation in the Sahara (20° N – 30° N and 15° W – 50° E) around 5500 years ago that is

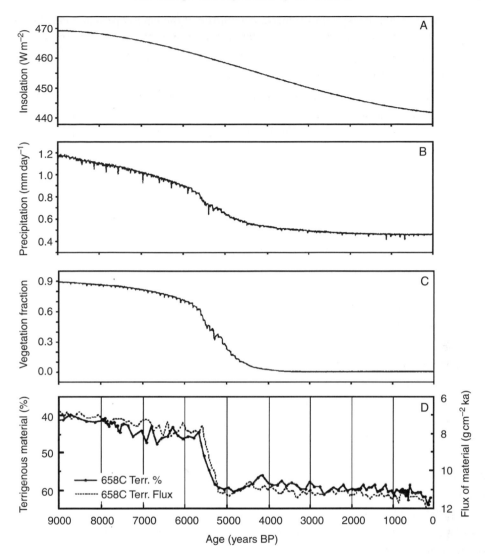

Figure 11.13 Simulation of transient development of precipitation (B) and vegetation fraction in the Sahara (C) as response to changes in insolation (A depicts insolation changes on average over the Northern Hemisphere during boreal summer). Results from Claussen et al. (1999) (B,C) are compared with data of terrigenous material and estimated flux of material in North Atlantic cores off the North African coast (D) by deMenocal *et al.* (2000). Reproduced from Kabat *et al.* (2004) with kind permission of Springer Science and Business Media.

supported by observations from sediment cores off the North African coast. The rapid change contrasts markedly with the slow decrease in insolation. Alterations in the African monsoon due to such abrupt desertification would be expected to have global atmospheric circulation (and therefore) global climate consequences.

Hayden, B. P., 1998. Ecosystem feedbacks on climate at the landscape scale. *Phil Trans. R.Y. Soc. B*, **353**, 5–18.
Pielke, R. A. Sr., 2005. Land use and climate change. *Science*, **310**, 1625–1626.

Additional reading

A summary of papers that present evidence of global climate impacts from land-use change and vegetation dynamics is given in Table 11.2.

Epilogue

Throughout this book, a number of common themes reappear. Some of these common themes were apparent to us before we began preparing the first edition of the book, and in fact, they were a major motivation for us to write the book. Others became more and more apparent as we performed the literature survey and analysis, and prepared the written text. These issues have become even better defined in the years since our first edition. Some of these common themes are:

- the importance and underappreciation of *natural variability*,
- the dangers of overselling,
- the capricious administration of science,
- scientific credibility and advocacy, and
- politics and science.

While these subjects have been discussed in depth by others including science policy experts (Pielke Jr., 2007), let us briefly overview each of these common themes. A very effective summary of several of these issue are presented at http://sciencepolicy.colorado.edu/prometheus/archives/climate_change/index.html

E.1 The importance and underappreciation of *natural variability*

We have seen that our ability of determining if human activity demonstrably *causes* some *observed or hypothesized effect*, such as changes in local rainfall or global climate, is strongly dependent upon the *natural variability* of the system. This has certainly been true of determining if cloud seeding actually is capable of causing increases of rainfall in specified target areas. However, it is also true in assessing the role of such human climate forcings as anthropogenic greenhouse gas emissions, or deforestation, or release of CCN on global climate. While the time and space scales are very different, nonetheless the bottom line in examining potential human-caused effects is: *are these effects large enough in magnitude to be extricated from the "noise" of the natural variability of the system?* As we have seen there are few, if any, cases in which we can answer this question affirmatively on the global scale.

Ice cores have shown, for example, that a switch from an ice age climate to a non-ice age environment can occur over only a few decades (La Brecque, 1989), without human intervention (e.g., see Fig. E.1), while more recently the hydrology of a region can suddenly switch (e.g., Figs. E.2 and E.3). Rial *et al.* (2004) present other examples of sudden climatic shifts on a variety of timescales.

The so-called "climate jump" of 1976/77 is another example of a sudden climate regime shift which is seen in many datasets (e.g., Changnon *et al.*, 1991; Seidel *et al.*, 2004). No climate model has been able to predict or even adequately explain these rapid transitions, which would likely have a much greater impact on society than gradual climate shifts, in which adaptation might be less difficult. Rial *et al.* (2004) present additional examples demonstrating that the Earth's climate system is highly non-linear, that inputs and outputs are not proportional (change is often episodic and abrupt, rather then slow and gradual), and that multiple equilibria are the norm. This limits the value of global models as forecast tools. Peters *et al.* (2004) show that spatial non-linearities also result in critical environmental thresholds, such that what may be an appropriate mitigation response for one spatial scale of a disturbance may be inappropriate for a different spatial scale.

E.2 The dangers of overselling

We have seen that funding of the science of weather modification underwent a period of rapid rise, followed by an abrupt crash. One of the leading causes of that crash, we believe, is that the program was oversold. The claims that only a few more years of research and development will lead to a scientifically proven technology that will contribute substantially to water management and severe weather abatement were either great exaggerations, or just false. This is largely because we greatly underestimated the complexity of the scientific and technological problems we were (and still are) faced with. This tendency to oversell can even be taken to the absurd such as the "pork-barrel" funding of the University of Alaska in Fairbanks of more than $57 million over several years through the efforts of Alaskan Senator Ted Stevens (Cohen, 1991). This funding has as one of its objectives the harnessing of the Aurora Borealis for electric power generation – a concept that has no scientific basis.

The same can be said about human impacts on global climate. There are many scientists who are claiming that the short-term (periods of year to year, or decades) variations in weather and climate are clear evidence that we are experiencing the effects of anthropogenic greenhouse emissions as given by the

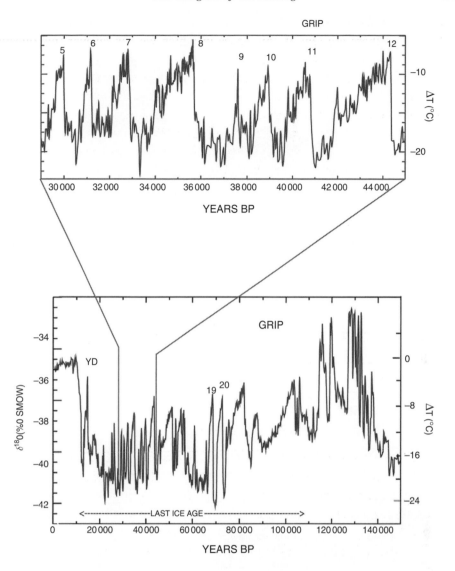

Figure E.1 Perhaps the most puzzling feature of recent paleoclimate records, highly relevant to understanding future global climate change, is the fast-warming/slow-cooling sequence found in the stable isotope fluctuations ($\delta^{18}O$) time series of Greenland's ice cores known as the Dansgaard–Oeschger (D/O) oscillations (Alley *et al.*, 1999). SMOW, Standard Mean Ocean Water. The D/O typically show very sudden, $6-10\,°C$ warming episodes lasting a few centuries or perhaps even a few decades, followed by millennia of relatively slow cooling. Remarkably, reconstructed SSTs in the tropical Atlantic mimic the D/O record in the 30 ka to 60 ka interval, and similar recordings are found in the subtropical Pacific and tropical Indian oceans. The longest period of the signal in the inset is a submultiple of the precession forcing and evidence of precession forcing exists elsewhere in the record. The ordinals near selected peaks correspond to numbered interstadials and YD is the Younger Dryas event. From Rial *et al.* (2004). Reprinted with kind permission of Springer Science and Business Media.

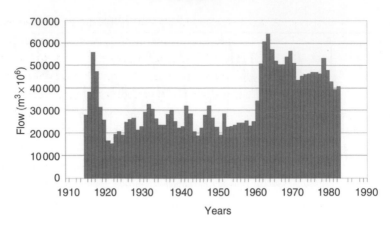

Figure E.2 Time series of annual outflows from the African equatorial lakes measured at the Mongalla station on the River Nile for the period 1915–83, showing an abrupt shift around 1961 and a slow decaying downward trend. Adapted from Salas *et al.* (1981), from Rial *et al.* (2004). Reprinted with kind permission of Springer Science and Business Media.

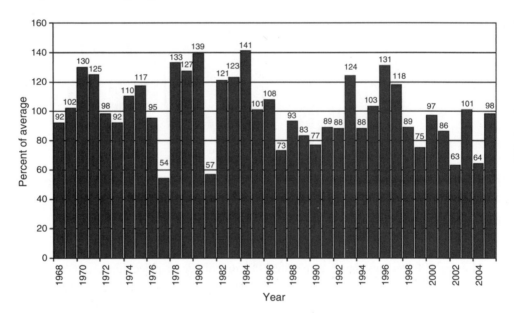

Figure E.3 Snowpack percent of average for Colorado by year from 1968 through present. Data available from NRCS Snow Survey Division (www.co.nrcs.usda.gov/snow/snow/).

example below. Moreover, many claim that the "forecasts" being made by global climate models represent realistic expectations of global averaged and regional changes in temperature and rainfall in the next decade or century. In our opinion, both of these claims represent overselling of the climate program.

These claims, as listed in the first edition of our book, appear and are discussed in both the professional literature (e.g., Schneider, 1990; Titus, 1990a,b; Intergovernmental Panel on Climate Change, 1991; Kellogg, 1991) and in the lay press (e.g., Brooks, 1989; Schneider, 1989; Thatcher, 1990; Bello, 1991; Luoma, 1991; UCAR/NOAA, 1991). These claims continue today in the professional literature (e.g., Barnett *et al.*, 2001) and lay press (e.g., Erickson, 2005).

As an example of such extreme claims to mitigate anthropogenically caused global warming, a National Academy of Sciences (1991) report considered the insertion of 50 000 100-km^2 mirrors in space to reflect incoming sunlight. Such gross global climate engineering represents a close analog to the exaggerated claims in weather modification that were made in the 1960s and 1970s. Short-term variations of weather and climate are clearly within the *natural variability* of climate to the extent that we can realistically assess it. Moreover, the models are not really "forecast" models (Pielke, 2002b). They are simply research models designed to simulate the responses of hypothesized anthropogenic changes to weather and climate, *other things being the same*. These model simulations should more appropriately be referred to as *process studies*, not *predictions*. With their many limitations in their description of climate forcings and feedbacks (National Research Council, 2003b, 2005), they are not capable of predicting climatic change. We simply do not know enough about all the processes of importance to climatic change to include them in any quantitative forecast system (Pielke, 1998, 2001a,b; National Research Council, 2005).

What it amounts to is that many scientists are grossly underestimating the complexity of interactions among the Earth's atmosphere, ocean, geosphere, and biosphere, and overstating the accuracy of the climate models to predict the future climates (e.g., see discussion by Pielke *et al.*, 2003; Pielke, 2004; Rial *et al.*, 2004). These problems are so complex that it may take many decades, or even centuries (if ever!), before we have matured enough as a scientific community to make *credible predictions* of long-term climate trends and their corresponding regional impacts. Widmann and Tett (2003) conclude, for example, that "even a perfect model with all forcings included will simulate only one of many possible climates consistent with the forcings." We may even find that, even by themselves, the uncertainty level of those predictions due to outside (the Earth) influences and volcanic eruptions are so large that those predictions are not useful for social planning.

E.3 The capricious administration of science and technology

In the United States, as well as many developed countries, science and technology are often poorly administered. As we have seen in weather modification, administration of many programs is fragmented among a number of basic and mission-oriented agencies, all of which compete for funding at national and state levels. This competition amongst the agencies often leads to the greatly exaggerated claims that many of the scientific and technological issues will be solved in the next 5 to 10 years.

In addition, because many of these agencies are mission-oriented, their job is to examine the impacts of human-induced changes on weather and climate on energy, air quality, water resources, or agriculture. Their job is not to advance the fundamental scientific issues regarding the behavior of the Earth system, but to get on with the business of evaluating the impacts of anthropogenic activity on their programs. As a result, they are often looking for short cuts to bottom line answers that can probably only be obtained through meticulous, often time-consuming scientific research.

Moreover, national governing bodies (legislatures, presidents, etc.) all work on timescales of 2, 4, or 6 years, and want to be able to identify impacts of their programs on timescales of their tenure. If significant progress is not made on those timescales, then often funding in those programs is reduced, if not curtailed, and new, competing programs are brought to the forefront. This results in short-sighted funding in science and technology in which programs are begun and before they reach maturity they are curtailed, then the rush is on to get on the bandwagon with the latest fad.

As we have discussed in this book, however, the scientific and technological problems associated with furthering our understanding of human impacts and *natural variability* and feedbacks of weather and climate are so complicated and multifaceted that many of the issues will not be resolved on timescales of decades or possibly centuries. *Thus, programs associated with the investigation of human impacts on weather and climate require sustained, stable national funding at a high level.* A view supporting this idea has been recommended in the Policy Statement of the American Meteorological Society on Global Climate Change (1991), which was produced by a group of climate scientists in the fall of 1990 who questioned the overselling and short-sighted perspective of current climate change government policy. In 2005, those recommendations were still not implemented.

E.4 Scientific credibility and advocacy

In this book, we have examined human impacts on weather and climate ranging from purposeful attempts at changing weather through cloud seeding, to inadvertent modification of weather and climate on regional or mesoscales, to human impacts on global climate. We have seen that the evidence indicates that human activity has significant impacts on weather and climate over a broad range of space and timescales. Furthermore, as long as the population of the human species continues to rise, the prospects for causing major changes in weather and climate becomes increasingly likely.

However, using modeling to assess regional or global climate processes is a distinct task from applying these models to skillfully predict the future climate. Claiming that models can accurately predict climate years and decades into the future (where we cannot yet test their predictive skill), and using this claim for the promotion of specific policy positions *is advocacy*.

With that in mind we ask, *should scientists be actively involved in advocating that we apply cloud seeding techniques to enhancing rainfall, or reducing emissions of greenhouse gases to alleviate greenhouse warming?* Pielke Jr. (2007) discusses in depth the very different functions of being an *honest broker*, who presents the spectrum of scientific understanding and suite of policy options, to being an *advocate* who cherrypicks from this information to promote a particular subset of scientific results and policy options.

Certainly the scientists are the best informed with regard to the consequences of human activity and, one could say, that if the informed scientist does not take an advocacy role in recommending that action be taken, then no one else will. Such a position is not without its dangers, however. For example, in the 1970s we worked for the former Experimental Meteorology Laboratory of NOAA in Miami, Florida, which was involved in developing and testing dynamic cloud seeding techniques to enhance rainfall. During that period, the State of Florida experienced a severe drought and asked the laboratory director to assist the State by directing a rainfall enhancement project. The director asked her three assistant administrators, one of which was Cotton, whether we should respond to the State's request by providing them operational cloud seeding support. I (Cotton) was the lone dissenter and argued that we were scientists who were involved in the objective assessment of the impacts of dynamic cloud seeding. As a result our participation in an operational program and obvious role as advocates of applying dynamic seeding would jeopardize our credibility as truly objective scientists and therefore adversely affect both the program and the individual scientists. The same can be said with regard to advocates of major disruptive societal changes with regard to greenhouse emissions.

Some might argue that the risk of losing one's scientific credibility is purely a personal one and must be weighed against the potential societal gains by taking immediate action to relieve drought or reduce greenhouse warming. In fact, the adverse impacts extend far beyond those affecting the individual scientist. Loss of scientific credibility is infectious and can, therefore, propagate through an entire scientific discipline and even to the scientific community as a whole. The fall of the science of weather modification by cloud seeding was almost certainly due, in part, to a loss of scientific credibility. The global climate change community must likewise be careful that a loss of scientific credibility does not propagate through their discipline, or the discipline of atmospheric science as well. Thus premature advocacy that a particular policy action be taken now, could, in the long run, destroy the prospects for obtaining solid scientific evidence on the role that human activity has on weather and climate.

E.5 Should society wait for hard scientific evidence?

Since we wrote our first edition, the Kyoto Climate Change Treaty has been signed, and on February 16, 2005, ratified by enough countries to put it into effect. This implementation occurred even though it is expensive and, if you accept the climate models used to rationalize the treaty, has hardly any effect on climate! It is a political placebo made to assure society that scientists are doing something on climate change. Cloud seeding provides an analogy to the rush for policy actions without firmly and convincingly establishing the scientific evidence. There are only a few limited examples of where cloud seeding has been scientifically shown to be effective in enhancing rainfall. Nonetheless, many nations are currently running operational cloud seeding projects.

Apparently, the decision has been made in those nations and states that the benefits outweigh the risks of applying the scientifically unproven technology of weather modification by cloud seeding. The major risks are the possibility of creating severe weather or floods, and to increasing rainfall in one local region at the expense of rainfall in a neighboring local region. Often the decision to apply cloud seeding technology in a particular country or state is a prescription of a *political placebo* or a decision that it is better to do something than to sit idly by and do nothing as reservoirs dry up and crops wither and die due to the absence of water. This is clearly where we are with respect to policies on applying changes to energy policy to influence climate.

More appropriately, however, the concerns on global climate change due to human activities should be recognized as a symptom of population growth. *One does not need strong scientific evidence that human activity is causing global climate change to recognize that a stable population on Earth will have long-term benefits; common sense is all that is needed!* Population growth is a problem that produces societal problems that are much more severe than any of the scenarios proposed to occur as a result of greenhouse gas warming. Catastrophic social upheavals are likely to result as the human density continues to increase and we see this today in the precarious state of a number of nations (Economist, 2005).

Population growth clearly increases our vulnerability to climate variability and change, as well as other environmental changes (e.g., see Vörösmarty *et al.*,

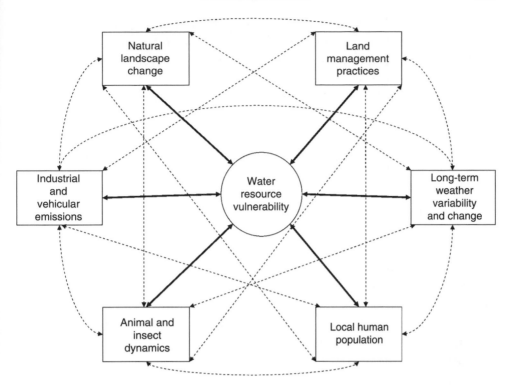

Figure E.4 Schematic of the relation of water resource vulnerability to the spectrum of environmental forcings and feedbacks. The arrows denote nonlinear interactions between and within natural and human forcings. Adapted from Pielke (2004), reprinted by permission of the International Geosphere–Biosphere Programme.

2000; Pielke, 2001). A focus on reducing vulnerability has been proposed to reduce this threat (Pielke, 2004; Pielke and Bravo de Guenni, 2004; Steffen *et al.*, 2004). Lomborg (2001) has discussed the relative risks and benefits of different environmental threats and finds climate change as a lower priority issue than other environmental threats. Such a focus, as illustrated schematically in Fig. E.4, should be adapted in lieu of focusing on prediction.

E.6 Politics and science

It is unfortunate but science and policy have become inextricably linked (see the insightful discussions of this subject in Pielke Jr. (2007), Pielke and Rayner (2004), with respect to Lomborg (2001)). This is nothing new as scientists like Galileo worked under the most extreme political pressures. But today science is often driven by political advocacy. Unfortunately, since the tenure of politicians is short, the science issues of interest during a given political era falls out of favor during a new era. The timescales for these political cycles is short being 4 to 8 years or so, yet to accomplish hard scientific results often takes timescales of decades or more. As a result, many fundamental issues never get fully resolved.

We have certainly seen this in weather modification where the scientific basis of hail suppression and rainfall enhancement still remains unresolved, due in part to lack of funding of research in those areas for over two decades. As the western United States has experienced a drought for the last 4 or 5 years cloud seeding operations have flourished and calls for renewed support in weather modification research by national committees like National Research Council (2003a) have come forth. But a few years of average or above-average precipitation and the push for renewed support of weather modification will wane, yet the fundamental scientific roadblocks will remain with us.

We have seen that scientists have become political activists pushing an agenda of research in nuclear winter and global warming. In their enthusiasm to push forward their political agendas using glossy pseudoscientific journals, newspaper articles, talk shows on television, and even Hollywood movies, uncertainties in the science are often glossed over and those scientists that question the science are labeled skeptics or, alternatively, contrarians. The term skeptic is often then associated with anti-environmentalists. An important component of the scientific method, debate of the fundamental issues, is thereby squelched.

The more vocal and respected of these scientists push their agenda with international organizations such as the International Panel on Climate Change where they seek the endorsement of a consensus of scientists. To quote Michael Crichton (2003), "Let's be clear: the work of science has nothing whatever to do with consensus. Consensus is the business of politics. Science, on the contrary, requires only one investigator who happens to be right, which means that he or she has results that are verifiable by reference to the real world. In science consensus is irrelevant. What is relevant is reproducible results. The greatest scientists in history are great precisely because they broke with the consensus. There is no such thing as consensus science. If it's consensus, it isn't science. If it's science, it isn't consensus. Period."

E.7 Conclusions

There are several conclusions we have reached with respect to the human influence on weather and climate.

(1) Humans clearly modify the weather and climate on the local scale through our influence on atmospheric composition and the landscape. The question is whether the human disturbance of the regional and global climate can be extracted from the large

natural variability that occurs on these scales. This book provides a discussion of the existing state of this assessment.

(2) Climate prediction is a much more difficult challenge than presented in the Intergovernmental Panel on Climate Change and US National Assessment reports. Despite claims to the contrary, skillful predictions (which are usually referred to as "projections"; MacCracken, 2002) years into the future have not been shown. The policy statement by the American Association of State Climatologists (2005) adopts this view. The publication of papers which use model predictions of the climate decades into the future as their scientific basis should be avoided (as there is no way to scientifically test their skill). We doubt, for instance, anyone would publish papers based on a weather forecast for a time period which had not yet occurred!

(3) The term "global warming" has become so politicized in its meaning so as to be worthless to advance the scientific debate of human-caused climate change. This term is applied to mean an increase in the globally averaged surface temperature in response to an increase in the concentration of well-mixed greenhouse gases, particularly carbon dioxide. The actual increase in the Earth's heat must occur in the oceans (Pielke, 2003).

(4) The main effect of humans on the global climate system, as it impacts society, is not from a globally averaged increase in surface temperature, or even a change in ocean heat content. One of the most important effects is if humans alter atmospheric and ocean circulation patterns, such as the position of the polar jet stream, and how the troughs and ridges of pressure are established, and if ocean conveyor belts are modified. Pielke *et al.* (2002) and Feddema *et al.* (2005b) illustrates how land-use change can accomplish large circulation changes, but without a change in the globally averaged surface temperature.

This perspective of focusing on global atmospheric circulation changes is recognized by the World Climate Research Programme (GEWEX) in the main objective of their Coordinate Enhanced Observing Period (CEOP) initiative: "To understand and model the influence of continental hydroclimate processes on the predictability of global atmospheric circulation and changes in water resources, with particular focus on the heat source and sink regions that drive and modify the climate system and anomalies." (from: Scientific Plan for the Coordinated Enhanced Observing Period (CEOP)– An Overview from a GEWEX Hydrometeorology Panel Perspective, available at: http://www.gewex.org/ceop/execsum.pdf).

The National Research Council (2005) also recognizes this major role of climate forcings in altering the large-scale atmospheric circulation. This perspective, however, is not widely adopted in the climate change community.

Additional readings

The following papers and books below are just a sample of the diversity of views on the issues presented in the Epilogue. These publications include discussions of the difference of scientific opinion on our ability to understand and, which is a separate question, our ability to predict future regional and global climate.

Foley, J. A., R. DeFries, G. P. Asner, *et al.*, 2005. Global consequences of land use. *Science*, **309**, 570–574.

Khandekar, M. L., 2004. Are climate model projections reliable enough for climate policy? *Energy Environ.*, **15**, 521–526.

Khandekar, M. L., T. S. Murty, and P. Chittibabu, 2005. The global warming debate: A review of the state of science. *Pure Appl. Geophys.*, **162**, 1557–1586.

Kondrat'yev, K. Ya., 2004. Key aspects of global climate change. *Energy Environ.*, **15**, 469–503.

Lomborg, B., 2001. *The Skeptical Environmentalist: Measuring the Real State of the World.* Cambridge, UK: Cambridge University Press.

MacCracken, M., 2002. Do the uncertainty ranges in the IPCC and US National Assessments account adequately for possibly overlooked climatic influences. *Climat. Change*, **52**, 13–23.

MacCracken, M., J. Smith, and A. C. Janetos, 2004. Reliable regional climate model not yet on horizon. *Nature*, **429**, 699.

Michaels, P. J., 2004. *Meltdown: The Predictable Distortion of Global Warming by Scientists, Politicians, and the Media.* Washington, DC: Cato Institute.

Pielke, R. A. Jr., 2007. *The Honest Broker: Making Sense of Science in Policy and Politics.* Cambridge, UK: Cambridge University Press.

Pielke, R. A. Sr., 2002b. Overlooked issues in the US National Climate and IPCC assessments. *Climat. Change*, **52**, 1–11.

Pielke, R. A. Jr., and S. Rayner (eds.), 2004. Science, policy and politics: Learning from controversy over *The Skeptical Environmentalist. Environ. Sci. Policy*, **7**, 355–433.

Sarewitz, D., R. A. Pielke Jr., and R. Byerly (eds.), 2000. *Prediction: Science Decision Making and the Future of Nature.* Covelo, CA: Island Press.

Steffen, W., A. Sanderson, P. D. Tyson, *et al.*, 2004. *Global Change and the Earth System: A Planet under Pressure.* Heidelberg, Germany: Springer-Verlag.

Weart, S. R., 2004. *The Discovery of Global Warming.* Cambridge, MA: Harvard University Press.

References

Ackerman, A. S., O. B. Toon, and P. V. Hobbs, 1995. Numerical modeling of ship tracks produced by injections of cloud condensation nuclei into marine stratiform clouds. *J. Geophys. Res.*, **100**, 7121–7133.

Ackerman, A. S., O. B. Toon, D. E. Stevens, *et al.*, 2000a. Reduction of tropical cloudiness by soot. *Science*, **288**, 1042–1047.

Ackerman, A. S., O. B. Toon, J. P. Taylor, D. W. Johnson, P. V Hobbs, and R. J. Ferek, 2000b. Evaluation of Twomey's parameterization of cloud susceptibility using measurements of ship tracks. *J. Atmos. Sci.*, **57**, 2684–2695.

Ackerman, T. P., K. -N. Liou, and C. B. Leovy, 1976. Infrared radiative transfer in polluted atmospheres. *J. Appl. Meteor.,* **15**, 28–35.

Adegoke, J. O., 2000. Satellite based investigation of land surface-climate interactions in the United States Midwest. Ph.D. dissertation, Department of Geography, The Pennsylvania State University, University Park.

Adegoke, J. O., R. A. Pielke Sr., J. Eastman, R. Mahmood, and K. G. Hubbard, 2003. Impact of irrigation on midsummer surface fluxes and temperature under dry synoptic conditions: A regional atmospheric model study of the US High Plains. *Mon. Wea. Rev.*, **131**, 556–564.

Adegoke, J. O., R. A. Pielke Sr., and A. M. Carleton, 2006. Observational and modeling studies of the impacts of agriculture-related land use change on climate in the central U.S. *Agric. Forest Meteor., Special Issue*, in press.

Akbari, H., A. Rosenfeld, and H. Taha, 1989. *Recent Developments in Heat Island Studies: Technical and Policy*. Berkeley, CA: Lawrence Berkeley Laboratory Technical Report.

Albrecht, B. A., 1989. Aerosols, cloud microphysics, and fractional cloudiness. *Science*, **245**, 1227–1230.

Alig, R. J., J. D. Kline, and M. Lichtenstein, 2004: Urbanization on the US landscape: Looking ahead in the 21st century. *Landsc. Urb. Plann.*, **69**, 219–234.

Alley, R. B., P. U. Clark, L. D. Keiwin, and R. S. Webb, 1999. Making sense of millennial scale climate change. In *Mechanisms of Global Climate Change at Millennial Time Scales*, AGU Geophysical Monograph No. 112 P. U. Clark, R. S. Webb, and L. D. Keigwin, eds., Washington, DC, American Geophysical Union, pp. 385–394.

Alpert P., P. Kishcha, Y. J. Kaufman, and R. Schwarzbard, 2005. Global dimming or local dimming?: Effect of urbanization on sunlight availability. *Geophys. Res. Lett.*, **32**, L17802, doi:10.1029/2005GL023320.

American Association of State Climatologists, 2005. *Policy Statement.* Available online at www.ncdc.noaa.gov/oa/aasc/aascclimatepolicy.pdf

American Meteorological Society, 1991. Policy Statement of the American Meteorological Society on Global Climate Change. *Bull. Am. Meteor. Soc.*, **72**, 57–59.

Amiro, B. D., J. I. MacPherson, and R. L. Desjardins, 1999. BOREAS flight measurements of forest-fire effects on carbon dioxide and energy fluxes. *Agric. Forest Meteor.*, **96**, 199–208.

Anthes, R. A., 1984. Enhancement of convective precipitation by mesoscale variations in vegetative covering in semiarid regions. *J. Climate Appl. Meteor.*, **23**, 541–554.

Arakawa, A., and W. H. Schubert, 1974. Interaction of a cumulus cloud ensemble with the large-scale environment. Part I. *J. Atmos. Sci.*, **31**, 674–701.

Arora, V., 2002. Modeling vegetation as a dynamic component in soil-vegetation-atmosphere transfer schemes and hydrologic models. *Rev. Geophys.*, **40**, 1006, doi:10.1029/2001RG000103.

Arrhenius, S., 1896. On the influence of carbonic acid in the air on the temperature of the ground. *Phil. Mag.*, **41**, 237–276.

Ashworth, J. R., 1929. The influence of smoke and hot gases from factory chimneys on rainfall. *Q. J. Roy. Meteor. Soc.*, **25**, 34–35.

Asner, G. P., and K. B. Heidebrecht, 2005. Desertification alters regional ecosystem–climate interactions. *Global Change Biol.*, **11**, 182–194.

Atlas, D., 1977. The paradox of hail suppression. *Science,* **195**, 139–145.

Aubreville, A., 1949. *Climats, forêts et désertification de l'Afrique tropicale.* Paris: Société d'Editions Géographiques et Coloniales.

Australia Conservation Foundation, 2001. *Australian Land Clearing, A Global Perspective: Latest Facts and Figures.* Fitzroy, Vic: Australia Conservation Foundation. Available online at www.acfonline.org.au/docs/publications/rpt0002.pdf

Avissar, R., and F. Chen, 1993. Development and analysis of prognostic equations for mesoscale kinetic energy and mesoscale (subgrid scale) fluxes for large-scale atmospheric models. *J. Atmos. Sci.*, **50**, 3751–3774.

Avissar, R., and Y. Liu, 1996. Three-dimensional numerical study of shallow convective clouds and precipitation induced by land surface forcing. *J. Geophys. Res.*, **101**, 7499–7518.

Avissar, R., and R.A. Pielke, 1989. A parameterization of heterogeneous land surfaces for atmospheric numerical models and its impact on regional meteorology. *Mon. Wea. Rev.*, **117**, 2113–2136.

Avissar, R., and T. Schmidt, 1998. An evaluation of the scale at which ground-surface heat flux patchiness affects the convective boundary layer using large-eddy simulations. *J. Atmos. Sci.*, **55**, 2666–2689.

Avissar, R., and D. Werth, 2005. Global hydroclimatological teleconnections resulting from tropical deforestation. *J. Hydrometeor.*, **6**, 134–145.

Baidya Roy, S., and R. Avissar, 2002. Impact of land use/land cover change on regional hydrometeorology in Amazonia. *J. Geophys. Res.*, **107**, 8037, doi:10.1029/2000JD000266.

Baidya Roy, S., G. C. Hurtt, C. P. Weaver, and S. W. Pacala, 2003. Impact of historical land cover changes on July climate of the United States. *J. Geophys. Res.*, **108**, **D24**, 4793, doi: 10.1029/ 2003JD003565.

Baik, J. -J., and Y. -H. Kim, 2001. Dry and moist convection forced by an urban heat island. *J. Appl. Meteor.*, **40**, 1462–1475.

Baker, R. D., B. H. Lynn, A. Boone, W. -K. Tao, and J. Simpson, 2001. The influence of soil moisture, coastline curvature, and land-breeze circulations on sea-breeze initiated precipitation. *J. Hydrometeor.*, **2**, 193–211.

Balling, R. C., 1988. The climatic impact of a Sonoran vegetation discontinuity. *Climat. Change*, **13**, 99–109.

Balling, R. C., and S. W. Brazel, 1987. Time and space characteristics of the Phoenix urban heat island. *J. Arizona–Nevada Acad. Sci.*, **21**, 75–81.

Banta, R. M., 1985. Nuclear firestorm results (preliminary) case: Standard atmosphere with moisture. Air Force Geophysics Laboratory/ LYC report.

Banta, R. M., and A. B. White, 2003. Mixing-height differences between land use types: Dependence on wind speed. *J. Geophys. Res.*, **108**, 4321, doi:10.1029/2002JD 002748.

Barkström, B. R., E. F. Harrison, and R. B. Lee, III, 1990. Earth Radiation Budget Experiment. *EOS*, **71**, 297, 304–305.

Barnett, T. P., D. W. Pierce, and R. Schnur, 2001. Detection of anthropogenic climate: Change in the World's oceans. *Science*, **292**, 270–274.

Barnston, A. G., and P. T. Schickedanz, 1984. The effects of irrigation on warm season precipitation in the southern Great Plains. *J. Climate Appl. Meteor.*, **23**, 865–888.

Barnston, A. G., W. L. Woodley, J. A. Flueck, and M. H. Brown, 1983. The Florida Area Cumulus Experiment's second phase (FACE-2). I. The experimental design, implementation, and basic data. *J. Appl. Meteor.*, **22**, 1504–1528.

Baron, J. S., M. D. Hartman, T. G. F. Kittel, *et al.*, 1998. Effects of land cover, water redistribution, and temperature on ecosystem processes in the South Platte basin. *Ecol. Appl.*, **8**, 1037–1051.

Barrie, L. A., D. Fisher, and R. M. Koerner, 1985. Twentieth century trends in Arctic air pollution revealed by conductivity and acidity observations in snow and ice in the Canadian high Arctic. *Atmos. Environ.*, **19**, 2055–2063.

Basara, J. B., and K. C. Crawford, 2002. Linear relationships between root-zone soil moisture and atmospheric processes in the planetary boundary layer. *J. Geophys. Res.*, **107**, doi:10.1029/2001JD000633.

Bastable, H. G., W. J. Shuttleworth, R. L. G. Dallarosa, G. Fisch, and C. A. Nobre, 1993. Observations of climate, albedo and surface radiation over cleared and undisturbed Amazonian forest. *Int. J. Climatol.*, **13**, 783–796.

Battan, L. J., 1969. Weather modification in the USSR: 1969. *Bull. Am. Meteor. Soc.*, **50**, 924–945.

Bauer, E., M. Claussen, V. Brovkin, and A. Hünerbein, 2003. Assessing climate forcings of the Earth system for the past millennium. *Geophys. Res. Lett.*, **30**, 1276, doi:10.1029/2002GL016639.

Beebe, R. C., 1974. Large scale irrigation and severe storm enhancement. *Proc. Symp. Atmospheric Diffusion and Air Pollution of the American Meteorological Society*, Co-sponsored by the World Meteorological Organization and the American Meteorological Society, Boston, MA, Sept 9–13, 1974, pp. 392–395.

Bello, M., 1991. Greenhouse warming threat justifies immediate action. *Natl Res. Council News Report*, **41**(4), 2–5.

Beltran, A., 2005. Using a coupled atmospheric-biospheric modeling system (GEMRAMS) to model the effects of land-use/land-cover changes on near-surface atmosphere. Ph.D. dissertation, Atmospheric Science Department, Colorado State University, Fort Collins, CO.

Ben-Zvi, A., 1997. Comments on "A new look at the Israeli cloud seeding experiments." *J. Appl. Meteor.*, **36**, 255–256.

Berger, W. H., 1982. Climate steps in ocean history-lessons from the Pleistocene. In *Climate in Earth History*, W. H. Berger and J. C. Crowell, panel co-chairmen. Washington, DC: National Academy Press, pp. 43–54.

Bergeron, T., 1965. On the low-level redistribution of atmospheric water caused by orography. *Suppl., Proc. Int. Conf. on Cloud Physics*, Tokyo, 96–100.

Betts, R. A., 1999. Self-beneficial effects of vegetation on climate in an Ocean–Atmosphere General Circulation Model. *Geophys. Res. Lett.*, **26**, 1457–1460.

Betts, R. A., 2000. Offset of the potential carbon sink from boreal forestation by decreases in surface albedo. *Nature*, **408**, 187–189.

Betts, R. A., 2001. Biogeophysical impacts of land use on present-day climate: near-surface temperature and radiative forcing. *Atmos. Sci. Lett.*, doi:10.1006/asle.2000.0023.

Betts, A. K., 2004. Understanding hydrometeorology using global models. *Bull. Am. Meteor. Soc.*, **11**, 1673–1688.

Betts, A. K., and M. J. Miller, 1986. A new convective adjustment scheme. II. Single column tests using GATE wave, BOMEX, ATEX and arctic air-mass data sets. *Q. J. Roy. Meteor. Soc.*, **112**, 693–709.

Betts, A. K., R. L. Desjardins, and J. I. MacPherson, 1992. Budget analysis of the boundary layer grid flights during FIFE 1987. *J. Geophys. Res.*, **97**, 18533–18546.

Betts, R. A., J. H. Ball, A. C. M. Beljaars, M. J. Miller, and P. Viterbo, 1996. The land-surface–atmosphere interaction: A review based on observational and global modelling perspectives. *J. Geophys. Res.*, **101**, 7209–7225.

Betts, R. A., P. M. Cox, S. E. Lee, and F. I. Woodward, 1997. Contrasting physiological and structural vegetation feedbacks in climate change simulations. *Nature*, **387**, 796–799.

Betts, R. A., P. M. Cox, and F. I. Woodward, 2000. Simulated response of potential vegetation to doubled-CO_2 climate change and feedbacks on near-surface temperature. *Global Ecol. Biogeog.*, **9**, 171–180.

Bibilashvili, N. Sh., I. I. Gaivoronski, G. G. Godorage, A. I. Kartsivadze, and R. N. Stankov, 1974. *Proc. World Meteorological Organization/International Association of Meteorology and Atmospheric Sciences Scientific Conf. on Weather Modification*, Tashkent, 1973, p. 333.

Bigg, E. K., 1988. Secondary ice nucleus generation by silver iodide applied to the ground. *J. Appl. Meteor.*, **27**, 453–457.

Bigg, E. K., 1990a. Long term trends in ice nucleus concentrations. *Atmos. Res.*, **25**, 409–415.

Bigg, E. K., 1990b. Aerosol over the southern ocean. *Atmos. Res.*, **25**, 583–600.

Bigg, E. K., 1997. A mechanism for the formation of new particles in the atmosphere. *Atmos. Res.*, **43**, 129–137.

Bigg, E. K., and E. Turton, 1988. Persistent effects of cloud seeding with silver iodide. *J. Appl. Meteor.*, **27**, 505–514.

Bigg, E. K., J. L. Gras, and C. Evans, 1984: Origin of Aitken particles in remote regions of the Southern Hemisphere. *J. Atmos. Chem.*, **1**, 203–214.

Biswas, K. R., and A. S. Dennis, 1971. Formation of rain shower by salt seeding. *J. Appl. Meteor.*, **10**, 780–784.

Biswas, K. R., R. K. Kapoor, and K. K. Kanuga, 1967. Cloud seeding experiment using common salt. *J. Appl. Meteor.*, **10**, 780–784.

Black, J. F., and B. H. Tarmy, 1963. The use of asphalt coating to increase rainfall. *J. Appl. Meteor.*, **2**, 557–564.

Bohren, C., 1989. The greenhouse effect revisited. *Weatherwise*, **42**, 50–54.

Bonan, G. B., 1997. Effects of land use on the climate of the United States. *Climat. Change*, **37**, 449–486.

Bonan, G. B., 1999. Frost followed the plow: Impacts of deforestation on the climate of the United States. *Ecol. Appl.*, **9**, 1305–1315.

Bonan, G. B., 2001. Observational evidence for reduction of daily maximum temperature by croplands in the midwest United States. *J. Climate*, **14**, 2430–2442.

Bonan, G. B., 2002. *Ecological Climatology: Concepts and Applications.* New York: Cambridge University Press.

Bonan, G. B., D. Pollard, and S. L. Thompson, 1992. Effects of boreal forest vegetation on global climate. *Nature*, **359**, 716–718.

Bonan, G. B., F. S. Chapin III, and S. L. Thompson, 1995. Boreal forest and tundra ecosystems as components of the climate system. *Climat. Change*, **29**, 145–167.

Bornstein, R. D., and G. M. LeRoy, 1990. Urban barrier effects on convective and frontal thunderstorms. *Proc. 4th American Meteorological Society Conf. Mesoscale Processes*, June 25–29, 1990, Boulder, CO, pp. 120–121.

Bornstein, R. D., and Q. Lin, 2000. Urban heat islands and summertime convective thunderstorms in Atlanta: Three case studies. *Atmos. Environ.*, **34**, 507–516.

Borys, R. D., and K. A. Rahn, 1981. Long-range atmospheric transport of cloud-active aerosol to Iceland. *Atmos. Environ.*, **15**, 1491–1501.

Borys, R. D., D. H. Lowenthal, and D. L. Mitchell, 2000. The relationships among cloud microphysics, chemistry, and precipitation rate in cold mountain clouds. *Atmos. Environ.*, **34**, 2593–2602.

Borys, R. D., D. H. Lowenthal, S. A. Cohn, and W. O. J. Brown, 2003. Mountain and radar measurements of anthropogenic aerosol effects on snow growth and snowfall rate. *Geophys. Res. Lett.*, **30**, 1538, doi:10.1029/2002GL016855.

Bosilovich, M. G., and S. D. Schubert, 2002. Water vapor tracers as diagnostics of the regional hydrologic cycle. *J. Hydrometeor.*, **3**, 149–165.

Bosilovich, M. G., and W. -Y. Sun, 1999. Numerical simulation of the 1993 midwestern flood: Land–atmosphere interactions. *J. Climate*, **12**, 1490–1505.

Boucher, O., and U. Lohmann, 1995. The sulfate-CCN-cloud albedo effect. *Tellus, Ser. B*, **47**, 281–300.

Boucher, O., G. Myhre, and A. Myhre, 2004. Direct human influence of irrigation on atmospheric water vapour and climate. *Climate Dynamics*, **22**, 597–603.

Bounoua, L., G. J. Collatz, S. O. Los, *et al.*, 2000. Sensitivity of climate to changes in NDVI. *J. Climate*, **13**, 2277–2292.

Bounoua, L., R. DeFries, G. J. Collatz, P. Sellers, and H. Khan, 2002. Effects of land cover conversion on surface climate. *Climat. Change*, **52**, 29–64.

Braconnot, P., S. Joussaume, O. Marti, and N. de Noblet-Ducoudre, 1999. Synergistic feedbacks from ocean and vegetation on the African monsoon response to mid-Holocene insolation. *Geophys. Res. Lett.*, **26**, 2481–2484.

Braham, R. R., 1981. Designing cloud seeding experiments for physical understanding. *Bull. Am. Meteor. Soc.*, **62**, 55–62.

Braham, R. R. Jr., 1986. Precipitation enhancement: A scientific challenge. *Meteor. Monogr.* **21**.

Braham, R. R., and P. A. Spyers-Duran, 1967. Survival of cirrus crystals in clear air. *J. Appl. Meteor.*, **6**, 1053–1061.

Braham, R. R. Jr., L. J. Battan, and H. R. Byers, 1957. Artificial nucleation of cumulus clouds, clouds and weather modification: A group of field experiments. *Meteor. Monogr.* **2**, 47–85.

Braslau, N., and J. V. Dave, 1975. Atmospheric heating rates due to solar radiation for several aerosol-laden cloudy and cloud-free models. *J. Appl. Meteor.*, **14**, 396–399.

Brasseur, G., W. Steffen, and K. Noone, 2005: Earth System Focus Geosphere-Biosphere Program. *EOS*, **86**, 209–216.

Brier, G., 1955. Seven-day periodicities in certain meteorological parameters during the period 1899–1951. *Bull. Am. Meteor. Soc.*, **36**, 265–277.

Brooks, W. T., 1989. The global warming panic. *Forbes*, **December**, 97–102.

Broström, A., M. Coe, S.P. Harrison, *et al.*, 1998. Land surface feedbacks and paleomonsoons in northern Africa. *Geophys. Res. Lett.*, **25**, 3615–3618.

Brovkin, V., M. Claussen, V. Petoukhov, and A. Ganopolski, 1998. On the stability of the atmosphere-vegetation system in the Sahara/Sahel region. *J. Geophys. Res.*, **103**, 31613–31624.

Brovkin, V., A. Ganopolski, M. Claussen, C. Kubatzki, and V. Petoukhov, 1999. Modeling climate response to historical land cover change. *Global Ecol. Biogeog.*, **8**, 509–517.

Brovkin, V., J. Bendtsen, M. Claussen, *et al.*, 2002. Carbon cycle, vegetation and climate dynamics in the Holocene: Experiments with the CLIMBER-2 Model. *Global Biogeochem. Cycles*, **16**, 1139, doi:10.1029/2001GB 001662.

Brovkin, V., S. Levis, M. F. Loutre, *et al.*, 2003. Stability analysis of the climate–vegetation system in northern high latitudes. *Climat. Change*, **57**, 119–138.

Brovkin, V., S. Sitch, W. Von Bloh, M. Claussen, E. Vauer, and W. Cramer, 2004. Role of land cover changes for atmospheric CO_2 increase and climate change during the last 150 years. *Global Change Biol.*, **10**, 1253–1266.

Brovkin, V., M. Claussen, E. Driesschaert, *et al.*, 2005. Biogeophysical effects of historical land cover changes simulated by six Earth system models of intermediate complexity. *Climate Dynamics*, **26**, 587–600.

Browning, K. A., 1968. The organization of severe local storms. *Weather*, **23**, 429–434.

Browning, K. A., and G. B. Foote, 1976. Airflow and hail growth in supercell storms and some implications for hail suppression. *Q. J. Roy. Meteor. Soc.*, **102**, 499–533.

Browning, K. A., C. W. Pardoe, and F. F. Hill, 1975. The nature of orographic rain at wintertime cold fronts. *Q. J. Roy. Meteor. Soc.*, **101**, 333–352.

Brubaker, K. L., and D. Entekhabi, 1995. An analytic approach to modeling land-atmosphere interaction. I. Construct and equilibrium behavior. *Water Resource Res.*, **31**, 619–632.

Brubaker, K. L., D. Entekhabi, and P.S. Eagleson, 1993. Estimation of continental precipitation recycling. *J. Climate*, **6**, 1077–1089.

Bruintjes, R. T., R. M. Rasmussen, W. Sukarnjanaset, P. Sukhikoses, and N. Tantipubthong, 1999. Variations of cloud condensation nuclei (CCN) and aerosol particles over Thailand and the possible impacts on precipitation formation in clouds. *Proc. 7th World Meteorological Organization Scientific Conf. on Weather Modification*, Chiang Mai, Thailand, pp. 33–36.

Bruintjes, R. T., D. W. Breed, V. Salazar, *et al.*, 2001. Overview and results from the Mexican hygroscopic seeding experiment. *Proc. 15th American Meteorological Society Conf. Planned and Inadvertent Weather Modification*, Jan 14–18, 2001, Albuquerque, NM, pp. 45–48.

Bryant, N. A., L. F. Johnson, A. J. Brazel, *et al.*, 1990. Measuring the effect of overgrazing in the Sonoran Desert. *Climat. Change*, **17**, 243–264.

Burke, E. J., W. J. Shuttleworth, and Z. -L. Yang, 2000. Heterogeneous vegetation affects GCM-modeled climate. *GEWEX WCRP News*, **10**, 7–8.

Burtsev, I. I., 1974. Hail suppression. *Proc. 2nd World Meteorological Organization Scientific Conf. Weather Modification*, Boulder, CO, pp. 217–222.

Byers, H. R., 1974. History of weather modification. In *Weather and Climate Modification*, W.N. Hess, ed. New York: John Wiley, pp. 3–44.

Cahill, T. A., K. Wilkinson, and R. Schnell, 1992. Composition analyses of size-resolved aerosol samples taken from aircraft downwind of Kuwait, Spring 1991. *J. Geophys. Res.*, **97**, 14513–14520.

Callendar, G. S., 1938. The artificial production of carbon dioxide and its influence on temperature. *Q. J. Roy. Meteor. Soc.*, **64**, 223–240.

Carleton, A. M., D. Travis, D. Arnold, *et al.*, 1994. Climatic-scale vegetation–cloud interactions during drought using satellite data. *Int. J. Climatol.*, **14**, 593–623.

Carleton, A. M., J. Adegoke, J. Allard, and D. L. Arnold, 2001. Summer season land cover–convective cloud associations for the Midwest US "Corn Belt." *Geophys. Res. Lett.*, **28**, 1679–1682.

Carlson, T. N., and S. G. Benjamin, 1980. Radiative heating rates for Saharan dust. *J. Atmos. Sci.*, **37**, 193–213.

Carrió, G. G., H. Jiang, and W. R. Cotton, 2005a. Impact of aerosol intrusions on Arctic boundary layer clouds I. 4 May 1998 case. *J. Atmos. Sci.*, **62**, 3082–3093.

Carrió, G. G., H. Jiang, and W. R. Cotton, 2005b. Impact of aerosol intrusions on Arctic boundary layer clouds Sea-ice melting rates. II. *J. Atmos. Sci.*, **62**, 3094–3105.

Carslaw, K. S., R. G. Harrison, and J. Kirby, 2002. Cosmic rays, clouds, and climate. *Science*, **298**, 1732–1737.

Castro, C. L., R. A. Pielke Sr., and G. Leoncini, 2005. Dynamical downscaling: Assessment of value restored and added using the Regional Atmospheric Modeling System (RAMS). *J. Geophys. Res.*, **110**, **D5**, 5108, doi:10.1029/2004JD004721.

Cathcart, R. B., and V. Badescu, 2004. Architectural ecology: A tentative Sahara restoration. *Int. J. Environ. Stud.*, **61**, 145–160.

Cess, R. D., G. L. Potter, S. J. Ghan, and W. L. Gates, 1985. The climatic effects of large injections of atmospheric smoke and dust: A study of climate feedback mechanisms with one- and three-dimensional climate models. *J. Geophys. Res.*, **90**, 12937–12950.

Cess, R. D., G. L. Potter, M. -H. Zhang, *et al.*, 1991. Interpretation of snow–climate feedback as produced by 17 general circulation models. *Science*, **253**, 888–892.

Chagnon, F. J. F., and R. L. Bras, 2005. Contemporary climate change in the Amazon. *Geophys. Res. Lett.*, **32**, L13703, doi:10.1029/2005GL022722.

Chagnon, F. J. F., R. L. Bras, and J. Wang, 2004. Climatic shift in patterns of shallow clouds over the Amazon. *Geophys. Res. Lett.*, **31**, L24212, doi:10.1029/2004Gl 021188.

Chameides, W.L., L. Xingsheng, T. Xiaoyan, *et al.*, 1999. Is ozone pollution affecting crop yields in China? *Geophys. Res. Lett.*, **26**, 867–870.

Chang, J. T., and P. J. Wetzel, 1991. Effects of spatial variations of soil moisture and vegetation on the evolution of a prestorm environment: A case study. *Mon. Wea. Rev.*, **119**, 1368–1390.

Changnon, D., T. B. McKee, and N. J. Doesken, 1991. Hydroclimatic variability in the Rocky Mountains. *Water Res. Bull.*, **27**, 733–743.

Changnon, D., M. Sandstrom, and C. Schaffer, 2003. Relating changes in agricultural practices to increasing dew points in extreme Chicago heat waves. Climate Res., **24**, 243–254.

Changnon, S. A., Jr. 1968. The La Porte anomaly: Fact or fiction? *Bull. Am. Meteor. Soc.*, **49**, 4–11.

Changnon, S. A., 1977. *Hail Suppression Impacts and Issues*, Final Report, Technology Assessment of the Suppression of Hail, ERP75–09980, RANN Program, National Science Foundation, Champaign, IL.

Changnon, S. A. Jr., 1981a. Midwestern cloud, sunshine and temperature trends since 1901: Possible evidence of jet contrail effects. *J. Appl. Meteor.*, **20**, 496–508.

Changnon, S. A. Jr., 1981b. *METROMEX: A review and summary*, Meteorological *Monographics*, No. **18**, Boston, MA.

Changnon, S. A. Jr., and F. A. Huff, 1977. La Porte again: A new anomaly. *Bull. Am. Meteor. Soc.*, **58**, 1069–1072.

Changnon, S. A. Jr., and F.A. Huff, 1986. The urban-related nocturnal rainfall anomaly at St. Louis. *J. Climate Appl. Meteor.*, **25**, 1985–1995.

Changnon, S. A. Jr., and J. L. Ivens, 1981. History repeated: The forgotten hail cannons of Europe. *Bull. Am. Meteor. Soc.*, **62**, 368–375.

Changnon, S. A. Jr., and W. H. Lambright, 1987. The rise and fall of federal weather modification policy. *J. Wea. Mod.*, **19**, 1–12.

Chappell, C. F., L. O. Grant, and P. W. Mielke Jr., 1971. Cloud seeding effects on precipitation intensity and duration of wintertime orographic clouds. *J. Appl. Meteor.*, **10**, 1006–1010.

Charlson, R. J., J. E. Lovelock, M. O. Andreae, and S. G. Warren, 1987. Oceanic phytoplankton, atmospheric sulfur, cloud albedo and climate. *Nature*, **326**, 655–661.

Charlson, R. J., S. E. Schwartz, J. M. Hales, *et al.*, 1992. Climate forcing by anthropogenic aerosols. *Science*, **255**, 423–430.

Charney, J. G., 1975. Dynamics of desert and drought in the Sahel. *Q. J. Roy. Meteor. Soc.*, **101**, 193–202.

Charney, J., and P. H. Stone, 1975. Drought in the Sahara: A biogeophysical feedback mechanism. *Science*, **187**, 434–435.

Charney, J. G., W. J. Quirk, S. -H. Chow, and J. Kornfield, 1977. A comparative study of the effects of albedo change on drought in semi-arid regions. *J. Atmos. Sci.*, **34**, 1366–1385.

Chase, T. N., R. A. Pielke, T. G. F. Kittel, R. Nemani, and S. W. Running, 1996. The sensitivity of a general circulation model to global changes in leaf area index. *J. Geophys. Res.*, **101**, 7393–7408.

Chase, T. N., R. A. Pielke Sr., T. G. F. Kittel, J. S. Baron, and T. J. Stohlgren, 1999. Potential impacts on Colorado Rocky Mountain weather due to land use changes on the adjacent Great Plains. *J. Geophys. Res.*, **104**, 16673–16690.

Chase, T. N., R. A. Pielke Sr., T. G. F. Kittel, R. R. Nemani, and S. W. Running, 2000. Simulated impacts of historical land cover changes on global climate. *Climate Dynamics*, **16**, 93–105.

Chase, T. N., R. A. Pielke Sr., T. G. F. Kittel, *et al.*, 2001. Relative climatic effects of landcover change and elevated carbon dioxide combined with aerosols: A comparison of model results and observations. *J. Geophys. Res.*, **D106**, 31685–31691.

Chen, C. H., and H. D. Orville, 1980. Effects of mesoscale on cloud convection. *J. Appl. Meteor.*, **19**, 256–274.

Chen, F., and R. Avissar, 1994a. The impact of land-surface wetness heterogeneity on mesoscale heat fluxes. *J. Appl. Meteor.*, **33**, 1323–1340.

Chen, F., and R. Avissar, 1994b. Impact of land-surface moisture variabilities on local shallow convective cumulus and precipitation in large-scale models. *J. Appl. Meteor.*, **33**, 1382–1394.

Chen, T. -C., J. -H. Yoon, K. J. St. Croix, and E. S. Takle, 2001. Suppressing impacts of the Amazonian deforestation by global circulation change. *Bull. Am. Meteor. Soc.*, **82**, 2209–2216.

Cheng, W. Y. Y., and W. R. Cotton, 2004. Sensitivity of a simulated mesoscale convective system to horizontal heterogeneities in soil moisture initialization. *J. Hydrometeor.*, **5**, 934–958.

Chisholm, A. J., and J. H. Renick, 1972. The kinematics of multicell and supercell Alberta hailstorms. In *Alberta Hail Studies* 1972, Hail Studies Report No. 72-2. Edmonton, Research Council of Alberta, pp. 24–31.

Chisnell, R. F., and J. Latham, 1976a. Ice multiplication in cumulus clouds. *Q. J. Roy. Meteor. Soc.*, **102**, 133–156.

Chisnell, R. F., and J. Latham, 1976b. Comments on the paper by Mason, "Production of ice crystals by riming in slightly supercooled cumulus." *Q. J. Roy. Meteor. Soc.*, **102**, 713–715.

Christy, J. R., R. W. Spencer, W. B. Norris, W. D. Braswell, and D. E. Parker, 2003. Error estimates of Version 5.0 of MSU/AMSU bulk atmospheric temperatures. *J. Atmos. Ocean. Tech.*, **20**, 613–629.

Christy, J. R., W. B. Norris, K. Redmond, and K. P. Gallo, 2006. Methodology and results of calculating central California surface temperature trends: Evidence of human-induced climate change? *J. Climate,* **19**, 548–563.

Chung, C. C., J. E. Penner, K. E. Taylor, A. S. Grossman, and J. J. Walton, 1997. An assessment of the radiative effects of anthropogenic sulfate. *J. Geophys. Res.*, **102**, 3761–3778.

Chung, C. E., and V. Ramanathan, 2003. South Asian haze forcing: Remote impacts with implications to ENSO and AO. *J. Climate*, **16**, 1791–1806.

Civerolo, K. L., G. Sistla, S. T. Rao, and D. J. Nowak, 2000. The effects of land use in meteorological modeling: Implications for assessment of future air quality scenarios. *Atmos. Environ.*, **34**, 1615–1621.

Clark, C. A., and R. W. Arritt, 1995. Numerical simulations of the effect of soil moisture and vegetation cover on the development of deep convection. *J. Appl. Meteor.*, **34**, 2029–2045.

Clark, D. B., Y. Xue, R. Harding, and P. J. Valdes, 2001. Modeling the impact of land surface degradation on the climate of tropical North Africa. *J. Climate*, **14**, 1809–1822.

Clark, D. B., C. M. Taylor, and A. J. Thorpe, 2004. Feedback between the land surface and rainfall at convective length scales. *J. Hydrometeor.*, **5**, 625–639.

Claussen, M., 1995. *Modeling Biogeophysical Feedback in the Sahel*, Report No. 163. Hamburg, Germany, Max-Planck Institut für Meteorologie.

Claussen, M., 1997. Modelling biogeophysical feedback in the African and Indian Monsoon region. *Climate Dynamics*, **13**, 247–257.

Claussen, M., 1998. On multiple solutions of the atmosphere–vegetation system in present-day climate. *Global Change Biol.*, **4**, 549–560.

Claussen, M., 2001a. Biogeophysical feedbacks and the dynamics of climate. In *Global Biogeochemical Cycles in the Climate System,* Schulze, E.-D., M. Heinman, S. Harrison *et al.* eds. San Diego, CA: Academic Press, pp. 61–71.

Claussen, M., 2001b. Earth system models. In *Understanding the Earth System: Compartments, Processes and Interactions*, E. Ehlers and T. Krafft, eds. Heidelberg, Germany: Springer-Verlag, pp. 145–162.

Claussen, M., and V. Gayler, 1997. The greening of Sahara during the mid-Holocene: Results of an interactive atmosphere-biome model. *Global Ecol. Biogeo., Lett.*, **6**, 369–377.

Claussen, M., V. Brovkin, A. Ganopolski, C. Kubatzki, and V. Petoukhov, 1998. Modeling global terrestrial vegetation–climate interaction. *Phil. Trans. Roy. Soc. Lond. B*, **353**, 53–63.

Claussen, M., C. Kubatzki, V. Brovkin, *et al.*, 1999. Simulation of an abrupt change in Saharan vegetation at the end of the mid-Holocene. *Geophys. Res. Lett.*, **26**, 2037–2040.

Claussen, M., V. Brovkin, V. Petoukhov, and A. Ganapolski, 2001. Biogeophysical versus biogeochemical feedbacks of large-scale land-cover changes. *Geophys. Res. Lett.* **26**, 1011–1014.

Claussen, M., V. Brovkin, and A. Ganopolski, 2002: Africa: Greening of the Sahara. In *Proc. Global Change Open Science Conference*, Amsterdam, The Netherlands, July 10-13, 2001, pp. 125–128.

Claussen, M., V. Brovkin, A. Ganopolski, C. Kubatzki, and V. Petoukhov, 2003. Climate change in northern Africa: The past is not the future. *Climat. Change*, **57**, 99–118.

Claussen, M., P. M. Cox, X. Zeng, *et al.*, 2004. The global climate. In *Vegetation, Water, Humans and the Climate: A New Perspective on an Interactive System*, P. Kabat, M. Claussen, M. Dirmeyer *et al.* eds. Heidelberg, Germany: Springer-Verlag, pp. 33–57.

Claussen, M., V. Brovkin, R. Calov, A. Ganopolski, and C. Kubatzki, 2005. Did humankind prevent an early glaciation? Comment on Ruddiman's hypothesis of a prehistoric anthropocene. *Climat. Change*, **69**, 409–417.

Coakley, J. A., and R. D. Cess, 1985. Response of the NCAR community climate model to the radiative forcing by the naturally occurring tropospheric aerosol. *J. Atmos. Sci.*, **42**, 1677–1692.

Coakley, J. A., R. L. Bernstein, and P. A. Durkee, 1987. Effect of ship-stack effluents on cloud reflectivity. *Science*, **237**, 1020–1022.

Cobb, W. E., and H. J. Wells, 1970. The electrical conductivity of oceanic air and its correlation to global atmospheric pollution. *J. Atmos. Sci.*, **27**, 814–819.

Cohen, S., 1991. Pork in the sky. *Washington Post Magazine*, **November 10**, 15–17, 36–39.

Collatz, G. J., J. T. Ball, C. Grivet, and J. A. Berry, 1991. Physiological and environmental regulation of stomatal conductance, photosynthesis and transpiration: A model that includes a laminar boundary layer. *Agric. Forest Meteor.*, **54**, 107–136.

Committee on Scholarly Communication with the People's Republic of China (CSCPRC), 1992. *Grasslands and Grassland Sciences in Northern China*. Washington, DC: National Academy Press.

Conant, W. C., A. Nenes, and J. H. Seinfeld, 2002. Black carbon radiative heating effects on cloud microphysics and implications for the aerosol indirect effect. I. Extended Kohler theory. *J. Geophys. Res.*, **107**, **D21**, 4604, doi:10.1029/2002JD 002094.

Cook, K .H., and E. K. Vizy, 2006. South American climate during the last glacial maximum: Delayed onset of the South American monsoon. *J. Geophys. Res.*, **111**, D02110, doi:10.1029/2005JD005980.

Cooley, H. S., W. J. Riley, M. S. Torn, and Y. He, 2005. Impact of agricultural practice on regional climate in a coupled land surface mesoscale model. *J. Geophys. Res.* **110**, D03113, doi:10.1029/2004JD005160.

Cooper, W. A., and R. P. Lawson, 1984. Physical interpretation of results from the HIPLEX-1 experiment. *J. Climate Appl. Meteor.*, **23**, 523–540.

Cooper, W. A., R. T. Bruintjes, and G. K. Mather, 1997. Calculations pertaining to hygroscopic seeding with flares. *J. Appl. Meteor.*, **36**, 1449–1469.

Copeland, J. H., R. A. Pielke, and T. G. F. Kittel, 1996. Potential climatic impacts of vegetation change: A regional modeling study. *J. Geophys. Res.*, **101**, 7409–7418.

Costa, M. H., and J. A. Foley, 2000. Combined effects of deforestation and doubled atmospheric CO_2 concentrations on the climate of Amazonia. *J. Climate*, **13**, 18–34.

Cotton, W. R., 1972a. Numerical simulation of precipitation development in supercooled cumuli, Part I. *Mon. Wea. Rev.*, **100**, 757–763.

Cotton, W. R., 1972b. Numerical simulation of precipitation development in supercooled cumuli, Part II. *Mon. Wea. Rev.*, **100**, 764–784.

Cotton, W. R., 1985. Atmospheric convection and nuclear winter. *Am. Sci.* **73**, 275–280.

Cotton, W. R., 1990. *Storms*, Geophysical Science Series, vol. 1. Fort Collins, CO: ASTeR Press.

Cotton, W. R., and R. A. Anthes, 1989. *Storm and Cloud Dynamics*, International Geophysics Series, vol. 44. San Diego, CA: Academic Press. 883 pp.

Cotton, W. R., R. A. Pielke Sr., R. L. Walko, *et al.*, 2003. RAMS 2001: Current status and future directions. *Meteor. Atmos. Phys.*, **82**, 5–29.

Couzin, J., 1999. Landscape changes make regional climate run hot and cold. *Science*, **283**, 317–319.

Cox, P. M., R. A. Betts, C. D. Jones, S. A. Spall, and I. J. Totterdell, 2000. Acceleration of global warming due to carbon-cycle feedbacks in a coupled climate model. *Nature*, **408**, 184–187.

Crichton, M., 2003. Aliens cause global warming. Caltech Michelin Lecture, Jan. 17, 2003. Available online at www.crichton-official.com/speeches/speeches_quote04.html

Crook, N. A., 1996. Sensitivity of moist convection forced by boundary layer processes to low-level thermodynamic fields. *Mon. Wea. Rev.*, **124**, 1767–1785.

Crutzen, P. J., and J. W. Birks, 1982. The atmosphere after a nuclear war: Twilight at noon. *Ambio*, **11**, 114–125.

Cullis, C. F., and M. M. Hirschler, 1980. Atmospheric sulphur: Natural and man-made sources. *Atmos. Environ.*, **14**, 1263–1278.

Cutrim, E., D. W. Martin, and R. M. Rabin, 1995. Enhancement of cumulus clouds over deforested lands in Amazonia. *Bull. Am. Meteor. Soc.*, **76**, 1801–1805.

Dalu, G. A., R .A. Pielke Sr., M. Baldi, and X. Zeng, 1996. Heat and momentum fluxes induced by thermal inhomogeneities. *J. Atmos. Sci.*, **53**, 3286–3302.

Davey, C. A., R. A. Pielke Sr., and K. P. Gallo, 2006. Differences between near-surface equivalent temperature trends for the Eastern United States: equivalent temperature as an alternative measure of heat content. *Global Planet. Change*, in press.

Deardorff, J. W., 1974. Three-dimensional numerical study of the height and mean structure of a heated planetary boundary layer. *Bound.-Layer Meteor.*, **7**, 81–106.

DeFries, R. S., L. Bounoua, and G. J. Collatz, 2002. Human modification of the landscape and surface climate in the next fifty years. *Global Change Biol.*, **8**, 438–458.

Delmas, R., and C. Boutron, 1980. Are the past variations of the stratospheric sulfate burden recorded in central Antarctic snow and ice layers? *J. Geophys. Res.*, **85**, 5645–5649.

Delworth, T., and S. Manabe, 1989. The influence of soil wetness on near-surface atmospheric variability. *J. Climate*, **2**, 1447–1462.

deMenocal, P. B., J. Ortiz, T. Guilderson, *et al.*, 2000. Abrupt onset and termination of the African Humid Period: Rapid climate response to gradual insolation forcing. *Q. Sci. Rev.*, **19**, 347–361.

DeMott, P. J., K. Sassen, M. R. Poellet, *et al.*, 2003. African dust aerosols as atmospheric ice nuclei. *Geophys. Res. Lett.*, **30**, 1732, doi:10.1029/2003GL017410.

Dennis, A. S., and A. Koscielski, 1972. Height and temperature of first echoes in unseeded and seeded convective clouds in South Dakota. *J. Appl. Meteor.*, **11**, 994–1000.

Dennis, A. S., and D. J. Musil, 1973. Calculations of hailstone growth and trajectories in a simple cloud model. *J. Atmos. Sci.*, **30**, 278–288.

Dennis, A. S., and H. D. Orville, 1997. Comments on "A new look at the Israeli cloud seeding experiments." *J. Appl. Meteor.*, **36**, 277–278.

Dennis, A. S, A. Koscielski, D. E. Cain, J. H. Hirsch, and P. L. Smith, Jr., 1975. Analysis of radar observations of a randomized cloud seeding experiment. *J. Appl. Meteor.*, **14**, 897–908.

deNoblet-Ducoudre, N., M. Claussen, and I. C. Prentice, 2000. Mid-Holocene greening of the Sahara: First results of the GAIM 6000 year BP Experiment with two asynchronously coupled atmosphere/biome models. *Climate Dynamics*, **16**, 643–659.

De Ridder, K., 1998. The impact of vegetation cover on Sahelian drought persistence. *Bound. -Layer Meteor.*, **88**, 307–321.

De Ridder, K., and H. Gallée, 1998. Land surface-induced regional climate change in southern Israel. *J. Appl. Meteor.*, **37**, 1470–1485.

Dessens, J., 1998. A physical evaluation of a hail suppression project with silver iodide ground burners in southwestern France. *J. Appl. Meteor.*, **37**, 1588–1599.

Dickinson, R. E., 1975. Solar variability and the lower atmosphere. *Bull. Am. Meteor. Soc.*, **56**, 1240–1248.

Dickinson, R. E. (ed.), 1987. *The Geophysiology of Amazonia*. New York: John Wiley.

Dickinson, R. E., and P. Kennedy, 1992. Impacts on regional climate of Amazonian deforestation. *Geophys. Res. Lett.*, **19**, 1947–1950.

Diem, J. E., and D. P. Brown, 2003. Anthropogenic impacts on summer precipitation in central Arizona, USA. *Profess. Geog.*, **55**, 343–355.

Diffenbaugh, N. S., 2005. Atmosphere-land cover feedbacks alter the response of surface temperature to CO_2 forcing in a topographically complex region. *Climate Dynamics*, **24**, 237–251.

Dirmeyer, P. A., 1994. Vegetation stress as a feedback mechanism in midlatitude drought. *J. Climate*, **7**, 1463–1483.

Dirmeyer, P. A., 1998. Land–sea geometry and its effect on monsoon circulations. *J. Geophys. Res.*, **103**, 11555–11572.

Dirmeyer, P. A., 1999. Assessing GCM sensitivity to soil wetness using GSWP data. *J. Meteor. Soc. Japan*, **77**, 367–385.

Dirmeyer, P. A., and J. Shukla, 1994. Albedo as a modulator of climate response to tropical deforestation. *J. Geophys. Res.*, **99**, 20863–20877.

Dirmeyer, P. A., and J. Shukla, 1996. The effect on regional and global climate of expansion of the world's deserts. *Q. J. Roy. Meteor. Soc.*, **122**, 451–482.

Dirmeyer, P. A., and F. J. Zeng, 1999. Precipitation infiltration in the simplified SiB land surface scheme. *J. Meteor. Soc. Japan*, **77**, 291–303.

Doherty, R., J. Kutzbach, J. Foley, and D. Pollard, 2000. Fully coupled climate/ dynamical vegetation model simulations over Northern Africa during the mid-Holocene. *Climate Dynamics*, **16**, 561–573.

Dolman, A. J., M. A. Silva Dias, J.-C. Calvet, *et al.*, 1999. Mesoscale effects of deforestation in Amazonia: Preparatory LBA modelling studies. *Ann. Geophys.*, **17**, 1095–1110.

Dolman, A. J., A. Verhagen, and C.A. Rovers, 2003. *Global Environmental Change and Land Use*. Dordrecht, The Netherlands: Kluwer.

Dolman, A. J., M. K. van der Molen, H. W. ter Maat, and R. W. A. Hutjes, 2004. The effect of forests on mesoscale atmospheric processes. In *Forests and the Land–Atmosphere Interface*, M. Mencuccini, J. Grace, J. Moncrieff, and K. McNaughton, eds. Oxford, UK: Oxford University Press, pp. 51–72.

Douglas E. M., D. Niyogi, S. Frolking, *et al.*, 2006. Changes in moisture and energy fluxes due to agricultural land use and irrigation in the Indian Monsoon Belt. *Geophys. Res. Lett.*, **33**, L14403, doi:10.1029/2006GL026550.

Douville, H., and J.-F. Royer, 1997. Influence of the temperate and boreal forests on the Northern Hemisphere climate in the Météo-France climate model. *Climate Dynamics*, **13**, 57–74.

Dudal, R., 1987. Land resources for plant production. In *Resources and World Development*, D. J. McLaren and B. J. Skinner, eds. New York: John Wiley, pp. 659–670.

Dunion, J. P., and C. S. Velden, 2004. The impact of the Saharan air layer on Atlantic tropical cyclone activity. *Bull. Am. Meteor. Soc.*, **85**, 353–365.

Durieux, L., L. A. T. Machado, and H. Laurent, 2003. The impact of deforestation on cloud cover over the Amazon arc of deforestation. *Remote Sens. Environ.*, **86**, 132–140.

Durkee, P. A., R. E. Chartier, A. Brown, *et al.*, 2000. Composite ship track characteristics. *J. Atmos. Sci.*, **57**, 2542–2553.

Dyson, F. J., 1988. *Infinite in All Directions*, Gillford Lectures given at Aberdeen, Scotland, UK, Apr.–Nov. 1985. New York: Harper and Row.

Eastman, J. L., M. B. Coughenour, and R. A. Pielke Sr., 2001a. The effects of CO_2 and landscape change using a coupled plant and meteorological model. *Global Change Biol.*, **7**, 797–815.

Eastman, J. L., M. B. Coughenour, and R. A. Pielke Sr., 2001b. Does grazing affect regional climate? *J. Hydrometeor.*, **2**, 243–253.

Eberbach, P. L., 2003. The eco-hydrology of partly cleared, native ecosystems in southern Australia: A review. *Plant and Soil*, **257**, 357–369.

Economist, 2005. From chaos, order. *Economist*, **March 5**, 45–47.

Eltahir, E. A. B., 1996. Role of vegetation in sustaining large-scale atmospheric circulations in the tropics. *J. Geophys. Res.*, **101**, 4255–4268.

Eltahir, E. A. B., and R. L. Bras, 1993a. On the response of the tropical atmosphere to large-scale deforestation. *Q. J. Roy. Meteor. Soc.*, **119**, 779–793.

Eltahir, E. A. B., and R. L. Bras, 1993b. A description of rainfall interception over large areas. *J. Climate*, **6**, 1002–1008.

Eltahir, E. A. B., and R. L. Bras, 1994a. Precipitation recycling in the Amazon Basin. *Q. J. Roy. Meteor. Soc.*, **120**, 861–880.

Eltahir, E. A. B., and R. L. Bras, 1994b. Sensitivity of regional climate to deforestation in the Amazon Basin. *Adv. Water Res.*, **17**, 101–115.

Eltahir, E. A. B., and C. Gong, 1996. Dynamics of wet and dry years in West Africa. *J. Climate*, **9**, 1030–1042.

Eltahir, E. A. B., and E. J. Humphries, 1998. The role of clouds in the surface energy balance over the Amazon forest. *Int. J. Climatol.*, **18**, 1575–1591.

Eltahir, E. A. B., and J. S. Pal, 1996. Relationship between surface conditions and subsequent rainfall in convective storms. *J. Geophys. Res.*, **101**, 26237–26245.

Eltahir, E. A. B., B. Loux, T. K. Yamana, and A. Bomblies, 2004. A see-saw oscillation between the Amazon and Congo basins. *Geophys. Res. Lett.*, **31**, L23201, doi: 10.1029/2004GL021160.

Emanuel, K. A., 1988. Toward a general theory of hurricanes. *Am. Sci.*, **76**, 371–379.

Emori, S., 1998. The interaction of cumulus convection with soil moisture distribution: An idealized simulation. *J. Geophys. Res.*, **103**, 8873–8884.

English, M., 1973. Alberta hailstorms. II. Growth of large hail in the storm. *Meteor. Monogr.*, **14**, 37–98.

Entekhabi, D., and D. L. Brubaker, 1995. An analytic approach in modeling land–atmosphere interaction. 2. Stochastic extension. *Water Resource Res.*, **31**, 633–643.

Entekhabi, D., I. Rodriguez-Iturbe, and R. L. Bras, 1992. Variability in large-scale water balance with land surface–atmosphere interaction. *J. Climate*, **5**, 798–813.

Erickson, J., 2005. Change in the air: Heating up the high country. *Rocky Mountain News*, Apr. 20, 2005. Available online at rockymountainnews.com/drmn/state/article/0,1299,DRMN_21_3712751,00.html

Farley, R. D., D. J. Musil, H. D. Orville, F. J. Kopp, and P. C. S. Chen, 1977. *Numerical Models of Hailstorms and Tests of Hail Suppression Concepts*, Report No. 77-5. Rapid City, SD: Institute of Atmospheric Science, South Dakota School of Mines and Technology.

Fast, J. D., and M. D. McCorcle, 1991. The effect of heterogeneous soil moisture on a summer baroclinic circulation in the central United States. *Mon. Wea. Rev.*, **119**, 2140–2167.

Feddema, J., K. Olson, G. Bonan, *et al.*, 2005a. A comparison of a GCM response to historical anthropogenic land cover change and model sensitivity to uncertainty in present day land cover representation. *Climate Dynamics*, **25**, 581–609.

Feddema, J. J., K. W. Olson, G. B. Bonan, *et al.*, 2005b. The importance of land-cover change in simulating future climates. *Science*, **310**, 1674–1678.

Federer, B., A. Waldvogel, W. Schmid, *et al.*, 1986. Main results of Grossversuch IV. *J. Climate Appl. Meteor.*, **25**, 917–957.

Feingold, G., W. R. Cotton, S. M. Kreidenweis, and J. T. Davis, 1999. The impact of giant cloud condensation nuclei on drizzle formation in stratocumulus: Implications for cloud radiative properties. *J. Atmos. Sci.*, **56**, 4100–4117.

Fennessy, M. J., and J. Shukla, 1999. Impact of initial soil wetness on seasonal atmospheric prediction. *J. Climate*, **12**, 3167–3180.

Ferranti, L., and F. Molteni, 1999. Ensemble simulations of Eurasian snow-depth anomalies and their influence on the summer Asian monsoon. *Q. J. Roy. Meteor. Soc.*, **125**, 2597–2610.

Fitzjarrald, D. R., O. C. Acevedo, and K. E. Moore, 2001. Climatic consequences of leaf presence in the eastern United States. *J. Climate*, **14**, 598–614.

Foley, J., J. E. Kutzbach, M. T. Coe, and S. Lewis, 1994. Feedbacks between climate and boreal forests during the Holocene epoch. *Nature*, **371**, 52–54.

Foley, J. A., S. Levis, I. Prentice, D. Colin-Pollard, and S.L. Thompson, 1998. Coupling dynamic models of climate and vegetation. *Global Change Biol.*, **4**, 561–579.

Foley, J. A., M. H. Costa, C. Delire, N. Ramankutty, and P. Snyder, 2003. Green surprise? How terrestrial ecosystems could affect Earth's climate. *Front. Ecol. Environ.*, **1**, 38–44.

Foley, J. A., R. DeFries, G. P. Asner, *et al.*, 2005. Global consequences of land use. *Science*, **309**, 570–574.

Fourier, J. B., 1827. Remarques générales sur les températures du globe terrestre et des espaces planétaries. *Mém. Acad. Sci. Paris*, **7**, 570–604.

Fournier d'Albe, E. M., and P. M. Aleman, 1976. A large-scale cloud seeding experiment in the Rio Nazas Catchment Area, Mexico. *Proc. 2nd World Meteorological Organization Scientific Conf. Weather Modification*, Boulder, CO, Aug 2–6, 1976, 143–149.

Fraedrich, K., A. Kleidon, and F. Lunkeit, 1999. A green planet versus a desert world: Estimating the effect of vegetation extremes on the atmosphere. *J. Climate*, **12**, 3156–3163.

Franchito, S. H., and V. B. Rao, 1992: Climatic change due to land surface alterations. *Climat. Change*, **22**, 1–34.

Friedlingstein, P., L. Bopp, P. Ciais, *et al.*, 2001. Positive feedback between future climate change and the carbon cycle. *Geophys. Res. Lett.*, **28**, 1543–1546.

Friis-Christensen, E., and K. Lassen, 1991. Length of the solar cycle: An indicator of solar activity closely associated with climate. *Science*, **254**, 698–700.

Gaertner, M. A., O. Christensen, J. A. Prego, *et al.*, 2001. The impacts of deforestation on the hydrological cycle in the western Mediterranean: An ensemble study with to regional climate models. *Climate Dynamics*, **17**, 875–887.

Gagin, A., 1965. Ice nuclei, their physical characteristics and possible effect on precipitation initiation. *Proc. Int. Conf. on Cloud Physics*, Tokyo and Sapporo, Japan, pp. 155–162.

Gagin, A., 1971. Studies of the factors governing the colloidal stability of continental cumulus clouds. *Proc. Int. Conf. on Weather Modification*, Canberra, Australia, *Amer.Med.Soc.* Sept. 6–11, 1971 pp. 5–11.

Gagin, A., 1975. The ice phase in winter continental cumulus clouds. *J. Atmos. Sci.*, **32**, 1604–1614.

Gagin, A., 1986. Evaluation of "static" and "dynamic" seeding concepts through analysis of Israeli II and FACE-2 experiments. *Meteor. Monogr.*, **43**, 63–70.

Gagin, A., and J. Neumann, 1974. Rain stimulation and cloud physics in Israel. In *Weather and Climate Modification*. W. N. Hess, ed. New York: John Wiley, pp. 454–494.

Gagin, A., and J. Neumann, 1981. The second Israeli randomized cloud seeding experiment: Evaluation of the results. *J. Appl. Meteor.*, **20**, 1301–1311.

Gagin, A., D. Rosenfeld, and R. E. Lopez, 1985. The relationship between height and precipitation characteristics of summertime convective cells in South Florida. *J. Atmos. Sci.*, **42**, 84–94.

Galloway, J. N., J. D. Aber, J. W. Erisman, *et al.*, 2003. The nitrogen cascade. *BioScience*, **53**, 341–356.

Galloway, J., F. Dentener, D. G. Capone, *et al.*, 2004. Nitrogen cycles: Past, present and future. *Biogeochemistry* **70**, 153–226, doi: 10.1007/s10533–004-0370–0.

Ganopolski, A., C. Kubatzki, M. Claussen, V. Brovkin, and V. Petoukhov, 1998. The influence of vegetation–atmosphere–ocean interaction on climate during the mid-Holocene. *Science*, **280**, 1916–1919.

Gao, X. J., Y. Luo, W. T. Lin, Z. C. Zhao, and F. Giorgi, 2003. Simulation of effects of land use change on climate in China by a regional climate model. *Adv. Atmos. Sci.*, **20**, 583–592.

Garrett, A. J., 1982. A parameter study of interactions between convective clouds, the convective boundary layer, and forested surface. *Mon. Wea. Rev.*, **110**, 1041–1059.

Garstang, M., S. Ulanski, S. Greco, *et al.*, 1990. The Amazon Boundary-Layer Experiment (ABLE 2B): A meteorological perspective. *Bull. Am. Meteor. Soc.*, **71**, 19–32.

Gash, J. H. C., and C. A. Nobre, 1997. Climatic effects of Amazonian deforestation: Some results from ABRACOS. *Bull. Am. Meteor. Soc.*, **78**, 823–830.

Gast, P. R., A. S. Jursa, J. Castelli, S. Basu, and J. Aarons, 1965. Solar electromagnetic radiation. In *Handbook of Geophysics and Space Environments*, S. L. Valley, ed. New York: McGraw-Hill, pp. 16-1–16-38.

Gates, D. L., and S. Liess, 2001. Impacts of deforestation and afforestation in the Mediterranean region as simulated by the MPI atmospheric GCM. *Global Planet. Change*, **30**, 305–324.

Gedney, N., and P. J. Valdes, 2000. The effect of Amazonian deforestation on the Northern Hemisphere circulation and climate. *Geophys. Res. Lett.*, **27**, 3053–3056.

Gentry, R. C., 1974. Hurricane modification. In *Climate and Weather Modification*, W. N. Hess, ed. New York: John Wiley, pp. 497–521.

Georgescu, M., C. P. Weaver, R. Avissar, R. L. Walko, and G. Miguez-Macho, 2003. Sensitivity of model-simulated summertime precipitation over the Mississippi River Basin to the spatial distribution of initial soil moisture. *J. Geophys. Res.*, **108**, 8855, doi:10.1029/2002JD003107.

Gero, A. F., A. J. Pitman, G. T. Narisma, C. Jacobson, and R. A. Pielke Sr., 2006. The impact of land cover change in the Sydney Basin, Australia. *Global Planet. Change*, in press.

Ghan, A. J., L. R. Leung, R. C. Easter, and H. Abdul-Razzak, 1997. Prediction of droplet number in a general circulation model. *J. Geophys. Res.*, **102**, 21777–21794.

Ghan, S. J., M. C. MacCracken, and J. J. Walton, 1988. Climatic response to large atmospheric smoke injections: Sensitivity studies with a tropospheric general circulation model. *J. Geophys. Res.*, **93**, 8315–8337.

Ghuman, B. S., and R. Lal, 1986. Effects of deforestation on soil properties and microclimate of a high rain forest in southern Nigeria. In *The Geophysiology of Amazonia*, R. E. Dickinson, ed. New York: John Wiley, pp. 225–244.

Giambelluca, T. W., J. Fox, S. Yarnasarn, P. Onibutr, and M. A. Nullet, 1999. Dry-season radiation balance of land covers replacing forest in northern Thailand. *Agric. Forest Meteor.*, **95**, 53–65.

Giorgi, F., and G. Visconti, 1989. Two-dimensional simulations of possible mesoscale effects of nuclear war fires. II. Model results. *J. Geophys. Res.*, **94**, 1145–1163.

Givati, A., and D. Rosenfeld, 2004. Quantifying precipitation suppression due to air pollution. *J. Appl. Meteor.*, **43**, 1038–1056.

Glantz, M. H., R. W. Katz, and N. Nicholls (eds.), 1991. *Teleconnections Linking Worldwide Climate Anomalies*. Cambridge, UK: Cambridge University Press.

Golding, B. W., P. Goldsmith, N. A. Machin, and A. Slingo, 1986. Importance of local mesoscale factors in any assessment of nuclear winter. *Nature*, **319**, 301–303.

Gong, C., and E. A. B. Eltahir, 1996. Sources of moisture for rainfall in West Africa. *Water Resources Res.*, **32**, 3115–3121.

Gopalakrishnan, S. G., M. Sharan, R. T. McNider, and M. P. Singh, 1998. Study of radiative and turbulent processes in the stable boundary layer under weak wind conditions. *J. Atmos. Sci.*, **55**, 954–960.

Gorbunov, B., A. Baklanov, N. Kakutkina, H.L. Windsor, and R. Toumi, 2001. Ice nucleation on soot particles. *J. Aerosol Sci.*, **32**, 199–215.

Gordon, L., M. Dunlop, and B. Foran, 2003. Land cover change and water vapour flows: Learning from Australia. *Phil. Trans. Roy. Soc. B*, **358**, 1973–1984.

Gordon, L. J., W. Steffen, B. F. Jönsson, *et al.*, 2005. Human modification of global water vapor flows from the land surface. *Proc. Nat Acad. Sci. USA*, **102**, 7612–7617.

Goutorbe, J.-P., T. Lebel, A. Tinga, *et al.*, 1994. HAPEX-Sahel: A large-scale study of land-atmosphere interactions in the semi-arid tropics. *Annales de Géophysique*, **12**, 53–64.

Govindasamy, B., P. B. Duffy, and K. Caldeira, 2001. Land use changes and Northern Hemisphere cooling. *Geophys. Res. Lett.*, **28**, 291–294.

Grabowski, W. W., 2001. Coupling cloud processes with the large-scale dynamics using the cloud resolving convection parameterization (CRCP). *J. Atmos. Sci.*, **58**, 978–997.

Grant, L. O., 1986. Hypotheses for the Climax wintertime orographic cloud seeding experiments. *Meteor. Monogr.*, **21**, 105–108.

Grant, L. O., and R. E. Elliot, 1974. The cloud seeding temperature window. *J. Appl. Meteor.*, **13**, 355–363.

Grant, L. O., and P. W. Mielke, 1967. A randomized cloud seeding experiment at Climax, Colorado, 1960–1965. *Proc. 5th Berkeley Symposium on Mathematical Statistics and Probability*, 115–131.

Grant, L. O., C. F. Chappell, L. W. Crow, *et al.*, 1969. *An Operational Adaptation Program of Weather Modification for the Colorado River Basin*, Interim Report to the Bureau of Reclamation, Contract No. 14-06-D-6467. Fort Collins, CO: Department of Atmospheric Science, Colorado State University.

Grasso, L., 1996. Numerical simulation of the May 15 and April 26, 1991 tornadic thunderstorms. Ph.D. dissertation, Department of Atmospheric Science, Colorado State University, Fort Collins, CO.

Grasso, L. D., 2000. The numerical simulation of dryline sensitivity to soil moisture. *Mon. Wea. Rev.*, **128**, 2816–2834.

Gray, W. M., 1990. Strong association between West African rainfall and US landfall of intense hurricanes. *Science*, **249**, 1251–1256.

Gray, W. M., 1991. Florida's hurricane vulnerability: Hostage to African rainfall. *Proc. 5th Governor's Hurricane Conference*, Tampa, FL, June 7, 1991, 24 pp.

Griffiths, J. F., and D. M. Driscoll, 1982. *Survey of Climatology*. Columbus, OH: Charles E. Merrill.

Grossman, R. L., D. Yates, M. A. LeMone, M.L. Wesely, and J. Song, 2005. Observed effects of horizontal radiative surface temperature variations on the atmosphere over a midwest watershed during CASES 97. *J. Geophys. Res.*, **110**, **D6**, D06117, doi:10.1029/2004JD004542.

Gu, L., D. D. Baldocchi, S. C. Wofsy, *et al.*, 2003. Response of a deciduous forest to the Mount Pinatubo eruption: Enhanced photosynthesis. *Science*, **299**, 2035–2038.

Guillevic, P., R. D. Koster, M. J. Suarez, *et al.*, 2002. Influence of the interannual variability of vegetation on the surface energy balance: A global sensitivity study. *J. Hydrometeor.*, **3**, 617–629.

Haberle, R. M., T. P. Ackerman, and O.B. Toon, 1985. Global transport of atmospheric smoke following a major nuclear exchange. *Geophys. Res. Letters*, **12**, 405–408.

Hadfield, M. G., W. R. Cotton, and R. A. Pielke Sr., 1991. Large-eddy simulations of thermally forced circulations in the convective boundary layer. I. A small-scale circulation with zero wind. *Bound.-Layer Meteor.*, **57**, 79–114.

Hahmann, A. N., and R. E. Dickinson, 1997. RCCM2-BATS Model over tropical South America: Applications to tropical deforestation. *J. Climate*, **10**, 1944–1964.

Hale, R. C., K. P. Gallo, T.W. Owen, and T. R. Loveland, 2006. Land use/land cover change effects on temperature trends at U.S. Climate Normals station. *Geophys. Res. Lett.*, **33**, L11703, doi:10.1029/2006GL026358.

Hallett, J., 1981. Ice crystal evolution in Florida summer cumuli following AgI seeding. *Proc. 8th Conf. Inadvertent and Planned Weather Modification*, Reno, NV, pp. 114–115.

Hallett, J., and S. C. Mossop, 1974. Production of secondary ice particles during the riming process. *Nature*, **249**, 26–28.

Haltiner, G. J., and R. T. Williams, 1980. *Numerical Prediction and Dynamic Meteorology*, 2nd edn. New York: John Wiley.

Hanamean, J. R., Jr., R. A. Pielke Sr., C. L. Castro, D. S. Ojima, B. C. Reed, and Z. Gao, 2003. Vegetation impacts on maximum and minimum temperatures in northeast Colorado. *Meteor. Appl.*, **10**, 203–215.

Hänel, R. A., B. J. Conrath, V. G. Kunde, *et al.*, 1972. The Nimbus 4 infrared spectroscopy experiment. I. Calibrated thermal emission spectra. *J. Geophys. Res.*, **11**, 2629–2641.

Hanesiak, J. M., R. L. Raddatz, and S. Lobban, 2004. Local initiation of deep convection on the Canadian Prairie provinces. *Bound.-Layer Meteor.*, **110**, 455–470.

Hansen, J. E., and L. Nazarenko, 2004. Soot climate forcing via snow and ice albedos. *Proc. Nat. Acad. Sci. USA*, **101**, 423–428, doi:10.1073/pnas.2237157100.

Hansen, J. E., W. Wong, and A. A. Lacis, 1978. Mount Agung eruption provides test of a global climatic perturbation. *Science*, **199**, 1065–1068.

Hansen, J. E., A. A. Lacis, P. Lee, and W.-C. Wang, 1980. Climatic effects of atmospheric aerosols. *Ann. N.Y. Acad. Sci.*, **338**, 575–587.

Hansen, J., I. Fung, A. Lacis, *et al.*, 1988. Global climate changes as forecast by Goddard Institute for Space Studies three-dimensional model. *J. Geophys. Res.*, **93**, 9341–9364.

Hansen, J., A. Lacis, D. Rind, *et al.*, 1986. The greenhouse effect: Projections of global climate change. In *Effects of Changes in Stratospheric Ozone and Global Climate*, J. G. Titus, ed. Washington, DC: US Environmental Protection Agency, pp. 199–218.

Hansen, J., M. Sato, and R. Ruedy, 1997a. Radiative forcing and climate response. *J. Geophys. Res.*, **102**, **D6**, 6831–6864.

Hansen, J., M. Sato, A. Lacis, and R. Ruedy, 1997b. The missing climate forcing. *Phil. Trans. Roy. Soc. B*, **352**, 231–240.

Hansen, J., M. Sato, R. Ruedy, *et al.*, 2005. Efficacy of climate forcings. *J. Geophys. Res.*, **110**, D18104, doi:10.1029/2005JD005776.

Hartmann, D. L., 1984. On the role of global-scale waves in ice–albedo and vegetation–albedo feedback. *Geophysi. Monogr.*, **5**, 18–28.

Harwell, M. A., and T. C. Hutchinson, 1985. *Environmental Consequences of Nuclear War*, Vol. 2, *Ecological and Agricultural Effects*. Chichester, UK: John Wiley.

Hasler, N., R. Avissar, and S. Glen, 2005. Issues in simulating the annual precipitation of a semi-arid region in Central Spain. *J. Hydrometeor.*, **6**, 409–422.

Havens, B. S., J. E. Jiusto, and B. Vonnegut, 1978. *Early History of Cloud Seeding*. Albany, NY: Project Cirrus Fund.

Hayden, B. P., 1998. Ecosystem feedbacks on climate at the landscape scale. *Phil Trans. Roy. Soc. B*, **353**, 5–18.

Heck, P., D. Lüthi, and C. Schär, 1999. The influence of vegetation on the summertime evolution of European soil moisture. *Phys. Chem. Earth B*, **24**, 609–614.

Heck, P., D. Lüthi, H. Wernli, and C. Schär, 2001. Climate impacts of European-scale anthropogenic vegetation changes: A sensitivity study using a regional climate model. *J. Geophys. Res.*, **106**, 7818–7835.

Henderson-Sellers, A., and V. Gornitz, 1984. Possible climatic impacts of land cover transformations, with particular emphasis on tropical deforestation. *Climate Change*, **6**, 231–257.

Henderson-Sellers, A., R. E. Dickinson, T. B. Durbridge, *et al.*, 1993. Tropical deforestation: Modeling local- to regional-scale climate change. *J. Geophys. Res.*, **98**, 7289–7315.

Herron, M. M, 1982. Impurity sources of F^-, Cl^-, NO_3, and SO_4^{2-} in Greenland and Antarctic precipitation. *J. Geophys. Res.*, **87**, 3052–3060.

Hindman, E. E., II, 1978. Water droplet fogs formed from pyrotechnically generated condensation nuclei. *J. Wea. Mod.*, **10**, 77–96.

Hindman, E. E., II, P. V. Hobbs, and L. F. Radke, 1977a. Cloud condensation nuclei from a paper mill. I. Measured effects on clouds. *J. Appl. Meteor.*, **16**, 745–752.

Hindman, E. E., II, P. M. Tag, B. A. Silverman, and P. V. Hobbs, 1977b. Cloud condensation nuclei from a paper mill. II. Calculated effects on rainfall. *J. Appl. Meteor.*, **16**, 753–755.

Hjelmfelt, M. R., 1980. Numerical simulation of the effects of St. Louis on boundary layer airflow and convection. Ph.D. dissertation, University of Chicago, Chicago, IL.

Hobbs, P. V., and L. F. Radke, 1969. Cloud condensation nuclei from a simulated forest fire. *Science*, **163**, 279–280.

Hobbs, P. V., and A. L. Rangno, 1985. Ice particle concentrations in clouds. *J. Atmos. Sci.*, **42**, 2523–2549.

Hobbs, P. V., L. F. Radke, and S. E. Shumway, 1970. Cloud condensation nuclei from industrial sources and their apparent influence on precipitation in Washington State. *J. Atmos. Sci.*, **27**, 81–89.

Hobbs, P. V., T. J. Matejka, P. H. Herzegh, J. D. Locatelli, and R. A. Houze, 1980. The mesoscale and microscale structure and organization of clouds and precipitation in midlatitude cyclones. I. A case study of a cold front. *J. Atmos. Sci.*, **37**, 568–596.

Hobbs, P. V., T. J. Garrett, R. J. Ferek, *et al.*, 2000. Emission from ships with respect to their effects on clouds. *J. Atmos. Sci.*, **57**, 2570–2590.

Hoffmann, W. A., and R. B. Jackson, 2000. Vegetation–climate feedbacks in the conversion of tropical savanna to grassland. *J. Climate*, **13**, 1593–1602.

Hoffmann, W. A., W. Schroeder, and R. B. Jackson, 2002. Positive feedbacks of fire, climate, and vegetation and the conversion of tropical savanna. *Geophys. Res. Lett.*, **29**, 2052, doi:10.1029/2002GL015424.

Holland, E. A., B. H. Braswell, J. Sulzman, and J.-F. Lamarque, 2005. Nitrogen deposition onto the United States and Western Europe: A synthesis of observations and models. *Ecol. Appl.*, **15**, 38–57.

Holt, T. R., D. Niyogi, F. Chen, *et al.*, 2006. Effect of land–atmosphere interactions on the IHOP 24–25 May 2002 convection case. *Mon. Wea. Rev.*, **134**, 113–133.

Hong, X., M. J. Leach, and S. Raman, 1995. A sensitivity study of convective cloud formation by vegetation forcing with different atmospheric conditions. *J. Appl. Meteor.*, **34**, 2008–2028.

Hou, A. Y., 1998. Hadley circulation as a modulator of the extratropical climate. *J. Atmos. Sci.*, **55**, 2437–2457.

Houghton, H. G., 1938. Problems connected with the condensation and precipitation processes in the atmosphere. *Bull. Am. Meteor. Soc.*, **19**, 152–159.

Houghton, J. T., G. J. Jenkins, and J. J. Ephraums, 1990. *Climate Change: The IPCC Scientific Assessment*. Cambridge, UK: Cambridge University Press.

Houghton, J. T., Y. Ding, D. J. Griggs, *et al.* (eds.), 2001. *Climate Change 2001: The Scientific Basis*, Contribution of Working Group I to the 3 Assessment Report of the Intergovernmental Panel on Climate Change. Cambridge, UK: Cambridge University Press.

Houghton, R., 1987. Biotic changes consistent with the increased seasonal amplitude of atmospheric CO_2 concentrations. *J. Geophys. Res.*, **92**, 4223–4230.

Houghton, R., 1988. Reply to S. Idso, Comment on "Biotic changes consistent with the increased seasonal amplitude of atmospheric CO_2 concentrations." *J. Geophys. Res.*, **93**, 1747–1748.

Huang, J., H. M. van den Dool, and K. P. Georgakakos, 1996. Analysis of model-calculated soil moisture over the United States (1931–1993) and applications to long-range temperature forecasts. *J. Climate*, **9**, 1350–1362.

Huang, X., T. J. Lyons, and R. C.G. Smith, 1995a. Meteorological impact of replacing native perennial vegetation with annual agricultural species. In *Scale Issues in Hydrological Modelling*, J. D. Kalma and M Sivapalan, eds. Chichester, UK: Advantar Communications, pp. 401–410.

Huang, X., T. J. Lyons, and R. C. G. Smith, 1995b. Meteorological impact of replacing native perennial vegetation with annual agricultural species. *Hydrol. Processes*, **9**, 645–654.

Hudson, J. G., and A. D. Clarke, 1992. Aerosol and cloud condensation nuclei measurements in the Kuwait plume. *J. Geophys. Res.*, **97**, 14533–14536.

Hulstrom, R. L., and T. L. Stoffel, 1990. Some effects of the Yellowstone fire smoke cloud on incident solar irradiance. *J. Climate*, **3**, 1485–1490.

Hung, H.-M., A. Malinowski, and S. Martin, 2003. Kinetics of heterogeneous ice nucleation on the surfaces of mineral dust cores inserted into aqueous sulfate particles. *J. Phys. Chem. A*, **107**, 1296–1306.

Hutjes, R. W. A., P. Kabat, and A. J. Dolman, 2003. Land cover and the climate system. In *Global Environmental Change and Land Use*, A. J. Dolman, A. Verhagen, and I. Rovers, eds. Dordrecht, The Netherlands: Kluwer, pp. 73–110.

Ichinose, T., 2003. Regional warming related to land use change during recent 135 years in Japan. *J. Global Environ. Engin.*, **9**, 19–39.

Idso, S. B, 1988. Me and the modelers: Perhaps not so different after all. *Climat. Change*, **12**, 93.

Idso, S., R. Jackson, R. Reginato, B. Kimball, and F. Nakayama, 1975. The dependence of bare-soil albedo on soil water content. *J. Appl. Meteor.*, **14**, 109–113.

iLEAPS, 2004. Integrated Land Ecosystem-Atmosphere Processes Study (iLEAPS): The IGBP – Land-Atmosphere Project. Science Plan and Implementation Strategy. Available online at www.atm.helsinki.fi/ILEAPS/index.php?page=draft

Imbrie, J., and K. P. Imbrie, 1979. *Ice Ages: Solving the Mystery*. Short Hills, NJ: Enslow Publishers.

Intergovernmental Panel on Climate Change, 1991. *Climate Change: The IPCC Scientific Assessment Intergovernmental Panel on Climate Change*. New York: World Meteorological Organization and United Nations Environmental Program.

Intergovernmental Panel on Climate Change, 2001. *Summary for Policymakers: A Report of the Intergovernmental Panel on Climate Change*. Available online at www.ipcc.ch/

Iribarne, J. V., and R. G. de Pena, 1962. The influence of particle concentration on the evolution of hailstones. *Nubila*, **5**, 7–30.

Irizarry-Ortiz, M. M., G. Wang, and E. A. B. Eltahir, 2003. Role of the biosphere in mid-Holocene climate of west Africa. *J. Geophys. Res.*, **108**, **D2**, 4042, doi: 10.1029/2001JD000989.

Isono, K., M. Komabayasi, and A. Ono, 1959. The nature and origin of ice nuclei in the atmosphere. *J. Meteor. Soc. Japan*, **37**, 211–233.

Jenkinson, D. S., D. E. Adams, and A. Wild, 1991. Model estimates of CO_2 emissions from soil in response to global warming. *Nature*, **351**, 304–306.

Jensen, E., D. Starr, and B. Toon, 2004. Mission investigates tropical cirrus clouds. *EOS*, **84**, 45–50.

Jiang, H., G. Feingold, and W. R. Cotton, 2002. Simulations of aerosol-cloud-dynamical feedbacks resulting from entrainment of aerosol into the marine boundary layer during the Atlantic Stratocumulus Transition Experiment, *J. Geophys. Res.*, **107**, **D24**, 4813, doi:10.1029/2001JD001502.

Jin, M., J. M. Shepherd, and M. D. King, 2005. Urban aerosols and their variations with clouds and rainfall: A case study for New York and Houston. *J. Geophys. Res.*, **110**, D10520, doi:10.1029/JD005081.

Johnson, D. B., 1982. The role of giant and ultragiant aerosol particles in warm rain initiation. *J. Atmos. Sci.*, **39**, 448–460.

Jones, A. S., I. C. Guch, and T. H. Vonder Haar, 1998. Data assimilation of satellite-derived heating rates as proxy surface wetness data into a regional atmospheric mesoscale model. II. A case study. *Mon. Wea. Rev.*, **126**, 646–667.

Joussaume, S., K. E. Taylor, P. Braconnot, *et al.*, 1999. Monsoon changes for 6000 years ago: Results of 18 simulations from the Paleoclimate Modeling Intercomparison Project (PMIP). *Geophys. Res. Lett.*, **26**, 859–862.

Kabat, P., Claussen, M., Dirmeyer, P. A., *et al.* (eds.), 2004. *Vegetation, Water, Humans and the Climate: A New Perspective on an Interactive System.* New York: Springer-Verlag.

Kallos, G., V. Kotroni, K. Lagouvardos, and A. Papadopoulos, 1998. On the long-range transport of air pollutants from Europe to Africa. *Geophys. Res. Lett.*, **25**, 619–622.

Kalnay, E., and M. Cai, 2003. Impact of urbanization and land-use change on climate. *Nature*, **423**, 528–531.

Kalnay, E., M. Kanamitsu, R. Kistler, *et al.*, 1996. The NCEP/NCAR 40-year Reanalysis project. *Bull. Am. Meteor. Soc.*, **77**, 437–471.

Kalnay, E., M. Cai, H. Li, and J. Tobin, 2006. Estimation of the impact land of surface forcings-on temperature trends in eastern United States. *J. Geophys. Res.*, **111**, D06106, doi:10.1029/2005JD006555.

Kanae, S., T. Oki, and K. Musiake, 1994. Game-tropics studies on deforestation effects in Indochina. *Global Energy and Water Cycle Experiment (GEWEX) News*, **9**, 1–4.

Kanae, S., T. Oki, and K. Musiake, 2001. Impact of deforestation on regional precipitation over the Indochina peninsula. *J. Hydrometeor.*, **2**, 51–70.

Karl, T. R., and P. D. Jones, 1989. Urban bias in area-averaged surface air temperature trends. *Bull. Am. Meteor. Soc.*, **70**, 265–270.

Karyampudi, V. M., and T. N. Carlson, 1988. Analysis and numerical simulations of the Saharan air layer and its effects on easterly wave disturbances. *J. Atmos. Sci.*, **45**, 3102–3136.

Kaufman, Y. J., and R. S. Fraser, 1997. The effect of smoke particles on clouds and climate forcing. *Science*, **277**, 1636–1639.

Kaufman, Y. J., C. J. Tucker, and R. L. Mahoney, 1991. Fossil fuel and biomass burning effect on climate: Heating or cooling? *J. Climate*, **4**, 578–588.

Keenan, T., G. Holland, S. Rutledge, *et al.*, 1994. *Science Plan: Maritime Continent Thunderstorm Experiment (MCTEX)*, BMRC Research Report No. 44. Melbourne, Vic: Bureau of Meteorology.

Keith, D. W., J. F. DeCarolis, D. C. Denkenberger, *et al.*, 2004. The influence of large-scale wind power on global climate. *Proc. Natl. Acad. Sci. USA*, **101**, 16115–16120.

Keller, V. W., and R. I. Sax, 1981. Microphysical development of a pulsating cumulus tower: A case study. *Q. J. Roy. Meteor. Soc.*, **107**, 679–697.

Kellogg, W. W., 1991. Response to skeptics of global warming. *Bull. Am. Meteor. Soc.*, **72**, 499–511.

Kerr, R. A., 1982. Cloud seeding: One success in 35 years. *Science*, **217**, 519–521.

Kerr, R. A., 1991. Could the sun be warming the climate? *Science*, **254**, 652–653.

Khain, A., D. Rosenfeld, and A. Pokrovsky, 2005. Aerosol impact on the dynamics and microphysics convective clouds. *Q. J. Roy. Meteor. Soc.*, **131**, 1–25.

Khandekar, M. L., 2004. Are climate model projections reliable enough for climate policy? *Energy Environ.*, **15**, 521–526.

Khandekar, M. L., T. S. Murty, and P. Chittibabu, 2005. The global warming debate: A review of the state of science. *Pure Appl. Geophys.*, **162**, 1557–1586.

Khvorostyanov, V., and K. Sassen, 1998. Cloud model simulation of a contrail case study: Surface cooling against upper tropospheric warming. *Geophys. Res. Lett.*, **25**, 2145–2148.

Kiang, J. E., and E. A. B. Eltahir, 1999: Role of ecosystem dynamics in biosphere–atmosphere interaction over the coastal region of West Africa. *J. Geophys. Res.*, **104**, 31173–31189.

Kiehl, J. T., V. Ramanathan, J. J. Hack, A. Huffman, and K. Swett, 1999. The role of aerosol absorption in determining the atmospheric thermodynamic state during INDOEX (abstract). *EOS*, **80**, 184.

Kiehl, J. T., T. L. Schneider, P. J. Rasch, M. C. Barth, and J. Wong, 2000. Radiative forcing due to sulfate aerosols from simulations with the National Center for Atmospheric Research Community Climate Model, Version 3. *J. Geophys. Res.*, **105**, **D1**, 1441–1457.

Kim, Y., and E. A. B. Eltahir, 2004. Role of topography in facilitating coexistence of trees and grasses within savannas. *Water Resources Res.*, **40**, W07505, doi:10.1029/2003WR002578.

King, D. A., G. E. Bingham, and J. R. Kercher, 1985. Estimating the direct effect of CO_2 on soybean yield. *J. Environ. Mgt.*, **20**, 51–62.

Kleidon, A., 2004. Beyond Gaia: Thermodynamics of life and Earth system functioning. *Climat. Change*, **66**, 271–319.

Kleidon, A., and M. Heimann, 1999. Deep rooted vegetation, Amazonian deforestation, and climate: Results from a modelling study. *Global Ecol. Biogeog.*, **8**, 397–405.

Kleidon, A., and M. Heimann, 2000. Assessing the role of deep rooted vegetation in the climate system with model simulations: mechanism, comparison to observations and implications for Amazonian deforestation. *Climate Dynamics*, **16**, 183–199.

Kleidon, A., and S. Lorenz, 2001. Deep roots sustain Amazonian rainforest in climate model simulations of the last ice age. *Geophys. Res. Lett.*, **28**, 2425–2428.

Kleidon, A., K. Fraedrich, and M. Heimann, 2000. A green planet versus a desert world: Estimating the maximum effect of vegetation on the land surface climate. *Climat. Change*, **44**, 471–493.

Klein Goldewijk, K., 2001. Estimating global land use change over the past 300 years: The HYDE Database. *Global Biogeochem. Cycles*, **15**, 417–433.

Knight, C. A., and P. Squires (eds.), 1982. *Hailstorms of the Central High Plains, Parts I and II*. Boulder, CO: Colorado Associated University Press.

Koenig, L. R., 1977. The rime-splintering hypothesis of cumulus glaciation examined using a field-of-flow cloud model. *Q. J. Roy. Meteor. Soc.*, **103**, 585–606.

Koenig, L. R., and F. W. Murray, 1976. Ice-bearing cumulus cloud evolution: Numerical simulations and general comparison against observations. *J. Appl. Meteor.*, **15**, 747–762.

Kondrat'yev, K. Ya., 2004. Key aspects of global climate change. *Energy Environ.*, **15**, 469–503.

Kondrat'yev, K. Ya., 2005. *Atmospheric Aerosol Formation Processes, Properties, and Climatic Impacts*. Saint Petersburg, Russia: Research Center for Ecological Safety of the Russian Academy of Sciences.

Kondrat'yev, K. Ya., V. I. Binenko, and O. P. Petrenchuck, 1981. Radiative properties of clouds influenced by a city. *Izv. Atmos. Ocean. Phys.*, **17**, 122–127.

Koster, R. D., and M. J. Suarez, 2004. Suggestions in the observational record of land–atmosphere feedback operating at seasonal time scales. *J. Hydrometeor.*, **5**, 567–572.

Koster, R. D., M. J. Suarez, and M. Heiser, 2000. Variance and predictability of precipitation at seasonal-to-interannual timescales. *J. Hydrometeor.*, **1**, 26–46.

Koster, R. D., P. A. Dirmeyer, Z. Guo, *et al.*, 2004. Regions of strong coupling between soil moisture and precipitation. *Science*, **305**, 1138–1140.

Kramer, M. L., D. E. Seymour, M. E. Smith, R. W. Reeves, and T. T. Frankenberg, 1976. Snowfall observations from natural-draft cooling tower plumes. *Science*, **193**, 1239–1241.

Kratzer, P. A., 1956: *The Climate of Cities*. Boston, MA: American Meteorological Society.

Krauss, T. W., and J. R. Santos, 2004. Exploratory analysis of the effect of hail suppression operations on precipitation in Alberta. *Atmos. Res.*, **71**, 35–50.

Kreidenweis, S. M., J. E. Penner, S.-C. Wang, *et al.*, 1991. The effects of dimethylsulfide upon marine aerosol concentrations. *Atmos. Environ. A*, **25**, 2501–2511.

Kubatzki, C., and Claussen, M., 1998. Simulation of the global biogeophysical interactions during the Last Glacial Maximum. *Climate Dynamics*, **14**, 461–471.

Kuhn, P. M., 1970. Airborne observations of contrail effects on the thermal radiation budget. *J. Atmos. Sci.*, **27**, 937–942.

Kuo, H. L., 1974. Further studies of the parameterization of the influence of cumulus convection on large-scale flow. *J. Atmos. Sci.*, **31**, 1232–1240.

Kutzbach, J. E., and P. J. Guetter, 1986. The influence of changing orbital parameters and surface boundary conditions on climate simulations for the past 18 000 years. *J. Atmos. Sci.*, **43**, 1726–1759.

Kuzelka, R. D., 1993. Introduction. In *Flat Water: A History of Nebraska and Its Water*, Resources Report No. 12, C. A. Flowerday, ed. Lincoln, NE: University of Nebraska Conservation and Survey Division, pp. 1–7.

Labitzke, K., 1987. Sunspots, the QBO, and the stratosphere temperature in the North Polar region. *Geophys. Res. Lett.*, **14**, 535–537.

La Brecque, M., 1989. Detecting climate change. I. Taking the world's shifting temperature. *Mosaic*, **20**, 2–9.

Lacaux, J. P., J.A Warburton, J. Fournet-Fayard, and P. Waldteufel, 1985. The disposition of silver released from Soviet OBLAKO rockets in precipitation during hail suppression experiment Grossversuch IV. II. Case studies of seeded cells. *J. Climate Appl. Meteor.*, **24**, 977–992.

Lamb, D., J. Hallet, and R. I. Sax, 1981. Mechanistic limitations to the release of latent heat during the natural and artificial glaciation of deep convective clouds. *Q. J. Roy. Meteor. Soc.*, **107**, 935–954.

Lambin, E. F., H. J. Geist, and E. Lepers, 2003. Dynamics of land-use and land-cover change in tropical regions. *Ann. Rev. Environ. Resources*, **28**, 205–241.

Lamptey, B. L., E. J. Barron, and D. Pollard, 2005. Simulation of the relative impact of land cover and carbon dioxide to climate change from 1700 to 2100. *J. Geophys. Res.*, **110**, D20103, doi:10.1029/2005JD005916.

Landsberg, H. E., 1956. The climate of towns. In *Man's Role in Changing the Face of the Earth*, W. L. Thomas, ed. Chicago, IL: University of Chicago Press, pp. 584–603.

Landsberg, H. E., 1970. Man-made climatic changes. *Science*, **170**, 1265–1274.

Langmuir, I., 1948. The production of rain by chain reaction in cumulus clouds at temperatures above freezing. *J. Meteor.* **5**, 175.

Langmuir, I., 1953. *Project Cirrus*, Final Report, Contract W-36–039-SC-32427, RL-785. Schenectady, NY: General Electric Company Research Laboratory.

Larson, V. E., R. Wood, P. R. Field, *et al.*, 2001. Small-scale and mesoscale variability of scalars in cloudy boundary layers: One-dimensional probability density functions. *J. Atmos. Sci.*, **58**, 1978–1996.

Larson, V. E., J.-C. Golaz, H. Jiang, and W. R. Cotton, 2005. Supplying local microphysics parameterizations with information about subgrid variability: Latin hypercube sampling. *J. Atmos. Sci.*, **62**. 4010–4026.

Lawton, R. O., U. S. Nair, R. A. Pielke Sr., and R. M. Welch, 2001. Climatic impact of tropical lowland deforestation on nearby montane cloud forests. *Science*, **294**, 584–587.

Lean, J., 2005. Living with a variable sun. *Physics Today*, **58**, 32–38.

Lean, J., and P. R. Rowntree, 1997. Understanding the sensitivity of a GCM simulation of Amazonian deforestation to the specification of vegetation and soil characteristics. *J. Climate*, **10**, 1216–1235.

Lean, J., and D. A. Warrilow, 1989. Simulation of the regional climatic impact of Amazon deforestation. *Nature*, **342**, 411–413.

Ledley, T. S., and S. L. Thompson, 1986. Potential effect of nuclear war smokefall on sea ice. *Climate Change*, **8**, 155–171.

Lee, R., 1978. *Forest Micrometeorology*. New York: Columbia University Press.

Lee, T. J., R. A. Pielke, and P. W. Mielke Jr., 1995. Modeling the clear-sky surface energy budget during FIFE87. *J. Geophys. Res.*, **100**, 25585–25593.

Leemans, R., 1999. Land-use change and the terrestrial carbon cycle: The International Geosphere-Biosphere Programme (IGBP). *Global Change Newslett.*, **37**, 24–26.

Lenton, T. M., 2000. Land and ocean carbon cycle feedback effects on global warming in a simple earth system model. *Tellus, Ser. B*, **52**, 1159–1188.

Lepers, E., E. F. Lambin, A. C. Jenetos, *et al.*, 2005. A synthesis of information on rapid land-cover change for the period 1981–2000. *BioScience*, **55**, 115–124.

Levi, Y., and D. Rosenfeld, 1996. Ice nuclei, rainwater chemical composition, and static cloud seeding effects in Israel. *J. Appl. Meteor.*, **35**, 1494–1501.

Levin, Z., E. Ganor, and V. Gladstein, 1996. The effects of desert particles coated with sulfate on rain formation in the eastern Mediterranean. *J. Appl. Meteor.*, **35**, 1511–1523.

Levis, S., J. A. Foley, and D. Pollard, 1999. Potential high-latitude vegetation feedbacks on CO_2-induced climate change. *Geophys. Res. Lett.*, **26**, 747–750.

Levis, S., J. A. Foley, and D. Pollard, 2000. Large-scale vegetation feedbacks on a doubled CO_2 climate. *J. Climate*, **13**, 1313–1325.

Levitus, S., J. I. Antonov, T. P. Boyer, and C. Stephens, 2000. Warming of the world ocean. *Science*, **287**, 2225–2229.

Levitus, S., J. I. Antonov, J. Wang, *et al.*, 2001. Anthropogenic warming of Earth's climate system. *Science*, **292**, 267–269.

Levitus, S., J. Antonov, and T. Boyer, 2005. Warming of the world ocean, 1955–2003. *Geophys. Res. Lett.*, **32**, L02604, doi:10.1029/2004GL021592.

Levy, G., and W. R. Cotton, 1984. A numerical investigation of mechanisms linking glaciation of the ice-phase to the boundary layer. *J. Climate Appl. Meteor.*, **23**, 1505–1519.

Lewis, W., 1951. On a seven-day periodicity. *Bull. Am. Meteor. Soc.*, **32**, 192.

Li, B., and R. Avissar, 1994. The impact of spatial variability of land–surface heat fluxes. *J. Climate*, **7**, 527–537.

Li, D., H. Komiyama, K. Kurihara, and Y. Sato, 2000. Case studies of the impact of landscape changes on weather modification in western Australia in summer. *J. Geophys. Res.*, **105**, 12303–12315.

Lian, M. S., and R. D. Cess, 1977. Energy balance climate models: A reappraisal of ice–albedo feedback. *J. Atmos. Sci.*, **34**, 1058–1062.

Liepert, B. G., J. Feichter, U. Lohmann, and E. Roeckner, 2004. Can aerosols spin down the water cycle in a warmer and moister world? *Geophys. Res. Lett.*, **31**, L06207, doi:10.1029/2003GL019060.

Lim, Y.-K., M. Cai, E. Kalnay, and L. Zhou, 2005. Observational evidence of sensitivity of surface climate changes to land types and urbanization. *Geophys. Res. Lett.*, **32**, L22712, doi:10.1029/2005GL024267.

Lindzen, R. S., 1990. Some coolness concerning global warming. *Bull. Am. Meteor. Soc.*, **71**, 288–299.

Linkletter, G. O., and J. A. Warburton, 1977. An assessment of NHRE hail suppression technology based on silver analysis. *J. Appl. Meteor.*, **16**, 1332–1348.

Liou, K. N., J. L. Lee, S. C. Ou, A. Fu, and Y. Takano, 1991. Ice cloud microphysics, radiative transfer and large-scale cloud processes. *Meteor. Atmos. Phys.*, **46**, 41–50.

Liston, G. E., J. P. McFadden, M. Sturm, and R. A. Pielke Sr., 2002. Modelled changes in arctic tundra snow, energy and moisture fluxes due to increased shrubs. *Global Change Biol.*, **8**, 17–32.

Liu, Y., and R. Avissar, 1999a. A study of persistence in the land–atmosphere system using a general circulation model and observations. *J. Climate*, **12**, 2139–2153.

Liu, Y., and R. Avissar, 1999b. A study of persistence in the land-atmosphere system with a fourth-order analytical model. *J. Climate*, **12**, 2154–2168.

Liu, Y., C. P. Weaver, and R. Avissar, 1999. Toward a parameterization of mesoscale fluxes and moist convection induced by landscape heterogeneity. *J. Geophys. Res.*, **104**, 19515–19553.

Liu, Z., M. Notaro, J. Kutzbach, and N. Liu, 2006. Assessing global vegetation-climate feedbacks from observations. *J. Climate*, **19**, 787–814.

Lofgren, B. M., 1995a. Sensitivity of land–ocean circulations, precipitation, and soil moisture to perturbed land surface albedo. *J. Climate*, **8**, 2521–2542.

Lofgren, B. M., 1995b. Surface albedo–climate feedback simulated using two-way coupling. *J. Climate*, **8**, 2543–2562.

Lohmann, U., 2002. A glaciation indirect aerosol effect caused by soot aerosols. *Geophys. Res. Lett.*, **29**, dio:10.1029/2001GL014357.

Lohmann, U., and J. Feichter, 2005. Global indirect aerosol effects: A review. *Atmos. Chem. Phys.*, **5**, 715–737.

Lohmann, U., J. Feichter, C. C. Chung, and J. E. Penner, 1999. Prediction of the number of cloud droplets in the ECHAM GCM. *J. Geophys. Res.*, **104**, **D8**, 9169–9198.

Lomborg, B., 2001. *The Skeptical Environmentalist: Measuring the Real State of the World*. Cambridge, UK: Cambridge University Press.

Loose, T., and R. D. Bornstein, 1977. Observations of mesoscale effects on frontal movement through an urban area. *Mon. Wea. Rev.*, **105**, 562–571.

Loveland, T. R., B. C. Reed, J. F. Brown, *et al.*, 2000. Development of a global land cover characteristics database and IGBP DISCover from 1-km AVHRR data. *Int. J. Remote Sens.*, **21**, 1303–1330.

Lowenthal, D. H., R. D. Borys, J. C. Chow, F. Rogers, and G. E. Shaw, 1992. Evidence for long-range transport of aerosol from the Kuwaiti oil fires to Hawaii. *J. Geophys. Res.*, **97**, 14573–14580.

Lu, L., R. A. Pielke, G. E. Liston, W. J. Parton, D. Ojima, and M. Hartman, 2001. Implementation of a two-way interactive atmospheric and ecological model and its application to the central United States. *J. Climate*, **14**, 900–919.

Luoma, J. R., 1991. Gazing into our greenhouse future. *Audubon*, **March**, 52–59, 124–125.

Lynn, B. H., D. Rind, and R. Avissar, 1995a. The importance of mesoscale circulations generated by subgrid-scale landscape-heterogeneities in general circulation models. *J. Climate*, **8**, 191–205.

Lynn, B. H., F. Abramopoulos, and R. Avissar, 1995b. Using similarity theory to parameterize mesoscale heat fluxes generated by subgrid-scale landscape discontinuities in GCMs. *J. Climate*, **8**, 932–951.

Lynn, B. H., W.-K. Tao, and P. J. Wetzel, 1998. A study of landscape-generated deep moist convection. *Mon. Wea. Rev.*, **126**, 828–942.

Lyons, T. J., 2002. Clouds prefer native vegetation. *Meteor. Atmos. Phys.*, **80**, 131–140.

Lyons, T. J., P. Schwerdtfeger, J. M. Hacker, *et al.*, 1993. Land–atmosphere interaction in a semiarid region: The bunny fence experiment. *Bull. Am. Meteor. Soc.*, **74**, 1327–1334.

Lyons, T. J., R. C. G. Smith, and H. Xinmei, 1996. The impact of clearing for agriculture on the surface energy budget. *Int. J. Climatol.*, **16**, 551–558.

Lyons, W. A., 1999: Lightning. In *Storms*, R. A. Pielke Sr., and R. A. Pielke Jr., eds. New York: Routledge, pp. 60–79.

MacCracken, M. C., 1985. Carbon dioxide and climate change: Background and overview. In *The Potential Climatic Effects of Increasing Carbon Dioxide*, M. C. MacCracken and F. M. Luther, eds. Washington, DC: US Department of Energy.

MacCracken, M. C., 2002. Do the uncertainty ranges in the IPCC and US National Assessments account adequately for possibly overlooked climatic influences, *Climat. Change*, **52**, 13–23.

MacCracken, M. C., and J. S. Chang, 1975. *Preliminary Study of the Potential Chemical and Climatic Effects of Atmospheric Nuclear Explosions*, Livermore, CA: Lawrence Livermore Laboratory. Report UCRL-51653.

MacCracken, M. C., J. Smith, and A. C. Janetos, 2004. Reliable regional climate model not yet on horizon. *Nature*, **429**, 699.

Mahfouf, J.-F., E. Richard, and P. Mascart, 1987. The influence of soil and vegetation on the development of mesoscale circulations. *J. Climate Appl. Meteor.*, **26**, 1483–1495.

Mahmood, R., and K.G. Hubbard, 2002. Anthropogenic land-use change in North American tall grass–short grass transition and modification of near surface hydrologic cycle. *Climate Res.*, **21**, 83-90.

Mahmood, R., and K. G. Hubbard, 2003. Simulating sensitivity of soil moisture and evapotranspiration under heterogeneous soils and land uses. *J. Hydrol.*, **280**, 72–90.

Mahmood, R., K. G. Hubbard, and C. Carlson, 2004. Modification of growing-season temperature records in the Northern Great Plains due to land-use transformation: Verification of modeling results and implication for global climate change. *Int. J. Climatol.*, **24**, 311–327.

Mahmood, R., S. Foster, T. Keeling, *et al.*, 2006. Impacts of irrigation on 20th century temperature in the North American Great Plains. *Global Planet. Change*, **52**, doi:10.1016/j.gloplacha.2005.10.004.

Mahrer, Y., and R. A. Pielke, 1976. The numerical simulation of the air flow over Barbados. *Mon. Wea. Rev.*, **104**, 1392–1402.

Mahrer, Y., and R. A. Pielke, 1978. The meteorological effect of the changes in surface albedo and moisture. *Israel Meteor. Soc. (IMS)*, 55–70.

Mahrt, L., J. S. Sun, D. Vickers, *et al.*, 1994. Observations of fluxes and inland breezes over a heterogeneous surface. *J. Atmos. Sci.*, **51**, 2484–2499.

Malm, W. C., 1989. Atmospheric haze: Its sources and effects on visibility in rural areas of the continental United States. *Environ. Monit. Assess.*, **12**, 203–225.

Malone, R. C., L. H. Auer, G. A. Glatzmaier, M. C. Wood, and O. B. Toon, 1986. Three-dimensional simulations including interactive transport, scavenging, and solar heating of smoke. *J. Geophys. Res.*, **91**, 1039–1053.

Mann, M. E., and P. D. Jones, 2003. Global surface temperatures over the past two millennia. *Geophys. Res. Lett.*, **30**, 1820, doi:10.1029/2003GRL017814.

Marland, G., R. A. Pielke Sr., M. Apps, *et al.*, 2003. The climatic impacts of land surface change and carbon management, and the implications for climate-change mitigation policy. *Climate Policy*, **3**, 149–157.

Marshall, C. H. Jr., R. A. Pielke Sr., and L. T. Steyaert, 2003. Crop freezes and land-use change in Florida. *Nature*, **426**, 29–30.

Marshall, C. H. Jr., R. A. Pielke Sr., L. T. Steyaert, and D. A. Willard, 2004a. The impact of anthropogenic land cover change on warm season sensible weather and sea-breeze convection over the Florida peninsula. *Mon. Wea. Rev.*, **132**, 28–52.

Marshall, C. H. Jr., R. A. Pielke Sr., and L. T. Steyaert, 2004b. Has the conversion of natural wetlands to agricultural land increased the incidence and severity of damaging freezes in south Florida? *Mon. Wea. Rev.*, **132**, 2243–2258.

Martin, G. M., D. W. Johnson, and A. Spice, 1994. The measurement and parameterization of effective radius of droplets in warm stratocumulus clouds. *J. Atmos. Sci.*, **51**, 1823–1842.

Marwitz, J. D., 1972. The structure and motion of severe hailstorms. I. Supercell storms. *J. Appl. Meteor.,* **11**, 166–179.

Mason, B. J., 1971. *The Physics of Clouds*, 2nd edn. Oxford, UK: Clarendon Press.

Mather, G. K., 1991. Coalescence enhancement in large multicell storms caused by the emissions from a Kraft paper mill. *J. Appl. Meteor.*, **30**, 1134–1146.

Mather, G. K., M. J. Dixon, and J. M. de Jager, 1996. Assessing the potential for rain augmentation: The Nelspruit randomized convective cloud seeding experiment. *J. Appl. Meteor.*, **35**, 1465–1482.

Matsui, T., and R. A. Pielke Sr., 2006. Measurement-based estimation of the spatial gradient of aerosol radiative forcing. *Geophys. Res. Lett.*, **33**, L11813, doi:10.1029/2006GL025974.

Matsui, T., H. Masunaga, R. A. Pielke Sr., and W.-K. Tao, 2004. Impact of aerosols and atmospheric thermodynamics on cloud properties within the climate system. *Geophys. Res. Lett.*, **31**, L06109, doi:10.1029/2003GL019287.

Matsui, T., H. Masunaga, R. A. Pielke Sr., *et al.*, 2006. Satellite-based assessment of marine low cloud variability associated with aerosol, atmospheric stability, and the diurnal cycles. *J. Geophys. Res.*, **111**, doi: 10.1029/2005JD006097.

Matthews, H. D., A. J. Weaver, K. J. Meissner, N. P. Gillett, and M. Eby, 2004. Natural and anthropogenic climate change: Incorporating historical land cover change, vegetation dynamics and the global carbon cycle. *Climate Dynamics*, **22**, 461–479.

Mayewski, P. A., W. B. Lyons, M. J. Spencer, *et al.* 1986. Sulfate and nitrate concentrations from a south Greenland ice core. *Science,* **232**, 975–977.

Maynard, K., and J.-F. Royer, 2004. Sensitivity of a general circulation model to land surface parameters in African tropical deforestation experiments. *Climate Dynamics*, **22**, 555–572.

McClatchey, R. A., R. W. Fenn, J. E. A. Selby, F. E. Volz, and J. S. Garing, 1972. *Optical Properties of the Atmosphere*, 3rd edn. AFCRL-72-0497. Hanscom, MA: Airforce Cambridge Research Laboratory.

McCumber, M. C., and R. A. Pielke, 1981. Simulation of the effects of surface fluxes of heat and moisture in a mesoscale numerical model. I. Soil layer. *J. Geophys. Res.*, **86**, 9929–9938.

McFadden, J. P., G. E. Liston, M. Sturm, R. A. Pielke Sr., and F. S. Chapin, III, 2001. Interactions of shrubs and snow in Arctic tundra: Measurements and models. In *Soil–Vegetation–Atmosphere Transfer Schemes and Large-Scale Hydrological Models*, A. J. Dolman *et al.*, eds. Wallingford, UK: Institute of Hydrology, pp. 317–325.

McGuffie, L., A. Henderson-Sellers, H. Zhang, T. B. Durbidge, and A.J. Pitman, 1995. Global climate sensitivity to tropical deforestation. *Global Planet Change*, **10**, 97–128.

McNider, R. T., and F. J. Kopp, 1990. Specification of the scale and magnitude of thermals used to initiate convection in cloud models. *J. Appl. Meteor.*, **29**, 99–104.

McNider, R. T., M. D. Moran, and R. A. Pielke, 1988. Influence of diurnal and inertial boundary layer oscillations on long-range dispersion. *Atmos. Environ.*, **22**, 2445–2462.

McPherson, E. G., and G. C. Woodard, 1989. The case for urban releaf: Tree planting pays. *Arizona's Economy*, **December**.

McPherson, E. G., and G. C. Woodard, 1990. Cooling the urban heat island with water-and-energy efficient landscapes. *Arizona Review*, **1990**, 1–8.

McPherson, R. A., D. J. Stensrud, and K. C. Crawford, 2004. The impact of Oklahoma's winter wheat belt on the mesoscale environment. *Mon. Wea. Rev.*, **132**, 405–421.

Mears, C. A., and F. J. Wentz, 2005. The effect of diurnal correction on satellite-derived lower tropospheric temperature. *Science Express*, August 11, doi:10.1126/science. 1114772.

Mears, C. A., M. C. Schabel, and F. J. Wentz, 2003. A reanalysis of the MSU channel 2 tropospheric temperature record. *J. Climate*, **16**, 3650–3664.

Meitín, J. G., W. L. Woodley, and J. A. Flueck, 1984. Exploration of extended-area treatment effects in FACE-2 using satellite imagery. *J. Climate Appl. Meteor.*, **23**, 63–83.

Menon, S., J. Hansen, L. Nazarenko, and Y. Luo, 2002a. Climate effects of black carbon aerosols in China and India. *Science*, **297**, 2250–2253.

Menon, S., A. D. Del Genio, D. Koch, and G. Tselioudis, 2002b. GCM simulations of the aerosol indirect effect: Sensitivity to cloud parameterization and aerosol burden. *J. Atmos. Sci.*, **59**, 692–713.

Mesinger, F., and N. Mesinger, 1992. Has hail suppression in eastern Yugoslavia led to a reduction in the frequency of hail? *J. Appl. Meteor.*, **31**, 104–111.

Meyers, M. P., P. J. DeMott, and W. R. Cotton, 1992. New primary ice nucleation parameterizations in an explicit cloud model. *J. Appl. Meteor.*, **31**, 708–721.

Miao, J. -F., L. J. M. Kroom, J. Vila-Guerau de Arellano, and A.A.M. Holtslag, 2003. Impacts of topography and land degradation on the sea breeze over eastern Spain. *Meteor. Atmos. Phys.*, **84**, 157–270.

Michaels, P. J., 2004. *Meltdown: The Predictable Distortion of Global Warming by Scientists, Politicians, and the Media*. Washington, DC: Cato Institute.

Michaels, P. J., D. E. Sappington, D. E. Stooksbury, and B. P. Hayden, 1990. Regional 500 mb heights and US 1000-500 mb thickness prior to the radiosonde era. *Theor. Appl. Climatol.*, **42**, 149–154.

Mielke, P. W., 1995. Comments on the "Climax I and II experiments including replies to Rangno and Hobbs". *J. Appl. Meteor.*, **34**, 1228–1232.

Mielke, P. W., L. O. Grant, and C. F. Chappell, 1970. Elevation and spatial variation effects of wintertime orographic cloud seeding. *J. Appl. Meteor.*, **9**, 476–488; Corrigenda: **10**, 842; **15**, 801.

Mielke, P. W., L. O. Grant, and C. F. Chappell, 1971. An independent replication of the Climax wintertime orographic cloud seeding experiment. *J. Appl. Meteor.*, **10**, 1198–1212; Corrigendum: **15**, 801.

Mielke, P. W., G. W. Brier, L. O. Grant, G. J. Mulvey, and P. N. Rosensweig, 1981. A statistical reanalysis of the replicated Climax I and II wintertime orographic cloud seeding experiments. *J. Appl. Meteor.*, **20**, 643-660.

Millán, M. M., M. J. Estrela, M. J. Sanz, *et al.*, 2005. Climatic feedbacks and desertification: the Mediterranean model. *J. Climate,* **18**, 684–701.

Miller, G., J. Mangan, D. Pollard, S. Thompson, B. Felzer, and J. Magee, 2005. Sensitivity of the Australian Monsoon to insolation and vegetation: Implications for human impact on continental moisture balance. *Geology*, **January**, 65–68.

Miller, W. F., R. A. Pielke, M. Garstang, and S. Greco, 1988. Simulations of the mesoscale circulation in the Able II region. *Proc. 5th Brazilian Congress of Meteorology*, Nov. 5-11, 1988, Rio de Janeiro, Brazil.

Milly, P. C. D., and K. A. Dunne, 1994. Sensitivity of the global water cycle to the water-holding capacity of land. *J. Climate*, **7**, 506–526.

Ming, Z., and Z. Xinmin, 2002. A theoretical analysis on the local climate change induced by the change of land use. *Adv. Atmos. Sci.*, **19**, 45–63.

Mitchell, J. F. B., C. A. Senior, and W. J. Ingram, 1989. CO_2 and climate: A missing feedback? *Nature,* **341**, 132–134.

Mohr, K. I., R. D. Baker, W. K. Tao, and J. S. Famiglietti, 2003. The sensitivity of West African convective line water budgets to land cover. *J. Hydrometeor.*, **4**, 62–76.

Mölders, N., 1998. Landscape changes over a region in East Germany and their impact upon the processes of its atmospheric water-cycle. *Meteor. Atmos. Phys.*, **68**, 79–98.

Mölders, N., 1999a. On the atmospheric response to urbanization and open-pit mining under various geostrophic wind conditions. *Meteor. Atmos. Phys.*, **71**, 205–228.

Mölders, N., 1999b. On the effects of different flooding stages of the Oder and different land-use types on the distributions of evapotranspiration, cloudiness and rainfall in the Brandenburg–Polish border area. *Contrib. Atmos. Phys.*, **72**, 1–25.

Mölders, N., 2000a. Similarity of microclimate as simulated over a landscape of the 1930s and the 1980s. *J. Hydrometeor.*, **1**, 330–352.

Mölders, N., 2000b. Application of the principle of superposition to detect nonlinearity in the short-term atmospheric response to concurrent land-use changes associated with future landscapes. *Meteor. Atmos. Phys.*, **72**, 47–68.

Mölders, N., and M. A. Olson, 2004. Impact of urban effects on precipitation in high latitudes. *J. Hydrometeor.*, **5**, 409–429.

Mölders, N., and W. Rühaak, 2002. On the impact of explicitly predicted runoff on the simulated atmospheric response to small-scale land-use changes: An integrative modeling approach. *Atmos. Res.*, **63**, 3–38.

Mooney, M. L., and G. W. Lunn, 1969. The area of maximum effect resulting from the Lake Almanor randomized cloud seeding experiment. *J. Appl. Meteor.*, **8**, 68–74.

Moore, N., and S. Rojstaczer, 2002. Irrigation's influence on precipitation: Texas High Plains, USA. *Geophys. Res. Lett.*, **29**, doi:10.1029/2002GL014940.

Mossop, S. C., and J. Hallett, 1974. Ice crystal concentration in cumulus clouds: Influence of the drop spectrum. *Science,* **186,** 632–634.

Muhs, D. R., 1985. Age and paleoclimatic significant of Holocene sand dunes in northeastern Colorado. *Ann. Assoc. Am. Geogr.*, **75**, 566–582.

Muhs, D. R., and V. T. Holliday, 1995. Evidence of active dune sand on the Great Plains in the 19th century from accounts of early explorers. *Quatern. Res.*, **43**, 198–208.

Murcray, W. B., 1970. On the possibility of weather modification by aircraft contrails. *Mon. Wea. Rev.,* **98**, 745–748.

Murray, F. W., and L. R. Koenig, 1979. *Simulation of Convective Cloudiness, Rainfall, and Associated Phenomena Caused by Industrial Heat Released Directly to the Atmosphere*, Report R-2456-DOE. Rand Corporation, Santa Monica, CA: Report R-2456-DOE. Santa Monica, CA: Rand corporation.

Murty, B. V. R., and K. R. Biswas, 1968. Weather modification in India. *Proc., 1st National Conference on Weather Modification,* Albany, NY, Apr. 28 – May 1, pp. 71–80.

Myhre, G., and A. Myhre, 2003. Uncertainties in radiative forcing due to surface albedo changes caused by land-use changes. *J. Climate*, **16**, 1511–1524.

Mylne, M. F., and P. R. Rowntree, 1992. Modelling the effects of albedo change associated with tropical deforestation. *Climat. Change*, **21**, 317–343.

Nair, U. S., R. O. Lawton, R. M. Welch, and R. A. Pielke Sr., 2003. Impact of land use on Costa Rican tropical montane cloud forests: Sensitivity of cumulus cloud field characteristics to lowland deforestation. *J. Geophys. Res.* **108**, 10.1029/2001JD001135.

Namias, J., 1978. Multiple causes of the North American abnormal winter 1976-77. *Mon. Wea. Rev.*, **106**, 279–295.

Narisma, G. T., and A. J. Pitman, 2003. The impact of 200 years land cover change on the Australian near-surface climate. *J. Hydrometeor.*, **4**, 424–436.

Narisma, G. T., and A. J. Pitman, 2004. The effect of including biospheric feedbacks on the impact of land cover change over Australia. *Earth Interactions*, **8**, 1–28.

Narisma, G. T., and A. J. Pitman, 2006. Exploring the sensitivity of the Australian climate to regional land cover change scenarios under increasing CO_2 concentrations and warmer temperatures. *Earth Interactions*, **10**, 1-27.

Narisma, G. T., A. J. Pitman, J. Eastman, *et al.*, 2003. The role of biospheric feedbacks in the simulation of the impact of historical land cover change on the Australian January climate. *Geophys. Res. Lett.*, **30**, 2168, doi:10.1029/2003GL018261.

National Academy of Sciences, 1975. *Long-Term Worldwide Effects of Multiple Nuclear-Weapon Detonations.* Washington, DC: National Academy Press.

National Academy of Sciences, 1980. *The Atmospheric Sciences: National Objectives for the 1980s.* Washington, DC: National Academy Press.

National Academy of Sciences, 1991. *Policy Implications of Greenhouse Warming: Synthesis Panel Report.* Washington, DC: National Academy Press.

National Agricultural Statistics Service, 1998. *Crop Values: Final Estimates by States, 1992-97*, Statistical Bulletin, Agricultural Statistics Board. Washington, DC: US Department of Agriculture. Available online at http://www.usda.gov/nass/pubs/agr98/acro98.htm

National Research Council, 1985. *The Effects on the Atmosphere of a Major Nuclear Exchange.* Washington, DC: National Academy Press.

National Research Council, 2000. *Reconciling Observations of Global Temperature Change*, Panel on Reconciling Temperature Observations, Climate Research Committee, Board on Atmospheric Sciences and Climate. Washington, DC: National Academy Press.

National Research Council, 2003a. *Critical Issues in Weather Modification Research.* Washington, DC: National Academy Press.

National Research Council, 2003b. *Understanding Climate Change Feedbacks*, Panel on Climate Change Feedbacks, Climate Research Committee, Board on Atmospheric Sciences and Climate, Division on Earth and Life Studies. Washington, DC: National Academy Press.

National Research Council, 2005. *Radiative Forcing of Climate Change: Expanding the Concept and Addressing Uncertainties*, Committee on Radiative Forcing Effects on Climate, Climate Research Committee, Board on Atmospheric Sciences and Climate (BASC). Washington, DC: National Academy Press. Available online at http://www.nap.edu/catalog/11175.html

Neftel, A., J. Beer, H. Oeschger, F. Zürcher, and R.C. Finkel, 1985. Sulfate and nitrate concentrations in snow from south Greenland. *Nature,* **314**, 611–613.

Negri, A. J., R. F. Adler, L. Xu, and J. Surrat, 2004. The impact of Amazonian deforestation on dry season rainfall. *J. Climate*, **7**, 1306–1319.

Nelson, L. D., 1979. *Observations and Numerical Simulations of Precipitation Mechanisms in Natural and Seeded Convective Clouds*, Technical Note No. 54. Chicago, IL: University of Chicago. Department of Geophysical Sciences.

Nemani, R. R., S. W. Running, R. A. Pielke, and T. N. Chase, 1996. Global vegetation cover changes from coarse resolution satellite data. *J. Geophys. Res.*, **101**, 7157–7162.

Nemani, R., M. White, P. Thornton, *et al.*, 2002. Recent trends in hydrologic balance have enhanced the terrestrial carbon sink in the United States. *Geophys. Res. Lett.*, **29**, 1468, doi:10.1029/2002GL014867.

Nenes, A., W. C. Conant, and J. H. Seinfeld, 2002. Black carbon radiative heating effects on cloud microphysics and implications for the aerosol indirect effect. II. Cloud microphysics. *J. Geophys. Res.*, **107**, **D21**, 4605, doi:10:1029/2002JD002101.

Neumann, J., and Y. Mahrer, 1974. A theoretical study of the sea and land breezes of circular islands. *J. Atmos. Sci.*, **31**, 2027–2039.

Neumann, J., and S. Parpola, 1987. Climatic change and the eleventh–tenth-century eclipse of Assyria and Babylonia. *J. Near East. Stud.* **46**, 161–182.

Neumann, J., and R. M. Sigrist, 1978. Harvest dates in ancient Mesopotamia as possible indicators of climatic variations. *Climat. Change*, **1**, 239–252.

Nicholson, S., 2000. Land surface processes and Sahel climate. *Rev. Geophys.*, **38**, 117–139.

Nicholson, S. E., C. J. Tucker, and M. B. Ba, 1998. Desertification, drought, and surface vegetation: An example from the west African Sahel. *Bull. Am. Meteor. Soc.*, **79**, 815–829.

Nickovic, S., G. Kallos, A. Papadopoulos, and O. Kakaliagou, 2001. A model for prediction of desert dust cycle in the atmosphere. *J. Geophys. Res.*, **106**, 18113–18129.

Niyogi, D., and Y. K. Xue, 2006. Soil moisture regulates the biological response atmospheric CO_2 concentrations in a coupled atmospheric biospheric model. *Global Planet. Change*, in press.

Niyogi, D., Dutta S., R. A. Pielke Sr., *et al.*, 2003. Challenges of representing land-surface processes in weather and climate models over the tropics: Examples over the Indian subcontinent. In *Weather and Climate Modeling*, S. V. Singh, S. Basu, and T. N. Krishnamurti, eds. New Delhi, India: New Age International Publishers, pp. 132–145.

Niyogi, D., H. -I. Chang, V. K. Saxena, *et al.*, 2004. Direct observations of the effects of aerosol loading on the net ecosystem CO_2 exchanges over different landscapes. *Geophys. Res. Lett.*, **31**, L20506, doi:10.1029/2004GL020915.

Niyogi, D. T., Holt, S. Zhong, P. C. Pyle, and J. Basara, 2006. Urban and land surface effects on the 30 July 2003 MCS event observed in the Southern Great Plains. *J. Geophys. Res.*, in press.

Nober, F. J., Graf, H. -F., and D. Rosenfeld, 2003. Sensitivity of the global circulation to the suppression of precipitation by anthropogenic aerosols. *Global Planet. Change*, **37**, 57–80.

Nobre, C. A., P. J. Sellers, and J. Shukla, 1991. Amazonian deforestation and regional climate change. *J. Climate*, **4**, 957–988.

Northern Eurasia Earth Science Partnership Initiative (NEESPI), 2004. *NEESPI Science Plan*, P. Y. Groisman and S. A. Bartalev, eds., Available online at NEESPI.gsfc. nasa.gov/

O'Brien, K. L., 1997. *Sacrificing the Forest: Environmental and Social Struggles in Chiapas*. Boulder, CO: Westview Press.

O'Brien, K. L., 2000. Upscaling tropical deforestation: Implications for climate change. *Climat. Change*, **44**, 311–329.

Odingo, R. S., 1990. The definition of desertification: Its programmatic consequences for UNEP and the international community. *Desertification Control Bull.*, Nov. 18, 1990.

Ohtake, T., and P. J. Huffman, 1969. Visual range of ice fog. *J. Appl. Meteor.*, **8**, 499–501.

Oke, T. R., 1987. *Boundary Layer Climates*, 2nd edn. New York: Routledge.

Ookouchi, Y., M. Segal, R. C. Kessler, and R. A. Pielke Sr., 1984. Evaluation of soil moisture effects on the generation and modification of mesoscale circulations. *Mon. Wea. Rev.*, **112**, 2281–2292.

Orme, A. R. (ed.), 2002. *Physical Geography of North America.* Oxford, UK: Oxford University Press.

Orville, H. D., 1986. A review of dynamic-mode seeding of summer cumuli. *Meteor. Monogr.*, **43**.

Orville, H. D., and J. -M. Chen, 1982. Effects of cloud seeding, latent heat of fusion, and condensate loading on cloud dynamics and precipitation evolution: A numerical study. *J. Atmos. Sci.,* **39**, 2807–2827.

Orville, H. D., and K. Hubbard, 1973. On the freezing of liquid water in a cloud. *J. Appl. Meteor.*, **12**, 671–676.

Orville, H. D., J. H. Hirsch, and L. E. May, 1980. Application of a cloud model to cooling tower plumes and clouds. *J. Appl. Meteor.,* **19**, 1260–1272.

Orville, H. D., P. A. Eckhoff, J. E. Peak, J. H. Hirsch, and F. J. Kopp, 1981. Numerical simulation of the effects of cooling tower complexes on clouds and severe storms. *Atmos. Environ.,* **15**, 823–836.

Otterman, J., 1974. Baring high-albedo soils by overgrazing: A hypothesized desertification mechanism. *Science*, **86**, 531–533.

Otterman, J., 1977. Anthropogenic impact on the albedo of the earth. *Climat. Change*, **1**, 137–155.

Otterman, J., and C. J. Tucker, 1985. Satellite measurements of surface albedo and temperatures in semi-desert. *J. Climate Appl. Meteor.*, **24**, 228–235.

Otterman, J., A. Manes, S. Rubin, P. Alpert, and D.O'C. Starr, 1990. An increase of early rains in southern Israel following land-use change? *Bound.-Layer Meteor.*, **53**, 333–351.

Oyama, M. D., and C. A. Nobre, 2003. A new climate–vegetation equilibrium state for tropical South America. *Geophys. Res. Lett.*, **30**, 2199, doi:10.1029/ 2003GL018600.

Oyama, M. D., and C. A. Nobre, 2004. Climatic consequences of a large-scale desertification in northeast Brazil: A GCM simulation study. *J. Climate*, **17**, 3203–3213.

Paluch, I. R., and D. H. Lenschow, 1991. Stratiform cloud formation in the marine boundary layer. *J. Atmos. Sci.*, **48**, 2141–2158.

Pan, Z., M. Segal, R. Turner, and E. Takle, 1995. Model simulation of impacts of transient surface wetness on summer rainfall in the United States midwest during drought and flood years. *Mon. Wea. Rev.*, **123**, 1575–1581.

Pan, Z., E. Takle, M. Segal, and R. Turner, 1996. Influences of model parameterization schemes on the response of rainfall to soil moisture in the central United States. *Mon. Wea. Rev.*, **124**, 1786–1802.

Pan, Z., R. W. Arritt, E. S. Takle, *et al.*, 2004. Altered hydrologic feedback in a warming climate introduces a "warming hole." *Geophys. Res. Letts*, **31**, L17109, doi:10.1029/2004GL020528.

Patterson, E. M., B. T. Marshall, and K. A. Rahn, 1982. Radiative properties of the Arctic aerosol. *Atmos. Environ.*, **16**, 2967–2977.

Pecker, J.-C., and S. K. Runcorn (eds.), 1990. *The Earth's Climate and Variability of the Sun over Recent Millennia*. London: The Royal Society.

Peel, D., A. J. Pitman, L. Hughes, G. Narisma, and R. A. Pielke Sr., 2005. The impact of realistic biophysical parameters for eucalyptus on the simulation on the January climate of Australia. *Env. Modelling and Software*, **20**, **5**, 595–612, doi:10.1016/j.envsoft.2004.03.004.

Penner, J. E., 1986. Uncertainties in the smoke source term for "nuclear winter" studies. *Nature,* **324**, 222–226.

Peters, D. P. C., R. A. Pielke Sr., B. T. Bestelmeyer, *et al.*, 2004. Cross-scale interactions, nonlinearities, and forecasting catastrophic events. *Proc. Nat. Acad. Sci. USA*, **101**, 15130–15135.

Peterson, T. C., 1991. The relationships between SST anomalies and clouds, water vapor, and their radiative effects. Ph.D. dissertation, Department of Atmospheric Science, Colorado State University, Fort Collins, CO.

Philander, S. G., 1990. *El Niño, La Niña and the Southern Oscillation*. London: Academic Press.

Pielke, R. A. Jr., 2001. Room for doubt. *Nature*, **410**, 151.

Pielke, R. A. Jr., 2007. *The Honest Broker: Making Sense of Science in Policy and Politics*. Cambridge, UK: Cambridge Universtiy press.

Pielke, R. A. Jr., and R. A. Pielke Sr., 1997. *Hurricanes: Their Nature and Impacts on Society*. Chichester, UK: John Wiley.

Pielke, R. A. Jr., 2004. when scientists Politicize Science: making sense of controversy over *The Skeptical Environmentalist. Environ. Sci. Policy*, **7** 405–417.

Pielke, R.A. Jr., and S. Rayner(eds)., 2004. science policy and politics: learing from controversy over The Sreptical Environmnetalist. Environ. Sci. Policy, **7**, 355–433.

Pielke, R. A. Sr., 1974. A three-dimensional numerical model of the sea breezes over South Florida. *Mon. Wea. Rev.*, **102**, 115–134.

Pielke, R. A. Sr., 1981. An overview of our current understanding of the physical interactions between the sea- and land-breeze and the coastal waters. *Ocean Mgt*, **6**, 87–100.

Pielke, R. A. Sr., 1984. *Mesoscale Meteorological Modeling*. New York: Academic Press.

Pielke, R. A. Sr., 1990. *The Hurricane*. London: Routledge.

Pielke, R. A. Sr., 1991. Overlooked scientific issues in assessing hypothesized greenhouse gas warming. *Environ. Software*, **6**, 100–107.

Pielke, R. A. Sr., 1998. Climate prediction as an initial value problem. *Bull. Am. Meteor. Soc.*, **79**, 2743–2746.

Pielke., R.A. Sr., 2001a. Influence of the spatial distribution of vegetation and soils on the prediction of cumulus convective rainfall. *Rev. Geophys.*, **39**, 151–177.

Pielke, R.A. Sr., 2001b. Carbon sequestration: The need for an integrated climate system approach. *Bull. Am. Meteor. Soc.*, **82**, 2021.

Pielke, R. A. Sr., 2002a. *Mesoscale Meteorological Modeling*, 2nd edn. San Diego, CA: Academic Press.

Pielke, R. A. Sr., 2002b. Overlooked issues in the US National Climate and IPCC assessments. *Climat. Change*, **52**, 1–11.

Pielke, R. A. Sr., 2003. Heat storage within the Earth system. *Bull. Am. Meteor. Soc.*, **84**, 331–335.

Pielke, R. A. Sr., 2004. A broader perspective on climate change is needed. *Int. Geosphere–Biosphere Programme Newsl.*, **59**, 16–19.

Pielke, R. A. Sr., 2005. Land use and climate change. *Science*, **310**, 1625–1626.

Pielke, R. A. Sr., and R. Avissar, 1990. Influence of landscape structure on local and regional climate. *Landsc. Ecol.* **4**, 133–155.

Pielke, R. A. Sr., and L. Bravo de Guenni (eds.), 2004. How to evaluate vulnerability in changing environmental conditions. In *Vegetation, Water, Humans and the Climate: A New Perspective on an Interactive System*, Global Change: The International Geosphere-Biosphere Programme Series, P. Kabat Berlin, Germany: Springer-verlag. *et al.* eds. pp. 483–544.

Pielke, R. A. Sr., and R. N. Chase, 2003. A proposed new metric for quantifying the climatic effects of human-caused alterations to the global water cycle. *Symp. Observing and Understanding the Variability of Water in Weather and Climate, 83rd American Meteorological Society Annual Meeting,* Feb. 9–13, Long Beach, CA.

Pielke, R. A. Sr., and J. R. Christy, 2005. Comment on "Earth's energy imbalance: Confirmation and Implications." Available online at num_on_blue.atmos.colostate.edu/publications/pdf/Hansen-Science.pdf

Pielke, R. A. Sr., and T. Matsui, 2005. Should light wind and windy nights have the same temperature trends at individual levels even if the boundary layer averaged heat content change is the same? *Geophys. Res. Lett.*, in **32**, 21813, doi: 10.1029/2005GL024407.

Pielke, R. A. Sr., and M. Segal, 1986. Mesoscale circulations forced by differential terrain heating. In *Mesoscale Meteorology and Forecasting*, P. Ray, ed. Boston, MA: American Meteorological Society, PP. 516–548.

Pielke, R. A. Sr., and M. Uliasz, 1993. Influence of landscape variability on atmospheric dispersion. *J. Air Waste Mgt.*, **43**, 989–994.

Pielke, R. A. Sr., and X. Zeng, 1989. Influence on severe storm development of irrigated land. *Natl. Wea. Digest*, **14**, 16–17.

Pielke, R. A. Sr., G. Dalu, J. S. Snook, T. J. Lee, and T. G. F. Kittel, 1991a. Nonlinear influence of mesoscale land use on weather and climate. *J. Climate*, **4**, 1053–1069.

Pielke, R. A. Sr., A. Song, P. J. Michaels, W. A. Lyons, and R. W. Arritt, 1991b. The predictability of sea-breeze generated thunderstorms. *Atmosfera*, **4**, 65–78.

Pielke, R. A. Sr., W. R. Cotton, R. L. Walko, *et al.*, 1992. A comprehensive meteorological modeling system: RAMS. *Meteor. Atmos. Phys.*, **49**, 69–91.

Pielke, R. A. Sr., T. J. Lee, J. H. Copeland, *et al.*, 1997. Use of USGS-provided data to improve weather and climate simulations. *Ecol. Appl.*, **7**, 3–21.

Pielke, R. A. Sr., G. E. Liston, J. L. Eastman, L. Lu, and M. Coughenour, 1999a. Seasonal weather prediction as an initial value problem. *J. Geophys. Res.*, **104**, 19463–19479.

Pielke, R. A. Sr., R. L. Walko, L. Steyaert, *et al.*, 1999b. The influence of anthropogenic landscape changes on weather in south Florida. *Mon. Wea. Rev.*, **127**, 1663–1673.

Pielke, R. A. Sr., G. Marland, R. A. Betts, *et al.*, 2002. The influence of land-use change and landscape dynamics on the climate system- relevance to climate change policy beyond the radiative effect of greenhouse gases. *Phil. Trans. Roy. Soc. A*, **360**, 1705–1719.

Pielke, R. A. Sr., H. J. Schellnhuber, and D. Sahagian, 2003. Nonlinearities in the Earth system. *International Geosphere–Biosphere Programme Newsl.*, **55**, 11–15.

Pielke, R. A. Sr., C. Davey, and J. Morgan, 2004a. Assessing "global warming" with surface heat content. *EOS*, **85**, 210–211.

Pielke, R. A. Sr., G. E. Liston, W. L. Chapman, and D. A. Robinson, 2004b. Actual and insolation-weighted Northern Hemisphere snow cover and sea ice: 1974–2002. *Climate Dynamics*, **22**, 591–595, doi:10.1007/s00382-004-0401-5.

Pielke, R. A. Sr., N. Doesken, O. Bliss, *et al.*, 2005a. Drought 2002 in Colorado: An unprecedented drought or a routine drought? *Pure Appl. Geophys.*, Special Issue in honor of Prof. Singh, **162**, 1455-1479, doi:10.1007/200024-005-2679-6.

Pielke, R. A. Sr., J. O. Adegoke, T. N. Chase, *et al.*, 2006a. A new paradigm for assessing the role of agriculture in the climate system and in climate change. *Agric. Forest Meteor.*, Special Issue, in press.

Pielke, R. A. Sr., T. Matsui, G. Leoncini, *et al.*, 2006b. A new paradigm for parameterizations in numerical weather prediction and other atmospheric models. *Natl. Wea. Digest*, in press.

Pilewskie, P., and F. P. J. Valero, 1992. Radiative effects of the smoke clouds from the Kuwait oil fires. *J. Geophys. Res.*, **97**, 14541–14544.

Pincus, R., and S. A. Klein, 2000. Unresolved spatial variability and microphysical process rates in large-scale models. *J. Geophys. Res.*, **105**, **D22**, 27059–27065.

Pinty, B., M. M. Verstraete, N. Bobron, F. Roveda, and Y. Govaerts, 2000. Do man-made fires affect Earth's surface reflectance at continental scales? *EOS*, **81**, 388–389.

Pitman, A. J., 2003. Review: The evolution of, and revolution in, land surface schemes designed for climate models. *Int. J. Climatol.*, **23**, 479–510.

Pitman, A. J., and G. T. Narisma, 2005. The role of land surface processes in regional climate change: a case study of future land cover change over South Western Australia. *Meteor. Atmos. Phys.*, **89**, 235–249.

Pitman, A. J., and M. Zhao, 2000. The relative impact of observed change in land cover and carbon dioxide as simulated by a climate model. *Geophys. Res. Lett.*, **27**, 1267–1270.

Pitman, A., R. A. Pielke Sr., R. Avissar, *et al.*, 1999. The role of the land surface in weather and climate: Does the land surface matter? *Int. Geosphere–Biosphere Programme Newslt.*, **39**, 4–11.

Pitman, A. J., G. T. Narisma, R. A. Pielke Sr., and N. J. Holbrook, 2004. The impact of land cover change on the climate of southwest Western Australia. *J. Geophys. Res.*, **109**, D18109, doi:10.1029/2003JD004347.

Pittock, A. B., T. P. Ackerman, P. J. Crutzen, *et al.*, 1986. *Environmental Consequences of Nuclear War*, vol. 1. *Physical and Atmospheric Effects*. Chichester, UK: John Wiley.

Polcher, J., and K. Laval, 1994a. A statistical study of the regional impacts of deforestation on climate in the LMD GCM. *Climate Dynamics*, **10**, 205–219.

Polcher, J., and K. Laval, 1994b. The impact of African and Amazonian deforestation on tropical climate. *J. Hydrol.*, **155**, 389–405.

Porch, W. M., C.-Y. J. Kao, and R. G. Kelley Jr., 1990. Ship trails and ship induced cloud dynamics. *Atmos. Environ.*, **24A**, 1051–1059.

Porporato, A., P. D'Odorico, L. Ridolfi, and I. Rodriguez-Iturbe, 2000. A spatial model for soil–atmosphere interaction: Model construction and linear stability analysis. *J. Hydrometeor.*, **1**, 61–74.

Prahm, L. P., U. Torp, and R. M. Stern, 1976. Deposition and transformation rates of sulfur oxides during atmospheric transport over the Atlantic. *Tellus*, **28**, 355–372.

Prentice, I. C., D. Jolly, and BIOME 6000 members, 2000. Mid-Holocene and glacial-maximum vegetation geography of the northern continents and Africa. *J. Biogeog.*, **27**, 507–519.

Prospero, J. M., and T. N. Carlson, 1972. Vertical and areal distributions of Saharan dust over the western equatorial North Atlantic Ocean. *J. Geophys. Res.*, **77**, 5255–5265.

Pruppacher, H. R., and J. D. Klett, 1978. *Microphysics of Clouds and Precipitation*. Dordrecht, The Netherlands: Reidel.

Quijano, A. L., I. N. Sokolik, and O. B. Toon, 2000. Radiative heating rates and direct radiative forcing by mineral dust in cloudy atmospheric conditions. *J. Geophys. Res.*, **105**, **D10**, 12207–12219.

Rabin, R. M., and D. W. Martin, 1996. Satellite observations of shallow cumulus coverage over the central United States: An exploration of land use impact on cloud cover. *J. Geophys. Res.*, **101**, 7149–7155.

Rabin, R. M., S. Stadler, P. J. Wetzel, D. J. Stensrud, and M. Gregory, 1990. Observed effects of landscape variability on convective clouds. *Bull. Am. Meteor. Soc.*, **71**, 272–280.

Raddatz, R. L., 1998. Anthropogenic vegetation transformation and the potential for deep convection on the Canadian prairies. *Can. J. Soil Sci.*, **78**, 657–666.

Raddatz, R. L., 1999. Anthropogenic vegetation transformation and maximum temperatures on the Canadian prairies. *Can. Meteor. Oceanogr. society. Bull.*, **27**, 167–173.

Raddatz, R. L., 2000. Summer rainfall recycling for an agricultural region of the Canadian prairies. *Can. J. Soil Sci.*, **80**, 367–373.

Raddatz, R. L., 2003. Aridity and the potential physiological response of C_3 crops to doubled atmospheric CO_2: A simple demonstration of the sensitivity of the Canadian Prairies. *Bound.-Layer Meteor.*, **107**, 483–496.

Raddatz, R. L., 2005. Moisture recycling on the Canadian Prairies for summer droughts and pluvials from 1997 to 2003. *Agric. Forest Meteor.*, **131**, 13–26.

Raddatz, R. L., and J. D. Cummine, 2003. Inter-annual variability of moisture flux from the prairie agro-ecosystem: Impact of crop phenology on the seasonal pattern of tornado days. *Bound.-Layer Meteor.*, **106**, 283–295.

Radke, L. F., J. A. Coakley Jr., and M. D. King, 1989. Direct and remote sensing observations of the effects of ships on clouds. *Science,* **246,** 1146–1148.

Ramanathan, V., 1981. The role of ocean–atmosphere interactions in the CO_2 climate problem. *J. Atmos. Sci.,* **38,** 918 – 930.

Ramanathan, V., and W. Collins, 1991. Thermodynamic regulation of ocean warming by cirrus clouds deduced from observations of the 1987 El Niño. *Nature,* **351,** 27–32.

Ramanathan, V., B. R. Barkstrom, and E. F. Harrison, 1989a. Climate and the earth's radiation budget. *Phys. Today,* **42,** 22–32.

Ramanathan, V., R. D. Cess, E. F. Harrison, *et al.*, 1989b. Cloud-radiative forcing and climate: Results from the Earth Radiation Budget Experiment. *Science,* **243,** 57–63.

Ramanathan, V., P. J. Crutzen, J. Lelieveld, *et al.*, 2001. Indian Ocean Experiment: An integrated analysis of the climate forcing and effects of the great Indo-Asian haze. *J. Geophys. Res.*, **106**, 28371–28398.

Ramankutty, N., and J. A. Foley, 1999. Estimating historical changes in global land cover: Croplands from 1700 to 1992. *Global Biogeochem. Cycles*, **13**, 997–1027.

Ramankutty, N., C. Delire, and P. Snyder, 2006. Feedbacks between agriculture and climate: An illustration of the potential unintended consequences of human land use activities. *Global Change Biol.*, in press.

Ramaswamy, V., 1988. Aerosol radiative forcing and model responses. In *Aerosols and Climate*, P. V. Hobbs and M. P. McCormick, eds. Hampton, VA: A. Deepak Publishing, pp. 349–372.

Ramaswamy, V., and J. T. Kiehl, 1985. Sensitivities of the radiative forcing due to large loadings of smoke and dust aerosols. *J. Geophys. Res.,* **90,** 5597–5613.

Ramírez, J. A., and S. Senareth, 2000. A statistical–dynamical parameterization of canopy interception and land surface–atmosphere interactions. *J. Climate*, **13**, 4050–4063.

Randall, D. A., and S. Tjemkes, 1991. Clouds, the earth's radiation budget, and the hydrologic cycle. *Paleogeog., Paleoclimatol., Paleoecol. (Global Planet. Change)*, **90**, 3–9.

Randall, D., T. Carlson, and Y. Mintz, 1984. The sensitivity of a general circulation model to Saharan dust heating. In *Aerosols and Their Climatic Effects*, H.E. Gerber and A. Deepak, eds. Hampton, VA: A. Deepak Publishing, pp. 123–132.

Randall, D. A., Harshvardhan, D. A. Dazlich, and T. G. Corsetti, 1989. Interactions among radiation, convection, and large-scale dynamics in a general circulation model. *J. Atmos. Sci.,* **46,** 1943–1970.

Randall, D. A., M. Khairoutdinov, A. Arakawa, and W. Grabowski, 2003. Breaking the cloud-parameterization deadlock. *Bull. Am. Meteor. Soc.*, **84**, 1547–1564.

Rangno, A. L., and P. V. Hobbs, 1987. A reevaluation of the Climax cloud seeding experiments using NOAA published data. *J. Climate Appl. Meteor.*, **26**, 757–762.

Rangno, A. L, and P. V. Hobbs, 1993: Further analysis of the Climax cloud-seeding experiment. *J. Appl. Meteor.*, **32**, 1837–1847.

Rangno, A. L., and P. V. Hobbs, 1995. Reply. *J. Appl. Meteor.*, **34**, 1233–1238.

Rangno, A. L., and P. V. Hobbs, 1997a. Reply. *J. Appl. Meteor.*, **36**, 253-254.

Rangno, A. L., and P. V. Hobbs, 1997b. Reply. *J. Appl. Meteor.*, **36**, 257-259.

Rangno, A. L., and P. V. Hobbs, 1997c. Reply. *J. Appl. Meteor.*, **36**, 272-276.

Rangno, A. L., and P. V. Hobbs, 1997d. Reply. *J. Appl. Meteor.*, **36**, 279.

Ray, D. K., U. S. Nair, R. M. Welch, *et al.*, 2003. Effects of land use in southwest Australia. 1: Observations of cumulus cloudiness and energy fluxes. *J. Geophys. Res.*, **108**, **D14**, 4414, doi:10.1029/2002JD002654.

Ray, D. K., R. M. Welch, R. O. Lawton, and U. S. Nair, 2006. Dry season clouds and rainfall in northern central America: Implications for the mesoAmerican biological corridor. *Global Planet. Change*, accepted.

Reynolds, D. W., 1988. A report on winter snowpack-augmentation. *Bull. Am. Meteor. Soc.,* **69**, 1290–1300.

Reynolds, D. W., and A. S. Dennis, 1986. A review of the Sierra Cooperative Pilot Project. *Bull. Am. Meteor. Soc.,* **67,** 513–523.

Rial, J., R. A. Pielke Sr., M. Beniston, *et al.*, 2004. Nonlinearities, feedbacks and critical thresholds within the Earth's climate system. *Climat. Change*, **65**, 11–38.

Richards, J. F., 1986. World environmental history and economic development. In *Sustainable Development of the Biosphere*, W. C. Clark and R. E. Munn, eds. Cambridge, UK: Cambridge University Press, pp. 53–74.

Riehl, H., and J. S. Malkus, 1958. On the heat balance in the equatorial trough zone. *Geophysica*, **6**, 504–537.

Riehl, H, and J. M. Simpson, 1979. The heat balance of the equatorial trough zone, revisited. *Contrib. Atmos. Phys.*, **52**, 287–305.

Rind, D., 1984. The influence of vegetation on the hydrologic cycle in a global climate model. In *Climate Processes and Climate Sensitivity.*, Geophysical Monograph No. 29, J. Hanson and C. Takahashi, eds. Washington, DC: American Geophysical Union, pp. 73–91.

Rind, D., E. W. Chiou, W. Chu, *et al.*, 1991. Positive water vapour feedback in climate models confirmed by satellite data. *Nature,* **349**, 500–503.

Rivers, A. R., and A. H. Lynch, 2004. On the influence of land cover on early Holocene climate in northern latitudes. *J. Geophys. Res.*, **109**, D21114, doi:10.1029/2003JD004213.

Roberts, P., and J. Hallett, 1968. A laboratory study of the ice nucleating properties of some mineral particulates. *Q. J. Roy. Meteor. Soc.*, **94**, 25–34.

Robock, A., 1978. Internally and externally caused climate change. *J. Atmos. Sci.,* **35**, 1111–1122.

Robock, A., 1979. The "Little Ice Age": Northern Hemispheric average observations and model calculations. *Science, 206*, 1402–1404.

Robock, A., 1981. A latitudinally dependent volcanic dust veil index, and its effect on climatic simulations. *J. Volcan. Geotherm. Res., 11*, 67–80.

Robock, A., 1984. Climate model simulations of the effects of the El Chichon eruption. *Geofis. Int., 23*, 403–414.

Rodríguez, S., X. Querol, A. Alastuey, G. Kallos, and O. Kakaliagou, 2001. Saharan dust contributions to PM 10 and TSP levels in southern and eastern Spain. *Atmos. Environ., 35*, 2433–2447.

Rodriguez-Iturbe, I., D. Entekhabi, and R. L. Bras, 1991a. Nonlinear dynamics of soil moisture at climate scales. I. Stochastic analysis. *Water Resources Res., 27*, 1899–1906.

Rodriguez-Iturbe, I., D. Entekhabi, J.-S. Lee, and R. L. Bras, 1991b. Nonlinear dynamics of soil moisture at climate scales. II. Chaotic analysis. *Water Resources Res., 27*, 1907–1915.

Rogers, D. C., P. J. DeMott, and S. M. Kreidenweis, 2001. Airborne measurements of tropospheric ice-nucleating aerosol particles in the Arctic spring. *J. Geophys. Res., 106*, 15053–15063.

Rosenfeld, D., 1997. Comments on "A new look at the Israeli cloud seeding experiments." *J. Appl. Meteor., 36*, 260–271.

Rosenfeld, D., 1999. TRMM observed first direct evidence of smoke from forest fires inhibiting rainfall. *Geophys. Res. Lett., 26*, 3105–3108.

Rosenfeld, D., 2000. Suppression of rain and snow by urban and industrial air pollution. *Science, 287*, 1793–1796.

Rosenfeld, D., and H. Farbstein, 1992. Possible influence of desert dust on seedability of clouds in Israel. *J. Appl. Meteor., 31*, 722–731.

Rosenfeld, D., and R. Nirel, 1996. Seeding effectiveness: The interaction of desert dust and the southern margins of rain cloud systems in Israel. *J. Appl. Meteor., 35*, 1502–1510.

Rosenfeld, D., and W. L. Woodley, 1989. Effects of cloud seeding in west Texas. *J. Appl. Meteor., 28*, 1050–1080.

Rosenfeld, D., and W. L. Woodley, 1993. Effects of cloud seeding in west Texas: Additional results and new insights. *J. Appl. Meteor., 32*, 1848–1866.

Rosenfeld, D., Y. Rudich, and R. Lahav, 2001. Desert dust suppressing precipitation: a possible desertification feedback loop. *Proc. Natl Acad. Sci. USA, 98*, 5975–5980.

Rotstayn, L. D., and U. Lohmann, 2002. Tropical rainfall trends and the indirect aerosol effect. *J. Climate, 15*, 2103–2116.

Rotstayn, L. D., B. F. Ryan, and J. E. Penner, 2000. Precipitation changes in a GCM resulting from the indirect effects of anthropogenic aerosols. *Geophys. Res. Lett., 27*, 3045–3048.

Roy, A. K., Bh. V. Ramana Murty, R. C. Srivastava, and L. T. Khemani, 1961. Cloud seeding trails at Delhi during monsoon months, July to Sept. (1957-1959). *Indian J. Meteor. Geophys., 12*, 401–412.

Rozoff, C. M., W. R. Cotton, and J. O. Adegoke, 2003. Simulation of St. Louis, Missouri, land use impacts on thunderstorms. *J. Appl. Meteor., 42*, 716–738.

Ruddiman, W. F., 2003. The anthropogenic greenhouse era began thousands of years ago. *Climat. Change, 61*, 261–293.

Rudolph, R. D., C. M. Sackiw, and G. T. Riley, 1994. Statistical evaluation of the 1984-1988 seeding experiment in northern Greece. *J. Wea. Mod., 26*, 53–60.

Ryan, B. F., and W. D. King, 1997. A critical review of the Australian experience in cloud seeding. *Bull. Am. Meteor. Soc., 78*, 239–254.

Sagan, C., 1983. Nuclear war and climatic catastrophe: Some policy implications. *Foreign Affairs,* **62** (Winter 1983/84), 257–292.

Saito, T., 1981. The relationship between the increase rate of downward long-wave radiation by atmospheric pollution and the visibility. *J. Meteor. Soc.,* **59,** 254–261.

Salas, J. D., J. T. B. Obeysekera, and D. C. Boes, 1981. Modeling of the equatorial lakes outflow. In: *Statistical Analysis of Rainfall and Runoff,* V.P. Singh, ed. Littleton, CO: Water Resources Publications, pp. 431–440.

Salati, E., J. Marques, and L. C. B. Molion, 1978. Origem e distribuição das chuvas na Amazônia. *Interciencia,* **3,** 200–205.

Sarewitz, D., R. A. Pielke Jr., and R. Byerly (eds.) 2000. *Prediction: Science Decision Making and the Future of Nature.* Covelo, CA: Island Press.

Sassen, K., 1997. Contrail-cirrus and their potential for regional climate change. *Bull. Am. Meteor. Soc.,* **78,** 1885–1903.

Sassen, K., P. J. DeMott, J. M. Prospero, and M. R. Poellet, 2003. Saharan dust storms and indirect aerosol effects on clouds: CRYSTAL-FACE results. *Geophys. Res. Lett.,* **30,** 1633, doi:10.1029/2003GL017371.

Sax, R. I., 1976. Microphysical response of Florida cumuli to AgI seeding. *Proc. 2nd World Meteorological Organization Scientific Conf. Weather Modification,* Boulder, CO, pp. 109–116.

Sax, R. I., and V. W. Keller, 1980. Water–ice and water–updraft relationships near $-10\,^{\circ}C$ within populations of Florida cumuli. *J. Appl. Meteor.,* **19,** 505–514.

Sax, R. I., J. Thomas, and M. Bonebrake, 1979. Ice evolution within seeded and nonseeded Florida cumuli. *J. Appl. Meteor.,* **18,** 203–214.

Schaefer, V. J., 1948a. The production of clouds containing supercooled water droplets or ice crystals under laboratory conditions. *Bull. Am. Meteor. Soc.,* **29,** 175–182.

Schaefer, V. J., 1948b. The natural and artificial formation of snow in the atmosphere. *Trans. Am. Geophys. Union,* **29,** 492.

Schaefer, V. J., 1949. The formation of ice crystals in the laboratory and the atmosphere. *Chem. Rev.,* **44,** 291.

Schaefer, V. J., 1954. The concentrations of ice nuclei in air passing the summit of Mt. Washington. *Bull. Am. Meteor. Soc.,* **35,** 310–314.

Schaefer, V. J., 1966. Condensed water in the free atmosphere in air colder than $-40\,^{\circ}C$. *J. Appl. Meteor.,* **1,** 481–488.

Schaefer, V. J., 1969. The inadvertent modification of the atmosphere by air pollution. *Bull. Am. Meteor. Soc.,* **50,** 199–206.

Schär, C., D. Lüthi, U. Beyerle, and E. Heise, 1999. The soil–precipitation feedback: A process study with a regional climate model. *J. Climate,* **12,** 722–741.

Schneider, S. H., 1989. The changing climate. *Sci. American,* **261,** 70–79.

Schneider, S. H., 1990. The global warming debate heats up: An analysis and perspective. *Bull. Am. Meteor. Soc.,* **71,** 1291–1304.

Schneider, S. H., and S. L. Thompson, 1988. Simulating the climatic effects of nuclear war. *Nature,* **333,** 221–227.

Schreiber, K., R. Stull, and Q. Zhang, 1996. Distributions of surface-layer buoyancy versus lifting condensation level over a heterogeneous land surface. *J. Atmos. Sci.,* **53,** 1086–1107.

Schumann, U., 1994. On the effect of emissions from aircraft engines on the state of the atmosphere. *Ann. Geophysicae,* **12,** 365–384.

Schwartz, M. D., 1994. Monitoring global change with phenology: The case of the spring green wave. *Int. J. Biometeor.,* **38,** 18–22.

Schwartz, M. D., and T. R. Karl, 1990. Spring phenology: Nature's experiment to detect the effect of "green-up" on surface maximum temperatures. *Mon. Wea. Rev.,* **118,** 883–890.

Schwartz, S. E., 1989. Sulphate aerosols and climate. *Nature,* **340**, 515–516.

Scientific Plan for the Coordinated Enhanced Observing Period (CEOP), 1999: An Overview from a GEWEX Hydrometeorology Panel Perspective. http://ioc.unesco.org/oceanteacher/resourcekit/Module1/GlobalPrograms/gewex/Gewex CeopSciPlan.html

Scorer, R. S., 1987. Ship trails. *Atmos. Environ.,* **21**, 1417–1425.

Scott, B. C., and P. V. Hobbs, 1977. A theoretical study of the evolution of mixed-phase cumulus clouds. *J. Atmos. Sci.,* **34**, 812–826.

Seaver, W. L., and J. E. Lee, 1987. A statistical examination of sky cover changes in the contiguous United States. *J. Climate Appl. Meteor.,* **26**, 88–95.

Seidel, D. J., J. K. Angell, J. Christy, *et al.,* 2004. Uncertainty in signals of large-scale climate variations in radiosonde and satellite upper-air temperature datasets. *J. Climate,* **17**, 2225–2240.

Segal, M., and R. W. Arritt, 1992. Non-classical mesoscale circulations caused by surface sensible heat-flux gradients. *Bull. Am. Meteor. Soc.,* **73**, 1593–1604.

Segal, M., R. A. Pielke, and Y. Mahrer, 1983. On climatic changes due to a deliberate flooding of the Qattara depression (Egypt). *Climatic Change,* **5**, 73–83.

Segal, M., R. Avissar, M. C. McCumber, and R. A. Pielke, 1988. Evaluation of vegetation effects on the generation and modification of mesoscale circulations. *J. Atmos. Sci.,* **45**, 2268–2292.

Segal, M., J. R. Garratt, G. Kallos, and R. A. Pielke, 1989. The impact of wet soil and canopy temperatures on daytime boundary-layer growth. *J. Atmos. Sci.,* **46**, 3673–3684.

Segal, M., J. H. Cramer, R. A. Pielke, J. R. Garratt, and P. Hildebrand, 1991. Observational evaluation of the snow-breeze. *Mon. Wea. Rev.,* **119**, 412–424.

Segal, M., R. W. Arritt, C. Clark, R. Rabin, and J. Brown, 1995. Scaling evaluation of the effect of surface characteristics on potential for deep convection over uniform terrain. *Mon. Wea. Rev.,* **123**, 383–400.

Segal, M., R. W. Arritt, J. Shen, C. Anderson, and M. Leuthold, 1997. On the clearing of cumulus clouds downwind from lakes. *Mon. Wea. Rev.,* **125**, 639–646.

Segal, M., Z. Pan, R. W. Turner, and E. S. Takle, 1998. On the potential impact of irrigated areas in North American summer rainfall caused by large-scale systems. *J. Appl. Meteor.,* **37**, 325–331.

Sellers, P. J., R. E. Dickinson, D. A. Randall, *et al.,* 1997. Modeling the exchanges of energy, water, and carbon between continents and the atmosphere. *Science,* **275**, 502–509.

Sen, O. L., Y. Wang, and B. Wang, 2004a. Impact of Indochina deforestation on the East Asian summer monsoon. *J. Climate,* **17**, 1366–1380.

Sen, O. L., B. Wang, and Y. Wang, 2004b. Impacts of re-greening the desertified lands in northwestern China: Implications from a regional climate model experiment. *J. Meteor. Soc. Japan,* **82**, 1679–1693.

Seth, A., and F. Giorgi, 1996. Three-dimensional model study of organized mesoscale circulations induced by vegetation. *J. Geophys. Res.,* **101**, 7371–7391.

Shabbar, A., B. Bonsal, and M. Khandekar, 1997. Canadian precipitation patterns associated with the Southern Oscillation. *J. Climate,* **10**, 3016–3027.

Shaw, B. L., R. A. Pielke, and C.Ł. Ziegler, 1997. A three-dimensional numerical simulation of a Great Plains dryline. *Mon. Wea. Rev.,* **125**, 1489–1506.

Sheets, R. C., 1981. *Tropical Cyclone Modification: The Project STORMFURY Hypothesis,* Technical Report ERL 414-AOML. Miami, FL: National Oceanic and Atmospheric Administration.

Shell, K. M., R. Frouin, S. Nakamoto, and R. C. J. Somerville, 2003. Atmospheric response to solar radiation absorbed by phytoplankton. *J. Geophys. Res.*, **108**, **D15**, 4445, doi:10.1029/ 2003JD003440.

Shen, S., and M. Y. Leclerc, 1995. How large must surface inhomogeneities be before they influence the convective boundary layer structure? A case study. *Q. J. Roy. Meteor. Soc.*, **121**, 1209–1228.

Shepherd, J. M., and S. J. Burian, 2003. Detection of urban-induced rainfall anomalies in a major coastal city. *Earth Interactions*, **7**, 1–14.

Shepherd, J. M., H. Pierce, and A. J. Negri, 2002. On rainfall modification by major urban areas. Observation from space borne rain radar on TRMM. *J. Appl. Meteor.*, **41**, 689–701.

Shepherd, J. M., O. O. Taylor, and C. Garza, 2004. A dynamic GIS-multicriteria technique for siting the NASA-Clark Atlanta urban rain gauge network. *J. Atmos. Ocean. Tech.*, **21**, 1346–1363.

Shinoda, M., and M. Gamo, 2000. Interannual variations of boundary layer temperature over the African Sahel associated with vegetation and the upper troposphere. *J. Geophys. Res.*, **105**, 12317–12327.

Shukla, J., C. Nobre, and P. Sellers, 1990. Amazon deforestation and climate change. *Science*, **247**, 1322–1325.

Silva Dias, M. A. F., S. Rutledge, P. Kabat, *et al.*, 2002. Cloud and rain processes in a biosphere–atmosphere interaction context in the Amazon Region. *J Geophys. Res.*, **107**, 8072, doi:10.1029/2001IDO00335.

Silverman, B. A., 1986. Static mode seeding of summer cumuli: a review. *Meteor. Monogr.* **21**, 7–24.

Silverman, B. A., 2001. A critical assessment of glaciogenic seeding of convective clouds for rainfall enhancement. *Bull. Am. Meteor. Soc.*, **82**, 903–923.

Silverman, B. A., and W. Sukarnjanaset, 1996. On the seeding to tropical convective clouds for rain augmentation. *Proc. 13th American Meteorological Society Conf. Planned and Inadvertent Weather Modification*, Atlanta, GA, pp. 52–59.

Simmonds, I., D. Bi, and P. Hopc, 1999. Atmospheric water vapor flux and its association with rainfall over China in summer. *J. Climate*, **12**, 1353–1367.

Simon, C., 1981. Clues in the clay. *Sci. News*, **Nov. 14**, 314–315.

Simpson, J. E., 1980. Downdrafts as linkages in dynamic cumulus seeding effects: Notes. *J. Appl. Meteor.*, **19**, 477–487.

Simpson, J. E., 1994. *Sea Breeze and Local Winds*. Cambridge, UK: Cambridge University Press.

Simpson, J. E., and A. S. Dennis, 1972. *Cumulus Clouds and Their Modification*, Technical Memorandum ERLOD-14. Washington, DC: National Oceanic and Atmospheric Administration.

Simpson, J. E., G. W. Brier, and R. H. Simpson, 1967. STORMFURY cumulus seeding experiment 1965: Statistical analysis and main results. *J. Atmos. Sci.*, **24**, 508–521.

Simpson, J. E., N. E. Westcott, R. J. Clerman, and R. A. Pielke, 1980. On cumulus mergers. *Arch. Meteor. Geophys. Bioklim., Ser. A.*, **29**, 1–40.

Simpson, J. E., Th. D. Keenan, B. Ferrier, R. H. Simpson, and G. J. Holland, 1993. Cumulus mergers in the maritime continent region. *Meteor. Atmos. Phys.*, **51**, 73–99.

Simpson, R. H., and J. S. Malkus, 1964. Experiments in hurricane modification. *Sci. American*, **211**, 27–37.

Simpson, R. H., M. R. Ahrens, and R. D. Decker, 1963. *A Cloud Seeding Experiment in Hurricane Esther, 1961*, National Hurricane Research Project Report No. 60. Washington, DC: US Department of Commerce, Weather Bureau.

Simpson, R. H., and co-authors, 1978. *TYMOD: Typhoon Moderation*, Final Report, Prepared for the Government of the Philippines. Arlington, VA: Virginia Technology.

Singer, S. F. (ed.), 1992. *Global Climate Change: Human and Natural Influences*. New York: Paragon House.

Skinner, W. R., and J. A. Majorowicz, 1999. Regional climatic warming and associated twentieth century land-cover changes in north-western North America. *Climate Res.*, **12**, 39–52.

Small, R. D., B. W. Bush, and M. A. Dore, 1989. Initial smoke distribution for nuclear winter calculations. *Aerosol Sci. Tech.*, **10**, 37.

Smith, P. L., L. R. Johnson, D. L. Priegnitz, B. A. Boe, and P. J. Mielke Jr., 1997. An exploratory analysis of crop hail insurance data for evidence of cloud seeding effects in North Dakota. *J. Appl. Meteor.*, **36**, 463–473.

Snyder, P. K., J. A. Foley, M. H. Hitchman, and C. Delire, 2004a. Analyzing the effects of complete tropical forest removal on the regional climate using a detailed three-dimensional energy budget: An application to Africa. *J. Geophys. Res.*, **109**, D21102, doi:10.1029/2003JD004462.

Snyder, P. K., C. Delire, and J. A. Foley, 2004b. Evaluating the influence of different vegetation biomes on the global climate. *Climate Dynamics*, **23**, doi:10.1007/s00382-004-0430-0.

Soon, W. -H., D. R. Legates, and S. Baliunas, 2004. Estimation and representation of long-term (>40 year) trends of Northern Hemisphere-gridded surface temperature: A note of caution. *Geophys. Res. Lett.*, **31**, L03209, doi:1029/2003GRL019141.

Souza, E. P., N. O. Rennó, and M. A. F. Silva Dias, 2000. Convective circulations induced by surface heterogeneities. *J. Atmos. Sci.*, **57**, 2915–2922.

Squires, P., 1966. An estimate of the anthropogenic production of cloud nuclei. *J. Rech. Atmos.*, **2/3**, 297–308.

Steffen, W., A. Sanderson, P. D. Tyson, *et al.*, 2004. *Global Change and the Earth System: A Planet under Pressure*. Heidelberg, Germany: Springer-Verlag.

Stephens, G. L., P. M. Gabriel, and P. T. Partain, 2001. Parameterization of atmospheric radiative transfer. I. Validity of simple models. *J. Atmos. Sci.*, **48**, 3391–3409.

Stevens, B., G. Feingold, W. R. Cotton and R. L. Walko, 1998. Elements of the microphysical structure of numerically simulated nonprecipitating stratocumulus. *J. Atmos. Sci.*, **55**, 980–1006.

Steyaert, L. T., F. G. Hall, and T. R. Loveland, 1997. Land cover mapping, fire disturbance regeneration, and scaling studies in the Canadian BOREAL forest with a 1-km AVHRR and Landsat TM data. *J. Geophys. Res.*, **102**, 29581–29598.

Stohlgren, T. J., T. N. Chase, R. A. Pielke, T. G. F. Kittel, and J. Baron, 1998. Evidence that local land use practices influence regional climate and vegetation patterns in adjacent natural areas. *Global Change Biol.*, **4**, 495–504.

STORMFURY, 1970. *Project STORMFURY Annual Report 1969*. Miami, FL: National Hurricane Research Laboratory, National Oceanic and Atmospheric Administration, ADML / Hurricane Research Division.

Stouffer, R. J., S. Manabe, and K. Bryan, 1989. Interhemispheric asymmetry in climate response to a gradual increase of atmospheric CO_2. *Nature*, **342**, 660–662.

Strack, J. E., G. E. Liston, and R. A. Pielke Sr., 2004. Modeling snow depth for improved simulation of snow–vegetation–atmosphere interactions. *J. Hydrometeor.*, **5**, 723–734.

Sud, Y. C., W. C. Chao, and G. K. Walker, 1993. Dependence of rainfall on vegetation: Theoretical considerations, simulation experiments, observations, and inferences from simulated atmospheric soundings. *J. Arid Environ.*, **25**, 5–18.

Sud, Y. C., K. M. Lau, G. K. Walker, and J. H. Kim, 1995. Understanding biosphere–precipitation relationships: Theory, model simulations and logical inferences. *Mausam*, **46**, 1–14.

Sud, Y. C., G. K. Walker, J.-H. Kim, *et al.*, 1996. Biogeophysical consequences of a tropical deforestation scenario: A GCM simulation study. *J. Climate*, **9**, 3225–3247.

Suh, M. -S., and D. -K. Lee, 2004. Impact of land use/cover changes on surface climate over east Asia for extreme climate cases using RegCM2. *J. Geophys. Res.*, **109**, D02108, doi:10.1029/2003JD003681.

Sulakvelidze, G. K., B. I. Kiziriya, and V. V. Tsykunov, 1974. Progress of hail suppression work in the U.S.S.R. In *Climate and Weather Modification*, W.N. Hess, ed. New York: John Wiley, pp. 410–431.

Sun, W.-Y., M. G. Bosilovich, and J. -D. Chern, 1997. Regional response of the NCAR CCM1 to anomalous surface properties. *Terrest., Atmos., Ocean. Sci.*, **8**, 271–288.

Super, A. B., 1974. Silver iodide plume characteristics over the Bridger Mountain Range, Montana. *J. Appl. Meteor.*, **13**, 62–70.

Super, A. B., 1986. Further exploratory analysis of the Bridger Range winter cloud seeding experiment. *J. Climate Appl. Meteor.*, **12**, 1926–1933.

Super, A. B., and B. A. Boe, 1988. Microphysical effects of wintertime cloud seeding with silver iodide over the Rocky Mountains. III. Observations over the Grand Mesa, Colorado. *J. Appl. Meteor.*, **27**, 1166–1182.

Super, A. B., and J. A. Heimbach, 1983. Evaluation of the Bridger Range winter cloud seeding experiment using control gages. *J. Climate Appl. Meteor.*, **22**, 1989–2011.

Super, A. B., and J. A. Heimbach, 1988. Microphysical effects of wintertime cloud seeding with silver iodide over the Rocky Mountains. II. Observations over the Bridger Range, Montana. *J. Appl. Meteor.*, **27**, 1152–1165.

Super, A. B., B. A. Boe, and E. W. Holroyd III, 1988. Microphysical effects of wintertime cloud seeding with silver iodide over the Rocky Mountains. I. Experimental design and instrumentation. *J. Appl. Meteor.*, **27**, 1145–1151.

Svensmark, H., and E. Friis-Christensen, 1997. Variation of cosmic ray flux and global cloud coverage: A missing link in solar–climate relationships. *J. Atmos. Solar-Terrest. Phys.*, **59**, 1225–1232.

Tanre, D., J. F. Geleyn, and J. Slingo, 1984. First results of the introduction of an advanced aerosol–radiation interaction in the ECMWF low resolution global model. In *Aerosols and Their Climatic Effects*, H. E. Gerber and A. Deepak, eds. Hampton, VA: A. Deepak Publishing, pp. 133–177.

Taylor, C. M., F. Saïd, and T. Lebel, 1997. Interactions between the land surface and mesoscale rainfall variability during HAPEX-Sahel. *Mon. Wea. Rev.*, **125**, 2211–2227.

Taylor, C. M., R. J. Harding, R. A. Pielke Sr., *et al.*, 1998. Snow breezes in the boreal forest. *J. Geophys. Res.*, **103**, 23087–23101.

Taylor, C. M., E. F. Lambin, N. Stephenne, R. J. Harding, and R. L. H. Essery, 2002. The influence of land use change on climate in the Sahel. *J. Climate*, **15**, 3615–3629.

Tegen, I., and I. Fung 1995. Contribution to the atmospheric mineral aerosol load from land surface modification. *J. Geophys. Res.*, **100**, 18707–18726.

Texier, D., N. de Noblet, and P. Braconnot, 2000. Sensitivity of the African and Asian monsoons to mid-Holocene insolation and data-inferred surface changes. *J. Climate*, **13**, 164–181.

Thatcher, M., 1990. On long term climate prediction. *J. Air Waste Mgt.*, **40**, 1086–1087.

Thielen, J., W. Wobrock, A. Gadian, P. G. Mestayer, and J. -D. Creutin, 2000. The possible influence of urban surfaces on rainfall development: a sensitivity study in 2D in the meso-gamma-scale. *Atmos. Research*, **54**, 15–39.

Thompson, D. W. J., and J. M. Wallace, 2000. Annular modes in the extratropical circulation. I. Month-to-month variability. *J. Climate*, **13**, 1000-1016.

Thompson, D. W. J., J. M. Wallace, and G. C. Hegerl, 2000. Annular modes in the extratropical circulation. II. Trends. *J. Climate*, **13**, 1018–1036.

Timbal, B., S. Power, R. Colman, J. Viviand, and S. Lirola, 2002. Does soil moisture influence climate variability and predictability over Australia? *J. Climate*, **15**, 1230–1238.

Timbal, B., and J. M. Arblaster, 2006. Land cover change as an additional forcing to explain the rainfall decline in the south west of Australia. *Geophys:Res.Lett.*, **33** L07717, doi:10.1029/2005GL025361.

Titus, J. G., 1990a. Strategies for adapting to the greenhouse effect. *J. am. Planning asso.*, **Summer**., 331–323.

Titus, J. G., 1990b. Greenhouse effect, sea level rise and barrier islands: Case study of Long Beach Island New Jersey. *Coastal Mgt.*, **18**, 65–90.

Tripoli, G. J., 1986. Nucleation scavenging in smoke plumes induced by large urban fires: Some preliminary results. Presentation to *Global Effects Technical Meeting*, NASA Ames Research Center, Moffett Field, CA, Feb. 25–27, 1986.

Tripoli, G. J., and W. R. Cotton, 1980. A numerical investigation of several factors contributing to the observed variable intensity of deep convection over south Florida. *J. Appl. Meteor.*, **19**, 1037–1063.

Tsonis, A. A., 2001. The impact of nonlinear dynamics in the atmospheric sciences. *Int. J. Bifurcation and Chaos*, **11**, 881–902.

Tukey, J. W., D. R. Brillinger, and L. V. Jones, 1978. *The Management of Weather Resources*, vol. 2, *The Role of Statistics in Weather Resources Management*, Report of the Statistical Task Force to the Weather Modification Advisory Board. Washington, DC: US Government Printing Office.

Turco, R. P., O. B. Toon, T. P. Ackerman, J. B. Pollack, and C. Sagan, 1983. Nuclear winter: Global consequences of multiple nuclear explosions. *Science,* **222**, 1283–1292.

Turco, R. P., O. B. Toon, T. P. Ackerman, J. B. Pollack, and C. Sagan, 1990. Climate and smoke: An appraisal of nuclear winter. *Science,* **247**, 166–176.

Turner, B. L. II, W. C. Clark, R. W. Kates, *et al.*, (eds.), 1990. *The Earth as Transformed by Human Action: Global and Regional Changes in the Biosphere over the Past 300 Years.* Cambridge, UK: Cambridge University Press.

Twine, T. E., C. J. Kucharik, and J. A. Foley, 2004. Effects of land cover change on the energy and water balance of the Mississippi River Basin. *J. Hydrometeor.*, **5**, 640–655.

Twomey, S., 1974. Pollution and the planetary albedo. *Atmos. Environ.*, **8**, 1251–1256.

Twomey, S., 1977. The influence of pollution on the shortwave albedo of clouds. *J. Atmos. Sci.,* **34**, 1149–1152.

Twomey, S., M. Piepgrass, and T. L. Wolfe, 1984. An assessment of the impact of pollution on global albedo. *Tellus, Ser. B*, **36**, 356–366.

Tzivion, S., T. Reisin, and Z. Levin, 1994. Numerical simulation of hygroscopic seeding in a convective cloud. *J. Appl. Meteor.*, **33**, 252–267.

UCAR/NOAA, 1991. *Reports to the Nation on our changing planet, winter 1991.* Boulder, CO: University Corporation for Atmospheric Research Office for Interdisciplinary Earth Studies and the NOAA Office of Global Programs.

Udelhofen, P. M., and R. D. Cess, 2001. Cloud cover variations over the United States: An influence of cosmic rays or solar variability? *Geophys. Res. Lett.*, **28**, 2617–2620.

Ulanski, S., and M. Garstang, 1978a. The role of surface divergence and vorticity in the lifecycle of convective rainfall. I. Observation and analysis. *J. Atmos. Sci.*, **35**, 1047–1062.

Ulanski, S., and M. Garstang, 1978b. The role of surface divergence and vorticity in the life cycle of convective rainfall. II. Descriptive model. *J. Atmos. Sci.*, **35**, 1063–1069.

van den Heever, S. C., and W. R. Cotton, 2004. The impact of hail size on simulated supercell storms. *J. Atmos. Sci.*, **61**, 1596–1609.

van den Heever, S. C., G. G. Carrió, W. R. Cotton, P. J. DeMott and A. J. Prenni, 2006. Impacts of nucleating aerosol on Florida storms. I. Mesoscale simulations. *J. Atmos. Sci.*, **63**, 1752–1775.

van den Hurk, B. J. J. M., P, Viterbo, and S. O. Los, 2003. Impact of leaf area index seasonality on the annual land surface evaporation in a global circulation model. *J. Geophys. Res.*, **108**, **D6**, 4191, doi:10.1029/2002JD002846.

van der Molen, M. K., H. F. Vugts, L. A. Vruijnzeel, F. N. Scatena, R. A. Pielke Sr., and L. J. M. Kroon, 2005. Mesoscale climate change due to deforestation in the Maritime tropics. *Proc. 2nd Int. Symp. Mountains in the Mist: Science for Conserving and Managing Tropical Montane Cloud Forest*, Waimea, HI, July 27–Aug 2, 2004.

Vidale, P. L., R. A. Pielke, A. Barr, and L. T. Steyaert, 1997. Case study modeling of turbulent and mesoscale fluxes over the BOREAS region. *J. Geophys. Res.*, **102**, 29167–29188.

Viterbo, P., and A. K. Betts, 1999. The impact of the ECMWF reanalysis soil water on forecasts of the July 1993 Mississippi flood. *J. Geophys. Res.*, **104**, 19361–19366.

Vitousek, P. M., H. A. Mooney, J. Lubchenco, and J. M. Melillo, 1997. Human domination of Earth's ecosystems. *Science*, **277**, 494–499.

Vizy, E. K., and K. H. Cook, 2005. Evaluation of Last Glacial Maximum sea surface temperature reconstructions through their influence on South American climate. *J. Geophys. Res.*, **110**, D11105, doi:10.1029/2004JD005415.

Vogelmann, A. M., A. Robock, and R. G. Ellingson, 1988. Effects of dirty snow in nuclear winter simulations. *J. Geophys Res.*, **93**, 5319–5332.

Vonnegut, B., 1947. The nucleation of icc formation by silver iodide. *J. Appl. Phys.*, **18**, 593.

Vörösmarty, C. J., P. Green, J. Salisbury, and R. B. Lammers, 2000. Global water resources: Vulnerability from climate change acid population growth. *Science*, **289**, 284–288.

Vukovich, F. M., J. W. Dunn III, and B. W. Crissman, 1976. A theoretical study of the St. Louis heat island: The wind and temperature distribution. *J. Appl. Meteor.*, **15**, 417–440.

Wahl, E., 1951. On a seven-day periodicity in weather in the United States during April, 1950. *Bull. Am. Meteor. Soc.*, **32**, 193.

Wallace, J. M., and D. S. Gutzler, 1981. Teleconnections in the geopotential height field during the Northern Hemisphere winter, 1981. *Mon. Wea. Rev.*, **109**, 784–812.

Wang, G., and E. A. B. Eltahir, 1999. *The Role of Vegetation Dynamics in the Climate of West Africa*, Department of Civil and Environmental Engineering Report No. 344. Cambridge, MA: MIT Press.

Wang, G., and E. A. B. Eltahir, 2000a. Ecosystem dynamics and the Sahel drought. *Geophys. Res. Lett.*, **27**, 795–798.

Wang, G., and E. A. B. Eltahir, 2000b. Biosphere–atmosphere interactions over west Africa. I. Development and validation of a coupled dynamic model. *Q. J. Roy. Meteor. Soc.*, **126**, 1239–1260.

Wang, G., and E. A. B. Eltahir, 2000c. Biosphere–atmosphere interactions over west Africa. II. Multiple climate equilibria. *Q. J. Roy. Meteor. Soc.*, **126**, 1261–1280.

Wang, G., and E. A. B. Eltahir, 2000d. The role of vegetation dynamics in enhancing the low-frequency variability of the Sahel rainfall. *Water Resources Res.*, **36**, 1013–1021.

Wang, G., and E. A. B. Eltahir, 2000e. Ecosystem dynamics and the Sahel drought. *Geophys. Res. Lett.*, **27**, 795–798.

Wang, G., and E. A. B. Eltahir, 2002. Impact of CO_2 concentration changes on the biosphere–atmosphere system of west Africa. *Global Change Biol.*, **8**, 1169–1182.

Wang, G., and D. Schimel, 2003. Climate change, climate modes, and climate impacts. *Ann. Rev. Environ. Resources*, **28**, 1–28, doi:10.1145/annurev.energy.28.050302. 105444.

Wang, G., E. A. B. Eltahir, J. A. Foley, D. Pollard, and S. Levis, 2004. Decadal variability of rainfall in the Sahel: Results from the coupled GENESIS-IBIS atmosphere-biosphere model. *Climate Dynamics*, **22**, 625–637.

Wang, H., A. J. Pitman, M. Zhao, and R. Leeman, 2003. The impact of historical land cover change on the June meteorology of China since 1700 simulated using a regional climate model. *Int. J. Climatol.*, **23**, 511–527.

Wang, J., R. L. Bras, and E. A. B. Eltahir, 1997. A stochastic linear theory of mesoscale circulation induced by the thermal heterogeneity of the land surface. *J. Atmos. Sci.*, **53**, 3349–3366.

Wang, J., R. L. Bras, and E. A. B. Eltahir, 1998. Numerical simulation of nonlinear mesoscale circulation induced by thermal heterogeneity of land surface. *J. Atmos. Sci.*, **55**, 447–464.

Wang, J., R. L. Bras, and E. A. B. Eltahir, 2000. The impact of observed deforestation on the mesoscale distribution of rainfall and clouds in Amazonia. *J. Hydrometeor.*, **1**, 267–286.

Wang, X. -P., R. Berndtsson, X. -R. Li, and E. -S. Kang, 2004. Water balances for re-vegetated xerophyte shrub area. *Hydrol. Sci.*, **49**, 283–295.

Warburton, J. A., L. G. Young, M. S. Owens, and R. H. Stone, 1985. The capture of ice nucleating and non ice-nucleating aerosols by ice-phase precipitation. *J. Recherche Atmosphérique*, **19**, 249–255.

Warburton, J. A., R. H. Stone, and B. L. Marler, 1995a. How the transport and dispersion of AgI aerosols may affect detectability of seeding effects by statistical methods. *J. Appl. Meteor.*, **34**, 1930–1941.

Warburton, J. A., L. G. Young, and R. H. Stone, 1995b. Assessment of seeding effects in snowpack augmentation programs: Ice nucleation and scavenging of seeding aerosols. *J. Appl. Meteor.*, **34**, 121–130.

Waring, R. H., and S. W. Running, 1998. *Forest Ecosystems: Analysis at Multiple Scales*, 2nd edn. San Diego, CA: Academic Press.

Warner, J., 1968. A reduction in rainfall associated with smoke from sugar-cane fires: An inadvertent weather modification? *J. Appl. Meteor.*, **7**, 247–251.

Warner, J., 1971. Smoke from sugar-cane fires and rainfall. *Proc. Int. Conf. on Weather Modification*, Canberra, Australia, Sept. 6–11, 1971.

Warner, J., 1973. The microstructure of cumulus clouds. V. Changes in droplet size distribution with cloud age. *J. Atmos. Sci.*, **30**, 1724–1726.

Warner, J., and S. Twomey, 1967. The production of cloud nuclei by cane fires and the effects on cloud droplet concentration. *J. Atmos. Sci.*, **24**, 704–706.

Warren, S. G., and W. J. Wiscombe, 1985. Dirty snow after nuclear war. *Nature*, **313**, 467.

Washington, W. M., and G. A. Meehl, 1984. Seasonal cycle experiment on the climate sensitivity due to a doubling of CO_2 with an atmospheric general circulation model coupled to a single mixed-layer ocean model. *J. Geophys. Res.*, **89**, 9475–9503.

Weart, S. R., 2004. *The Discovery of Global Warming*. Cambridge, MA: Harvard University Press.

Weaver, C. P., and R. Avissar, 2001. Atmospheric disturbances caused by human modification of the landscape. *Bull. Am. Meteor. Soc.*, **82**, 269–281.

Weaver, C. P., and R. Avissar, 2002. Reply to: "Comments on 'Atmospheric disturbances caused by human modification of the landscape'" by J. C. Doran and S. Zhong. *Bull. Am. Meteor. Soc.*, **83**, 280–283.

Weaver, C. P., R. Avissar, and Y. Liu, 2000. On the parameterization of convective precipitation generated by land cover change/land use in large-scale atmospheric models. *Proc. 15th Conf. Hydrology, American Meteorological Society*, Long Beach, CA, pp. 289–291.

Wei, H., and C. Fu, 1998. Study of the sensitivity of a regional model in response to land cover change over northern China. *Hydrol. Proc.*, **12**, 2249–2265.

Weickmann, H. K., 1964. The language of hailstorms and hailstones. *Nubila*, **6**, 7–51.

Welch, R. M., and W. G. Zdunkowski, 1976. A radiation model of the polluted atmospheric boundary layer. *J. Atmos. Sci.*, **33**, 2170–2184.

Welch, R. M., J. Paegle, and W. G. Zdunkowski, 1978. Two-dimensional numerical simulation of the effects of air pollution upon the urban–rural complex. *Tellus*, **30**, 136–150.

Werth, D., and R. Avissar, 2002. The local and global effects of Amazonian deforestation. *J. Geophys. Res.*, **107**, **D20**, 8087, doi:10.1029/2001JD000717.

Werth, D., and R. Avissar, 2005. The local and global effects of African deforestation *Geophys. Res. Lett.*, **32**, L12704, doi:10.1029/2005GL022969.

Wetzel, P. J., S. Argentini, and A. Boone, 1996. Role of land surface in controlling daytime cloud amount: Two case studies in the GCIP-SW area. *J. Geophys. Res.*, **101**, 7359–7370.

Wexler, H., 1951. Periodicity of weather. *Chem. Engin. News,* **29**, 3933.

Whitten, R. C., W. J. Borucki, and R. P. Turco, 1975. Possible ozone depletions following nuclear explosions. *Nature,* **257**, 38–39.

Widmann, M., and S. Tett, 2003. Simulating the climate of the last millennium. *Global Change Newslett.*, **56**, 10–13.

Wigley, T. M. L., 1988. The climate of the past 10 000 years and the role of the Sun. In *Secular Solar and Geomagnetic Variations in the Last 10 000 Years*, F. R. Stephenson and A. W. Wolfendale, eds. Hingham, MA: Kluwer, pp. 209–244.

Wigley, T. M. L., and P. M. Kelly, 1990. Holocene climatic change, ^{14}C wiggles and variations in solar irradiance. *Phil. Trans. Roy. Soc.*, **330**, 547–559.

Wigley, T. M. L., P. D. Jones, and P. M. Kelly, 1986. Empirical climate studies. In *The Greenhouse Effect, Climate Change and Ecosystems*, B. Bolin, B. R. Döös, J. Jäger, and R. A. Warrick, eds. Chichester, UK: John Wiley,.

Wild, M., 2005. Solar radiation budgets in atmospheric model intercomparisons from a surface perspective. *Geophys. Res. Lett.*, **32**, L07704, doi:10.1029/ 2005GL022421.

Williams, J. H., and S. Murfield, 1977. *Agricultural Atlas of Nebraska*. Lincoln, NE: University of Nebraska Press.

Williams, K. D., A. Jones, D. L. Roberts, C. A. Senior, and M. J. Woodage, 2001. The response of the climate system to the indirect effects of anthropogenic sulfate aerosol. *Climate Dynamics*, **17**, 845–856.

Williams, M., 2003. *Deforesting the Earth: From Prehistory to Global Crisis*. Chicago, IL: University of Chicago Press.

Willis, J. K., D. Roemmich, and B. Cornuelle, 2004. Interannual variability in upper ocean heat content, temperature, and thermosteric expansion on global scales. *J. Geophys. Res.*, **109**, C12036, doi:10.1029/2003JC002260.

Willson, R. C., and H. S. Hudson, 1988. Solar luminosity variations in solar cycle 21. *Nature,* **332**, 810–812.

Willson, R. C., H. D. Hudson, C. Frohlich, and R. W. Brusa, 1986. Long-term downward trend in total solar irradiance. *Science,* **234**, 1114–1117.

Wolff, G. T. 1986. Measurements of SO_x, NO_x and aerosol species off Bermuda. *Atmos. Environ.,* **20**, 1229–1239.

Woodcock, A. H., and R. H. Jones, 1970. Rainfall trends in Hawaii. *J. Appl. Meteor.,* **9**, 690–696.

Woodley, W. L., 1970. Precipitation results from a pyrotechnic cumulus seeding experiment. *J. Appl. Meteor.,* **9**, 242–257.

Woodley, W. L., 1997. Comments on "A new look at the Israeli cloud seeding experiments." *J. Appl. Meteor.,* **36**, 250–252.

Woodley, W. L., and D. Rosenfeld, 2004. The development and testing of a new method to evaluate the operational cloud-seeding programs in Texas. *J. Appl. Meteor.,* **43**, 249–263.

Woodley, W., J. Flueck, R. Biondini, *et al.*, 1982a. Clarification of confirmation in the FACE-2 experiment. *Bull. Am. Meteor. Soc.,* **63**, 273–276.

Woodley, W., J. Jordan, J. Simpson, *et al.*, 1982b. Rainfall results of the Florida Area Cumulus Experiment, 1970–1976. *J. Appl. Meteor.,* **21**, 139–164.

Woodley, W., A. Barnston, J. A. Flueck, and R. Biondini, 1983. The Florida Area Cumulus Experiment's Second Phase (FACE-2). II. Replicated and confirmatory analyses. *J. Climate Appl. Meteor.,* **22**, 1529–1540.

Woodley, W. L., B. A. Silverman, and D. Rosenfeld, 1999a. *Final Contract Report to the Ministry of Agriculture and Cooperatives.* Littleton, CO: Woodley Weather Consultants Report, 110 pp.

Woodley, W. L., D. Rosenfeld, W. Sukarnjanaset, *et al.*, 1999b. The Thailand cold-cloud seeding experiment. II. Results of the statistical evaluation. *Proc. 7th World Meterological Organization Conf. Weather Modification,* Chaing Mai, Thailand, pp. 25–38.

World Meteorological Organization, 1996. Meeting of experts to review the present status of hail suppression. *Programme on Physics and Chemistry of Clouds and Weather Modification Research.*, Golden Gate Highlands National Park, South Africa, November 6-10, 1995. WMO Technical Document No. 764, WMP Report No. 26 Geneva, Switzerland: WMO.

World Resources Institute, 1994. *World Resources 1994–1995: A Guide to the Global Environment.* New York: Oxford University Press.

Wu, Z. -X., and R. E. Newell, 1998. Influence of sea surface temperature of air temperature in the tropic. *Climate Dynamics,* **14**, 275–290.

Xue, Y., 1996. The impact of desertification in the Mongolian and the inner Mongolian grassland on the regional climate. *J. Climate,* **9**, 2173–2189.

Xue, Y., 1997. Biosphere feedback on regional climate in tropical north Africa. *Q. J. Roy. Meteor. Soc.,* **123**, 1483–1515.

Xue, Y., and M. D. Fennessy, 2002. Under what conditions does land cover change impact regional climate? In *Global Desertification: Do Humans Cause Deserts?* J. F. Reynolds and D. M. Stafford Smith, eds. Berlin, Germany: Dahlem University Press, pp. 59–74.

Xue, Y., and J. Shukla, 1993. The influence of land surface properties on Sahel climate. I. Desertification. *J. Climate,* **6**, 2232–2245.

Xue, Y., and J. Shukla, 1996. The influence of land surface properties on Sahel climate. II. Afforestation. *J. Climate,* **9**, 3260–3275.

Xue, Y., K. N. Liou, and A. Kasahara, 1990. Investigation of the biogeophysical feedback on the African climate using a two-dimensional model. *J. Climate*, **3**, 337–352.

Xue, Y., M. J. Fennessy, and P. J. Sellers, 1996. Impact of vegetation properties on U.S. summer weather prediction. *J. Geophys. Res.*, **101**, 7419–7430.

Xue, Y., F. J. Zeng, K. E. Mitchell, Z. Janjic, and E. Rogers, 2001. The impact of land surface processes on simulations of the U.S. hydrological cycle: A case study of the 1993 flood using the SSiB land surface model in the NCEP Eta regional model. *Mon. Wea. Rev.*, **129**, 2833–2860.

Xue, Y., H.-M. H. Juang, W.-P. Li, *et al.*, 2004a. Role of land surface processes in monsoon development: East Asia and West Africa. *J. Geophys. Res.*, **109**, D03105, doi:10.1029/2003JD003556.

Xue, Y., R. W. A. Hutjes, R. J. Harding, *et al.*, 2004b. The Sahelian climate. In *Vegetation, Water, Humans and the Climate*, P. Kabat, M. Claussen *et al.*, eds. Berlin, Germany: Springer-Verlag, pp. 59–77.

Xuejie, G., L. Yong, L. Wantao, Z., Zongci, and F. Giorgi, 2003. Simulation of effects of land use change on climate in China by a regional climate model. *Adv. Atmos. Sci.*, **20**, 583–592.

Young, K. C., 1977. A numerical examination of some hail suppression concepts. *Meteorol. Monogr.*, **16**, 195–214.

Yum, S. S., and J. G. Hudson, 2001. Vertical distributions of cloud condensation nuclei spectra over the springtime Arctic Ocean. *J. Geophys. Res.*, **106**, 15045–15052.

Zeng, N., 1998. Understanding climate sensitivity to tropical deforestation in a mechanistic model. *J. Climate*, **11**, 1969–1975.

Zeng, N., and J. D. Neelin, 1999. A land–atmosphere interaction theory for the tropical deforestation problem. *J. Climate*, **12**, 857–872.

Zeng, N., R. E. Dickinson, and X. Zeng, 1996. Climatic impact of Amazon deforestation: A mechanistic model study. *J. Climate*, **9**, 859–883.

Zeng, N., J. D. Neelin, K.-M. Lau, and C. J. Tucker, 1999. Enhancement on interdecadal climate variability in the Sahel by vegetation interaction. *Science*, **286**, 1537–1540.

Zeng, N., K. Hales, and J. D. Neelin, 2002. Nonlinear dynamics in a coupled vegetation–atmosphere system and implications for desert–forest gradient. *J. Climate*, **15**, 3474–3487.

Zeng, X., and R. A. Pielke, 1995a. Landscape-induced atmospheric flow and its parameterization in large-scale numerical models. *J. Climate*, **8**, 1156–1177.

Zeng, X., and R. A. Pielke, 1995b. Further study on the predictability of landscape-induced atmospheric flow. *J. Atmos. Sci.*, **52**, 1680–1698.

Zeng, X., Y.-J. Dai, R. E. Dickinson, and M. Shaikh, 1998. The role of root distribution for climate simulation over land. *Geophys. Res. Lett.*, **25**, 4533–4536.

Zeng, X., S. S. P. Shen, X. Zeng, and R. E. Dickinson, 2004. Multiple equilibrium states and the abrupt transitions in a dynamical system of soil water interactions with vegetation. *Geophys. Res. Lett.*, **31**, L05501, doi:10.1029/2003GL018910.

Zhang, H., A. Henderson-Sellers, and K. McGuffie, 1996a. Impacts of tropical deforestation. I. Process analysis of local climatic change. *J. Climate*, **9**, 1497–1517.

Zhang, H., K. McGuffie, and A. Henderson-Sellers, 1996b. Impacts of tropical deforestation. II: The role of large-scale dynamics. *J. Climate*, **9**, 2498–2521.

Zhang, H., A. Henderson-Sellers, and K. McGuffie, 2001. The compounding effects of tropical deforestation and greenhouse warming on climate. *Climat. Change*, **49**, 309–338.

Zhang, J., U. Lohmann, and P. Stier, 2004. A microphysical parameterization for convective clouds in the ECHAM5 Climate Model. I. Single column model results evaluated at the Oklahoma RM site. *J. Geophys. Res.*, **110**, D15S07, doi:10.1029/2004JD005128.

Zhao, M., and A. J. Pitman, 2002a. The impact of land cover change and increasing carbon dioxide on the extreme and frequency of maximum temperature and convective precipitation. *Geophys. Res. Lett.*, **29**, 2-1–2-4.

Zhao, M., and A. J. Pitman, 2002b. The regional scale impact of land cover change simulated with a climate model. *Int. J. Climatol.*, **22**, 271–290.

Zhao, M., and A. J. Pitman, 2005. The relative impact of regional scale land cover change and increasing CO_2 over China. *Adv. Atmos. Sci.*, **22**, 58–68.

Zhao, M., A. J. Pitman, and T. N. Chase, 2001a. The impact of land cover change on the atmospheric circulation. *Climate Dynamics*, **5/6**, 467–477.

Zhao, M., A. J. Pitman, and T. Chase, 2001b. Climatic effects of land cover change at different carbon dioxide levels. *Climate Res.*, **17**, 1–18.

Zheng, X., and E. A. B. Eltahir, 1997. The response to deforestation and desertification in a model of west African monsoons. *Geophys. Res. Lett.*, **24**, 155–158.

Zheng, X., and E. A. B. Eltahir, 1998. The role of vegetation in the dynamics of west African monsoons. *J. Climate*, **11**, 2078–2096.

Zheng, X., E. A. B. Eltahir, and K. A. Emanuel, 1999. A mechanism relating tropical Atlantic spring sea surface temperature and west African rainfall. *Q. J. Roy. Meteor. Soc.*, **125**, 1129–1164.

Zheng, Y., G. Yu, Y. Qian, *et al.*, 2002. Simulations of regional climate effects of vegetation change in China. *Q. J. Roy. Meteor. Soc.*, **128**, 2089–2114.

Zhou, L., R. E. Dickinson, Y. Tian, *et al.*, 2004. Evidence for a significant urbanization effect on climate in China. *Proc. Natl Acad. Sci. USA*, **101**, 9540–9544.

Ziegler, C. L., T. J. Lee, and R. A. Pielke, Sr., 1997. Convective initiation at the dryline: A modeling study. *Mon. Wea. Rev.*, **125**, 1001–1026.

Zuberi, B., A. Bertram, C. Cassa, L. Molina, and M. Molina, 2002. Heterogeneous nucleation of ice in $(NH_4)2SO_4$-H_2O particles with mineral dust immersions. *Geophys. Res. Lett.*, **29**, doi:10.1029/2001GL014289.

Index

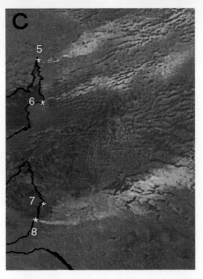

Plate 1 Satellite visualization of NOAA AVHRR images, showing the microstructure of clouds for three cases over three different continents with streaks of visibly smaller drops due to ingestion of pollution originating from known pollution sources that are marked by white numbered asterisks. (A) A 300×200 km cloudy area containing yellow streaks originating from the urban air pollution of Istanbul (*1), Izmit (*2), and Bursa (*3) on 25 December 1998 at 12:43 UT. (B) A 150×100 km cloudy area containing yellow streaks showing the impact of the effluents from the Hudson Bay Mining and Smelting compound at Flin-Flon (*4) in Manitoba, Canada ($54° 46' N 102° 06' W$), on 4 June 1998 at 20:15 UT.(C) An area of about 350×450 km containing pollution tracks over South Australia on 12 August 1997 at 05:25 UT originating from the Port Augusta power plant (*5), the Port Pirie lead smelter (*6), Adelaide port (*7), and the oil refineries (*8). All images are oriented with north at the top. The images are color composites, where the red is modulated by the visible channel; blue is modulated by the thermal infrared; and green is modulated by the solar reflectance component of the 3.7μm channel, where larger (greener) reflectance indicates smaller droplets. The composition of the channels determines the color of the clouds, where red represents cloud with large drops and yellow represents clouds with small drops. The blue background represents the ground surface below the clouds. From Rosenfeld (2000). Reprinted with permission from D. Rosenfeld, © 2000 American Association for the Advancement of Science. See also Figure 4.3.

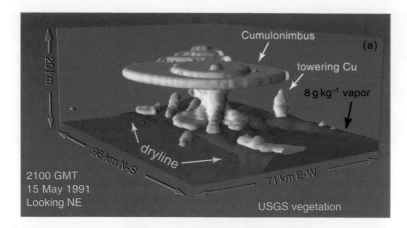

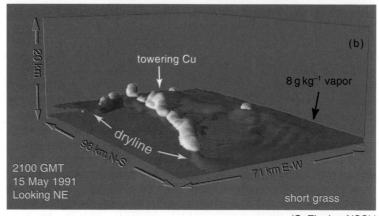

(C. Ziegler, NSSL)

Plate 2 (a) and (b) Model output cloud and water vapor mixing ratio fields on the third nested grid (grid 4) at 21:00 UT on 15 May 1991. The clouds are depicted by white surfaces with $q_c = 0.01$ g kg^{-1}, with the Sun illuminating the clouds from the west. The vapor mixing ratio in the planetary boundary layer is depicted by the shaded surface with $q_v = 8$ g kg^{-1}. The flat surface is the ground. Areas formed by the intersection of clouds or the vapor field with lateral boundaries are flat surfaces, and visible ground implies $q_v < 8$ g kg^{-1}. The vertical axis is height, and the backplanes are the north and east sides of the grid domain. Reproduced from Pielke *et al.* (1997) with permission from *Ecological Applications* and the Ecological Society of America. See also Figure 6.10.

Plate 3 South of the South Platte River, south and west of North Platte, Nebraska, looking north on 1 June 2004 at approximately 1300 LST. A number of pivot irrigators are not watering. Photo courtesy of Kelly Redmond. See also Figure 6.16.

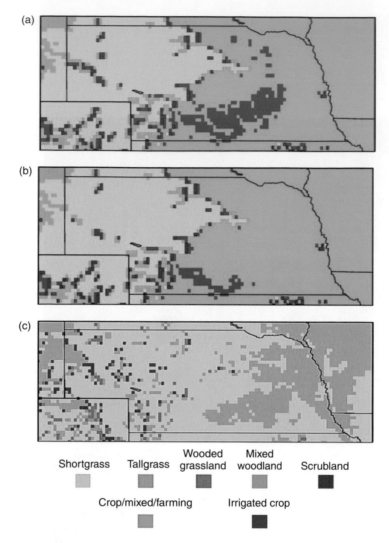

Plate 4 Land-cover datasets used for RAMS simulations for (a) 1997 Landsat and ancillary data irrigation, (b) OGE, and (c) Küchler potential vegetation. From Adegoke *et al.* (2003) reproduced with permission from the American Meteorological Society. See also Figure 6.17.

Plate 5 This MODIS view shows the denuded high albedo regions of the Sinai and Gaza Strip, in contrast to the darker western Negev. Sensor: Terra/MODIS; Datastart: 2000-09-10; Visible Earth v1 ID: 5606; Visualization date: 2000-10-12. Courtesy of NASA Visible Earth and Jacques Descloitres, MODIS Land Science Team. See also Figure 6.24.

Plate 6 Examples of clear-cutting of the tropical forest in two areas of the Amazon. Photos provided by Carlos Nobre of the Center for Weather Prediction and Climate Studies – CPTEC, INPE, Brazil. See also Figure 6.25.

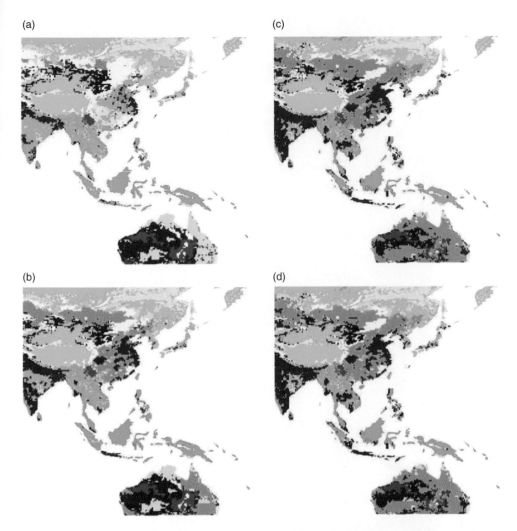

Plate 7 Examples of land-use change from (a) 1700, (b) 1900, (c) 1970, and (d) 1990. The human-disturbed landscape includes intensive cropland (red), and marginal cropland used for grazing (pink). Other landscape includes, for example, tropical evergreen and deciduous forest (dark green), savanna (light green), grassland and steppe (yellow), open shrubland (maroon), temperate deciduous forest (blue), temperate needleleaf evergreen forest (light yellow), and hot desert (orange). Of particular importance is the expansion of the cropland and grazed land between 1700 and 1900. Data obtained from the Hyde Database available at www.mnp.nl/hyde/. Reproduced with permission from Kees Klein Goldewijk. See also Figure 11.1.

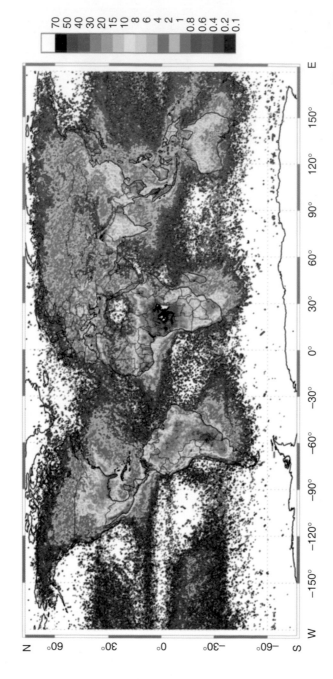

Plate 8 Global distribution of lightning from April 1995 through February 2003 from the combined observations of the NASA Optical Transient Detector (OTD) (4/95-3/00) and Lightning Imaging Sensor (LIS) (1/98-2/03) instruments. From http://thunder.nsstc.nasa.gov/images/HRFC_AnnualFlashRate_cap.jpg. See also Figure 11.2.

(a) Radiative forcing due to carbon sequestration (nW m^{-2} ha^{-1})

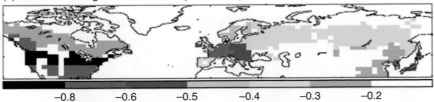

| −0.8 | −0.6 | −0.5 | −0.4 | −0.3 | −0.2 |

(b) Radiative forcing due to albedo change (nW m^{-2} ha^{-1})

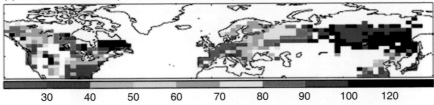

| 0.05 | 0.1 | 0.15 | 0.2 | 0.25 | 0.3 | 0.35 | 0.4 |

(c) Carbon emissions equivalent to albedo change (t C ha^{-1})

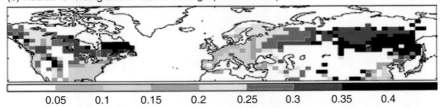

| 30 | 40 | 50 | 60 | 70 | 80 | 90 | 100 | 120 |

(d) Net radiative forcing by afforestation (nW m^{-2} ha^{-1})

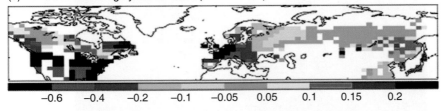

| −0.6 | −0.4 | −0.2 | −0.1 | −0.05 | 0.05 | 0.1 | 0.15 | 0.2 |

Plate 9 Radiative forcing of climate by afforestation, considering illustrative 1-ha plantations in the temperate and boreal forest zones. Calculations apply to the time at the end of one forestry rotation period, relative to the start of the rotation period with plantation areas unforested. (a) Global mean long-wave radiative forcing due to carbon dioxide removal through sequestration (n W m^{-2} ha^{-1}). (b) Global mean shortwave radiative forcing due to albedo reduction (n W m^{-2} ha^{-1}). (c) Carbon emissions that would give the same magnitude of radiative forcing as the albedo reduction (t C ha^{-1}). (d) Net radiative forcing due to afforestation, found by summing (a) and (b) (n W m^{-2} ha^{-1}). Positive forcing implies a warming influence; where (d) shows positive values, afforestation would warm climate rather than cooling it as would be expected by considering carbon sequestration alone. After Betts (2000); from Pielke *et al.* (2002). See also Figure 11.9.

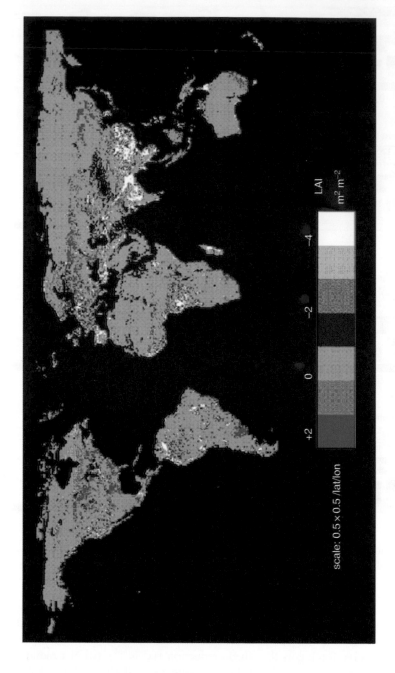

Plate 10 Effect of land-use changes on plant canopy density (potential LAI - actual LAI). Scale 0.5 latitude × 0.5 longitude. From Pielke *et al.* (2002), based on Nemani *et al.* (1996). See also Figure 11.10.

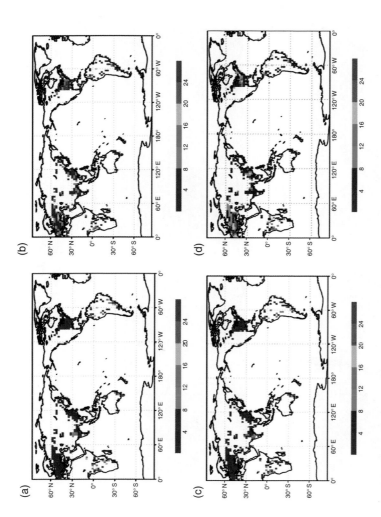

Plate 11 The 10-year average absolute value change in (a) January surface latent turbulent heat flux, (b) July surface latent turbulent heat flux, (c) January surface sensible heat flux, and (d) July surface sensible heat flux in W m^{-2} at the locations where land-use change occurred. Based on Chase *et al.* (2000); from Pielke *et al* (2002). See also Figure 11.11.

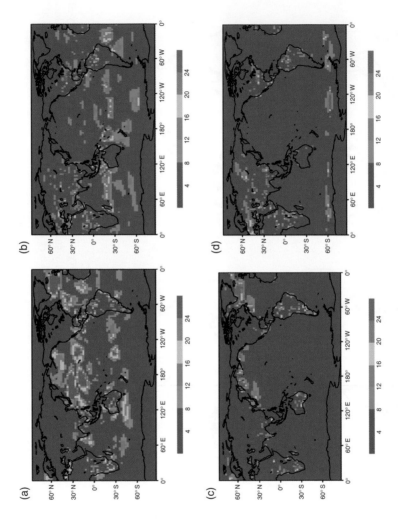

Plate 12 The 10-year average absolute value change in surface sensible and latent turbulent heat flux in W m^{-2} worldwide as a result of the land-use changes. (a) January surface latent turbulent heat flux, (b) July surface latent turbulent heat flux, (c) January sensible turbulent heat flux, and (d) July sensible turbulent heat flux. Based on Chase *et al.* (2000); from Pielke *et al.* (2002). See also Figure 11.12.

ASHES, ASHES

JO TREGGIARI

ASI

ASI

ES,

ES

SCHOLASTIC INC.

NEW YORK TORONTO LONDON AUCKLAND

SYDNEY MEXICO CITY NEW DELHI HONG KONG

ISBN 978-0-545-43656-4

12 11 10 9 8 7 6 5 4 3 2 1   12 13 14 15 16 17/0

Printed in the U.S.A.   40

First Scholastic paperback printing, January 2012

Book design by Elizabeth B. Parisi

FOR BETTY FRANKLIN,
SUSAN TREGGIARI,
AND MY FEARLESS, INTREPID LUCY PARRIS

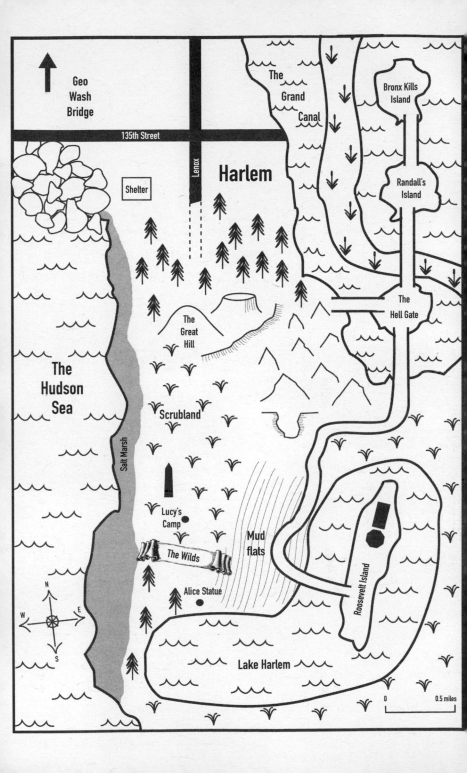

Ring around the roses
A pocketful of posies
Ashes, ashes
We all fall down.

CHAPTER ONE
TURTLE

Lucy hunched over the corpse and felt a tiny bubble of hysterical laughter gurgle up. But as she stared at the lifeless turtle stretched out on the rough plank, the laughter died abruptly. The tang of fresh blood was unpleasant. She should have butchered it outside, by the shore, but the hour was getting late and she'd felt exposed on the sand. Besides, she had never actually done a turtle before, never noticed how the

wizened face and papery eyelids made it look like a very old person.

She positioned the knife edge along the thinnest section of gray, wrinkled neck and pushed down, fixing her gaze steadily in front of her. The knife stuck. She tried to stop her brain from screaming thoughts of sinew and bone, and leaned her weight on her hand. The flesh resisted, then suddenly gave way. The knife slammed into the hard wood underneath, and the head rolled off onto the ground with an audible *thump*.

Her stomach heaved. Fortunately, it was empty. Lucy put her knife down and dragged the woven screen away from the entry hole to her shelter, letting a breeze sweep in and clear the stench from her nose.

She closed her eyes and breathed in deeply, sinking to her knees. She could smell the scent of impending rain. She wondered whether she could survive it for another year. Two days of steady rain had already turned the ground outside her camp into a series of muddy pools threaded by soggy grassland, and since her shelter lay in a hollow, there was now what amounted to a narrow moat right outside the front entrance.

The floods had first come about five years ago, when she was eleven years old. Melting polar caps; rising sea levels; increased rainfall; a steady battering of hurricanes, tornadoes, and earthquakes weakening the land: everything the scientists had warned them about. And the world mapped in her geography books had changed with a frightening rapidity;

continents shifting shape, coastlines altered. San Francisco, Los Angeles, Venice, Thailand, Spain, her beloved Coney Island, Japan, had all but vanished beneath the waves. Australia was half the size it had been, shrinking like an ice cube in a warm drink, and New York City had become a clump of six or seven scattered islands connected to the mainland by a few big bridges — the Geo Wash, the RFK, the Will Burg. Some were only accessible during the Long Dry.

Small but fast-moving canals flowed over the same routes as the old roads. Lexington Avenue, Fifth Avenue, 42nd Street, were all underwater now. But people had rallied and rebuilt. They'd stretched suspension bridges strong enough to hold a dozen people and a few bicycles at a time across the swollen canals that now ran in a crisscrossing grid over what had been Manhattan. Thousands of sandbags shored up the dikes along the smaller waterways, and a massive wall of masonry and detonated high-rises had been built in an attempt to keep the inland sea back from the edges of Harlem and Washington Heights. Cheap plywood houses sprang up on stilts, altering the cityscape. Deep, wide gutters were cut into the ground, and cars were banned from the city, except on the outskirts and the few roads that had survived the earthquakes.

They'd called it "New Venice" jokingly, and it had seemed okay then. Lucy, living in the solidity of northern New Jersey, miles away from the shores of the sea, had felt safe, and she'd kept on taking the train into the city or hitching, kept on

cruising the vintage stores for cool clothes. It rained frequently, making whole neighborhoods inaccessible for months out of the year, and the summers were more sweltering than ever, but the streets were still packed with people buying and selling or just hanging out. And then, as if all that had been just a dress rehearsal for some disaster movie, four years later the plague had arrived.

She could almost hear the newspeople again, like something out of one of those cheesy old sci-fi television shows: the warnings to stay inside; the rising panic; the video of gaunt, red-eyed survivors, their skin seemingly charred; the doomsayers with their sandwich boards, black robes, and crazy talk about disease-carrying birds and God's wrath. Seeing anchorwomen, who normally looked like airbrushed mannequins, seriously freaking out was scary. Lucy still had nightmares. She still woke up certain that her skin was covered in scabs and she was bleeding to death from the inside out.

Instead of globe-eyed aliens or a gigantic meteorite headed straight for Earth, it was the resurgence of a killer disease that had reduced the global population to less than 1 percent of what it had been within three short months. Eating healthy, exercising, living in a big house, driving a fancy car—none of that mattered at all. The pox took almost everybody, and it seemed that people between the ages of thirty and sixty died faster and harder than anyone.

And who would have guessed that Lucy would have turned out to be luckier than her entire family?

Lucy "Lucky" Holloway. She used to hate her nickname, but now—now it was different.

It was weird to think that her younger brother, Rob, had started calling her that as a joke. He'd also nicknamed the dog, Rex, "Tex Mex" after it had been discovered that the golden retriever could scarf down a dozen frozen burritos without vomiting, and he'd renamed their older sister, Susan, "Maggie" (short for "Maggot") because she liked to eat rice pudding while bundled up on the couch in an old blanket.

Maybe Lucy had gotten off easy. *Lucky* instead of *Lucy* wasn't too bad, and most people didn't realize it was meant sarcastically because of her ability to trip over her own two feet, break dishes, and knock books off of shelves merely by walking past them. As a preteen, she'd managed to run through the glass French doors that led into the kitchen from the pool, not once, but twice, necessitating visits to the emergency room and eleven stitches in her calf the first time and then, six under her chin.

She knew her clumsiness annoyed her parents. She'd always felt as if she were a changeling dumped into their magazine-perfect midst. She didn't even look like them, having inherited some recessive gene from an ancient Welsh ancestor. She was slim and gray-eyed, with wildly curly black hair—shocking

compared to their pink and blond athletic good looks. She was awkward and she was ugly. And worse than that, she wasn't a superjock like her brother or a brainiac like her sister. She was something her happy-homemaker mom and her big-lawyer dad just couldn't understand: good at nothing in particular.

In her journal she'd written long, angry, tearful diatribes about feeling out of step and alone at home and at school, where the cliques were ruled by people just like her brother and sister, until she'd convinced herself she didn't care, forced herself to tune out when Rob's latest game score or Maggie's newest scholarship was being discussed over the dinner table. At least she hadn't been nicknamed after something that pulsed and wriggled on rotten food.

Poor Maggie. Gone, and her old blanket gone, too, burned in a useless attempt to get rid of the sickness. Together with the quilts Grandma Ferris had painstakingly pieced together and Lucy's threadbare teddy bear and the embroidered sofa cushions and everything else it seemed had made life soft and comfortable. Great piles of sheets and bedspreads, mattresses and pillows were piled sky-high in every neighborhood, then doused in gasoline and coaxed into infernos that burned for weeks. When the wind was coming from the east, Lucy imagined she could still smell the acrid fumes like burnt hair, could still see the towers of black smoke billowing against the blazing blue sky. It was not until the Long Dry was over and the Long Wet began again that the fires were finally quenched

by the pelting rains. The parched dirt liquefied into slow-moving rivers of sludge and covered everything in mud, including the pits—the deep trenches used once the cemeteries were full, where bodies were stacked in rows like cut logs and scattered with quick lime to hurry the decay in an attempt to prevent reinfection. Then the controlled bombings began, turning high-rises into massive concrete cairns over the sites where thousands had died within days of one another. The skeletal bodies, the livid marks on blackened skin, were buried under tons of rock, and all the crushing details of life as it used to be were erased.

The memories she tried to preserve were of her life before the plague descended, and in her mind it was like those times were lit by a gentler sun, all Technicolor blurry and beautiful. She remembered the smallest things: her mother's buttermilk pancakes and homemade blackberry jam, the smell of fabric softener, the feel of socks without holes. Now, looking at her grimy fingernails and dirt-encrusted skin, she was amazed at how much she had changed. How things like homework, a daily shower, and a hot breakfast on the table seemed so unimaginable to her now.

She was certainly not lucky. She was Lucy, plain and simple.

She leaned back on her heels, staring at the flattened pile of dry grasses where she slept, her sleeping bag with her vintage leather motorcycle jacket scrunched up for a pillow; the

crooked shelf she'd hung from a couple of branches holding a dented tin plate and bowl; a camping knife, fork, and spoon strung on a piece of string so she wouldn't lose them; her backpack with the essentials; a change of clothing. All she had left in the world besides her knife, the survival manual she'd scooped off the floor of a bookshop with all the windows smashed out, and a few other personal items. The book was battered and stained, pages escaping from the cracked binding, but it was precious.

She scraped her hair back off her sweaty forehead. It was too short to stay tucked behind her ears and just long enough to fall into her eyes constantly. She felt an oozing wetness on her cheek and looked at her hands. Mud, blood, and who knew what else. Tears? Lucy didn't cry too often. She figured she'd used them all up by now. She bit her lip hard between her teeth and stood up clumsily. She'd been crouched over for so long that her right foot had gone to sleep. She dragged the screen back over the doorway, then limped over to the bucket she kept filled with rainwater and rinsed her hands, drying them roughly on the legs of her jeans.

The turtle wasn't getting any deader, and she had a lot to do before she lost the last of the daylight. She walked back over to the rough table she'd made out of a few pallets and peered down at the manual, held flat under a couple of rocks. The instructions had seemed simple enough. The capture had been easier than she had expected: sneaking up on the creature

while it sunbathed on a mud bar on the shore of the Hudson Sea, grabbing it by the thin leather whip of a tail, and holding it well away from her body until she could shove a stick between the snapping jaws. And she hadn't felt much sympathy, not after it tried to bite her; hadn't felt so much as a twinge, even though before everything that had happened she'd been a strict vegetarian. No, she'd held it flat against the ground with the pressure of her knee, waggled another stick in front of the cruel, predatory-bird mouth until the neck was stretched taut, and then bopped it hard on its little old lady head with a handy rock.

She checked the book again. There was a page missing; there must be. She flipped backward and forward, looking for the sequence of actions that would yield four slabs of pink meat, as pristine and antiseptic as anything you could have bought off a refrigerated shelf in a grocery store. If there had still been any around. She stabbed at the creature in a sudden fury. The knife turned on the shell. She yelped and tossed it from her in disgust. She'd gouged her left palm — a long, wide gash that instantly welled blood. She sucked on her hand, not really enjoying the coppery taste. She pulled her bandanna from her neck and wrapped it around the wound, pulling the ends tight in a knot with her teeth. Then she kneeled down and picked up the knife, rubbing the dirt from it and checking the blade for damage. She heaved a sigh of relief. It seemed okay. She ran her thumb over the edge, feeling a burr of

roughness, the smallest of nicks. It would need to be re-honed before she could continue.

"Stupid, stupid, stupid," she told herself.

At the bottom of her backpack was a narrow rectangle of gray stone. It felt like fine sandpaper. Five sweeps of the blade against it and the edge was sharp enough to draw a thin line of blood across the fleshy part of her thumb. She turned the knife over to sharpen the other side.

Lucy walked back over to the book and pushed her hair back behind her ears again with force. The gory remains of the turtle were laid out on some broad leaves. It looked nothing like the neat illustration. The vibrantly colored picture showed tidy quarters of pale rosy meat—not this mutilated lumpy mass leaking muddy water and blood. The shell was the problem. Bony, hard as granite, it just wouldn't come off. She'd followed the instructions, tossing the corpse into a saucepan of water over the fire. She'd even left the turtle in the pot for longer than the ten minutes specified, but now she had to wonder if perhaps the water hadn't been hot enough.

The words danced in front of her eyes. She stared so long that they stopped making any kind of sense. The sun was setting, and the light leaking through the willow screens was dim. Lucy inserted the knife between the shell and the carcass and jimmied it around. There was a snap and the tip of the blade broke off. She stared at the knife for a moment, disbelieving, and then with a cry of rage, she picked up and

threw the book with all her strength, sending it skidding across the dirt floor.

"Crap!" she yelled, and instantly was aware of the frustration welling up in her throat and the hot tears coming. She bit down hard on her lower lip until the pain pushed back the angry sobs catching in her throat. *Deep breath*. You did not waste food. Not when it was so scarce. Not when the birds were poison and squirrels were skittish. Carefully she checked over the knife. The main part of the blade, about six inches, was still good. She could use it for most jobs. With a sigh, Lucy bent down and picked up the book, shuffling the pages back into the binding and smoothing the cover.

She leaned over the body, poked at it with her finger. The turtle's legs flopped like a rag doll. She couldn't imagine anything less appetizing, but there was no way she was going to give up now. She hadn't eaten anything since that morning, and then it had only been a scoop of porridge and a handful of dried, shriveled raspberries, which had tasted moldy. She took a couple of pieces of wood from the scanty pile stacked beside her and added them to the fire. She held her hand over the mouth of the cooking pot. It was hardly steaming. The wood was too green, the fire still not hot enough; the cooking stones barely sizzled when she aimed a gobbet of spit at them, and the dented saucepan of water refused to boil.

She sighed. Her fists were clenched and she could feel the pinch of her nails against her palms. It was already too late in

the day to build up a good fire. The turtle's mottled skin, the ragged ruin of its neck, were taking on an unhealthy gray appearance. And she could smell something swampy and briny, like stagnant water. It was already cooler than it had been for the last six months, but still warm enough to turn meat bad fast. If the last four hours weren't going to be a complete waste of time, she'd have to do something. Lucy hefted one of the heavy, river-smoothed rocks she kept nearby and smashed it into the shell, which broke into irregular pieces, some large enough to dig out with her fingers, some small bits, like yellow pottery shards, embedded into the leathery skin of the turtle's underside. She went to work picking out the pieces until she could make a long incision in the belly. She shoved her hand in under the tough hide and scooped out the stomach and intestines, careful to breathe through her mouth. She'd gutted plenty of fish in the last year, and the looping entrails didn't bother her too much anymore. They were neat little parcels as long as she was careful not to puncture them. She piled them on a few broad dock leaves and covered them up against the flies. Later she'd bait her fishing lines with them and see if the catfish and eels liked innards better than the night crawlers she normally used. She flipped the turtle over, smashed the upper plate with the rock, and picked out as much of the shell as she could. After consulting the book again, she made four slits down the inside of each leg and cut away the skin. It slipped back easily, sort of like

peeling a banana, and with only a little bit more cutting she was able to pull it free from the turtle's feet.

Lucy ran her finger over the hide, wondering if it was tough enough to patch the many holes in her boots. *Nothing wasted,* she thought, putting it aside to deal with later. She'd cured a rabbit pelt and a couple of squirrel skins before and ended up with serviceable but stinky leather, too stiff to work with easily but good enough to mend holes. She looked down at the oozing carcass, casting her mind back to tenth-grade biology, trying to remember anything useful. That had been frogs, in any case. Rubbery, fake-looking, and smelling overwhelmingly of formaldehyde. If she had a frog in front of her now, she'd have been able to skin and fillet it in two minutes flat, like one of those Japanese chefs.

She had a wild impulse to just dump the turtle and eat acorn porridge and dried berries for the fourth day in a row, but her acorn flour store was getting low, fresh meat was rare, and she needed the protein. She suspected, too, that there was more squirming weevil than powdered acorn at the bottom of the old coffee can. Perhaps if she just shoved the meat back into the saucepan, put the lid on, and left it to sit for a while over the bedded embers, the flesh would fall from the bones and she'd have turtle soup or turtle tea. There were a couple of shriveled wild onions left, some woody mushrooms she could toss in. She'd eaten far worse.

Lucy stuck the turtle in the pot and covered it, piling the

smoking wood around it. She rinsed her right hand in the bucket of water, getting most of the blood off, though not the dark matter stuck deep under her nails, and wiped it dry on her jeans. Her left hand throbbed, and she wondered if the bandage was on too tight. Her skin felt greasy with sweat. She stripped off her thick sweatshirt. Underneath she wore a tank top. Lucy sniffed at her armpits, wrinkling her nose, and then quickly sluiced her upper body. Now that the waters were rising, she'd be able to bathe again. It had been far too long. Weird how she hadn't noticed the odor of stale sweat and grime that permeated the shelter. Time to change the bedding grasses and air out the rush mats she had pieced together during the long nights. Lucy had ended up with cuts striping her palms and fingers and looking as if she'd lost a fight with a bramble bush. She would light some sage bundles later to clear out the musty stink and the smell of dead turtle.

She pulled her sweatshirt back on. Her neck felt tight, her hands shaky, and the wound beat in time with her heart. She wrapped a shawl and then her sleeping bag around her shoulders and sat as close as she could to her small fire. The smoke burned her eyes. She was procrastinating. There were things to do before nightfall, but she had checked the calendar notches she had cut into the bark of one of the four support trees and knew the moon would be full, which would give her

more light than usual. She could get a later start, and a few minutes' rest would do her good.

Beyond the cracks in the interlaced willow screens she had made to disguise the small clearing where she lived, she could see the huge, red sun setting above waters that looked as thick and black as molasses.

Lucy dreaded this time of day, when there was a pause and her thoughts rose up and threatened to submerge her. As long as she was busy doing, she could keep the loneliness at bay. She drew the edge of the sleeping bag up around her ears, the shawl over her head, and nuzzled into them, smelling the nose-tickling mustiness of leaf mold, ground-in dirt, and the dried grasses she slept on. Her mind buzzed at her like an annoying mosquito.

She needed to walk the circumference of her camp, check the snares she'd set, the trip wires, the bundles of grass she had laid down on the ground that would show her if anyone heavy-footed had come near. Lucy groaned. She was so tired. Her days were always long, but sometimes it seemed as if it didn't matter how early a start she got.

She thrust the sleeping bag aside, bundled up the shawl, and got to her feet, popping her neck and shrugging her shoulders up and down a few times to work out the tightness. She checked a few more things off her mental list: She needed drinking water, so she'd have to make a wide arc and pass by

the lake, and she needed as much wood as she could carry, now that the rains were gathering force. There'd been two torrential downpours lasting ten or twelve hours already, and it was only the beginning of June by her rough monthly calendar. She'd been less than careful about keeping track of the days and nights. The Long Wet brought monsoons, riptides, flash floods, and sudden lightning fires—the worst of them falling roughly in the middle of the cycle, but if anything was true these days, it was that the weather was erratic.

She'd check her snares, of course, hoping for a ground squirrel or rat, and her fishing lines, although during the Long Dry the lake waters had receded, leaving about twenty feet of dry, cracking mud before the first dribbles accumulated in shallow pools. A mudskipper maybe, a newt, or a salamander, though she didn't like the gluey taste much. It was too dark to go digging for shellfish by the sea. She'd plan on doing that tomorrow.

First she listened. But there were no noises except for the rhythmic hum of insects. Next she peered through a hole in the mesh of supple willow limbs that screened the front entrance to her camp. Lucy knew every tree, every bush, every grassy hump silhouetted in the gathering dusk. It was a landscape she had peered at and studied night after night. She had counted the weird hummocks carved out of the earth after the last quake—there were twenty-three of them standing guard like silent sentinels.

Nothing seemed different, but lately she'd had the unsettling feeling of eyes on her. She checked for movement. The air was so still, the grasses didn't stir. She pulled up the black sweatshirt hood. Then she picked up a couple of plastic gallon jugs for the water, looping a length of rope through the handles, slung a woven grass bag over her shoulder, checked that her knife was snug against her hip, and lifted the front door screen out of the way.

A long puddle of water lapped against the piled sticks and brush she'd stacked against the outer walls to keep the rainwater from seeping in. She splashed through, feeling the cold wetness through the thick leather of her boots and a double layer of socks, ducked her head slightly, and replaced the screen. She backed up about five feet, making sure her small fire pit was invisible from the outside. It was. *Good*. She'd spent a lot of time stuffing most of the larger chinks with moss and dried grass recently. Plus, now that the rains were coming, the willow sticks she'd shoved into the ground to make thicker walls would begin to grow and leaf-out. Willow was amazing! A cut stick would take root easily. The four slender, flexible trees she had bent down and bound together at the top to make the sloping roof were already bushy with new growth.

If you didn't know the camp was there, it was almost invisible against the surrounding foliage and shrubs, like the snug, domed nests the field mice made themselves out of grass stalks. She glanced at the sky. The moon was beginning its

rise, full as she had hoped. Purple clouds boiled; the wind had suddenly picked up and the scent of rain was heavy. It would mask the smells of smoke and cooking turtle, she thought. Taking one last look around, she set off toward the lake, her nerves stretched tight and jumping.

The terrain was already changing. There were splashes of green leaves within the dusty gold. And the ground was spongy underfoot, treacherous with puddles and sinkholes. There were pretty much only two seasons now — drought and flood.

Her boots squelched a bit, but so far they were not leaking too badly. It was so quiet — save for the scritch of small claws scrabbling up tree trunks and the angry, explosive noises of disturbed squirrels. She always thought they sounded as if they were cussing her out. On the way, she inspected various snares she'd concealed under bushes and by likely holes, crossing back and forth along the narrow spit of land, her senses in hyperdrive. They were all empty. One showed signs that a predator had gotten there first. Tufts of silver fur snagged in the branches, a few driblets of blood. She kneeled down, touched the soft, downy clumps. Rabbit, she thought, rather than squirrel. Too bad. Rabbit was a delicacy these days, but she couldn't help but be glad that there were still foxes and coyotes around. So many animal species had been wiped out in the plague.

As if to echo her thoughts, a howl rent the air. She stiffened. She knew the clear belling and crystal sharp barks of the foxes and coyotes as they called to one another. This was deeper, urgent—the sound of a hunting pack of dogs. Her head swiveled in the direction of the baying. She thought it was at some distance yet. But behind her. She shuddered, fighting the urge to break into a panicked run. Not just behind her, but between her and her camp. Most predators were still scared around humans; her smell was enough to keep them at a distance. But the packs of feral dogs were large and hungry, and they had no fear of people.

She considered. She'd work her way to the lake and circle around, giving them wide berth. The land rose slightly just beyond the water's edge, and she'd be able to get a better look. And she could check the water levels at the same time. There was the tarnished bronze statue of a girl sitting on a large toadstool surrounded by an assortment of strange characters, and Lucy used this to keep track, scratching lines into the metal every second full moon. The last time she'd looked, the water had been barely lapping at the girl's toes, but by the middle of the Long Wet it would be up to her shoulder level. Lucy couldn't remember the girl's name now, although when she was a child her mother used to bring her here to climb on the statue. She recalled jumping from toadstool to toadstool, feeling the smooth, sun-hot metal, playing king of the castle

with other kids. The bravest of them leapt from the hare to the man in the top hat or perched on the girl's head, gripping the long locks of her flowing hair. That wasn't Lucy, though. She never made it higher than the girl's lap—broad and solid and safe.

Now she moved quickly. There was no cover but scrubby grasses and spindly bushes. The ground underfoot had changed from loose, sandy earth to cracked, oozing mud. The lake was to her left. It had dwindled over the hot season to a series of small, murky pools surrounded by rings of soft, slippery sludge. A larger expanse of water lay far beyond her reach, as smooth as glass. Her fishing lines were marked by twists of bark. Lucy pulled them up, and, finding the hooks empty, tossed them back into the shallow water. All around her was the plopping sound of frogs, as they woke to her presence and alerted one another. The splashes they made sounded like a string of tiny firecrackers going off. She needed her spear to catch frogs. They were too quick, too alert.

The dogs had stopped barking. The night was silent again except for the small animal noises. Lucy crouched and submerged her water bottles to fill them. The flow of water gurgled gently. Her eyes darted around, her head lifted. She pushed her hood back so that she could see better. The quiet was unnerving after the cacophony of howls and barks. The hairs on the back of her neck rose. She was being watched. Slowly, she got to her feet, capped the jugs, and hung them

from her neck, easing the rope into position so it lay across her shoulders. Then she loosened the knife in its sheath. She strained her ears, listening hard. Suddenly there were small, ominous noises coming from all around. A rat snake rustled past, its heavy black body as thick around as her wrist. There was the squeal of something just caught.

Lucy pulled her hood back over her face, trying to blend into the inadequate shadows. She froze. Directly across from her, at the edge of a pool of fresh rainwater, belly flat to the ground, was a cougar. So close she could see the pink tongue lap. They locked eyes. Lucy barely breathed. She tried to remember if the manual said she should play dead or make a racket. The cougar didn't move. Lucy's fingers fumbled at the hilt of the knife, trying to prepare herself for an attack if it came; quietly telling herself to slash a volley of cuts; reminding herself that the blade was broken, that stabbing would have no effect. But behind that voice, the knowledge that she'd be helpless against two hundred pounds of lithe muscle and bone, a natural killer, and the hope that death would be quick and the pain numbed by fear and shock. Maybe she shouldn't be making eye contact? Perhaps that was a threat? She closed her eyes and murmured a quick prayer. Her thigh muscles quivered. She ducked down, trying to move smoothly. Her feet slid awkwardly in the mud. She slipped and fell backward, the weight of the water bottles pulling her off balance. Quickly she was back up on her feet, knife in her hand. Her

jeans were so coated in mud, they looked like a statue's legs. The cougar was gone, soundlessly, no movement of grasses even to mark its passing. And now Lucy realized that the dogs were yelping again, an excited chorus of barks, much closer, and she heard the crash and thud of many paws trampling the earth.

There was an ominous rumble overhead. Immediately, as if the sky had ripped open, the rain began, a torrent drenching her to the skin and plastering her hood to her skull. The ground was instantly hammered into sogginess. Lucy looked to her right. She saw hillocks of flattened grass too low to conceal a ground squirrel, and the tossing sea beyond. To her left was a series of muddy pools fast expanding and the shifting sludge that would slow her down, sucking at her boots, and beyond it the rain-shattered lake. She could make out the silhouette of the statue. The rainwater had already pushed the level up above the top of the toadstool, much higher already, she thought, than at this time last year. Directly in front of her, past a patch of soggy scrubland and up a slight rise, was a thick stand of trees, shadowed and dark. Behind her, she saw the first dog loping in her direction. Its muzzle grazed the ground, plumed tail up, fur raised in a spiky ridge over its back. Through the sheets of rain it looked like an illustration from a children's fairy tale cut out of black construction paper. Wolflike.

Without hesitation she sprang forward toward the grove, dodging around the hummocks of slick, sharp grasses, running, like a panicked rabbit, in a crooked line, until she was pushing through dense and prickly bushes, ignoring the barbs that caught and tore her skin and snagged her clothes. She secreted herself behind the nearest tree — a pine, wind-battered and salt-poisoned, with rough, shaggy bark, and no branches low enough or strong enough to hoist herself up on. The rain drove into her eyes. She wiped a streaming hand across her face. Her water bottles tugged at her neck. She lifted the rope over her head and hurriedly stowed the bottles under a nearby shrub. Her grip on her knife was slippery, and she rubbed her hand uselessly on her wet pants to dry the moisture from it. She tightened her grasp and leaned her forehead against the tree, trying to catch her breath. She had a cramp in her side and she kneaded it with one bunched fist. Pressing her body against the coarse bark, she squinted her eyes against the drizzle to make out the shapes of the dogs.

The throng broke apart, dozens of dogs fanning out and then coming back together as they caught a trace of her along the lakeshore. The moonlight made shadows everywhere. They had definitely found her trail. The rain might slow them down a little, the puddles she had sloshed through would mask her scent, but they were serious about tracking her and unlikely to give up. She could hear the heavy panting

and excited bursts of barking as they called to one another, like the high-pitched yelps of puppies scuffling over a bone. They were so close.

Lucy forced herself to leave the comforting solidity of the tree and move backward, as quietly as she could, sliding her feet through the mush of wet leaves. She took shallow breaths, darting quick glances over her shoulder, making for the place where the trees grew thickest. Black shapes wove back and forth, just beyond the pines in front of her, as the dogs tried to pick up her scent on the wet ground. She crept toward a cluster of pine, elm, willow, and leggy maples. The tall trees stood trembling; water cascaded down from their branches. She backed against the smooth trunk of an elm, the biggest tree in the glade. Too high overhead, wide branches spread out against the dark, fractured sky. The moon was directly above her. She hunkered down, listening to the sounds of the dogs coming ever closer. She held her knife in both hands, the blade pointing straight out in front of her. She'd kill at least one or two before they savaged her. The cramp was back again, jabbing into her side with a ferocity that made her wince; her lungs felt starved of oxygen; her heartbeat echoed in her ears. Then the crack of a branch snapping, loud as a gunshot, made her look up.

CHAPTER TWO

THE DOGS

What are you waiting for? Come on, grab it!" hissed a low voice. A hand dangled a few feet above Lucy's head. She blinked against the rain streaming into her eyes. Long fingers waggled. The rest of the person was shrouded in shadow. She stumbled backward, brandishing the knife. Behind her, at the perimeter of the small wood, the barks coalesced into a uniform baying and delighted howls, and she

heard the sound of many bodies plowing through the under-growth. They had found her.

The person made an annoyed explosive sound halfway between a curse and a grunt. "Well?"

She could tell it was a male voice. The hand gestured impa-tiently. "I can't hold on much longer. Are you coming or what? Do you want to be dog food?" She took one last look over her shoulder and jumped for the hand. Her fingers grabbed and slipped. The branch bucked under their weight. For a brief moment before she dropped to the ground, she saw his eyes— light-colored, not the bloody red eyes of the S'ans.

He grunted again. "And put that pig-sticker away before you cut off my nose. You're going to have to help yourself get up here, you know. I'm barely hanging on." Lucy hesitated; she didn't think the branch would hold them both. He leaned far-ther forward. His arms were bare and his skin was tanned, unblemished save for the silvery puckered scar of the vaccina-tion on his biceps. She thrust the knife into her sweatshirt pocket, unsheathed. Dangerous, but she wanted it close at hand. Heavy bodies thudded through the underbrush. She turned and saw two dogs angling in, mouths open, black lips peeled back from their long, spittle-flecked fangs. They cov-ered the ground with terrifying speed. She could see the lean muscles bunching as they prepared to leap.

She turned back to the tree and the hand that was still held out toward her and jumped for it. His grasp caught,

slipped again, and then his fingers tightened around her wrist. She scrabbled at the trunk with the blunt toes of her heavy boots, reaching out for a branch or something to hold on to. The gash across her palm stung. She felt the wound split open again under the bandage. For one horrible moment she hung suspended just a few feet from the ground, and she imagined a dog's sharp teeth grinding into her ankle. There was a volley of awful noise: crashing, panting, and a chorus of snarls.

He heaved, she kicked desperately, her left foot hit some-thing with a solid *thwack* and a yelp, and all of a sudden her momentum carried her up to the high branch he straddled, so quickly that she almost went over it and back down to the ground, but he kept hold, jerking her back. Lucy threw out her free arm and clasped it around the branch, then swung her leg across. Looked down. Ten or twelve dogs were clustered around the bottom of the tree. One blew a froth of bloody bubbles from its shattered nose. More dogs were coming, the pack, rearing up on their hind legs, jostling for position, black claws digging into the bark.

The earth tilted, pulled at her, and she squeezed her eyes shut, feeling her tenuous hold loosen and her equilibrium leave her with a stomach-twisting suddenness. Lucy fell against the stranger, a boy. Her head spun, and for a moment she thought she might vomit. Her stomach cramped. She bit down on her tongue until the nausea passed.

"I lost my balance," she muttered, even though he hadn't said anything. Her voice was gruff, and she was conscious of his hand still clasping hers, the warm solidity of his chest. Lucy scuttled back against the trunk, pulling her hand away and clutching the bark tightly. She had never liked heights, not even the swings or slides in the playground growing up. There was no room on the branch to move very far away. She carefully avoided looking down again. She could hear the dogs milling around, snapping and growling at one another. She tried to pretend she was not fifteen feet up. The boy stared at her with an amused expression that she longed to slap from his face. She cleared her throat. Her hand went to the knife inside her pocket.

"What?" she blurted out. She was painfully aware of the grime on her face and hands, the dry sweat that stiffened her hair, the stale, dirty stink coming off her mud-soaked jeans. He smelled clean. Soap—a memory so sharp, it hurt. His clothes were worn and patched, but not as filthy as hers. They were brightly colored, too— not the best choices for blending into the dull, beige landscape and shadows—as if he didn't care who could spot him from a mile away, and he'd cut the sleeves from his red sweatshirt as though he didn't feel the damp chill. She shot a look at him from under her eyelashes. He was about her age, with green eyes, dirty blond hair, and a generous mouth that was smirking at her. The grin slipped a little as her tone of voice registered.

"I'm waiting for you to thank me," he said. Her cheeks flamed.

It had been months since Lucy had been around another human being. She went out of her way to avoid them, to hide. She felt acutely uncomfortable; sort of like the feeling she used to get on the first day of school after the freedom of the summer holidays. She looked away from the intensity of his gaze. The dogs were sniffing around the tree. A couple of them had plopped down and begun licking themselves. They looked almost friendly, except that whenever she or the boy shifted just a little bit on the branch, they leapt to their feet and started growling and snarling again. Farther out in the brush she could see more shadowy sentinels waiting for her to make a break for it. How long would they wait for a meal? How soon before they'd give up and try for a mouse in a burrow or a rat on the garbage heaps? How long before she could climb down and go home? The boy was still looking at her, almost as if he could tell what she was thinking. The smirk was back. She dropped her eyes, busied herself with tightening the bandanna around her left hand. It was wet with new blood. She'd left a bright smear on the bark on her way up.

Clasping his hands to his chest and adopting a high-pitched voice, he said, "Oh, thank you, Aidan, for saving me from that pack of vicious dogs! That was *so* great of you to hang out of the tree like that and risk your life or possibly a serious

accident for a complete stranger!" She scowled, wondering if she could jump off the far side of the tree, avoid the dogs, and get the heck out of there. She looked down at the new hole torn in the knee of her jeans.

"Thanks," she said, after a long moment. "I'm Lucy." Her voice sounded raspy, and she was aware of how dry her throat was. "Do you have any water?"

He shook his head. "Didn't plan on being here that long. Just came out to relax. See what I could see . . ." He stared at her and she wondered if her hair was bushing out.

"Are you scouting?" Lucy asked. She knew, of course, that there were others out there, loners like herself, but most people kept to their safe places and didn't wander. She saw campfires sometimes, heard voices from a long way off, but Aidan was the first person she'd seen in a while. As far as she was concerned, the streets belonged to the S'ans—survivors of the plague who were horribly scarred and sick in the brain.

He shook his head. The sarcastic curl was back in the corner of his mouth, and she decided it was just something he couldn't help, but it didn't exactly make her warm to him.

She looked at him properly. As far as she could tell, he carried no collecting bags, no blade, not even a big stick.

"What do you mean?" she said. "You're not scouting?" Lucy straightened up; her fingers felt for her knife again. "Are you a spy?" she blurted out. "Are you spying on me?" Her greatest fear was that someone would force her back to the shelter.

Aidan's eyes flicked to her face and then away again. He stared at his hands. She waited for him to say something. He cleared his throat. "Not spying," he said. "But I've seen you before."

She remembered the disquieting feeling that she was being watched and waved the knife in front of his face. "You've been following me."

He looked up. "No!" he said, as if horrified. She set her teeth.

"Your camp is visible from here if you know where to look. That's all. I noticed . . ." Now it was *his* cheeks that reddened. He stopped in mid-sentence, then shrugged his shoulders up and down and said in a louder voice, which set the dogs below whining and snapping, "It's lucky for you that I was here; otherwise you'd be dog meat. You ran to this tree. I didn't make you come here."

That was true enough. She eyed him, fingering her blade. "Sort of creepy, though," she muttered. "So what were you doing here, then?" she asked, lifting her chin and staring hard at him. "Are you just . . . hanging out?" The words felt odd on her lips.

"Yeah," he said easily. "I guess you could call it that. I just like to climb trees, and the view from here is pretty much three hundred and sixty degrees." He gestured wildly with one outstretched arm. Just watching the sweep of his hand made Lucy feel dizzy again, and she clutched at her tree

branch, trying to do it inconspicuously. She was appalled. There hadn't been a moment in the last twelve months, except for when she was sleeping, when she wasn't doing something. If she wasn't gathering food, she was plugging gaps, collecting water, or baiting hooks. And in the evenings she'd plait coarse grasses into rough lengths for ropes or mats, cure skins, smoke meat, pound acorns, or mend tears and patch holes in her clothing and shoes. She definitely didn't have time to *hang out*.

Lucy stared at the boy thinking he was insane, but the *really* crazy thing was that he was staring back at her with the exact same expression mirrored in his eyes.

"Hello?" she blurted out now, slapping the branch so hard, it stung her palm. "Why risk everything for no reason except that you wanted to look at the view!" She pointed to the dogs. "This isn't a park anymore."

Aidan froze for a moment and then leaned back against the tree limb, his arms crossed behind his head. She had no idea how he was balancing himself, but he looked as comfortable as if he were lying on a couch.

"I think you *think* you know more than you do, wild girl," he said.

"What does that mean?" Lucy said, bristling.

"How long have you been out here?"

"Long enough to know how dangerous it is. The S'ans! The Sweepers! The scavengers!"

"I'm careful," he said after a pause. "And the scavengers aren't all bad."

"You're nuts." *And stupid*, she added silently. "The scavengers will rob you blind, the Sweepers will lock you up, and the S'ans will give you the pox. Or, if you're lucky, just plain kill you," she added, digging her knife into the tree trunk.

He was looking amused again, and her hand itched to slap him. A little snort of laughter escaped from his mouth.

She carefully swiveled her torso so that she was facing away from him. Less than a mile away, past the trees and the scrubland, was her camp. It might as well have been on the other side of the world. Aidan whistled a tuneless song under his breath and she did her best to ignore him. The dampness soaked into her skin, chilling her bones. She stunk of swampy mud. Her fingers cramped on the hilt of her knife, but she kept it out and ready.

The rain finally fizzled to a stop. Mist rolled in from the sea and wreathed the ground below. There was the tinkling splash all around them of drops falling from leaves onto the earth. Lucy's hand crept up to pat her head. Moisture made her hair frizz out. She probably looked a mess.

She scowled, shifting on the branch. Her butt was falling asleep and she longed to move, but there was no escape. The dogs panted and grumbled and prowled below. One, a terrier, Lucy thought, the sort of dog she'd once have thought was cute, just sat and whined pathetically at the bottom of the

tree as if it was starving. A tussle broke out between two of them, a black pit bull with fur so short and slick it looked spray-painted on and a burly rottweiler. Smaller dogs darted in, nipping at flanks, and the chorus of barks was deafening. It was a short, vicious fight that ended with lacerated ears and bleeding muzzles. Tufts of fur floated in the air. The two dogs collapsed, chests heaving, licking their wounds. The audience of dogs lay down as well, as if exhausted by the excitement. Some of them seemed to fall asleep. Lucy carved out a chunk of bark and tossed it down onto one of the sprawled bodies.

The animal was up in an instant, growling ferociously and clawing at the tree. More dogs rushed in from every direction, baying in excitement. Their eyes reflected the moonlight, and thick strands of saliva sprayed from their jaws. Lucy wondered if they were rabid.

"Smooth," said Aidan. He'd been so quiet, she'd half-suspected he had fallen asleep.

She glared at him.

"Come on, seeing as we're stuck here for a while," he said, leaping to his feet. He was standing on the branch, perfectly balanced, one hand stretched out toward her. Below, the dogs were going crazy again, catapulting themselves up into the air, scrabbling at the tree trunk.

"Uh, no . . ." The thought of moving made her head swim.

"I want to show you something. Up there," Aidan said, shifting easily on the branch, his arms relaxed by his sides. He was wearing brightly painted high-top sneakers. His feet seemed to grip the bark. Lucy's heavy boots felt like weights at the ends of her legs. Her wounded hand twinged when she clenched it experimentally.

"Scared?" he asked.

She imagined pushing him down or kicking his legs out from under him.

"I'm not," she said, clenching her jaw. She got carefully to her feet, holding on to a branch above her head with one hand and tightening her grip on the knife with the other. Lucy ignored Aidan's outstretched arm, and eventually he shoved his hand in his jeans pocket as if to mock her, and started climbing. He moved with an ease that made her face flush red with annoyance. She rubbed her thumb over the bone handle. Her stomach twisted and she felt a rush of bile in her mouth. She bit her lip hard and forced herself to look ahead to where he was standing with his head crowned by new bright green leaves. *Not down, don't look down,* Lucy told herself fiercely. He was a jerk, and there was no way she'd let him see how terrified she was. She remembered how she'd followed her little brother, Rob, across a fallen tree in the park once, although her knees had turned to water, just because he'd taunted her. When she'd caught up to him, she

had wrestled him down to the ground and stuffed handfuls of rotting leaves down his shirt.

Aidan walked casually to the end of the branch and then pulled himself up to the one above it. It was about chest-high on Lucy. She watched to see how he swung his leg over and then stood up. There were plenty of small branches overhead to hang on to, and she was pretty pleased with her performance. Just the slightest wobble on the way up, a misstep, forced her to drop to her knees and cling to the branch before continuing. But she'd sprung up again quickly before he'd noticed, not realizing until she was moving again that her fear of heights was being suppressed by feelings of irritation and a burning desire to prove to him that she was tougher than he would ever be. The tree was solid and broadly branched, and the bark smooth enough not to snag her feet but rough enough to give her some purchase. Aidan climbed and she clambered after him until they were near the top. The branches thinned out. Lucy gripped a handlelike pair of limbs and felt a little more secure. The air was much colder up here, and she pulled her sweatshirt hood forward, annoyed, too, that he didn't seem to feel the cold at all.

"So what's so special—" she began, and then she caught her breath. They were above the fog bank. Below it to the west lay the scrubby wasteland, the mudflats, the salt marsh, and just beyond, the vast waters of the Hudson Sea. To the south, under the low-slung moon, on a narrow wedge of rock

and soil were the toppled skyscrapers—row upon row of fallen dominoes—and the ridges of pulverized concrete and steel girders like jagged, broken teeth. Such a strange skyline now, full of odd angles and deep chasms with no symmetry; it no longer seemed like something built by humans. The new wooden structures that bristled from every area of high ground looked like they would blow over in a stiff breeze. And to the east, Lake Harlem took the shape of a bulging Christmas stocking, the misshapen toe cupping the southern-most part of the promontory they perched above.

Lucy's rib cage felt suddenly too small for her lungs. The devastation was overpowering seen at this distance. An entire city leveled. Some structures brought down by the gale force winds and the earthquakes, others by friendly bombing. And buried deep within the mortar and brick and sheets of steel were millions of people who had sickened and died in a matter of hours, many dropping where they stood in the first and second waves of the plague.

"They're like giant gravestones."

"Easy to forget living out here, I bet," Aidan said.

"Yeah," she said slowly. "I had begun to forget." She shifted her weight without thinking and grabbed a branch to steady herself, ignoring the pain as her left hand flexed. In her camp, on her spit of land bordered by mud and water, it had felt as if she lived in a wilderness, when in fact the remnants of her old life were only a few miles away.

He pointed east over the mudflats. "Check it out."

Lucy could just make out the blurred shape of a landmass in the middle of the lake, a narrow island not more than a few miles long. Just visible against the blue-black sky was a darker silhouette. A tower, strangely shaped—an octagon or hexagon. At the very top blinked a red light.

"What's that?"

"That's Roosevelt Island."

The name stirred a memory in her, but she couldn't place it.

"That's where your Sweepers come from. The Compound."

There was another large, low, rectangular building attached to it. Just the outline was discernible, and it was unlit, but she could tell that it was solid, massive.

"The hospital," Aidan said.

And suddenly Lucy remembered.

She gripped her knife, feeling a chill creep up her spine. She remembered the dozens of newscasts, the mass hysteria that each one brought. The island was where the smallpox hospital was. In the early days of the plague, notices and warnings had originated from there, but as the epidemic had spread, the status reports had ceased. Anyone with common sense could just look around and see that most of the people they knew and saw every day were sick, no matter what the television might be telling them about vaccine supplies and control. The live footage of calm, white-coated doctors and pretty,

smiling nurses had ceased, replaced by pretaped public service announcements, and the hospital had become just another derelict building. The little she knew about the S'ans and the dangers of the world she lived in now had come from those early news reports—a mixture of public service announcement and disinformation. "Stay in your homes. Avoid crowded places. Inform your doctor of any symptoms." And flashing across the screen 24/7, the plague hotline number to report your infected friends and neighbors. The hazard squads, they were told, patrolled constantly, seeking out pockets of infection, affected birds and animals, and those too sick to get themselves to the hospital. The white vans touring the neighborhoods and the white-suited men became a frequent sight, but they always gave Lucy the creeps. Once the disease took hold, most people had stopped believing that they were getting anything approaching the truth and ignored the reports and the government orders to remain in quarantine. People had left the cities in droves and the sickness had left with them, spreading like a wildfire.

There was something unsettling about the building, Lucy thought, taking a deep breath. In a landscape without any other artificial lighting, the red beacon at the top of the tower seemed like the baleful eye of some giant beast. There must be people inside, but whether they were doctors or government people or squatters, she couldn't tell.

Aidan grabbed her arm. "But look," he said, turning her to

face north. His face was lit up with excitement. He stood on the highest branch capable of holding his weight. The wind rustled the leaves. Far beyond the Hudson Sea—out where, Lucy knew from childhood Sunday drives with her family, there had once been farmland and apple stands, cows and pumpkin patches and the sweet cider donuts Maggie had loved and eaten by the dozen, and where the land was now given over to wilderness—a flickering light had appeared, followed by another farther away, and then another, strung out like shimmering golden beads on a necklace. A crooked line of fires. Aidan's fingers dug into her arm. She would have pushed him away, but she was afraid of falling.

"Oww," she said.

He barely took any notice of her.

"There are people out there who don't have to hide. Who are just . . . living. That's where I'm heading someday."

"So why don't you go, then? What are you waiting for?" She tried unsuccessfully to keep the sneer out of her voice. He was probably one of those people who didn't move without someone telling him it was okay.

He looked at her, taking in her expression. He let go of her arm. "Oh, so you're happy, right? You're having fun hiding from everyone, playing survivor out here with the dogs, eating frogs and acorns, being cold and wet or hot and itchy? Bathing once every few months?" His voice was scathing. His nose wrinkled, and once again she became aware of the smell

wafting from her grimy clothes and her hair, which must resemble a bird's nest.

Her face reddened.

"I'm not leaving here."

"The Sweepers will find you sooner or later."

"I thought they were just looking for diseased people. Or the S'ans?"

Aidan shook his head. "No. Now they're looking for whoever they can find. And who's to say the plague won't come back again? Another wave that'll take out more survivors? Maybe it's breeding in the sewers. In the rats. Or in the birds again."

She pressed her lips together. "Sounds like a good reason to stay away from people."

"Nothing will get rebuilt without people," he said.

"*People*"—she put a snotty emphasis on the word—"are the reason we're in this mess in the first place. Too many people, and most of them are a waste of oxygen."

She glared at the tree limb, dug her fingers into the cracks in the bark. He snorted. She could just tell without looking that his mouth was twisted in a sneer again. She felt heat flood the back of her neck. *What a jerk!*

"The waters are rising," he said. "Every season, they creep a little higher. You can tell by Alice."

She looked at him uncomprehendingly.

"The Alice statue. Isn't that how you keep track of the lake levels?"

"How did you know about that?" She tightened her hold on her knife. "You have been watching me!"

He rolled his eyes. "You don't own the park, you know. It's not yours."

Eyeing him suspiciously, she thought about what he had said.

Alice! That was the name of the girl sitting on the mushroom. Now that she heard it again she had no idea how she could have forgotten. She'd seen the animated movie, and her mother had read the book to her, stopping when it gave Lucy nightmares about having her head chopped off. There were so many things she had forgotten or blocked out, as if she couldn't help squashing down all the memories with the ones that hurt still. She stared at the tower. And then at the water, knowing that what Aidan said was true. The rains would come again with staggering ferocity like something out of the Bible; the oceans would swell, devouring the new brittle edges of coastline; rivers would spill over; and the lakes would grow until they swallowed the land. They were due for something big and devastating, she could feel it. Locusts, maybe.

"Where do you live?" Lucy asked, shrugging the crawling feeling from between her shoulder blades.

He turned to look at her, one hand rumpling his shaggy hair until it stuck out in all directions. The smirk was gone. He pointed into the darkness. "See?"

She shook her head.

He grabbed her shoulder, turned her a few degrees to the east. Past the hollow where her camp was, a tall silhouette loomed. She recognized the Egyptian-style marble column which stood there, as out of place as a camel, and to the northeast of it a plateau and a series of gorges where a massive earthquake had caused the concrete slabs of a big road to slide and sink and bunch upward like a swathe of gray ribbon.

Aidan pointed with his finger, and she followed the invisible path with her eyes. "See the plateau? If you keep going across the escarpment about three or four miles as the crow flies, you'll come to the canals. It's pretty hard going." She could just make out the slender silhouettes of rope bridges slung like webs above the cement-veined crevasses, and clusters of stilt houses sticking up along the slopes like bunches of strange flowers.

"There," he said, stabbing the air with his finger, "the Hell Gate." He sounded proud and embarrassed at the same time. "The camp was actually part of Wards Island before the floods."

"What's with the name?" she asked, thinking it sounded overly dramatic. "I thought the Hell Gate was a bridge or something?"

"We adopted it because it seemed appropriate."

"Sounds homey," she said sarcastically.

A dog howled suddenly from outside the thicket. Under their tree the pack lurched to its feet, barking raucously. The

howl came again, a long, sustained cry like a signal of some kind, and the pack, jostling one another and snapping at the air, scrambled about in excitement, tearing up the mossy ground with their thick claws. Lucy tracked them as they milled and broke apart, never moving more than a few yards away from the tree. Something had gotten them riled up again. She sensed his eyes on her.

"You can't just hide in your hollow like a mouse."

She stared at him. "I'm not hiding," she snapped. "I'm surviving. And I've been doing just fine on my own."

His gaze flicked away. She felt him tense beside her.

"Those are not feral dogs," he said. "They're hunting dogs."

"So what are they hunting?"

"Well, I'm pretty sure it's not me. They didn't appear until you did. They're trackers. They're looking for something."

She felt her jaw drop open. "What do you mean?" Her voice was a croak. "What are they looking for?"

"I don't know exactly." He frowned. "But something makes them go crazy. I've watched them before," he said. "They're sent out from the Compound. I've seen them around, out on the Great Hill, on the Cliff, in the Hell Gate, down in the Village. They go out, they find people who are hiding, and then the Sweepers come."

She blinked. Her brain felt fuzzy. Her knife was in her hand again. It felt clumsy in her hand, as if she couldn't will her fingers to hold it properly.

"So they're just keeping us here until . . ."

"Until the Sweepers arrive."

"How did the dogs know?" she asked.

He shrugged. "I guess they smelled you."

She shot a swift glance at him, but he wasn't smirking. His eyebrows were drawn across his forehead and one hand raked through his hair.

She judged the jump to the ground. Maybe she could push off from a branch before dropping, get some distance from the dogs before running. She thought there were maybe a couple of dozen of them. And more beyond her sight, out there in the gloom with the dog that had howled the announcement of their location. She squinted into the gathering darkness, straining to see a sign that the Sweepers were coming. Could she kill a dog? If she had to. But would that stop the others? Or would the blood drive them into a killing frenzy?

"Give me that bandanna," he said, pointing to her wounded hand.

"Why?"

"Come on!" He made an impatient gesture when she remained frozen. She held out her arm and he untied the knot from the bandage. Dark, fresh blood clotted the blue and white paisley design, and there were older, rusty stains where it had dried. He shoved it in the back pocket of his jeans and pulled two large, smooth rocks and a slingshot from the pouch

of his sweatshirt. "Stay here until they're gone, then run as fast as you can," he said.

"What are you going to do?"

He grinned, his teeth very white.

The dogs were huddled beneath the tree in a solid mass of resting bodies. Aidan fitted a rock into the rubber pocket of his sling and then drew it back between two fingers. The stone whistled through the air, hitting the trunk of a tree at the edge of the grove with a sharp *thunk*. Furry heads came up, and the dogs bolted toward the sound. He quickly aimed and shot the second rock at a tree farther on, and then, with one easy motion, swung out from the branch catching the limb below him. Quickly he made his way down before Lucy could even gasp out a word. He jumped the last ten feet, landing softly on his feet. The bandanna was out and in his hand, and he ran in the opposite direction from the pack, ducking every few feet to trail the material along the ground. Once he'd cleared the woods, he stopped and turned. He stood for a moment at the top of a small grassy hill, and then a wild, almost joyful cry burst from his lips. It rang through the trees and was answered by the dogs. Outlined against the sky, he raised the bandanna like a flag and waved it. He whooped and hollered. Lucy watched him disappear toward the lake. A chorus of barks rang out, and then the pack rushed back in a boiling frenzy. The small black-and-white terrier she had noticed before fought to get to the front. Its nose was down to

the ground, a steady whine rising from its throat, its stump of a tail wagging furiously. It sent up an excited yapping, which was echoed almost immediately by another chorus of mad barks, and the small dog sped away. The other dogs hurtled after it, shoulder to shoulder in a melee of bristling fur, passing underneath her tree and onward in the direction the boy had taken.

CHAPTER THREE
THE TIME BEFORE

Lucy half climbed, half fell out of the tree. Her knees were shaking and her muscles felt stiff and cold. Her hand was sore and so caked with congealed blood that she could barely close her fingers. She clambered backward until she reached the crotch of the tree, grabbed hold of a branch, and slipped, wrenching her shoulder. Her feet skidded against the wet bark. Her heart in her mouth, Lucy inched her way down,

staring fixedly at her boots and letting the ground fuzz out at the edges of her vision. Any hope she had of natural coordination abandoned her, and the memory of Aidan's confidence frayed her nerves even more. She dropped the last twelve feet gracelessly, slithering down against the trunk and scraping the side of her left arm and the length of her ribs against the bark. She came down hard on one foot, jarring her ankle.

By the time she had hobbled all the way to the clearing that surrounded her camp, she was panting and hunched over. Her ankle had swollen like a golf ball; her hair was glued to her face by a combination of sweat and moist air. Every shadow cast by the moon, every whisper in the grasses, sent a surge of panic through her body. So much adrenaline was coursing through her that she felt physically sick with it.

But there had been no sign of the Sweepers, no sign of the dogs beyond the occasional echoing howl carried over the mudflats from the lake. Lucy stopped in the middle of a barren patch of ground where she could see in all directions and listened hard, forcing her breathing to slow so she could pay attention. Would she hear if Aidan was being torn to pieces? Could she tell if the dogs had caught up to him? A lone insect stilled its buzz-saw melody as she slowly turned in a circle. There was no other sound.

The night wrapped around her, and with a soft sigh the rain began to fall again, passing from sprinkle to gushing torrent in a few seconds. She moved forward at a jog, then an

all-out run, frantic suddenly to reach the shelter of her camp. But then she paused, fiercely reminding herself to be cautious and check the silhouetted hummocks. She added them up and felt her heartbeat calm. Twenty-three hunched shapes, and that one fallen tree stump that looked like the curved back of a breaching whale—everything as usual. Lucy started forward again. It was impossible to move quietly. The ground was covered in puddles, some treacherously deep. She splashed forward, hands out in case she fell on the slippery grass, muddy streams on either side of her feet. Around the circumference of her camp, barely twenty yards from her concealed front door, dozens of paw prints were gouged into the wet soil.

Casting one final look around, she bolted the last ten feet to her door, ignoring the pain in her ankle. She dragged the willow screen aside and ducked inside. Her shelter was smoke-filled and thick with the briny smell of stewing turtle. She pulled her sweatshirt sleeve over her hand, picked up the pot by the scalding handle, and moved it from the smoldering fire, setting it on the ground. She raked the meager embers with a stick and added the last of the wood to the pile, then sank down, holding her hands out to the pitiful warmth. She was soaking wet. Her waterlogged sweatshirt dragged at her body; her filthy jeans were pasted to her legs. Her toes squelched in the two inches of water inside her boots, and she could smell the stink of her sodden socks even without taking her boots off. Her hands shook, and then the tremors

traveled swiftly up her arms and down her spine. She knew the smart thing would be to strip, dry herself off as best she could, and then put on a change of clothing, but she was too exhausted to do more than stare at the flames and shiver. Her fingers were slathered with mud and blood and covered in scratches. Her whole left side, where she had fallen from the tree, felt bruised and raw. She peeled her sweatshirt and tank top up. Furrows of skin were scraped from her ribs and along her forearm. Her shoulder was bruised; purple bloomed above the bone. Lucy looked at her upper arm: smooth, unblemished except for four freckles set in a line as if someone had pierced her with the tines of a fork. No scars. She pulled her shirts back down, gasping as the cold, wet material touched her body. She wrapped her arms around her chest and rocked back and forth, slitting her eyes against the tendrils of smoke that wreathed the floor.

The rain pounded the branches above her. Occasionally a drop would force its way through the densely woven wood and fall on her head. She had a plastic tarp she could sleep under, but it was splotched with mildew and it crackled and rustled and slid away from her sleeping bag. She always had nightmares when she used it, waking in a confused tangle and feeling as if she were being suffocated.

Seeing another person, talking to Aidan, had thrown Lucy off balance. She was perfectly fine living on her own, relying on no one else, but she'd almost forgotten that there were

other people out there. It was easier if she could pretend that she was the only survivor. Then her mind was completely occupied with foraging and hunting and all the small problems she had to solve, and she'd crawl into her sleeping bag at the end of another long day with no troubling thoughts. But now she was remembering how things used to be, and it was almost like a part of her, the human part, which was social and — she hated to admit it — craved conversation and interaction, had awoken again.

Lucy hobbled over to her backpack, unbuckled it, and pulled it open. She pushed her hands down to the very bottom, letting her fingers dance over her flint and tinderbox, her journal, a dead flashlight, her transistor radio, one last precious book of matches, until she felt the smooth leather cover. She hardly knew why she had kept it when so much of her life before was strewn in piles on the floor of her New Jersey home. The last weeks there were a blur in which only endless phone conversations with her parents' doctors and the countless forms to be signed stood out in her memory. A jumble of decisions were made while she could scarcely remember her own name, until at last the bodies were packed into the ambulance and taken away, leaving a silence that felt heavy and buzzed in her ears. She'd scanned her mother's phone book, called women she remembered as being kind, but the phone rang and rang and no one ever picked up.

And after that the house was almost unbearable, and the neighborhood she'd grown up in felt empty and forlorn, like a ghost town. She had become increasingly nervous, jumping at sounds, scared of the lights that came on in the adjacent houses in the middle of the night, the strange, silent men in hazard suiting who seemed to be looking for something, the white vans they drove. Lucy had taken to sleeping on the cold linoleum floor in the mudroom, which had no windows but did have a door that double-bolted and let out into the yard with its thick screen of cedar hedges. She'd listened to reports on the solar-powered radio her dad had kept on a shelf by the cellar door with stubs of emergency candles and freeze-dried camping meals. The college stations she was familiar with were not transmitting, and one by one the big news stations stopped, until finally there'd been nothing but a pirate channel, fuzzy and frustrating to pick up strongly. But in the early days, she'd lie on an inflatable mattress with the radio pressed to her ear, happy to be hearing another human voice. The host, who called himself Typhoid Harry, had been the first and only person to explain the plague in words she could understand. From him she'd learned that most people had contracted the plague in the first wave of contagion. Out of every one million people, 999,999 had died. Most of the survivors were picked off by the second wave. However, there were a scant few who seemed protected by the routine childhood vaccine

given out with those for chicken pox, measles, polio, and bird flu, and an even fewer number who somehow survived the disease, horribly scarred and insane—the S'ans.

On the day she'd left for good, she'd run from room to room, breathless, crying jagged sobs that hurt her chest, careful not to look at too much, but becoming transfixed by the sights of her mother's faded dressing gown still hanging on its hook on the bedroom door, her shawl draped on her favorite armchair, her father's coffee mug on the draining board in the kitchen. She'd spent most of the time in her dad's home office searching for she didn't know what, catching the lingering scent of his aftershave, and finding the hunting knife and sheath in the bottom drawer of the desk.

Lucy had taken the knife not so much for defense. At that point everything was odd, surreal, but she had no notion of any physical danger to herself. She'd slipped it into her bag with her mother's shawl, a box of assorted freeze-dried food, and a bottle of spring water, because it was so unlike her father to own a weapon. He was all about leather attachés and legal briefs and dark, perfectly pressed suits. It was a puzzle to be gnawed on.

And she had taken her tenth-grade yearbook, too, even though she'd hated school, never infiltrating the groups of popular kids. The yearbook was a superficial slice of high school life that completely ignored the pain and boredom of it. She couldn't help thinking that Aidan would have fit in

perfectly at her school, although she had to admit there was an edge to him that was different from the preppy, stuck-up boys she used to have classes with.

She opened the yearbook. The blank pages in front and back were empty of those insipid *Have a great summer!* messages. Inside she'd scrawled over pictures of the hair-sprayed, shiny lip-glossed, made-up girls in her class with a big, thick, black pen, giving them punk hairdos and raccoon eyes and thought bubbles that said stuff like "Do you think I'm pretty?" Somehow their deaths had changed it all. The yearbook touched on the life before. It had become something to remind her that things had been normal once.

Lucy flipped the pages with difficulty. They'd swollen from the damp and stuck together, and the red cover was warped. Past the graduating seniors' portraits, where everyone was posed like they were selling wristwatches; carefully avoiding the formal photo of Maggie, who was smiling so widely, happy and secure in the knowledge that she had her pick of Ivy League schools; past Rob and the rest of the ninth graders who looked like little kids and always would be. She got to her class picture. Ran her eyes over the list of names: Julie, Scott, Chad, Angie—people who'd barely noticed she was alive even though they'd known one another since kindergarten. In the class roster she'd been marked absent, but she'd been there. It was like a bad joke that even her teachers seemed unaware of her existence. She stood at the end of the

row toward the back, shoulders hunched and hair pulled forward across her pale face, which appeared to float like the moon above the unrelenting black of her combat boots, jeans, T-shirt, and zippered hoodie.

Chad was standing next to her, but he'd squeezed over so that there were at least a couple of feet between them. God, she had hated him! He'd always acted as if she were diseased or something.

Lucy chewed her thumbnail, remembering how strange life had been that spring. The flyers with the lists of symptoms had appeared, plastered all over school, and it seemed as if everyone visited the nurse's office complaining of headaches and muscle cramping and fever. A few girls had fainted in class. Lucy had felt perfectly fine. She turned the pages of the yearbook slowly, flicking past photos of football teams and teachers and school staff. She paused at the picture of the nurse, Mrs. Reynolds, looking so neat and trim and motherly in her white outfit.

But she hadn't been so calm the last time Lucy had seen her, when she was called into the health office for yet another blood test. Mrs. Reynolds had seemed distracted. Even her smooth blond hair, normally pinned in a neat bun, was messily tucked behind her ears, and she'd had dark circles under her eyes. There'd been none of the usual chatter, the casual questions about Lucy's health or how the school year was going. She'd been nervous, preoccupied. And she'd flubbed

the test somehow. Instead of blood squirting into the needle, it had dribbled all over Lucy's arm and the black-and-white tiled linoleum floor, and quite a lot of it had spattered onto the woman's white brogues. And although Lucy knew from sex ed class the previous year that the nurse could field the most embarrassing questions lobbed at her by Chad and his idiot posse, she had mumbled when Lucy asked her how many kids were sick and if it was contagious.

"What is it?" Lucy had said. "Strep? Or is it mono?" For some reason there was a coolness factor associated with mono. It meant you'd been kissing someone. Julie Reininger's rep had been cemented by having mono and being out of school for a whole month last winter.

"Maybe that bird flu they were talking about on the news?" Lucy had continued, and she'd been almost mesmerized by the weird spasm that quivered across Mrs. Reynolds's fingers and the way her eyes skittered away. And then she'd bitten her lip, as the nurse jabbed the needle into her arm again. Shortly afterward, Mrs. Reynolds had left the room, clasping the full tube of blood and closing the door firmly behind her. Lucy had heard the sound of the lock clicking shut. She had waited, until her sweaty thighs had stuck to the paper covering the gurney and she realized that she had to go to the bathroom. Finally, after looking at the closed door and the frosted glass window, she got up and walked around the small room, sliding drawers open and checking out the

plastic-wrapped syringes, the tongue depressors flavored with cinnamon, the model of the female reproductive system all shiny purple and pink plastic — she'd wondered if the colors were anatomically accurate — and blowing balloons with a couple of powdery surgical gloves. She tried not to think about how full her bladder was. One of the bottom drawers held a thick stack of folders. Lucy was about to close it when she noticed Chad Grey's name and casually flipped the cover open. Chad had been absent for the last few days, and Lucy couldn't say she missed him. He always had some lame comment to make when she walked past his locker in the hallway, and he liked coming up with stupid words to rhyme with her name. Being called "Goosey" or "Moosey" might not have been exactly insulting, but it was almost impossible to walk to your desk with any kind of poise when a crew of boys was hissing it under their breaths. Maybe he had an STD or something. . . .

A wallet-sized student photo was clipped to the top of the page. A black bar was slashed across his eyes in marker, and the letter *D* was carefully marked next to his name. Lucy would have liked to believe that it stood for *dumb* but even then she was afraid it meant something much more terminal. She read: "Student complains of abdominal pain, fever, headache, backache, nausea. No lesions. Subconjunctival bleeding, subcutaneous bleeding. Hemorrhagic variant suspected. Sent to Dr. Lessing/R. Island for confirmation."

And there was a folder for Hilly Taylor and one for Samantha Barnes and that massive jerk AJ Picard, and, come to think of it, she hadn't seen any of them around for a while. She had trained herself to ignore them for so long, but now it seemed crazy that she could go from class to class and sit there doodling in the margins of her notebooks without noticing the empty desks. And each photo had been altered in the same way, black bars slashed across their eyes, the letter *D* written in thick, black lines, and the same listing of symptoms.

Suddenly she had to pee so badly, she squeezed her legs together like a toddler. She kept rifling through the papers and there, almost at the bottom of the pile, was one marked "Lucy Holloway." It was thicker than the rest. Seeing it made the latest needle hole in her arm twinge. They'd turned her into a pincushion these last few weeks. And there'd been no explanation. Just more tests following the first physical exam, when Mrs. Reynolds had run her fingers over the smooth skin of Lucy's upper arms, looking for the puckered scar of a vaccination that wasn't there.

"My parents didn't believe in them," Lucy had whispered when the nurse had finally thought to ask her and she'd begun to feel afraid that there was something really wrong with her. It was one of the only ways her straight parents deviated from the norm. She remembered when Maggie had told her, in a hushed voice, about their older brother, who'd died

when he was barely two years old from an allergic reaction to a shot. "That's when Mom and Dad moved out of New York to Sparta, here in New Jersey," Maggie had said breathlessly, her eyes round with delicious horror. "Because it's less crowded and people are healthier, so it doesn't matter." Then she'd added, "Alex's face swelled up like a pumpkin, and his hands looked like shiny pink balloons, and then his tongue turned black." And even though there was no way Maggie could have known all that, the image had given Lucy nightmares for years.

Now she opened the folder slowly, half-afraid she'd see the black bar and the *D*, but then thinking that if she did then that would indicate it meant something other than *deceased*. And that would be good, right? Her startled face looked back at her from the photo. It was a copy of the picture on the student ID she was supposed to wear clipped to her backpack but never did. Her thick, curly bangs obscured her left eye completely, and her mouth was pressed into a thin line that almost made her lips disappear, but there was no black bar, no letter *D*.

She'd picked up her folder and stumbled backward to the gurney, spilling pages covered in weird symbols, rows of numbers and decimal points, percentages and charts. Too much information for someone who had suffered many scraped knees and cuts and broken bones but had never contracted anything worse than a head cold. One thing you could say about Lucy Holloway was that she had near perfect attendance.

She wasn't able to make heads or tails of the science stuff, and her bladder had finally demanded that she do something about it. She peed in a blue plastic cup and then dropped it in the biohazard receptacle, trying not to tip it over. The can was filled with used syringes and marked with a yellow and black skull.

And then Mrs. Reynolds had come back; the clicking of the lock opening gave Lucy just enough time to stuff her folder back into the drawer and vault onto the examining table. The nurse had jotted down a few notes and then made the calls that brought a legion of white-coats in, their faces blank behind their masks, and the battery of testing had begun again, until finally her father had shown up. He'd seemed twelve feet tall standing in the doorway, swinging his briefcase like an axe, his face purple with rage. She had never seen him with his cuffs undone, his tie unknotted, his carefully combed hair bristling. Something stony in his face had stopped her from asking any questions when he'd dragged her out of the room. Afterward, there'd been no time. Lucy never went back to school. Missed final exams, never picked up her report card, and soon there had been no reason to think about school. She'd received her yearbook in the mail a month or so later, sent directly from the printer with a computer-generated mailing label affixed.

A few pages further and there Lucy was again, a shot of her hunched over her journal, her hair a tangled curtain drawn

across her face, scribbling away furiously. All around her were people caught in the midst of laughing and talking, their hands a blur of motion, moving around her as if she weren't there. And she kind of hadn't been. In her mind she'd been traveling and thinking about the day she could escape, and she had written it all down in her journal. Even to herself, though, she had to admit she looked like a strange, dull girl. Lucy closed the book, thinking not for the first time that she should burn it or throw it away, and ended up stowing it safely in a fold of the orange tarp against the wall. She sat back down next to the fire and clasped her arms around her shins, resting her chin on one bony kneecap. Her body thrummed with exhaustion.

If Aidan had gone to her school, she wondered if he'd have talked to her or if he'd have gravitated toward the in crowd. She could picture him: confident, easy, and relaxed. She could see Julie and Hilly hanging on his muscular arms, imagine him in a letterman jacket or a numbered basketball jersey. What was his deal, anyway? He reminded her of the boys at school. He had that swagger, that confidence, which she could only suppose came from having things always go your way. And from looking like he did, like something out of a preppy sportswear catalog. But he was different, too, in a way she couldn't put her finger on.

Two things bothered Lucy: One was that Aidan seemed to know quite a lot about her and where she lived. She wondered

if he was spying on her after all. She'd felt as if she was being watched for a long time now. And if so, that begged the question *why*? And the other, far more important, was, what if he was right about the dogs? She could think of no reason why someone would be tracking her, but it made her feel unsafe for the first time since she'd found the hollow and the willow grove. She'd built her camp. It was as warm and dry as she could manage. It was comforting in a weird way, maybe because it was completely hers and proof that she could survive on her own, away from other people.

At first the thought of being alone had been terrifying. It was what had driven her to leave her family house and venture back into the city, searching for someone to tell her what to do next. She'd wandered blindly, attempting to mesh her memories of various streets with the rubble-strewn desolation around her, and eventually followed a cluster of scared-looking teenagers who seemed to be heading somewhere.

The shelters that had been organized after the first and second waves of the plague had passed were depressing and crowded with survivors looking for answers or authority figures that weren't there. There were maybe five of them set up in gutted churches and schools and sports arenas around what remained of the city, and they all looked the same: long rows of camp beds, flickering tube lights, and huddled bodies under thin blankets. People bundled under the covers like they were children afraid of the dark. She couldn't help but

be reminded of the last time she'd seen her parents in the hospital, lying on gurneys side by side in the hallway with sheets pulled up over their faces.

It had been impossible to sleep. Every three hours there was the grinding roar of old generator-powered fans starting up and pushing around the warm air and the thick smell of unwashed clothing, unwashed skin, and instant noodles. And always, from one bed or another, a constant keening, like a wounded animal had crawled inside to die, and strangled sobs sometimes exploding into rage. The woman in the bed next to her, her gray face sagging with exhaustion, had never stopped crying and moaning, "The Sweepers took my boy away, the Sweepers took my boy away," until it started to sound like the lyrics to a sad song. Lucy had slept in her clothes with her boots on and her backpack clasped to her chest, too scared even to visit the bathrooms at night, when grimy, wild-eyed people gathered for secret reasons of their own.

One day, as she was coming back from a solitary walk around a swampy, mosquito-infested neighborhood that used to have the best used record stores, she'd seen a squad of people in white hazard suits come out of the shelter with the sad woman and a few others—mostly children—bundle them into a white van with darkened windows, and speed away. The men's faces were covered with blue surgical masks and their hands were gloved. The lady had left her purse behind, pushed halfway under her pillow, as if she was planning on

coming back in a few minutes; as if she hadn't meant to leave. By the next morning it was gone, and the pillow, too. Pillows were in short supply. After that, Lucy had gotten out of there as soon as she could. She'd been better away from people and among the trees, where she felt like she could breathe.

A sudden flurry of raindrops forced their way through the roof and dripped onto her head and neck. She blinked. She'd set bundles of sage burning in the corners and the purple smoke was thick on the ground, the spicy fug strong enough to mask the briny scent of cooked turtle. Her clothes were still damp, but beginning to stiffen. She had been sitting for an hour at least, staring at nothing. She peeled off her wet things, scrubbed her skin with a scratchy towel, and put on dry clothes. The thick woolly socks on her feet felt like heaven, even though her big toes poked through. She wrapped herself in her mother's shawl, and then slipped on her leather jacket, pulling the collar up around her ears. She pushed her waterlogged boots close to the fire. Then she peered into the depths of the cooking pot. It looked like a thick soup, greenish-brown, and it smelled salty and wild. Chunks and strings of indefinable matter floated on the top. Lucy's stomach turned an unhappy somersault, but from nausea or hunger, she wasn't sure. It had been at least sixteen hours since she'd choked down a heap of lukewarm acorn mush, and she dipped a bowl in now, being careful not to stir up the murk too much. It was stronger tasting than she expected. As salty as boiled seaweed,

and although she was careful to sip with a pursed mouth, straining it through her teeth, there was plenty of sand and little bits of turtle shell floating around with the gluey wild onions and the chewy dried mushrooms. It was slightly less repulsive than the salamander stew she'd made before finding out it was better to skin them first, and she reminded herself that the survival book had praised turtle meat as being high-protein and low-fat. However, she would not be recommending it to anyone.

She forced the food down, and then sat determinedly not thinking about what she had just eaten for a few minutes until she could be sure that it was going to stay down. Instead she found her thoughts returning to Aidan. Lucy decided that she was pretty sure she disliked him intensely, his attitude, his annoying self-assured way. The fire wheezed and snapped and sent out tiny wavering flames that occasionally puffed gouts of smoke as if they concealed a small dragon. The flicker of rain falling beyond the walls reminded her of snow on a television screen. Lucy fell asleep, sitting up, her jacket pulled tight around her, the smell of worn leather comforting.

In her dream there were dogs swimming in the lake, their pelts dark and streaming water like seals, and they were herding the small boat she was in, pushing it toward shore. There was something hidden in the pitch-black that terrified her. Was Aidan somewhere? She could hear him, but the sound of

his voice echoed all around her, and she couldn't tell where it was coming from, and it was too dark to see him. Suddenly she was certain that the dogs were pushing her away from the safety of land, into the open waters.

She came awake in a rush, not sure what had roused her. Her eyes felt as if they were filled with grit. The camp was flooded with a soft gray light. It was too quiet, and after a moment Lucy realized that the storm had blown itself out and that it was the encompassing stillness that had wakened her. She could hear the trickle of water sheeting down the walls of her shelter, but other than that there was a deep silence, muffled, as if she still had her head under her arm, or she were still asleep. It was eerie. She got up, forced on her boots without tying them, and moved the screen aside. She was definitely awake. Her boots were clammy, the leather stiff. It was not quite dawn. Droplets of moisture glistened on the grass stems. The rain must have just stopped. The trees about her shook as though a giant had flicked their tops carelessly as he walked past, and Lucy realized that the roof of her shelter was swaying as if blown by a strong breeze. But there was not a breath of wind now. It was still and so, so hushed, it seemed the entire world was frozen between moments.

Lucy stumbled out toward the shore. Wet reeds slapped against her hands. Her jeans were already soaked to the knee.

The air was warm. All the myriad sounds of animals waking up were missing. No frogs. No birdsong. No rustle of mouse or vole in the long grass. The sun was rising now, just cresting the purple edge of the horizon behind her. She felt the heat on the back of her neck and shrugged out of her leather jacket and the shawl, carrying both under one arm. She checked to make sure her knife was sheathed at her waist. Everything seemed crystal clear, the curious quality of light so sharp it hurt her eyes. Her booted feet squelched and slid in the sand, the loose laces clumpy with mud. A sound like the *flip-flop* of a car's windshield wipers in a rainstorm reached her ears, but magnified a hundredfold. Ahead of her, the surface of the sea appeared to be seething, like molten silver at a boiling point. She stopped and slit her eyes, shading them against the brilliance of the light. She'd seen the ocean just before a sudden storm, with a blazing sun overhead, when the waves seemed picked out in metal wires and the sky was almost black, but this was different. This was like a spilling of gleaming coins.

She realized that she was looking at fish flopping on the beach, thousands of silver bodies leaping like dancers. The tide was out so far, she could see nothing but the fish and the brown sugary sand, the water drained away as if someone had pulled a giant bath plug. Far beyond, reflecting flickers and flashes of sun, she could see the ocean. It was retreating, the waters drawing back like a tide in reverse.

She turned and began to run. Panic spilled into her mouth like bile. The waterlogged sand tugged at her feet, slowing her down and threatening to trip her. She pushed on, forcing her knees higher. No time to bend and tie her laces. Only two thoughts yammered in Lucy's brain and she grabbed hold of them: *Get my stuff. Get to the highest ground I can find.*

CHAPTER FOUR

SEA

After the first disasters, they'd had emergency drills at school: what to do in case of earthquake, cyclone, and flash flood. They'd watched countless hours of video footage, of Maui engulfed by lava and the devastating eruptions of Mount St. Helens and Mount Vesuvius, massive explosions that buried all of Portland, Oregon, in ash and molten rock, and tilted the city of Naples into a boiling sea. Even the

youngest kids knew to find a doorway or a desk, a cellar or the highest ground.

So before the thought had hit her brain, Lucy had turned and started running. She had twenty minutes if she was lucky, ten if she was not, and considering how things usually played out in her life, she'd better not count on having enough time.

She had to abandon her home. The thought of it was a physical pain in her chest. Lucy was past the sands now, resisting the urge to turn around and look behind her, fearing the sight of that wave building as it rolled back in. She'd seen films of tsunamis towering a thousand feet, waters so high and fierce you expected to see Godzilla charging through them with tiny destroyers and navy boats bobbing around his leathery ankles. And she'd seen the footage of what was left behind: miles of wreckage, houses splintered, buildings mowed down and crushed, and the drowned bodies of humans and animals flung on the shore like driftwood.

Time seemed to slow down and then speed up again. Lucy felt like she was watching herself in a movie. Short, flickering scenes, as if the film were old and missing frames, the whole thing spliced together badly. She found herself in the salt marsh with no idea how long it had taken her to get there. It seemed mere moments. The ground was firm under her feet; she ran faster, and then the bristly grasses gave way to low shrubs and spindly bushes, and she skirted some and leapt over others, letting the panic take the lead. Ahead of her was

71

the clump of supple trees that marked her camp. And the ground was wetter, slippery as oil, where it had flooded from the rains. She dodged hummocks of greasy grass, her breath coming in heaving gasps. Sweat trickled down her back. Just before the entrance she slid in a foot of water, but was up on her feet again before she felt the wet soak through her jeans. Lucy pulled the screen aside, hurled it from her, ducked down, and was in, casting her eyes around.

What should she take? No time to think. She unbuckled her backpack, pulled at the laces until it gaped open, stuffed the shawl inside, and jammed her arms into the sleeves of her leather jacket. Her brain was taking snapshots of each corner of her camp. Sleeping bag; the survival manual from the table; her clothes from yesterday, a damp, dirty pile on the ground. She shoved everything in, pushing it down as much as she could, feeling to make sure her journal was there, and then the bag was buckled and slung over her shoulder. She paused to kick dirt over the smoldering fire, then berated herself for wasting time. Tons of water were about to crash down on her, but it was a habit learned during the Long Dry when a wayward spark could destroy everything. One last look around. She didn't have much. The pots and pans were an unnecessary weight. What food stores she had left were not worth taking. She grabbed a half-full water bottle, not sure if she'd find a stream or a spring safe to drink from. She hung her spoon and fork around her neck. The hammer of her heart

seemed to be counting off the seconds. Was there anything else? She turned to leave, then suddenly remembered and ran to the place where her sleeping bag had been spread on a flattened pile of dried grasses, shoved her hand against the wall, and pulled out her yearbook. She clasped it to her chest, took one last look around, and ducked outside.

She bent and tied her laces, fumbling for a moment and finally settling for two tight knots which would be impossible to undo later. She stowed her yearbook in the bag and shrugged it back over her shoulders. The beach was still empty, the fish a thin layer of throbbing silver at this distance, with the deep blues of water and sky above. Choosing which way to go was a nonissue. West was the sea. East and south ended in water as well. North would take her up a slope and eventually to the Great Hill, and from there she could make a decision. A small voice in her head piped up and reminded her that the Hell Gate, Aidan's camp, also lay in that direction, but she pushed it down. From the Great Hill she could journey on for a few days and cross the Geo Wash Bridge if she wanted or loop back around. Maybe come home in a day or two and try to salvage something, rebuild. She told herself she could completely avoid the Hell Gate if she wanted to.

Lucy hurried along the narrow track—a muddy animal trail worn into the grass by sharp deer hooves when they came down from the heights to drink from the lake. Beyond the

scrublands the ground rose sharply. She went straight up, taking it at a run, her backpack bouncing with every step, reaching forward with her hands, low to the ground, ready to catch herself if she fell. The terrain became loose, crumbling earth and pebbles, spiked with rocky outcrops and straggling trees. Stones rolled under her feet, threatening to bring her down. She pulled herself up, grabbing at slender branches and roots to keep her balance. A few hundred yards up, she paused for breath. Her sprained ankle was a hot ball of pain. Her throat was raw. Her ribs hurt. Her fingers were scratched and bleeding. The wound on her palm had opened again. She'd left a trail of blood on the stones. The thought crossed her mind that the dogs would have no trouble tracking her this time. Lucy felt a jolt of fear and suppressed it. Drowning in a monstrous wave would fix that problem. Just ahead was a thicket of wind-twisted fir and pine clinging tenaciously to the slope, and beyond, she knew, was a bare cap of gray rock at the summit of the hill. And surely that would be high enough. She ran on, limping now, her leg muscles trembling with exhaustion. There were pine needles underfoot; it smelled mossy, pleasant. Dappled light filtered down. She paused, her breath hitching in her throat, and drank the water in her bottle in a few, panicked gulps. She felt safe under the canopy of trees, but her fear pushed her onward. She had just reached the far edge of the wood when she heard

a roar like a subway train hurtling through a tunnel. It seemed frighteningly close.

Lucy broke through the line of trees, clawed her way up to a rocky ledge, and looked down from the height. She had a panoramic view of the drained beach, so peaceful at this distance. The thin slice of land where she'd lived for more than a year fell away beneath her only a mile or two from where she stood. She could see the green dome of her camp, the line of grass hummock sentinels, the black trunks of salt-burned trees by the shore, the wide swathe of sand. And then the wave came. Suddenly there was water everywhere, rushing in as fast as a jet plane. The waves jostled to fill every available space. The bowl where her home nestled was an upended snow globe shaken with a ferocity that robbed the breath from her lungs. Trees were uprooted and flung into the air; bushes and slabs of earth were ripped loose, rolled and tossed into the seething mass of water. The stone needle was completely submerged. The wave grew higher as it came, a cataclysmic wall of water dwarfing everything before it, taller than her father's office building. It smashed against the hill like a massive fist, and she felt the tremor vibrate through her body. It broke less than a quarter mile from where she stood. A quarter mile was only 1,320 feet, she remembered from some math class long ago, and yet it seemed closer. If she hadn't forced herself to take more than 1,320 steps, it would

have caught up to her. She looked into the wave, a dizzying swirl of stormy blue and emerald green, darkening to purple at the depths and exploding with foam at the crest. It was near enough that Lucy felt the soaring spray hit her face and her nose filled with the smell of salt. Her eye was caught by a splash of bright orange within the brown swampy swirl of pulverized tree and bush and earth, and she recognized the tarp from her camp. When the wave rolled back out with a sucking sound that she felt as a pressure around her throat, it left nothing behind but a thick sludge. The ground steamed in the morning sun. It was quiet and nothing moved.

Lucy realized that she had bitten her lip. Blood trickled down her chin. She rubbed it away, staring at the bright red smear on her fingers before wiping them on her jeans. She looked down at the devastation, trying to will her brain to comprehend it. The splintered trees, the slick layer of mud and pools of water. Nothing remained of her shelter. Even the tarp had been dragged back to sea. There were shapes left sprawled in the mud. Rabbits, groundhogs, other small animals, drowned in their burrows. Bile flooded her mouth and she vomited. Turtle soup. And that brought on more heaving until her stomach was empty.

After some minutes she got up, moved away from the steaming pile of puke, and sat down with her back to the wreckage. She peeled her sock back from her ankle. It was soft and puffy to the touch, but she could rotate her foot

and flex her toes. She stripped the sock off and tied it around her ankle and then put her boot back on. The sole and heel of her foot were covered in calluses about a centimeter thick — she could walk without a sock for a while. Next, the wound on her palm, split open again and weeping a little blood. She wrapped it with the only bandanna she now owned, pulling the ends tight and securing them with a knot. Lucy's fingers were shredded from the rocks and the tips throbbed, but at least it was a distraction from the pain in her ankle. She leaned back against her backpack, listening to the thud of her heart. The slope ahead was a gentler rise topped with cracked and weathered gray stone. Tiny pink-flowered plants anchored themselves in the crannies. In the sky, so brilliant a blue that it seemed unreal, a hawk climbed in ever-tightening circles. *It must be wonderful to be so free,* she thought, *to be able to travel away from everything.*

It was the yucky taste in her mouth more than anything else that propelled her to her feet eventually. She walked up to the crest of the hill, favoring her ankle and working the stiffness out of her legs, and scanned the area in front of her, wondering if she could find a spring or a small stream where she could refill her bottle, maybe soak her ankle. The hill dropped off into a gorge, but it was not so deep that she couldn't scramble down into it and up the other side. It was what lay beyond it that made her pause and begin chewing on her thumbnail: a long expanse of buckled highway, driven up

into a series of concrete ridges by the powerful earthquake that had collapsed the Empire State Building three years ago and pulverized most of Midtown. Strewn with rubble, the road dropped twelve feet in places and climbed twenty feet in others. The concrete was crumbling and pierced with weeds. Dandelions bobbed their yellow heads from every crack. She'd always liked dandelions. They seemed like free spirits, growing wherever they wanted, and springing back no matter how often her mother dug them up. Lucy started walking toward the first crevasse.

CHAPTER FIVE
THE HELL GATE

Lucy wiped her mouth. After three hours of steady hiking, climbing, and risking severe bodily injury crawling in and out of crevasses, she'd found a puddle of rainwater that tasted of tarmac but wasn't too gritty. The water made her stomach cramp and she realized how hungry she was. The sun climbed in the sky. It looked huge and more orange than yellow. She guessed the time must be close to noon. She wanted to be off

the ridge before night fell. She felt exposed and vulnerable with no foliage above her, and although the sky was cloudless, Lucy knew that a vicious storm could move in with unnatural speed. The day had become humid, still, as if the tsunami had driven out most of the oxygen when it took the trees. Her bangs hung in limp ringlets over her eyes, and she could tell by touch that her hair had frizzed up. She wished for an elastic band or a piece of string to tie it back with, but she had nothing. She touched the hilt of her knife, rubbing her thumb over the smooth bone. She could hack off the mass of hair, cutting it close to the nape of her neck, but then she'd have the same problem in another month or two, and in the meantime she would look like a freak, or a boy. She wasn't sure which was worse, but she did know that she didn't want Aidan to see her looking like a head-injury victim.

Aidan was an uncomfortable thought. Lucy pushed it away. She wasn't going to see Aidan. She was going to stock up, rest, and figure out where she would live now. Aidan was where people were, and where food was, that was all. She cupped her hands, scooped up more lukewarm water, and dribbled it over her head and neck, then smoothed her hair down as best she could. The road was flat for a few hundred yards. Beyond that it dropped off again, but she couldn't tell how far. She walked, watching out for loose rubble. In places the mangled tarmac was marked with a broken white line, but it was no longer straight. It deviated from the middle and twisted

suddenly and disappeared. She estimated that she was around Second Avenue and 92nd Street, although acres of road and earth had been shifted in the big quake, the landscape completely reconfigured. Sometimes she thought it looked as if a toddler had built a city out of blocks and then knocked them all down in a rage.

Lucy had reached a gorge that was as big as a canyon. It went down about forty feet and then climbed back up nearly the same distance in a series of trenches like giant steps. There was no way around it—it crossed the entire width of the ridge. When she finally pulled herself up the last craggy slope, bruising her knees in the process, she found herself on top of a plateau. Straight ahead of her was a deep, wide ravine, and stretched across it, ridiculously fragile, a suspension bridge. It swung in a gentle rhythm, although there was no breeze. This must be the Grand Canal. For a minute or two Lucy looked across the chasm. She chewed on her lip. Sweat trickled down her back and her heart thumped painfully against her breast bone. It was so high. The bridge was anchored on her side by several loops of rough-looking braided rope attached to an outcrop of rock. Lucy tugged on it and then stepped onto the bridge, which dipped with her weight. Each step created vibrations that traveled the length of the bridge and then bounced back, throwing her off balance. She crept forward, holding on to the rope supports with both hands, her arms outstretched to their full length. She tried to keep her eyes on

the far side, but she couldn't control her gaze. It was drawn to the ground far below. The channel bed was almost completely dry. The two downpours they'd had at the beginning of the Long Wet were not enough to flood it yet. Sharp rocks and rubble were strewn on the bottom, along with mounds of garbage. She saw a baby stroller, a dented refrigerator with its door hanging loose, wads of rain-soaked paper, tattered clothes and blankets, the twisted wreck of an old metal bed—the kind they used to have in hospitals, with wheels and coiled springs.

The rung she shuffled onto snapped with a sharp crack, half of the wood breaking off jaggedly and spinning out into the air. Her already weakened ankle twisted. Her foot went through the hole; the weight of her body threw her forward onto her knees, and the bridge swung crazily from side to side, tilting so that she was no longer on a level surface. Now one edge was vertical. She was being tipped off. She grabbed at the ropes, burning red stripes across her hands, and halted the fall. For several minutes she didn't move. She lay there sideways with her head hanging over the edge, waiting for the bridge to stop swaying and right itself again. Lucy squeezed her eyes shut, trying to erase the image of the rocks sticking up like spearheads at the bottom of the canal. Slowly she shifted her weight toward the middle. The bridge leveled out. Once her heart had stopped pounding, she pulled her foot from the hole. Like a bear trap, splinters of wood had pierced

her jeans and the sock she'd tied as a bandage over the bone. Her ankle was ringed with scrapes like tooth marks. She moved from her knees to her feet and began to inch her way forward again. Her teeth chattered so hard, her skull hurt and her jaw ached. By the time she got halfway across there was a sheen of sweat across her face, which she dared not wipe off, and her legs were trembling. She forced herself to keep moving. When she stepped off the bridge onto firm ground, her legs gave way beneath her.

After a few long moments with her head down around her knees, Lucy got up again. Her hair was plastered to the back of her neck with sweat and her damp arms clung to the lining of her leather jacket. Her throat was parched and her stomach growled with hunger. In the forefront of her brain was the fervent hope that wherever Aidan was, it would be straight ahead and not across any more suspension bridges. She looked around at the dilapidated buildings, the mountains of pulverized concrete and twisted girders. This may have been a neighborhood before, but now it was just the shell of one. A path, barely discernible, snaked through the rubble, disappearing a dozen yards ahead between the remnants of two brownstones, their roofs missing, their foundations sagging so that they almost touched at the top. The Hell Gate. The question was, were you entering hell going in or coming out? As far as she was concerned, the jury was still out on that one.

The terrain was unpredictable, and in most places sharply inclined on crumbling slopes made up of equal parts soil and man-made materials. Cinder blocks, sandbags, and planks of wood shored up the various levels like a humongous ladder. She followed the track—so narrow a goat would have had a problem with it. She went slowly, testing the ground, which was loose and studded with rocks. She kept her eyes open for people. Scavengers. Bands of roaming thieves who scoured the streets for anything that could be reused or resold. Rumor was they stole the fillings out of the mouths of corpses.

Suddenly Lucy was conscious of a hum not far ahead, down the next hill. She unclasped her knife, making sure it slid freely in the sheath, and pulled her leather jacket tighter around her body. It was too hot for leather, but it gave her confidence. She hoped it made her look tough. She walked slowly toward the noise, unable to tell if it was machinery, music, or the buzz of human voices. A guide rope was fastened to stakes where the edge of the hill dropped precipitously, with white flags of cloth tied onto wires to make the way clear. Wooden pallets were laid over deep puddles. She stopped. A curve in the trail along the edge of a crag revealed a view of the settlement below: tents clustered like mushrooms, lean-tos made of rough pieces of plywood. She was barely fifty feet above the source of the jumble of noise. She ducked down, feeling nervous all of a sudden. Lying on her belly in the loose dirt, Lucy peered over the edge. A few pebbles

rattled down the slope. Just ahead, the path dropped down and opened onto a crowded square.

A wide road, which had somehow escaped devastation, rose high over the canal and ran southward; small walking alleys radiated in all directions, leading to more plywood shacks and, farther up, to the other suspension bridges she had seen from a distance. The central area had been part of the big street. You could tell because it was relatively flat and by the broken white line running down the middle, but the surface of the tarmac was cracked and uneven, giving it the appearance of large black paving stones. Reappropriated awnings and large lengths of canvas were slung on poles around the edges as protection from the sun's heat and the rains, but the middle circular area was clear.

More people than she'd seen since she left the emergency shelter massed in small groups. They seemed to be mostly children and teenagers, with a sprinkling of gray heads, which didn't surprise Lucy. It was the middle-aged population that had suffered from the plague the most. People like her parents.

She heard the hubbub of human voices. They sounded excited, happy. And unexpectedly she heard music. A guitar, she thought, and a few singers. People jostled and bantered; some pushed wheelbarrows piled high with broken appliances, and others lounged cross-legged on long benches. Smoke gusted from a massive fire pit. A large black pot steamed above it. Lucy scanned the crowd for Aidan, unable

to stop herself from feeling a jolt of excitement at the thought of seeing him again and ruthlessly reminding herself that she didn't like him.

And there he was, taller than she remembered. His shaggy blond head, his red sweatshirt. He leaned against a crumbling wall that was covered in faded posters and graffiti. As she watched he threw his head back, laughing at something his companion, a girl standing very close to him, said. The girl reached up and smoothed her hand across his face. Even from this distance she was striking. Her thick black hair so sleek it looked oiled and a jumble of silver bracelets on her tanned arms that caught the light.

Lucy wasn't sure what to do. She'd crossed miles of treacherous ground. She'd lost everything but what she carried on her back. And now she just felt like crawling away. She couldn't imagine walking downhill into that crowd of people. Knowing her, she'd probably trip and fall. The buzz of dialogue almost hurt her ears. She wasn't even sure if she remembered how to start a conversation. "Hi," she said experimentally, and her voice cracked.

On the other hand, they had water, and whatever was cooking above the fire smelled good. Dusk was approaching, and the thought of sleeping out here was daunting. She could cut through the settlement to link up with the Geo Wash Bridge farther north if she meant to keep on going. Or backtrack the

way she had come, across the rope bridge again, and then go miles around, and that was an unbearable thought.

She stood up, brushed the dirt from her clothes. Her hands crept up to her hair. The humidity had matted it into the corkscrew curls she despised. She spat on her fingers and dragged them through the unruly mass, but it was no good. She scowled. This was stupid. She didn't need anybody. There was no one down there whose opinion meant anything to her. She squared her shoulders, shrugged the backpack into position, checked her knife, and took one last look around.

Suddenly she stopped in her tracks. She saw a billow of dust coming from the south along the road. Not a cyclone. This hugged the tarmac, and it moved fast and low. The cloud dispersed, and now she could see a line of vans speeding toward the square. Four white vehicles like delivery trucks, but unmarked. The same type of van she'd seen crawling through her neighborhood sixty days after the plague arrived, searching out the sick and dead, dragging people from their homes. "Sweepers" was what the TV anchors had called them. Cleaning up the mess. Her eyes darted to the thronging crowd. She remembered what Aidan had said, that the Sweepers were hunting survivors now. She was gripped by a fear so strong, it cramped her belly. She was still too far away; the vehicles were moving too fast. How could she alert them?

She waved her arms in the air. No one noticed, not even

when she jumped up and down. She could yell, but her voice would be drowned by the tumult of voices below. A sound. Something unexpected. Something that would carry across the square. She pursed her lips. She was pretty good at wolf whistling, a skill she'd mastered to annoy Maggie. But her throat was too dry, and her tongue felt thick in her mouth. She couldn't whistle, but she had an idea. She filled her lungs and howled, a long wail that cut through the air like a knife.

CHAPTER SIX
SWEEPERS

The mournful cry seemed to echo. Down below people snapped to attention, froze for a long moment, and then the jumble of noise started up again. There was some laughter and excited chatter, as if it were a prank. Heads turned this way and that looking for the source of the howling. Then someone screamed. She heard shouts: "The Sweepers are coming!" and a dozen arms pointed at the speeding vans barely

one hundred yards away now. So close that Lucy could hear the rev of the accelerators, smell the sharp odor of gasoline. They arrived in a column, the exhaust fumes and the dust boiling up along the road behind them.

And then, like an anthill kicked open, people were running everywhere, making for the alleyways, melting into the shadows. It seemed like everyone was yelling. Kids disappeared under tarps and into tents. It was chaotic, but in a way it seemed rehearsed. Aidan was lost in the tumult. She leaned forward, crouched against the ground, searching the crowd for his bright sweatshirt, and found him bent over an old woman who was frantically trying to tie the corners of her blanket together around a pile of fruits and vegetables. Shriveled apples rolled in the dirt. A child stumbled and fell, screaming when he scraped his palms on the rough surface. Aidan scooped him up. A boy and girl, eight or nine years old with identical rats' nest hair, scabby knees, and dingy undershirts, squatted under an awning with their arms around each other. Two older kids threaded their hands through a column of bicycle tires. They could barely walk with their load. The dark-haired girl yelled at them and they dropped the stuff and scuttled off. Lucy pressed her body into the earth, lifting her head to see. She had a clear view. The square had emptied out. About fifteen or twenty people remained, and most of them seemed to be Lucy's age. A few of them picked up rocks and sticks from the ground. Some pulled short knives and

slingshots from their pockets. They spread out in a thin line. Their faces were set and grim.

Brakes squealed. The clamor of the engines seemed incredibly loud. One of the vans sideswiped the edge of a caved-in building, dislodging pieces of brick. Another plowed through a heap of pots and pans, sending them flying into the air. The vehicles slowly pulled up in a wedge and came to a stop, although the engines continued to roar. They effectively blocked the road. The front and rear windows of the vans were tinted. Heavy steel bars were welded to the bumpers. Huge truck wheels lifted them up four feet from the ground. The engines cut off simultaneously. The back doors were flung open, and a dozen figures in white hazmat suits spilled out. They wore shiny headgear and heavy, laced boots and carried small black boxes. Their hands looked like they were made of marble, and Lucy realized they were wearing white surgical gloves. Someone else appeared around the side of a van, holding the ends of several thick leashes in his black leather-clad hands. Four vicious-looking dogs struggled to free themselves from their trainer. They were powerfully built, with barrel-chested black and tan bodies. Rottweilers and German shepherds, Lucy thought, watching their noses scent the air, ears pressed flat against their skulls.

Aidan stood with the dark-haired girl and an older man, who was muscular with a shaved head and the glimmer of gold in both ears. He looked five or six years younger than

Lucy's father had been. Mid-thirties, she guessed. His bulky arms were inked with swirling blue tattoos, his calves bulged, and his back was ramrod straight. There was something military about him, as if he'd been trained for conflict. Lucy glanced from the teenagers standing in their thin line to the Sweepers who had spread out in a solid row. The Sweepers stood shoulder to shoulder, their helmets reflecting the sunlight. They looked as impenetrable as a steel wall. The teenagers didn't stand a chance.

Lucy fought the urge to run. Adrenaline scurried up and down her spine. Things seemed to be happening in slow motion, but she knew it had barely been a minute or two since the vans arrived. The Sweepers moved forward. Aidan and the others faced them, flimsy weapons ready. The bald man raised his hand. Chunks of stone flew through the air. Lucy heard the clatter as they connected with the Sweepers' headgear. Aidan yelled something indecipherable, but the anger was clear, and a second volley of rocks flew. One of the Sweepers broke ranks and took a wild swing at him — the blow missed his nose but connected with the side of his head. Aidan staggered and threw a punch back. The Sweeper dodged it and darted back to his line. Aidan pressed his hand to his cheek. Lucy winced. She could see the scarlet welt.

The black-haired girl erupted into a barrage of curses. Another volley of stones flew through the air, followed by a hail of garbage: old tin cans and bottles. It had no effect from

what Lucy could see. Slowly Aidan and the others were forced backward against a wall. Lucy stood up. No one was paying any attention to anything that was happening outside the square. Anxiety coursed through her; she felt ill and riled up all at the same time. She had a clear view from where she stood. The Sweepers looked like white chess pieces in their close ranks. Two of them were sidling to the right, trying to flank the kids. Lucy pursed her lips and howled again. Aidan shouted something and another volley of garbage spattered against the dark visors. The Sweepers reformed into their solid line. Lucy couldn't understand why the kids didn't just rush them. They seemed to approach no closer than ten feet. She narrowed her eyes, trying to see more clearly. The slim black boxes that the Sweepers carried looked like old-fashioned transistor radios. She didn't notice any weapons.

Suddenly the bald man rushed forward with a roar. Eight of the Sweepers closed around him. The remaining four faced Aidan and the rest of the teenagers. They stood locked in some kind of staring contest. Lucy couldn't understand it. Only the bald man did anything, and he was a blur of motion. He jabbed a punch, ducking low and driving up with an uppercut that struck a Sweeper under the chin and sent his head snapping back on his neck. Another Sweeper rushed him from behind. He pivoted, kicking out with one leg. His booted foot connected with an arm. Lucy heard a *crack* like the snapping of a twig. His momentum carried him forward.

He spun again and slammed his boot into the chest of the nearest Sweeper who fell to his knees. The bald man skipped backward, keeping a safe distance. The teenagers cheered.

Aidan picked up a discarded length of metal pipe. "Leo," he shouted, darting into a clear space. Leo ducked a blow and raised his arm high. Aidan threw the steel bar. Leo caught it and swept it around in a circle. The men fell back. Leo swung the bar as if it were an ax, catching a Sweeper behind the knees and bringing him down. Moving so quickly that his hands were a blur, he brought up the pipe like a spear and jabbed a second man in the back and another in the belly. His breath was coming in gusts now, and his broad chest was heaving. His movements slowed, but they were no less deadly or accurate. The five or six Sweepers who were still on their feet circled him warily. He held the bar easily in one hand, his head swiveling to clock their movements, but there were too many of them.

Aidan and the teenagers moved forward with their rocks and pieces of wood. The Sweepers moved back.

And then, all of a sudden, the dogs were among them, loosed from their leashes, advancing slowly, the fur raised along their spines, tails curled against their flanks. Even from where she crouched, Lucy could hear the rumbling growls, and, terrified, she buried her face in her arms.

Could they smell her from here?

The Sweepers broke ranks, two remaining in place while the dogs circled, growling continuously, keeping the teenagers

at bay. A red-haired boy, not much older than Lucy's brother, darted forward, screaming and stabbing the air with a heavy stick. The dog trainer made a gesture with one hand, and immediately one of the German shepherds was on the kid, hurling him backward, its jaws clamped on his forearm. The boy screamed, an awful high-pitched noise. The trainer shouted something and the dog released the arm, but settled its weight across the kid's chest, pinning him on the ground. Its jaws were inches from the boy's pale face.

Leo dropped his weapon and stepped back. His eyes were on the boy and the dog. None of the teenagers moved. It was as if they were frozen.

The hazmat-suited men dispersed; only the dog trainer remained where he was. Making their way around the perimeter of the square, the rest of them conducted a search, shining powerful flashlights into the tents and plywood huts.

Lucy heard a patter of voices and an excited yell from inside one of the shelters.

"Hold your ground," Aidan shouted. At his side, the dark-haired girl tossed a stick from hand to hand. Her face was turned toward his, and Lucy could see the rage disfiguring her features as she argued with him.

"No, Del," he yelled, putting out a hand, which she smacked away.

Leo shook his head at her, barked a command and then held up one arm, keeping the teenagers back. A bottle flipped

end over end through the air and smashed in front of one of the dogs. A volley of growls erupted, but the dogs stayed in position as though they were tethered by invisible lines.

None of the teenagers moved, as two grimy children were pulled from their hiding places and thrown over broad backs like they were bags of potatoes. A woman with drooping shoulders was shoved forward with such force, she staggered. An old man with a shapeless cardigan and a fringe of mousy hair followed, as though he were sleepwalking. A cluster of seniors with their arms around one another were forced through the open double doors and pushed down onto the floor of a van. The men Leo had hurt were helped to their feet and bustled away. The dogs, summoned by a whistle or gesture that Lucy didn't notice, returned to the trainer, who placed his black leather-clad hands on top of their rough heads before ordering them into a vehicle. The German shepherd was the last dog to be summoned. The dog moved slowly off the prone redhead and backed away. Del rushed to the boy, kneeled on the ground, and helped him up. Blood dripped from his arm, and it hung limply at his side.

A fat drop of rain splattered on Lucy's cheek as she watched in horror. In the next second, the sky had cracked open with an explosive blast and the rain poured down in a solid sheet. Lucy wiped the water from her eyes in vain. She was almost blinded. Engines started up, wheels screeched, and the column

of vehicles drove off. And now people moved. They ran after them, yelling and throwing stones.

The rain fell in heavy sheets, reducing everything to slippery mush. The path was a treacherous mess of mud and rushing rivulets of water. Lucy didn't know what to do. Part of her wanted to disappear. She stood there, shifting from foot to foot, trying to decide. Aidan looked up, spotting her. Lucy cringed — too late to duck down. She should have remembered that she'd be silhouetted against the sky and that he'd be wondering who had made the signal. He frowned, but then his expression cleared. He raised his hand, and after a moment she waved, too. She felt the rocky soil slide under her boots as she made her way carefully down to the square. She pulled the collar of her leather jacket tight around her throat. Why hadn't she joined the fight? What would Aidan think? Chances were he already considered her some kind of coward, hiding in her camp, ignoring the reality of life outside her safe acres.

But when Lucy finally reached the bottom of the path and stood there, water streaming down her neck and filling her boots, at a loss for anything to say, he pulled her into a hug. She buried her nose in the shoulder of his sweatshirt, smelling a clean, fresh scent, like lemons mixed with his sweat. Her heart gave a little skip before she realized that all his concern was for those who had been taken. He probably just needed

someone to hold. His arms tightened around her, but she made herself step away, ignoring the confused expression in his eyes.

"Hi," she said. Her voice only cracked a little. Inside her pocket, she dug her fingernails into her palm. She felt herself flush. She had to stop this. It was ridiculous. He clearly had feelings for that Del girl.

"Lucy. I should have guessed it was you." A hint of the crooked smile appeared on his face, but it vanished quickly. He scraped his hand through his wet hair and steered her under an awning. Others slowly appeared from their hiding places, assembling in small groups under the shelters. Lucy glimpsed more people who hovered in the shadows at the perimeter. *Too scared to come out even now*, she thought. The murmur of subdued voices rose. Brightly striped umbrellas went up around the fire pit, shielding it from the rain. They seemed too cheerful in contrast to the stunned atmosphere, a splash of color in a scene as monochromatic as an old postcard.

They stood quietly as the rain poured down, washing the dust from the road. The frown was back on Aidan's face, but he didn't talk. He rubbed the knuckles of one hand. Lucy saw that the skin was broken and bleeding, and the flesh over the bones looked swollen. His cheek was red and bruised. The viciousness of the attack had been shocking. Now that it was over, she felt a weariness that threatened to submerge her. If

she'd been able to, she'd have crawled into her sleeping bag, drawn her jacket over her head, and slept for two weeks. The pain of losing the safety and comfort of her home squeezed her heart.

Aidan looked past her. His arm went to hers, cradling her elbow, and he turned her to face the bald man striding toward them. He was huge. Tall, and as broad as a brick wall.

"Lucy, this is Leo." Leo nodded but did not smile. His shirt was damp with sweat, and beads of moisture flecked his scalp and upper lip. He wiped a ham-like hand on his cargo pants and then held it out, pumping her own in a bone-crushing grip. His blue eyes studied her with an intensity that made her nervous.

"Leo just needs to check you out," Aidan said, giving her a little pat and then a push on the back.

"Wha-a-t?"

Simultaneously Leo's hand gripped her forearm and she was directed toward one of the larger army green tents. There was no question of breaking loose. His hold was bruising. She tried to grab her knife, but it was out of reach. She threw a panicked look at Aidan over her shoulder. He nodded at her reassuringly, but his eyebrows were bunched and his expression was worried.

CHAPTER SEVEN
EXAMINATION

Inside, the tent was dark and smelled of mildew. A bench heaped with clothing stood in the middle of the packed dirt floor along with a table, two chairs, and some milk crates stuffed with wads of material. A few sheets hung from rings on the ceiling, making a small enclosure. A large bucket of water rested nearby. A hurricane lamp smoked gently and gave off a pungent odor.

And by the wall — her heart started beating quickly — stood a wheeled hospital gurney, like the ones her parents had died on.

Leo had finally let go of her arm. She stood rubbing it, eyeing the doorway flap and his bulk in front of it. She wondered if she could squeeze through the tiny gap under the tent where it was pinned to the earth with stakes, tried to decide if he was as slow as his height and weight suggested. But then she remembered the grace with which he had fought and resigned herself miserably to being his prisoner.

"You can put your backpack down and take a seat," he said, motioning toward the chair. His voice was brisk, impersonal. It gave away nothing.

He busied himself at the small table. There were small glass bottles and a few odd-looking metal implements on a steel tray, which he pushed to one side.

Lucy remained standing, balanced on the balls of her feet so she could run if she had to. Her hand went to the knife. She wasn't sure what to think. She tried to read him. It didn't seem as if he was about to attack her, but her nerves were zinging anyway.

What was she in for? Torture? Execution? And the more nagging thought: Why had she ever trusted Aidan?

Leo now turned to the heap of clothes and rummaged through them with his broad back to her. It was such a target.

She half drew the knife and prepared to pounce. If she could get the knife to his neck, she could make him let her go.

His words distracted her.

"Henry's not here, but I've learned enough from him not to cause you too much discomfort. Let's see, might as well get this sorted. A medium should do, I think."

Henry? Medium what?

He swiveled suddenly, as coordinated as the cougar she'd seen at the lake, and tossed something at her. She almost dropped the knife in her attempt to catch the soft bundle.

His eyes widened as he caught sight of the blade. And then a grin spread across his ruddy face.

"It's not what you think."

She held up two pieces of clothing. Worn, black faded to gray—a pair of loose drawstring pants and a baggy thermal shirt.

"For you to change into. After."

He nodded at the tarps hanging behind her. "There's a makeshift shower in there. Water's cold, I'm afraid. No disrespect, but I'm thinking it's been a while." Her cheeks flamed. Then he pointed to a box spilling more clothes on the floor. "Underwear and so on in there." And now she could have sworn *he* blushed, but the light was pretty poor. "They're secondhand, but clean."

He took a step toward her, his fingers spread out in a nonthreatening pose.

She held her ground. "What do you want?"

"Can you put the knife down, Lucy?" He had stopped moving toward her, and his voice was gentle. She felt tears pricking at her eyelids. He sounded like her dad. The same burr in his voice.

She got a grip on her emotions and did not lower the knife.

He picked up something small and metallic from the table and flicked a switch. A small dot of light came on. She recognized the scope doctors used. Like the one they'd used on Rob when he was four and stuck orange pips up his nose.

He showed it to her, moving slowly, as if she were a little kid. The circle of light bobbed around.

"I need to ask you to trust me. Just for a little while."

She thought about her choices. Surprise! She didn't really *have* any. Seemed like that was the way it was recently.

She scowled and nodded.

"I'll trust you, too," he said, his eyes on her weapon. "I just need to look in your mouth and ears. Check your glands for swelling, your fingernails for blackening."

With the worst cases of the plague, bleeding started under the skin, a darkness spreading like crude oil on water, and a high fever boiled the blood. In the first few months she'd been obsessive about checking every bruise, every lump, but she was a klutz, and she always had some cuts or contusions sprinkled across her legs and arms.

He shone the light in her eyes and grunted. "Your eyes are clear."

"Are you afraid I'll infect you?" she asked sarcastically.

"Frankly, right now I'm more worried about your blade." He shifted around so he could peer into her ears. She hoped they were moderately clean.

He checked her fingernails, pressing along the edges. He turned her palm over. The knife cut on it still oozed, and the edges were raw.

"Nasty," he said. "There's a salve here somewhere." He placed her hand palm up on her knee and rummaged through the clutter on the table, emerging with a flat tin and a rectangular piece of material. He opened the box, revealing a paste which resembled brown Vaseline. It had a pungent smell like oregano.

"Goldenseal, echinacea, and comfrey," he said, as though that meant anything to her. "Grammalie makes it."

He smeared some over the wound, wrapped it tightly in a cloth bandage, and used some thin strips of cloth to bind it in place. The edges of the wound stung briefly and then stopped. She clenched her fist experimentally. The pain was numbed.

"Here." He handed her a surgical glove. "To keep your hand dry."

She was oddly reluctant to take it. The Sweepers wore gloves like that. She was reminded of a question she'd wanted to ask. "Do they always send dogs?"

He considered. "Lately." He squared his shoulders and rubbed his chin thoughtfully. "Lately it seems like they're looking for something in particular."

She couldn't control the shiver that snaked up her spine.

"How can you live like this, not knowing if they're going to come back?"

"We try to prepare as best we can. Look out for one another." He glanced at her with narrowed eyes. "You were living alone? Out in the Wilds?"

She nodded.

"Easier, I bet. But lonely, maybe?"

She shrugged, feeling the sudden prickle of tears. She rubbed vigorously at her nose.

"People just naturally cluster together, you know. Everyone's got a version of the same story." He cocked an eyebrow at her. "Probably for the first time ever, we have an understanding, a compassion for one another, you know? Everyone has lost someone."

She said nothing, though a part of her wanted to. Alone, she could squash down all the emotions. He was making it hard.

"Say *ahhh*."

She wondered how bad her breath smelled.

He put the scope down and reached behind him. When he turned around again, he held a thermometer. "Open again."

She opened her mouth and he placed it beneath her tongue. The thermometer was uncomfortable in her mouth. She

moved her tongue. He frowned slightly and repositioned it. "Hold still for two minutes." She exhaled through her nose.

"Family maybe means something else these days," he said. "It's not about blood ties anymore."

She grunted and shifted on the chair. She ducked her head so she didn't have to meet his eyes.

After a seemingly endless time he said "Open" again and removed the thermometer. He shook it a couple of times and squinted at it, trying to read the numbers.

"People are scared. They fear that the disease is just dormant, that it might mutate again, resurge. We have to face the possibilities," he continued, holding the thermometer toward the light. "Normal." He placed the thermometer back on the table and faced her. "Good."

Lucy ran her dry tongue across her lips. The thought that the plague could appear again was terrifying.

"I could have told you that. I'm not sick."

"It's hard to tell. By the time the bleeding and fever appear, it's usually too late. And contagion usually occurs before the symptoms show themselves. We're barely hanging on here. We can't let you into the camp if there's even the smallest chance that you could bring infection."

"Aidan's the first person I've seen in six months. None of this is necessary." She stared at him, her chin thrust out. He looked amused. "I don't know if I'm staying past tonight," she said.

"Even so. We'll have to dispose of the clothes you're wearing, too. We ran out of bleach a month ago, and none of the herbal concoctions do the job."

She remembered her mother burning their family blankets and pillows on the pyres.

"You can't take my leather jacket. Or my boots! I'll leave." She pulled her jacket around her. She'd had the boots so long that, ripped and shredded as they were, they felt like old friends.

He shot them a glance, then looked at her stony face. "You can keep them. It's the plant fibers that hold the disease. Let's finish up." He moved slowly, holding his hands out where she could see them.

Then he pressed his thumbs in under her jawline and behind her ears. His hands were quick and firm. She closed her eyes and tried not to think about how her father had smoothed her hair away from her face or tweaked her nose when she was little and didn't want to take her fish oil gel tabs. "Anything hurt?"

She shook her head impatiently. He exhaled and wiped his sweating forehead. She wondered if he was more nervous than he admitted.

"Henry will ask you some questions when he gets back. He's out on hunt detail right now."

"Hunting animals?" Lucy asked.

He shot her an amused glance. "What else would it be?"

She shrugged. "Who's Henry?"

"He's our resident medical expert." He sat back on the stool, spreading his large hands on his knees and leaning so that only one chair leg still touched the floor.

Lucy suffered another twinge. Her dad used to sit like that at his desk. Even though her mom always complained that it scuffed the floor.

"Any more clothing in your bag?" he asked.

She nodded, unlaced the opening, and pulled out the sodden mass of her clothes. Her nose wrinkled. They smelled of mold and ancient sweat and the iron tang of blood from her wounded hand. She dropped them on the ground. They were torn and disgusting and probably unwearable anyway. She continued to dig, dropping her dead flashlight, tinderbox, journal, yearbook, survival manual, and her musty, polyester sleeping bag in a heap. Her fingertips touched soft wool at the bottom of the bag and her heart leapt. Her mother's shawl! Surely he wouldn't take it from her? He had said plant fibers, like cotton. This was wool. Wool was okay, right?

She withdrew her hand and raised her eyes. "That's it," she said firmly, indicating the pile of things. His glance passed over them slowly, and then he nodded and she shoved everything but the clothing back in and tied the laces tightly.

"That's it?"

"Yeah. Yes," she said hugging the bag to her. Could he tell that she was lying?

A furrow appeared across his forehead. "How long were you out on your own?"

She exhaled.

"About a year." His eyebrows went up, but all he said was, "There are more clothes over there if you need anything. No towels, but you can use them to dry off with, too." He got up heavily and pointed toward the shower stall. "You've got about three gallons of water there. If you use it all before you rinse off, you'll have to hike a ways to get more." He handed her a slab of rough soap. It smelled overpoweringly of peppermint and lemons and felt greasy against her palm.

"So? Okay?" he said, preparing to go. "I'll be outside if you need anything."

Wait. Now that he was leaving, she felt the familiar lump of dread settle in her stomach. Funny how she felt safer when she was out in the open and could see her surroundings. Anyone could approach the tent and she wouldn't know until it was too late.

"You can leave your old clothes there on the ground. I'll be right outside." He met her eyes, nodded, then ducked out the tent flap. She heard his deep voice as he greeted someone. It was comforting to think of him so close by.

The water was not as cold as she had feared. She made a washcloth out of her tank top and paid particular attention to her armpits and the back of her neck. The soap was gritty and

so pungent, it made her want to sneeze. She gave up trying to work it into a lather after a couple of minutes, doused her head, and tried to work through the worst of the tangles. She washed her mouth out and ran her finger over her teeth to clean them. When she was done, her skin tingled and she could bear to smell herself.

It was a relief to kick her old clothes to the side. She'd been wearing the same two pairs of jeans for almost a year, the same T-shirts and tank tops and hoodie, washing them in the lake when she could. She'd tried to make her own detergent from soapwort and the fat layer from the belly of a dead squirrel, but it had been a disaster. The stink of cooking lard had driven her from her camp for a few hours, and she'd ruined one of her only saucepans. She sniffed her sweatshirt before tossing it onto the discard pile in disgust. It was funny how she hadn't really smelled her stink before. She'd gotten so used to it.

She dragged her fingers through her curls one last time, both wishing for and glad there wasn't a mirror.

The new clothes smelled strongly of bleach and were rough and slightly itchy against her newly scrubbed skin, but they fit okay. She rolled the pant legs up a little, laced her boots, and then dug through the pile looking for a sweatshirt. She needed something with a hood, preferably dark-colored, so she could vanish if she had to. *Aha!* She pulled out a sweatshirt. It was

faded with washing and too big, but she slipped it on, instantly comforted by the fleece lining. Over that went her leather jacket. Now she could rough it outside for a few nights if she had to. She also grabbed another change of clothes, underwear, socks, and a couple of tank tops and stuffed them into her bag.

She shouldered the backpack and ducked outside. The rain had stopped, and the ground steamed slightly in the blazing sun.

Lucy shaded her eyes. The hospital tent stood in its own little area apart from the other lean-tos and awnings she could see scattered on the outskirts of the big square. People clustered together, exchanging worried glances and talking in low voices. None of the young kids were unaccompanied. Each had an older guardian, grim-faced and wary. Some of the teenagers were gathering piles of rocks; some stood along the path Lucy had traveled down, acting as sentries.

Feeling shy and awkward, she spotted Aidan a dozen yards away. He was standing close to *that Del girl.* Funny how she'd just started calling her that in her mind. Petty and sort of mean, actually, but there was something in the way the other girl held herself, as if she knew that she was beautiful and expected attention for it, that was really annoying.

Aidan leaned into her. Their heads were almost touching. His hand was on her sleeve. She yanked her arm away. A torrent of angry words spilled from her lips. He frowned and

made a series of exaggerated gestures with his hands, and suddenly she laughed and pulled him close, her left arm slung around his shoulder. His arm slipped around her waist. It was an intimate gesture, and it halted Lucy in her tracks.

Lucy fumbled with the too-long sleeves of her sweatshirt. She must look like an elephant. And it was way too hot to be wearing all her clothes. Del was in a tank top and a pair of faded cargo shorts.

Slowly, Lucy walked in their direction, her eyes fixed on the pebbly ground. She tried to look as if she had a destination, a purpose. She kicked a rock. A minute ago she'd felt clean, refreshed; now she was sweating. She touched her hair, pushing the riot of damp curls back without success.

"Lucy!" Aidan said, and waved.

Del moved even closer to him. She didn't smile. Lucy had never been so conscious of tripping as she was now, covering the ground that separated them. She prayed she wouldn't stumble in front of Del's piercing blue eyes. And if she did, she hoped she'd be knocked unconscious or something.

"Hi," she said, reaching them. She was striving for unconcerned and cool, but it came out sounding like a question. Del smirked.

"Del Flowers, this is Lucy . . . ?"

"Holloway," Lucy said. "Lucy Holloway." *Man, even the girl's name is exotic.*

They shook hands. Del's eyes slid away from hers, and as soon as she'd given Lucy's hand the expected up-down shake, she dropped it like it was a snake. Her fingers crept around Aidan's forearm.

Lucy put her backpack down and shrugged her arms out of her leather jacket. The sun was beating down. The glare beating off the broken tarmac was giving her a headache. She remembered how long it had been since she'd eaten. And most of that she'd puked up. She felt suddenly dizzy.

Del was tiptoeing her fingers along Aidan's biceps now. He stepped away and bent down to tighten his shoelace. "How'd it go with Leo?" he asked.

Lucy was instantly angry. She remembered the fear she'd felt. "You could have warned me."

"Would you have stuck around?"

"I almost knifed him."

Del snickered. "Leo is a black belt. I think he'd probably manage to defend himself against you."

"Not if he wasn't expecting an attack," Lucy fired back.

Del rolled her eyes. "Oh *come* on! He took on six guys today." She tugged at Aidan's arm. "Tell her!"

Aidan shook his head and mumbled something incomprehensible. Del glared at him, and then turned a poisonous gaze on Lucy.

"Whatever," she said, and stormed off.

After a long moment, Aidan said, "It's been a bad day. She's upset."

You think? Lucy barely stopped herself from voicing the thought.

Without speaking, they walked to the center of the square. Although it was still midafternoon and bright, the shadows were creeping forward. The sun was suddenly obscured behind boiling black clouds. The air felt heavy.

Rain again, thought Lucy, and then the fat drops fell. In just a few seconds, they became a torrent. Pools of already saturated mud surged under her boots. She felt the weight of the water in the weave of her clothes. Aidan's shaggy hair was plastered against his scalp. It seemed as if the weather never did anything by half measures anymore.

He pulled her under a pale blue awning, but he released her arm far too soon. At a loss for anything to say, Lucy stared at her feet. Aidan looked toward the wide road by which the Sweepers had come. His face was set. She followed his gaze.

"Where does the road go?" Lucy asked.

"It dips down and follows the shoreline for a few miles and ends up at the island."

"So they've got a straight route from here to there?"

"Yeah, it's one of the only routes still accessible. They keep it clear for the vans. Otherwise they'd be on foot."

An older woman, her head covered by a black scarf, dragged

a cover over the big pot on the fire and then joined Lucy and Aidan, who instantly made room for her. She walked slowly, as if her joints were stiff. Lucy's Grandma Ferris had moved like that. Her solid body was swathed in black shawls. Her nose was curved like a beak and she wore heavy gold hoops in her ears, which had elongated the lobes. Lucy recognized her as the woman with the fruits and vegetables. Her black eyes flashed. "They took the priest, Walter, and sad Olive?" she asked Aidan. Her voice was accented, the consonants thickly pronounced. "My little *zabkos*, too?"

Aidan nodded. "And some others I didn't know." She made a guttural noise in her throat and then sighed. "At least Emi and Jack are still together. They had barely settled in." She sighed again.

She turned toward Lucy. "And who is this?"

Lucy tried to meet her gaze but failed. Water dripped from her hair into her eyes. Her nose was running like a faucet. She thought about wiping it on her sleeve but didn't. Not in front of this fierce woman.

She pulled her sweatshirt hood up, but it was too late. She was already soaked.

"This is Lucy. She gave the signal," Aidan said. "And this is Grammalie Rose," he told Lucy.

The woman stared at her for a long moment. Her dark eyes were framed by thick, black brows. They gave her face a

strength that made Lucy nervous. She felt like a mouse pinned by a hawk.

"The howl?"

Lucy cleared her throat. "Yes," she said in a raspy whisper, and then, louder, "Yeah." She shot a quick glance at the stern old woman, wondering if that had sounded sort of smart-alecky.

"You are a wolf, perhaps?" She made a dry coughing sound which Lucy realized with surprise was a laugh.

"I just thought the sound would carry. And people would notice."

The old woman stared at her openly. Her eyes were very black. There was no definition between iris and pupil. It made it hard to look away.

"So," she said eventually, nodding her head. "Good. We need people like her."

"What?" said Lucy, glancing at Aidan. The corner of his mouth twisted and then flattened into a thin line again. "I'm not much use in a fight."

Aidan touched the welt on his cheek. "Yeah, well, neither am I." He looked down the road and frowned. "Especially when we're up against Tasers and a plan, and we've got nothing but some teenagers and senior citizens with sticks and stones."

"Tasers?" Lucy echoed. Those were the black boxes she'd seen the Sweepers holding. Stupidly she'd thought they were radios. No wonder the kids had held back.

"They don't always use them. Not against the young ones at least. It's as if they don't want to injure them or something," Aidan said.

Grammalie Rose said, "They will have their attention on Leo now."

Aidan nodded.

"So you don't really know what they're doing with the people they take?" Lucy asked.

"No idea, but I doubt it's a spa treatment," he said.

"Nothing good," Grammalie said heavily.

They both fell silent.

After a few seconds, Aidan loosened his shoulders. "Grammalie Rose, do you think . . . ?" He paused. She swiveled those piercing eyes toward him.

"Do I think they will come back?" Grammalie Rose exhaled. "Do I think we should try to find them?"

Aidan nodded. His hands were clenched in fists, but Lucy thought he was unaware of it.

"I think that would be both dangerous and foolhardy." Aidan made an impatient gesture. The old woman raised her hand and pointed her forefinger at his chest. "And I think we will have a meeting soon and hear from everyone."

"Soon? Tonight?"

She shook her head. "Feelings are running high. Not everyone is here."

Aidan grunted.

Her black eyebrows bunched. "Okay?" she asked.

"Okay," said Aidan.

She glanced at Lucy. A quirk appeared in the corner of her mouth.

"I will see you soon, *zabko*. There are still a few hours of light left."

"Ummm. Okay." Immediately Lucy berated herself. Why hadn't she said she had no intention of hanging around? That she was just passing through?

Grammalie Rose walked away, and Lucy watched her make slow progress, pausing to speak to one person, lay a hand on a bowed shoulder, give a swift hug to a small child who ran up to her, chattering away.

Lucy turned to Aidan, who was flexing his bruised hand. "What's . . . *jabco*?"

"I think it means 'little frog.' She calls everybody under sixty that."

"Oh. So should I be worried? She scares me."

"She sort of scares me, too, but don't be nervous." Aidan stared out into the rain. "I'm glad you came."

Lucy glared at the ground. She pressed the backs of her hands against her hot cheeks.

"No choice," she mumbled, and then wished she'd kept her mouth closed.

He shot her a quick smile which turned to a frown. "Why?"

She told him briefly about the tsunami, skipping over the

details in case she burst into tears at the thought of her lost camp. The frown got deeper.

"Well," he said after a long pause. She looked back at him. She'd been focusing with all her might on a cloud shaped like a teapot. "Now you can join us. We all pitch in together. No one is alone."

She wasn't at all sure about this. She felt nervous surrounded by people, and there was the danger of the Sweepers. She'd decide later. She could always sneak off in the middle of the night.

Finally she cleared her throat. "Why do you think they take them?"

"I honestly don't know."

"Well, where —"

"To the hospital on the island. That's where the white vans come from. That's where the answers have always come from." He frowned. "And the lies."

"That makes sense," Lucy said slowly. She dreaded asking the next question, so she asked a different one. "How many times has this happened?"

"Twice before. They used to grab the older folk, the ones who didn't move as quickly. But now they're taking anyone who is healthy. Mostly the kids. Today there were a bunch of people from . . ." He paused, searching for the correct word. "From elsewhere. Come for the trading. They didn't know the drill."

"The people," she said through a tight throat. She desperately needed a drink of water. Funny that she could be so wet and so thirsty at the same time. "Do they ever come back?"

He stared at her, and then his face sort of went blank. When he finally spoke, she could barely hear him.

"No."

CHAPTER EIGHT
GRAMMALIE ROSE

The old woman glowered at her. Lucy dropped her eyes. A thin trickle of sweat crept down her neck. The rain had stopped again as suddenly as it had started, and now the sun was blazing once more. She'd taken off the sweatshirt and dumped it and her leather jacket in a pile with her backpack. Even in just the thermal shirt she was hot, and also grimy

with mud and what she had a sneaking suspicion was manure. Whatever Lucy had been expecting, it wasn't this. She leaned against her shovel and looked out over the straight rows of vegetables, the vines rambling over wrought iron gates planted in the earth like trellises, the greenhouses made from old storm windows. In the distance she glimpsed others working. They moved slowly and their faces, from what she could see, were sort of strange. There were bumps and ridges where there shouldn't have been, and their skin was smooth but strangely colored. They wore hoods and long robes. Monks, maybe? But she was too exhausted to ponder it for long.

This mean old woman had made her dig potatoes, carrots, and beets; pick beans and zucchini; stake straggling tomato plants; and remove slugs and beetles from the leaves of Swiss chard and spinach until her fingers were covered with a gluey black residue of insect slime and guts. She hadn't complained, though, mostly because she couldn't. Grammalie Rose, who must have been in her seventies, had worked right alongside her, silently, one of her black shawls pulled over her hair and tied under her chin. Lucy had stared at her tough, reddened hands with their short, blunt nails as she squished bugs between her fingers and brushed dirt off of waxen potatoes before piling them into plastic buckets and tubs. She pulled rows of peas with a minimum expenditure of energy. None of the wrestling with tough stalks that Lucy was doing. Another one of those situations where her awkwardness wasn't

doing her any favors. She could almost hear her brother, Rob, pipe up, "Lucy loses to a plant!"

Lucy kept waiting for the woman to ask her questions or just make small talk, but she didn't. It was like she conserved energy. She marched along the rows, picking or digging, her back hunched, keeping up a solid pace while Lucy had trailed a few yards behind, *her* back screaming and *her* knees aching, barely able to drag the full container behind her. Finally, Lucy had casually asked where Aidan was.

"Out scouting, probably. Hunting or foraging. He likes to roam, that one," the old woman said before pointing out a cluster of bright green caterpillars on a head of lettuce.

And now that they were finally sitting down at the edge of the lot, Grammalie Rose was sticking with the silent treatment *and* giving her the evil eye. It was making her feel uncomfortable. The thought that Aidan had probably known what was in store for her and had uttered no warning *again!* made her stutter with rage, but she shoved it down to her belly where it simmered and spat. She ground her teeth and shot the old lady her best under-the-bangs-slit-eye stare. Grammalie Rose just looked amused and lit another one of the foul-smelling brown cigarettes she liked. The threads of black smoke it gave off stunk like burning hair.

"Have you never shelled beans, *zabko*?"

As she said this, Grammalie Rose was stripping the leathery pods from the dried beans and tossing them into a pail

where they rattled like marbles. She made it look really easy. Snap the end, pull off the string, split the shell with her thumbnail before spilling the purple and white beans into the palm of her hand, and throwing the shriveled ones onto a compost heap. Lucy had tried to copy her and ended up cutting her fingers to shreds and losing most of her beans in the dirt. They rolled everywhere, and who would have known that the dry pods sliced flesh like the edges of thick envelopes? Lucy hunched her shoulders, ignored the pain in her fingers, and yanked on a stubborn pod string.

"Did you eat nothing but meat and acorn mush?"

"Cattail bulbs. Chicory," Lucy said. "Wild onions."

"No wonder you are so skinny. And without energy."

"I outran a tsunami today, and then I hiked over a couple of mountains," Lucy said, feeling her ears go red. She hurled a handful of beans into the bucket with force. Her legs were falling asleep and it was impossible to find an inch of soil without rocks to sit on. "Shouldn't we be preparing or something, in case the Sweepers come back?" she asked the old woman.

"They will be back."

Lucy stared at her. "So?"

"So, people must eat. Life goes on."

"You're saying that we shouldn't do anything?"

Grammalie Rose just nodded and kept shelling. Her plastic bucket was full already, and she hooked Lucy's with the toe of

her clog and drew it closer. Her hands dipped and rose and dipped again.

Lucy bit her tongue. She felt like she might explode.

"But . . . that's . . . just . . . grrrrr!" she shouted finally, leaping to her feet and pacing back and forth.

After a few seconds, Grammalie Rose thumped the ground beside her. The frown was back on her face. Despite her annoyance, Lucy marveled at the bushy blackness of Grammalie Rose's eyebrows and the wrinkles fanning across her crumpled-tissue-paper face. How old was she?

"Sit down," she said.

There was no arguing with her tone, and Lucy could hardly pull her knife on the old woman. She exhaled through her nose and sank down ungracefully, crossing her legs and shifting until she found a comfortable spot. Grammalie Rose thrust a bucket of bean pods at her.

"We *are* doing something, *zabko*," she said.

"I'm not a frog!" Lucy said, turning a hot stare on her.

Grammalie Rose snorted out one of her dry laughs again. "*Wilcze*, then," she said, seeming amused.

"What does that mean?" Lucy snapped, suspecting that she was being teased. She felt like she was being treated like a three-year-old and struggled to control her temper.

"Wolf cub," she said, gesturing to the full bucket. "Full of snarls and bites." She chuckled and held Lucy's gaze until she sat and began work again.

Lucy slid her nail into the tough bean skin and split it, finding a rhythm that was missing before. Gradually she relaxed. *I could still leave*, she told herself, *anytime I want.*

She pushed her hand into the bucket and let the smooth beans sift through her fingers. She played with a pod, crunching it, and cast her mind about for something to say.

"How many people live in this settlement?"

"About thirty-five now. There were close to seventy-five when I first came, but some chose to move on. North. Some prefer to live on the outskirts and come in on market days. And others were taken."

Less than forty, Lucy thought. *And most of them kids.*

She cleared her throat and reached for a bottle of stale-tasting water. She'd drunk about a gallon already, and it seemed that she'd never get enough.

"Did you build all of this?" She waved her arm. Around the periphery of the lot were tumbled dinner plate—size slabs of concrete, rubble, and mounds of garbage big enough to climb. A chain-link fence sagging and busted through snaked around the edge.

"Not me personally."

Lucy stared at her. The black eyes gleamed through their veil of smoke. Was she joking with her?

"It was an old landfill. A dump, literally, and next to it, a cement parking lot which did not fare too well in an earthquake. Pickaxes and a lot of sweat did the rest." She pointed.

"See the low walls over there?" Lucy nodded. Gray walls sectioned off various rectangular areas.

"Corn and herbs. We're trying wheat and barley for the first time, now that flour can't be had for a song or a prayer. We built those out of blocks of concrete we dug up out of the parking lot. It is backbreaking work, but a creative use of salvage."

Her hands stopped working for a moment and she gazed over the furrows of sandy earth.

"It is not good soil, but it is good enough for what we grow here. The manure helps."

"So there are animals?" The thought of eating meat that was not newt or squirrel made her mouth water.

"Not anymore. We lost our last five goats just last week. Poachers." She scowled blackly. "The cows and the chickens died in the second wave." She sighed. "What I wouldn't give for a good honest egg."

"And the vegetables?" asked Lucy. "I mean, these are like what you used to get in a store. Not foraged daylily bulbs and wild greens."

"This was a neighborhood once. The kind that existed before all the troubles. People owned their homes and they grew extra food for their families. We adopted their gardens and their sheds. We used whatever we could find. That's what scavengers do." She picked up a trowel and turned it over in her hand. The clunky handle had obviously once belonged

to some other implement and was kept in place with coils of wire.

Scavengers. Grammalie Rose said it with pride, but Lucy had always thought that scavengers were no better than thieves. She remembered what she had said to Aidan up in the tree and she blushed. Fortunately, the sharp-eyed old woman didn't notice. She had risen to her feet, uttering small complaining noises as her knees creaked, and picked up the two full buckets of beans. She jerked her head at Lucy and then at the other tubs overflowing with produce. Lucy slid her arms through the handles, two on each side. She was balanced, but she felt the tug across the back of her shoulders and the promise of pain to come. Not for the first time in the last year, she thought longingly of a hot bath.

They walked back toward the square. A few people were sweeping the grimy puddles of rainwater away. Others were making piles of cans, rocks, and chunks of brick and sharpening sticks. More were unrolling large carpets woven from bright strips of plastic or squatting on the ground making repairs or dismantling pieces of machinery. Lucy could only guess where the heaps of gears and chains and oddly-shaped metal bits had come from and what use they could be now.

After Grammalie Rose's explosion of conversation, she'd reverted to silence again. When they reached a long awning, she grunted and stopped. A narrow table of pine planks

stretched at least twenty-five feet beneath the canvas tent. It was supported in five or six places with sawhorses. Knives of varying sizes gleamed on the rough surface. There were pots and pans, wooden cutting boards, colanders, and more of the plastic tubs. A fire crackled and smoked at the far end.

Lucy dropped her containers on the ground with a relieved groan and eased her backpack off. She worked her arms around, trying to loosen her shoulders. Maybe now she could go sprawl out somewhere, enjoy the last of the sun's warmth, and take a nap.

Grammalie Rose raised an eyebrow as if she knew what was passing through Lucy's mind. They locked eyes for a long moment. Lucy was on the verge of walking away when Grammalie Rose said in a mild tone, "You don't work, you don't eat."

Lucy nodded shortly. Her stomach felt like an empty balloon. She'd had no food for more than twenty-four hours. The old woman thumped her on the back. "Not so sullen, *wilcze*. The preparers get to eat first," she said, "and everyone must take a turn. This is not some cruel sort of punishment dreamt up especially for you." She made a croaking noise that sounded suspiciously like a chuckle and pushed her toward the table. There were people down at the far end scrubbing potatoes. They were surrounded by mounds of dirty yellow spuds, and yet they chattered and laughed together.

"Henry!" Grammalie Rose called, and pointed to one of the tubs Lucy had hauled.

Henry was small and dark, maybe in his early twenties, with brown hair that stuck up in a duck's tail over his forehead and twinkling brown eyes.

"Henry, this is Lucy."

Henry grinned at Lucy and stuck out his hand. She shook it, conscious of her filth-encrusted fingers. "Leo told me about you," he said. "He said you seemed okay."

"I am."

"Hmm." He ran his eyes over her in a not entirely clinical way.

"You don't look like a doctor," Lucy said, trying to cover her embarrassment.

"Oh, I'm not a doctor, but I do my best."

She felt vaguely unsettled.

He checked out the rest of their containers. "Looks like bean soup tomorrow. We'll have to soak them first."

Grammalie Rose grunted. "I've been making bean soup since your father was your age. Let tomorrow take care of itself." Henry made a face and skipped backward to avoid a slap. "First things first," the old woman said in a louder voice. "Take the onions and the carrots, *malpa*."

Henry staggered off, bowed under the weight of two tubs, and Grammalie Rose showed Lucy where a large bowl of tepid water was. Next to it was another slimy block of soap, this

one off-white and shapeless. "Wash well, *wilcze*," the old woman said, sinking down on a bench. Even though her eyes had closed as if she were resting, Lucy still felt them on her back.

She rolled her sleeves up above her elbows and scrubbed her hands with the soap. It smelled of industrial strength solvent and lard. Her fingers tingled now as if they'd been sprayed with an acid solution, and the bandaged wound across her palm stung. She also took the opportunity to wash her arms and the back of her neck, too, so although she smelled of paint stripper, she felt a little cleaner.

There were six skinned dead things on the table that needed their innards removed, and she was pretty sure that was next on the list of chores. Whoever had skinned them had left their little white tails attached. "Letting us know they are not cats," Grammalie Rose said, touching a cotton ball tuft.

"Do you hunt cats often?" Lucy asked. Meat was meat, but cats were carnivores and never tasted good. Most carnivores didn't. She'd caught a weasel in a snare once and boiled it up, and the meat had been stringy, gamey, and really tough.

"Not often, but when the hunters bring them, they remove the fur *and* the tails so we can pretend it is something else."

Looking away from the sad little carcasses, Lucy's attention was captured by a hooded, cloaked figure that approached the table and carefully picked up the tub Lucy had used to wash

her hands. The water was almost black with grime and bits of insect wings floated in it. His hands were gloved.

"Good," said Grammalie Rose. "Sammy, please bring us some fresh water if you can."

The man nodded. His hood slipped back a little and Lucy gasped. His features were covered completely by a mask. It was smooth and beautifully ornate, painted a luminescent white with gold filigree and flourishes around the cutout ovals where the eyes should be. His own eyes burned red behind the mask. A bow-shaped mouth was molded half-open and colored with more of the gold paint. The skin of his neck where the mask ended and his cloak gaped was blackened and cracked, as if it had been charred in a terrible fire.

Lucy stepped backward. Her heart was hammering in her chest, and she was unaware that her arm had knocked over a bowl and sent garlic and beets spilling onto the ground. He was a S'an. He was infected. A walking time bomb.

Without realizing it, she must have spoken out loud. Grammalie Rose pinched her elbow. Lucy half turned, and her fists came up defensively. The old woman batted them down. Lucy tried to form words but was unable to. Her brain was speeding. Those other strange people in the fields must be S'ans, too. She didn't understand how this could be. Everything she had ever been told said that they were carriers of the disease, as much as the urban birds were. They were

to be avoided, and yet they were here. Working and mingling as if they were regular people.

The pinch became increasingly painful. She dragged her gaze from the S'an and stared at Grammalie Rose.

"I will slap you if I have to," the old woman said in a ferocious voice. "Are you going to faint?"

Lucy shook her head. Her legs felt weak, but her head was clear.

"Listen to me. You hate because you are scared, and you fear because you don't understand."

A shiver of horror ran up Lucy's spine.

"The S'ans are to be pitied, not feared," Grammalie Rose said. "They have survived the disease, but they are damaged. Their skin, their bodies, are ravaged. Do you hate him because he is not so pretty as you?"

"No, I . . . I . . ." Incredibly, through the fright, she felt a flush of shame. The S'an was three feet away and Grammalie Rose was carrying out this inquisition in front of him. "No," she said again, feeling like she was being hauled up in front of the class and reprimanded for cheating on a test. She knew that everyone under the awning was staring at her.

"Do you know where the name came from?" Grammalie Rose asked, relaxing her grip a little. Lucy felt the tips of her fingers buzz as the blood flooded back in. She thought back to the news reports.

"They are escapees from the sanatoriums, right? Driven insane by the second wave of the plague? After the disease mutated, people started losing their minds and their legs fell off and they craved fresh brains and stuff. . . ." Her voice trailed off. It sounded sort of stupid when she said it out loud.

Grammalie Rose snorted. "Zombies, huh? I used to love those old movies." She let go of Lucy's arm, nodded to the S'an who was still standing there, holding the sloshing tub in both arms. He shambled off, spilling water in a trail behind him, but before he went he winked at Lucy.

She stared after him, her mouth open. She closed it with a snap.

"That is an urban myth arising out of fear; because they were infected, but against all odds survived with their skin burned and cracked and their eyes bloody. They are not crazy people."

"But don't they carry the contagion still?" Lucy asked. Her hand had gone to her knife; she rubbed her thumb over the hilt, tried to control the shaking.

"No."

This was contrary to everything Lucy had heard.

She rallied her thoughts. "I thought the symptoms meant the sickness is still there. His eyes are red. He's still bleeding under his skin. He was wearing gloves."

"No. His immune system remains weak, but they are no more sick than we are. The gloves and robes shield his

damaged skin from the sun. He wears the mask to protect *our sensibilities*." Grammalie Rose put stress on the words. Her black eyes were flashing with anger. Lucy stepped backward, tripping over a beet. She bent down to pick it up, noticed the other vegetables and the fallen bowl, and started piling them together.

It seemed impossible, but surely if the S'ans had been living with the settlers for all this time then they must be all right. Leo had said that people were scared so they had to be careful. Somehow Sammy and the others had checked out. She got to her feet. She needed time to get her head around this.

Grammalie Rose bent with a grunt and picked up a garlic bulb that had rolled near her foot. She handed it to Lucy, who put it in the bowl with the rest.

"You look to be good with a knife, *wilcze*. Think you can deal with the rabbits? Small pieces for a stew." Grammalie Rose spoke calmly, as if nothing had happened.

"Uhh." Lucy was startled.

"Good. I'll be over helping to peel that mountain of potatoes. Otherwise we'll never eat today." She nodded grimly at her and made her slow way to the far end of the table where Henry and the others stood. After a few seconds more of staring at Lucy, they got back to work and the hum of conversation started up again.

Lucy rubbed her nose with the back of a hand covered in blood. She'd already gutted and filleted three of the rabbits. The chef's knife she was using was very sharp, the edge honed beautifully. The S'an — Sammy — had not come back again, but the expectation of seeing him was making her jumpy. Part of her wondered where Aidan was. He was the only person she sort of knew in the camp, and he'd disappeared. She cut the pieces of glistening red meat into chunks and pulled another rabbit toward her. Using a cleaver, she chopped off the cotton ball tail. The knife whacked through the bone and into the board. She rocked it back and forth to loosen it.

"Whoa!" Henry jumped backward, his eyes wide in mock alarm. "Do you think you can put the knife down for a second?"

After a moment, Lucy recognized the teasing note in his voice. She smiled back reluctantly. The wood in front of her was covered in deep cuts. She hadn't noticed how much pressure she'd been exerting.

Henry slouched against the table. Freckles sprayed across his nose and cheeks. His eyes were dark brown and round like a child's. The spike of hair made him look like a cartoon character. He held a plate. On it was a loaf of bread and a small bowl filled with green-tinged oil.

Lucy's belly rumbled audibly, but she was so hungry, she didn't care.

He pushed the food toward her. "I thought you might be hungry. It came out of the oven about fifteen minutes ago."

Hastily wiping her filthy hands on her pants, Lucy tore off a chunk of bread and shoved it into her mouth. It was still warm. Henry watched her with an amused expression. "You can dip it in the oil, if you like. We have to make our bread with water, no milk you know, and it makes it sort of dense . . . and hard to swallow."

She stopped chewing for a minute. "Tastes pretty good to me." She took his advice, though, and swirled the next chunk in the oil. It was fruity and rich, and absolutely delicious. From the far end of the awning, the mouthwatering smell of onions, garlic, and carrots simmering in oil wafted into the air. She hadn't realized how much she'd missed fried food until now. The thought of fried potatoes made her giddy.

Henry pointed to the other side of the square where the remains of a building stood. Someone had attached frayed lengths of canvas from the two remaining corner uprights to make a rough roof. "Used to be an Italian deli," he said. "Nothing survived in the shop, but they had a cellar filled with wine and bottles of oil. The wine's gone now, unfortunately," he concluded. He met her eyes with a wry look. "But we've got enough oil to fry a mountain of potatoes." He rummaged in his back pocket and brought out a stub of

pencil and a tattered notebook. "Can I ask you a couple of questions?"

"Sure, I guess," said Lucy, brushing bread crumbs off the front of her shirt.

"We keep an informal sort of census now. So many people coming and going," Henry said. "Name?"

"Lucy Holloway."

"Age?"

"Sixteen. I'll be seventeen in a . . . few months." She realized that she wasn't sure exactly how soon her next birthday was.

He nodded. "Nice to have another mature person here. It's mostly little kids and the DAs."

Henry answered her querying look, lowering his voice: "Doddering Ancients, but don't let Grammalie Rose hear it."

"What did she call you? *Malpa?*"

"Polish for *monkey*. She thinks she's so funny!"

He jotted Lucy's responses down, and then swept his gaze over her. His eyes widened in appreciation. "I'd say you're healthy. Very healthy." Henry scribbled something else.

She blushed. There was no mistaking the fact that he was checking her out. He must be, what, at least twenty-one? Her hand crept up to her messy hair. Boys were so weird. Working in the fields all afternoon had covered her skin with a fresh layer of stink. Plus manure. And blood. And still, he was flirting.

Henry put his notepad away and leaned on the table.

"So. Grammalie Rose was a bit rough on you?"

"Yeah, because of the . . ." She corrected herself. "Because of Sammy."

"Hey, it was a shock for me, too, when I first came here, but soon enough you realize that they are just regular people."

"I guess," Lucy said. "So how many are there?"

"Three in this settlement. But there are more out there." He waved his hand in a vague way.

"And they help with the chores?" She fought to keep her voice neutral.

Henry shot her a look. "Yeah, everything but the cooking. Bits of them, you know, fingers and the like, kept falling into the stew, so we put a stop to that."

Lucy gasped, and then caught the wide grin spreading across his face. She went red. Henry put up his hands in a conciliatory gesture.

"Sorry—couldn't help myself. Corny as it sounds," he continued, "we're like a family. Literally, in some cases."

She looked at him.

"Emi and Jack are siblings." His face fell, and Lucy remembered that these were the names of the little kids who'd been grabbed earlier. "And Sammy is Aidan's brother."

"Really?" Lucy said. "I mean, how could that be? That one of them is fine and the other is . . ." She broke off. No one else in her family survived.

Henry raised an eyebrow and she got the feeling he'd read her mind. "There's no clear answer. Most people died if they got sick. Sammy's lucky to be alive," he said.

She nodded, and tried to swallow the lump in her throat.

"If you think about it, Sammy and Aidan beat the odds. Two brothers in the same family."

"But Aidan didn't even get sick. He has no scars . . . does he?"

Henry's mouth twisted. "Luck again, I guess. Of those who contracted the mutated hemorrhagic smallpox in the second wave, maybe one in a million survived, even with the vaccine. Most died within seventy-two hours. Those are some bad numbers. The regular pox left about one in one hundred thousand alive, so if you look at it like that, you and I, and everyone else here are blessed. Right? A few scars here and there, maybe, but nothing like what the S'ans have to bear. Pretty soon you won't even notice a difference." He shot her a grin, and she couldn't help but grin back at him.

"How do you know so much about it?"

"I was a premed student before."

"So how many unvaccinated people survived?"

He looked startled. "None. Big fat goose egg." He made a zero with his thumb and forefinger.

"No, seriously," she began before noticing his face. The smirk was gone. He shook his head.

"Basically that's why the majority of the deaths were adults aged thirty to sixty. The kids and teenagers were okay 'cause they were up to date on their shots including the reinstated ones."

Lucy nodded. She remembered her classmates back in grade school complaining that they'd had to get a whole slew of new injections after the first bird flu cases had been diagnosed.

". . . and the older people like Grammalie had been given live smallpox inoculations during the War, but the rest of them . . . Nope. One hundred percent mortality."

Her mouth shaped itself into an O. Her hand crept up to her left shoulder and pulled the rolled sleeve of her shirt down so that it covered her upper arm.

None. Zero. That made her an even bigger freak than the S'ans. She remembered the thick folder in the nurse's office. The countless blood tests. What exactly was wrong with her?

"Are you all right?" he asked.

"Yeah, yeah, I just ate too fast."

"Okay, well give a holler when you're done," Henry said, pointing to the cutting board, and sauntered away.

CHAPTER NINE
CAMP SCAVENGER

After Henry had gone back to his potatoes, Lucy forced her fingers to continue cutting up bunnies. Her brain was yammering away at full speed. If what Henry had said was true, she should not be alive. Unless she *had* been vaccinated when she was a kid and Maggie had been wrong or lying. Her sister wasn't the type to play tricks on her

siblings—that was more like something Rob would have done—but maybe she just hadn't had the facts straight.

Could she have forgotten? The first time she'd run through the glass doors had completely faded from memory, so maybe something in her just preferred to ignore unpleasant events.

Lucy pressed her fingers against the skin of her upper arm, trying to feel for the raised scar of a smallpox vaccine. She felt nothing. She needed to go somewhere where she could look. Find a mirror. Examine every inch of her skin. But she couldn't do it with all these people around. She wondered about bathrooms: Did they have them? Were there latrines out in the fields or something? Surely someone here owned a mirror. Perhaps Henry? He looked like he spent time getting his hair just right.

Another part of her brain reminded her that if she did have a vaccine scar, she would surely have noticed it before now.

Her stomach twisted. She *had* eaten too much too fast. Lucy took a sip of water and tried to think. Maybe she should leave? Go back out in the Wilds? But the Sweepers . . . Now that she'd seen them in action, she was scared. No one seemed able to stand against them, and by herself she'd be totally helpless. And the dogs—they were hunting people with the dogs.

She wished there were someone she could talk to. She might be sick and not showing symptoms. She could be a carrier like Typhoid Mary, who'd shown no symptoms but had infected people just by cooking their meals. She looked down at the chunks of rabbit glistening on the chopping block, the pile of cotton ball tails. Her stomach heaved again. Cooking would kill the disease, right? *If* she had it. Lucy imagined her body swarming with virus. She grabbed the edge of the table and pressed her fingers into it until her gut settled. Maybe she could tell Aidan. Or maybe she shouldn't say a word.

As if the thought had summoned him, Aidan appeared behind her and hopped up onto the table. "Hey," he said casually. "I was looking for you."

Lucy fought to control her panic. She made a noncommittal noise and stared at the table. She drew a bowl of water toward her and sluiced the blood from her hands, scrubbing at them with her nails. Slowly, her heart stopped racing. She peeked at Aidan under her eyelashes.

His sweatshirt was damp with sweat; there were mud stains on the knees of his jeans and a few dry leaves caught in his hair. She stared at his fingers, thinking how strong they seemed. That made her heart race again and distracted her from morbid thoughts. He tore off a piece of bread, swooshed it in the oil, and popped it into his mouth. He eyed the portions of chopped-up meat. "Cat?" he asked sadly. Lucy

flicked a rabbit tail at him. "Oh good," he said and flicked it back at her.

"So where have you been?" she asked, looking at Aidan's dirty fingernails.

"Out," he said.

Lucy felt a surge of irritation. Which was good. It banished the last of the fear and made it possible for her to meet his eyes without blushing. "With Del?" she asked before she could curb her tongue. Aidan looked at her and then jumped down, ignoring the question. Maybe he picked up on the sneer in her voice? She vowed to keep her mouth shut about the other girl.

He picked up the cutting board and transferred the meat into a plastic serving bowl. "Come on," he said, grabbing her arm. He led her to where the others were standing around a large pot on the fire. It was blackened iron and big enough to bathe a child in. About forty pounds of carrots and onions simmered at the bottom. The smell that rose was heady. Henry stirred the mixture occasionally with a long wooden spoon. A young girl, maybe eleven years old, with long blond pigtails was cutting up the last of the potatoes, helped by two small kids and a gray-haired old man with a walrus mustache. Grammalie Rose was no longer there.

"Ready for this?" Aidan asked Henry, hefting the bowl of meat.

"Sure, pile it on," Henry said. "Potatoes next, Sue," he told

the pigtailed girl. He lifted a bucket of water and held it poised for a few seconds before upending it into the sizzling pot.

The good smells were making Lucy woozy. She sat down on a bench and closed her eyes, letting the fragrant steam wash over her.

Aidan sat down beside her. "About fifteen minutes," he said with an amused tone in his voice. "Can you bear it?"

"Possibly not."

"Well, at least we eat first," he said.

"*We?*" Henry said, waggling his eyebrows at Aidan. "And how exactly have you helped with this fine meal?"

"I believe I hauled that water," Aidan said. "And I washed up last night."

Henry spread his fingers. "All right, all right." He turned to Lucy. "Can you help Aidan with the bowls and spoons?" He pointed toward the stacks of mismatched kitchenware. "We need thirty, forty of everything."

Lucy grabbed a handful of spoons and shoved them in her back pocket. She stacked bowls along the length of her arm and anchored them with her chin. It was a brave move. One clumsy step and she'd drop everything; but amazingly she made it to the table safely. She set the places. Aidan put down plates, water jugs, and three more loaves of the crusty bread, and scattered a few bread knives along the length of the rough pine table. He started cutting slices and, grabbing a knife and

another loaf of bread, she did the same. With half an ear she listened to the surrounding chatter.

People materialized from the corners of the kitchen tent, wiping their wet hands on their pant legs, removing stained aprons, and stretching sore neck muscles. They each grabbed a bowl and lined up for a few ladles of thick stew. Dishes clattered. A dozen conversations were going at once. Lucy felt the familiar shyness creep into her bones. It was like the high school cafeteria. She'd always eaten alone, outside in the quad or in the library. Aidan pulled her to her feet and shoved a bowl into her hands. "If you don't get in there, you'll never eat," he said, elbowing a space for her.

Henry grinned as he served her. He leaned forward and winked. "I gave you a little extra."

She smiled shyly and sat down at the far end of the table, away from the little clusters of people. For the next few minutes, she concentrated on eating. It wasn't until she looked up that she realized Aidan was sitting right across from her. He was smirking like a maniac.

"Never saw anyone actually inhale food before," he remarked.

"Oh God, I . . ." She put her hand up to her mouth and wiped it clean. A few spots of stew were speckled across the front of her shirt. There may even have been some caught in her hair. She wished she could just sink through the floor.

He pointed to the corner of his lip and tapped.

"What?"

"You missed a spot."

He half stood up and reached across the table. Was he going to touch her? And then suddenly he jerked away.

"Make room," Del barked, squeezing in next to him. The kid she'd forced farther along the bench glared at her but said nothing. He just picked up his bowl and turned to his neighbor, Sue. They bent their heads together, whispering furiously and darting covert glances at Del.

"Hi," Lucy said, determined not to get into a situation with her again. She even managed a small smile before returning her attention to her bowl.

Del stared, then nodded.

Score one maturity point for Lucy! She carefully scooped up a minute amount of stew and carried it to her mouth. The food had reached her stomach, and she felt a soothing warmth spread to her limbs. *I could fall asleep right here on the table,* she thought. She listened sleepily to Del.

"Did you scout today?" she asked Aidan. "I couldn't find you."

He shook his head. "I went a mile or two up the road. I thought about setting permanent sentries, but there's too much area to cover."

"What about pit traps? We could dig some around the periphery and then just keep watch by the road," Del said.

"Too dangerous for the little kids. You know they run around everywhere."

"Well, how about blocking it off? Then they couldn't drive the vans up to the camp." Del banged her spoon against the table for emphasis.

Aidan rubbed his forehead. He looked as if he hadn't slept for a week. "Leo figured out that it would take a ton or more of rubble to block it off. We can try, but it'll take weeks. And it's harvest time. We can't really spare people from the fields right now."

"There must be something!" Del stared at the table. "What'll they do to them?" she asked suddenly in a gentle voice, sounding completely unlike herself. "To Emi and Jack?"

"They'll be okay."

She gripped his arm hard. Aidan winced. "Promise me."

He shook his head, looking uncomfortable.

Del subsided into a stormy silence. The conversation along the table had dwindled, the sound of cutlery against bowls quieted; people finished and left. Others hauled water for washing up. Lucy was thankful not to be summoned for more chores. She rested her head on her arm.

Things were bad. She was scared, but she was also not hungry for the first time in months. She felt strong and revitalized. If she had to, she could run for miles, cross New

Venice, and find somewhere safe on the outskirts. Or travel farther. North, perhaps, like Aidan had said.

She didn't realize she'd fallen asleep until someone shook her gently by the shoulders. She opened drowsy eyes and looked into Aidan's face. Directly behind him stood Del, her arms crossed tightly, her mouth flattened in a line.

"Del said you could share her sleeping tent," Aidan said, helping her to her feet.

"Scout's been sleeping somewhere else anyway," Del said with a bitter laugh. She didn't look as if she'd offered freely, but at this point Lucy didn't care. She swayed and shook her head, trying to drive the cobwebs from her brain.

Aidan picked up her backpack and handed it to her. She clasped it to her chest. He led the way inside a small tent tucked between the gutted remains of two buildings. Trickles of the day's fading light fell through gaps between the canvas panels. Lucy caught a glimpse of three tarps on the ground arranged around an unlit hurricane lamp, two covered with blankets and clothing, before Aidan pushed her gently toward the third. She bumbled forward, tripping over her feet, and let the backpack slide to the ground as she sank down to her knees. Too tired to even dig out her sleeping bag, she pulled her sweatshirt and jacket firmly around her, tucked her knees in, and curled up on her side. She was asleep again almost immediately.

CHAPTER TEN

ATTACK!

Lucy was awakened by an excruciating pain in her wounded hand. She opened her mouth to scream but someone's fingers pinched it shut before she could emit a sound.

"Quiet!" Del hissed viciously. Lucy gripped Del's forearm and squeezed as hard as she could. "Don't yell," the girl said, slowly removing her hand from Lucy's mouth.

"Get off me," she replied between clenched teeth. The lamp in the middle of the tent threw a small circle of light. It was enough to make out Del, fully clothed, balanced on the balls of her feet as if she was expecting an attack.

Del shifted her weight and the boot lifted. Lucy decided it had been an accident; the opposite was too much to consider. She clutched her squashed fingers to her chest. She wiggled them. They seemed bruised, but not broken.

She wondered if it was close to dawn, but then through the gap in the roof she glimpsed the moon, half-hidden behind clouds. Standing above Lucy's bed, Del was a still form against dark shadows. She could see the other girl's rib cage move in and out with her breathing. She seemed to be waiting for something.

"What's going on?" Lucy asked. She'd only slept a few hours and her body was stiff and achy.

"I heard someone outside the tent," Del murmured. Her head darted around. She cocked it to one side like a dog. Lucy listened, too, straining to hear.

Distantly she heard a rumble. Thunder? Or could it be car engines? She sat up. She'd gone to bed with all her clothes and her boots on. Under the triple layer of jacket-sweatshirt-thermal, her skin was clammy with sweat. Her head felt groggy, her eyelids rimmed in sand.

Del was as immobile as a statue.

A crescendo of rumbling rose, followed by a shout and, closer still, the thud of running feet. And now she could hear voices raised in panic coming from all around and the sound of tires on sandy soil.

Cars!

"It's them. The Sweepers. They came back," Del said.

Lucy was instantly awake. Her heart pounded as if it would spring out of her chest; every muscle twitched. She scrambled to her feet. Through the canvas, she could see the dim shapes of figures moving outside. They had lights. Maybe flashlights or torches. She couldn't tell if they were scavengers or the enemy.

"Stay low," Del breathed. "They might have brought the dogs." She was frozen in a half crouch. Lucy mimicked her posture.

"If they brought the dogs, we should get out of here!" Lucy said, trying to breathe normally.

"Don't move," Del said with an imperious hand gesture that made Lucy bite her lip in annoyance.

The other girl moved slowly, her eyes on the hurricane lamp in the middle of the floor.

Lucy suddenly realized that their silhouettes must be visible from the outside.

Before she could say anything there was a thump as Del kicked the lamp over. The flame went out. Lucy blinked, trying to accustom her eyes to the sudden gloom.

"Let's go," Del muttered.

Lucy fumbled for her knife. It wasn't at her waist. For the first time, she had forgotten to fasten it on her belt before she went to bed. She cursed herself and dug into the sweatshirt pocket. Not there, either.

It must have fallen out.

She bent to the tarp she'd been sleeping on, searched the surface with trembling fingers. The thick layer of clothing made her clumsier than usual.

"Come on," said Del. "And be quiet, would you?" Lucy thought but couldn't be sure that she'd muttered something about a "buffalo."

"My knife."

"Leave it."

"No!"

Del snorted with impatience. "Hurry," she said. Lucy felt around the edges of the tarp and was finally rewarded with the hard outline of the hilt under her fingers. She picked up the knife, instantly feeling more confident.

A motor revved nearby. It sounded as if it was right in front. The tent flap was tied shut, but it was flimsy. They couldn't go out that way.

"We've got to get out of here," Del whispered, making for the back of the tent. She pulled on the bottom where it was pinned to the earth by metal hooks.

"Help me get this loose. Quietly."

Lucy joined her and, stowing her knife in her pocket, grabbed a handful of canvas and heaved. The ground was hard, compacted mud; the stakes had been pounded in, and they couldn't pull it loose. Del muttered a few choice curse words under her breath.

"Wait," Lucy said and pulled her knife back out. She stabbed at the heavy material. The sound of tearing canvas seemed incredibly loud. Del stifled an angry exclamation, which Lucy ignored. Once she had a big enough hole, she held it open and Del clambered through it, swearing as her boots caught in the folds of material.

Lucy started to follow her, and then she remembered her backpack. First rule: Always carry what you need with you. She hesitated. She could imagine what Del would say, but the habit was too ingrained. For over a year she had survived on her own because she was always prepared for the worst, and because her backpack held everything necessary for her survival.

She couldn't leave it.

Del was just about through, but it would just take a second.

She turned back to the tarp, scooped up her backpack, and shrugged it over her shoulders. Then she ran back to where Del's foot had just disappeared through the gap.

"What are you doing?" Del hissed, sticking her head through the hole.

"I'm coming!"

Someone burst into the tent. A heavy body struck Lucy in a tackle. She fell to the ground, biting her tongue hard on the way down. The taste of blood was in her mouth. She was pinned by strong arms attempting to grab her own and the weight of someone's body across her legs. She flailed, striking out wildly, and twisted around so she was lying on her back. Lucy kept struggling and kicked out with both feet. A sharp *crack*. She had hit something hard enough to make her ankles throb. An explosive grunt, the sound of fumbling, and then a helmet thudded to the ground near her head. The visor was smashed. Before she could struggle to her feet, the man threw himself forward. Lucy brought her knees up, trying to force him off of her. She lashed out with her fist, feeling the vibration in her elbow as she connected with his face. His breath was hot against her cheek. She felt a slick wetness on her forehead. Her blood or his?

A thick arm pushed against her neck. She tried to land another punch, but it was hard to breathe. Black dots danced in front of her eyes, and her pulse pounded in her temples.

Lucy attempted to scream, but the sound was choked off in her throat. With one last burst of energy she raised her head and bit down as hard as she could. It wasn't much — the man's arm was covered in thick material, denim or heavy cotton, but it was enough. He shifted and she arched her back, simultaneously rolling to one side, and managed to push him

off. She scrambled to her feet, panting, and aimed a kick at him, not caring where it landed. She heard the satisfying *thump* of impact.

The man groaned and threw out a hand, closing his fingers around her ankle, and suddenly she was pulled off her feet. She landed hard on her back, the blow cushioned by the bag across her shoulders, although she felt a sharp pain radiate up her spine. She must have landed on her dead flashlight. The breath left her body in one involuntary gasp. And then he was dragging her toward him. She dug her fingers into the dirt, but it was useless.

Lucy twisted her body, trying to break free, and then from the side and slightly behind him, a shadowy figure appeared. Del raised the hurricane lamp high, then brought it down. Some instinct must have warned him because he moved slightly, and rather than hit him on the head, the lamp smashed into his shoulder. Still, it was enough to break his hold on Lucy's leg. Shards of glass flew everywhere. Lucy felt a chunk sting her face. Del hauled her to her feet and dragged her out the front of the tent.

They ran toward the shouts and screams, then stopped, blinded. The square was lit up. The vans were positioned in a half circle, their engines idling. On top of each one was a powerful searchlight.

Lucy shielded her eyes. She could see little pockets of people to either side, mostly younger kids clutching one another

and crying; the old man with the walrus mustache she'd noticed at dinner. No one who looked like they could defend themself. Many of the shelters had been destroyed, the wooden supports smashed, canvas tarps trodden into the dust. There were perhaps a dozen Sweepers scattered around.

She could hear yelling coming from other parts of the camp. They must be everywhere. The scavengers were divided. She wondered where Aidan was. Henry, Leo, Grammalie Rose? Had they been captured already?

Del swore again under her breath.

Lucy cast a glance behind them. A Sweeper stood there, blocking the way back into the maze of alleyways. Why hadn't they run in that direction?

"Get ready to fight," Del said.

Lucy pulled her knife out and held it ready.

Del moved so she was standing with her back to Lucy's. She bent and scooped up a length of twisted metal. They circled, taking small steps and trying to look in every direction at once, to find a hole they could break through. A searchlight was moved so that it pointed directly at them. Lucy tried to see past it, but the intense light threw everything else into deep shadow. She caught a glimpse of white-suited figures rushing toward them, flanking them.

Del shouted and was suddenly yanked away from her. Lucy moved her knife, blade edge out, in a sweeping motion. She was grabbed from behind, her knife hand pinned. A

muscular arm settled around her neck, restricting but not cutting off her breathing. "Easy," a voice breathed in her ear. She felt a pinching sensation in her wrist, and her fingers opened. She dimly heard the thud as the knife fell to the ground. She was lifted off her feet. There was a flurry of movement to her left. And then Del was free. She swung the metal bar, aiming at her captor's groin. The blow connected and the man dropped to his knees. She swung again with a ferocity that was terrifying, catching him across the back, and he fell forward, groaning. Lucy's Sweeper was distracted, and she took the opportunity to stamp on his instep with all her might. With a roar of rage he pushed her away with so much force that she stumbled, falling to her knees.

"Duck!" Del yelled, swinging the metal rod like a madwoman. Lucy made herself as small as possible.

Her knife gleamed in the dust a few yards away. It might as well have been on Mars. She began to crawl toward it. She could hear Del cursing, panting, and, amazingly, mocking the Sweepers. But there were too many of them.

"Del!" Leo bellowed, appearing from an alleyway at a run and beating his way to the girl's side. He was armed with a two-by-four studded with nails. The Sweepers fell back for an instant, and then, as if obeying an order, advanced in a solid line on the two of them.

Two more Sweepers joined them, effectively containing the scavengers within a small area. Leo circled.

And then, surprisingly, one of the Sweepers moved forward alone. It didn't make sense. The scavengers were totally outnumbered. Lucy had reached her knife. She picked it up. No one was paying any attention to her. She crouched in the shadows, looking for an opportunity to help.

Leo pushed Del behind him and faced off with the Sweeper. A grin spread across his face. "Come on then," he yelled, advancing. He slapped the wood against his palm.

The Sweeper moved in closer and suddenly lunged forward, his arm outstretched. Concealed within his hand was a small black box. Lucy screamed a warning, but her cry was drowned out.

Leo raised his club.

And then there was a flash of electric blue light and a sound like meat being seared on a hot grill.

Leo collapsed to the ground like a felled tree. His body jerked spasmodically, and then he was still.

"Leo!" Del shouted, running toward his prone form. She'd taken only a few steps when another Sweeper stepped into her path. The blue light flashed again, and she crumpled.

Lucy shrank back into the shadows, drawing her hood forward around her face.

The Sweepers were gathered around Leo and Del, who were as still as corpses.

Were they dead? Leo groaned as one of the Sweepers nudged him roughly with a boot. Lucy felt a surge of relief.

160

Another man picked up Leo's nail-studded two-by-four, studied it, then hurled it to the ground in disgust. It bounced, landing not far from where Lucy lay hidden.

She thought she could see smears of blood on the board. What could she do? If she tried to help, they would capture her, too. She was safe for now. She felt a sense of relief mingled with the shame of escaping capture. If she had not paused for her backpack, then Del might still be free.

Slowly, Lucy forced herself to crawl backward toward the collapsed shelter behind her. Once she was concealed under the tarpaulin, she curled into a ball and tried to control the tremors that racked her body. She listened to the heavy sounds and grunts of the Sweepers, picturing them as they loaded Del's and Leo's unconscious forms into the nearest van. She heard the low rumble of voices and the piteous sound of a child sobbing. The doors slammed shut, the engines roared, and heavy tires crunched through the debris.

The rumble of the vans slowly died away, and afterward there was silence, which felt oppressive and filled with threat.

It was hours before she moved from her hiding place.

Lucy sat on a small hill above the camp and watched the sun rise. She was cold and cramped, but she could see everything from this vantage point. As soon as she had dared, she'd scuttled out from under the tarp and run, heading for higher ground.

Her thoughts went around and around in her head. She'd been arguing with herself for hours, unable to sleep.

She'd figured out this much. One: She was scared. Two: She badly wanted to leave. Three: That was the one thing she could not do.

The reality was, she was involved. Not only because she'd been right there when Del and Leo were taken, but because probably—*definitely*—Del wouldn't have been caught if not for her.

She untied her backpack and scanned the contents. A journal, a few pieces of clothing, her sleeping bag, a yearbook, a broken radio, a flashlight without batteries, a sharpening stone, and a tinderbox with a book of soggy matches in it.

Nothing she owned was worth someone else's life.

She looked down on the settlement. All the shelters around the square had been ripped apart. Flimsy supports lay twisted and snapped in two, plywood lean-tos were scattered in splintered heaps. The soft, packed earth where the trading market had been held just a day ago was torn up, and the deep tire tracks of the vans snaked through the devastation.

She looked south in the direction the road curved. The road that ended at Roosevelt Island. From here she couldn't see the red light flash at the top of the tower, but she knew it was there.

She got to her feet with a sigh.

A small group had assembled in the square. She recognized Aidan, Henry, Grammalie Rose, and a few others.

Lucy paused. Turned and looked in the opposite direction. She could make her way inland, find another perfect place to build a home, go back to life as she knew it. Alone.

A shout wafted up from below.

She gazed down. Henry had lifted his hand in a wave. He flourished it back and forth as if he were signaling an airplane.

Slowly, she waved in return and started down the path.

CHAPTER ELEVEN
AIDAN

The scavengers stood in a close group near the kitchen. Lucy shuffled her feet, not sure if she should interrupt, and was grateful when Henry hollered at her to join them. His left eye was blackened and puffy, the whites shot through with red. His hair was plastered to his head with sweat and filth and he bounced from foot to foot with nervous energy. He moved over to make room for her in the huddle.

"Lucy," Aidan said. His eyes went to the cut on her cheek and she touched it, shaking her head. It stung, but it was little more than a scratch. She'd collected a few more bruises, but she'd been lucky. Most of the blood on her belonged to other people. "I'm fine," she said.

Aidan's knuckles were scabbed and raw. A deep gash ran across his forehead, and there was dried blood in his hair. Lucy looked away from the expression of anguish on his face.

She stood apart and he made no attempt to move closer to her.

There were two teenagers, a boy and a girl, Lucy hadn't seen before. They both appeared to be about nineteen.

"*Wilcze*," Grammalie said, and nodded as if she was pleased she was still around. "This is Connor and Scout."

Connor was tall and rangy with red hair and a very direct gaze. Scout was tiny and had a pixie cut and worried brown eyes. They exchanged awkward greetings.

"We were out hunting," Connor told the group. "Didn't get back until this morning."

"How many this time?" Scout asked, wringing her fingers together. Connor grabbed her hand and enfolded it in his own.

"Five," Henry said. "Two more kids, Lottie and Patrick, and Hank—you know, walrus mustache, who helps out in the kitchen? And . . ." He took a deep breath. "Del and Leo."

Scout groaned.

"We're trying to decide if we should keep everyone together or disperse," Aidan told her.

"We cannot decide anything until we have a camp meeting," Grammalie Rose said.

"Fine. We're *discussing* it, then," Aidan rejoined. Lucy was surprised at the anger in his voice.

Grammalie Rose shrugged. "We all need to calm down first."

Aidan nodded curtly. He stalked a few paces away and stood with his back to them. Lucy understood how he felt. She wanted to be doing something. Right now she felt like she was just waiting for the next awful thing to happen.

"There are shelters farther out, near the bridges, that are harder for vehicles to reach. There are hiding places. Bombed-out buildings. We can lie low for a while," Henry urged. "I spent last night in one of the canals."

That explained the mud encrusting his clothing.

"The Long Wet is just beginning. There is the danger of flash floods," Connor pointed out.

"Well, high ground then," Scout suggested. "The plateau, maybe?"

"No protection," Henry said. "Gales, lightning storms, fire." He checked off the points on the fingers of one hand.

Connor glared at him. It was turning into a shouting match.

"How will people be fed if we are spread to the winds? The little ones? It is mostly little ones now," Grammalie said. "Here we have shelter. Supplies. Water."

"They're just picking us off," Henry said. "We're like sitting ducks!"

Lucy silently cheered him.

Grammalie Rose put up a hand, forestalling any further argument. "We may indeed end up moving, but nothing can happen until everyone is here to decide what is best. Sammy and Beth are still out foraging. They have a voice in this as well."

Her brows drew across her forehead, giving her sharp eyes a hooded appearance. Lucy thought she looked older suddenly. She noticed how bowed the woman's back was, and how swollen the knuckles on her work-reddened hands. Suddenly she wanted to offer Grammalie Rose a chair, but there were none.

"In the meantime," she continued, "there is food to gather, washing to be done, injuries to tend." She looked pointedly at Aidan's forehead and Henry's eye.

Henry muttered something unintelligible.

"We need water," Grammalie Rose said. "That is paramount."

Aidan whirled around.

"We'll get that," he said, scooping up a few large plastic jugs and thrusting them at Lucy. She took them, too surprised to

say anything. He grabbed four more and steered her out of the square. She pulled her arm away and stood still. She hated how he just grabbed her and started moving. Like she was a kid who couldn't cross the road by herself. He stopped, surprised.

"Listen, I've been working my butt off ever since I got here," she said. "Maybe I don't want to haul water!"

"I wanted to talk to you alone. I've been thinking about stuff," Aidan said.

"Okay, so talk."

"If Grammalie Rose sees us just sitting around, she'll put us both on latrine duty, and you don't want that, believe me."

Lucy had already caught the earthy odor coming from the row of narrow tents on the west end of the camp.

"Okay," she said slowly. It would actually feel good to be moving. "Listen, all this talking stuff is kind of irritating, though. Why can't someone just make a decision and then we act on it?"

"You mean like someone in charge?" He shook his head. "That's not how it is here. Sure, Grammalie Rose is kind of the boss, because she's the oldest and she's had experience living in a commune. And Leo—" Aidan's voice hitched. "He was a natural leader, but everyone is equally important here. That's the point."

"But doesn't it drive you crazy? I mean, how does anything ever happen quickly?"

"It doesn't."

At least he sounded as frustrated as she felt. Lucy began to feel a little thrill of excitement.

They started walking again. Aidan took a narrow path between two rows of old brick houses. The second floors were still mostly intact, but the foundations were crumbling and the roofs leaned together like two people about to kiss. Wooden scaffolding pressed up against the masonry on both sides, keeping it all standing, but Lucy couldn't help but be glad when they came out on a demolished area filled with rubble. Some kids were playing kick-the-can in the dust. They hollered when they saw Aidan and he waved back. A green garden hose coiled on the ground like a big snake. It was attached to a pipe that stuck up out of the debris. Aidan turned on the spigot. Water, rust-colored and full of debris, started to flow in a series of jerks and spurts.

"Why don't we get fresh water from a spring?" Lucy asked.

"I've found a couple of sources out in the woods, but it's a long trek. This is more convenient, at least until the cistern dries up."

The water cleared. He fitted the lip of a jug over the hose end and stood back up.

"So, what's the big plan?" asked Lucy, feeling a shiver race along her spine. Standing around waiting for bad things to happen was worse than actually doing something. "Were you trying to find a way to get onto the island?"

"Are you crazy?" he said, looking thunderstruck. "I'm not James Bond. Where would I even start?" Aidan scratched his head with his free hand, completely missing a leaf that was trapped there. "No, I was just working out ways we could post lookouts, cut off entrances to the camp, prepare for the next raid. And . . ." He seemed uncomfortable. "I was thinking that this isn't really your battle. You can leave. No harm done."

She was instantly furious. It didn't matter that she had been considering the same thing. That was her choice to make, not his.

"Don't be stupid," Lucy said. "I was there when they grabbed Leo and Del." She choked on the words, she was so mad.

She met his gaze. "I hid." He dropped his eyes and stared at the gushing water. "They could easily have taken me, too," Lucy said.

He looked flabbergasted.

She was pretty surprised herself. Saying it out loud made it all clearer in her head. She could try to make things right, and she could always leave afterward.

"I get the whole 'camp is the safest place' thing," she said. "And it is for the little kids, but there are the rest of us."

Aidan nodded.

"I was thinking rescue mission," she continued, trying to sound convincing.

"Some day, sure; but we know nothing about how things are set up, the layout of the place. We don't even know what

kind of weapons they have. Or how many of them there are. Believe me, I want to get Del and Leo and the others back worse than anybody, but if we go charging in there, we'll get caught."

Lucy stared at him. This talk didn't seem like him. It wasn't what someone who climbed to the tops of trees should be saying.

He must have read the surprise in her eyes. "What?"

"You said no one has ever come back. We don't know what's happening to them. So what are you waiting for? An invitation?" She bit her lip. This was one of her worst faults, speaking without thinking, but she couldn't help it. The words just kept bubbling out of her.

He looked as if she had slapped him in the face. She tried a gentler approach.

"You're so calm about it. The Sweepers took your friends, and it's like you've given up without even trying."

"I've been thinking about this all night!" Aidan yelled. "There's no easy solution. You think me and maybe two others can just barge in there like ninjas?!"

They glared at each other. The water overflowed the container, but he made no move to replace it with another. He paced back and forth, savagely kicking chunks of brick out of his way.

"You don't know. It's not the first time," he said in a calmer voice. "Leo was the fighter, the planner, and even *he*

couldn't think of a way to get the others back. With him gone . . ." Aidan shrugged his shoulders. "We're just not that organized."

"What about all of this?" she said, pointing to the shelters, the cultivated fields in the near distance.

He scowled. "It took us about six months to get all the shelters up, the water situation figured out, and the vegetable gardens planted. It was mostly Grammalie Rose's doing. And it takes the best part of every day to keep it going."

She understood that. It had taken all her energy to keep herself fed, warm, and dry.

"Yeah, but we can't just do nothing," she said.

"You're a lot like Del," he said. "Hot-tempered. What's the nice word? Impetuous." His mouth twisted into a wry grin.

Lucy wasn't too pleased to be compared to Del. She'd caught the admiring note in his voice. She pressed her lips firmly together. He was looking off into the distance— probably thinking of *her*. Lucy lifted the hose and stuck the end into another jug. The water was flowing more freely now and was less brown. She resisted the temptation to spray him.

"You said Leo was the fighter. We might not be able to throw a right hook or a roundhouse kick, but we could arm ourselves and go after them," she said. Aidan jumped a little, as if he'd forgotten she was there.

"You saw the kids in the square?"

She nodded. "The ones collecting projectiles?"

"They were so upset about Lottie and Patrick and the rest of them that they begged for something to do. I couldn't think of anything else. We've got rocks and cans, but they've got stun guns and chloroform gas and masks. The best thing for us to do is to keep watch and to hide."

"Watching and hiding didn't do much good last night."

"We weren't expecting it."

Lucy felt a surge of frustration. "We'll never be able to expect it. They've got the element of surprise on their side. They'll just pick us off."

He looked so miserable; she felt awful, but she couldn't stop talking. "We don't even know what they're doing to them." She was remembering the sad lady and the children from the shelter. The vans coming and taking them away. This had been going on for months.

Aidan shook his head. "We have to think about everyone in the camp. How would it be if most of the adults were gone on some disastrous rescue mission? There are kids here who can't make it on their own."

"Grammalie Rose," she said feebly.

"She's eighty, did you realize?" Aidan shook his head.

Lucy kicked out at a rock as big as a bowling ball. Her toes stung but she ignored the pain. Aidan bent down and placed another bottle under the tap.

She stared at his thick blond hair with the leaf caught there. "I thought she was your friend." It was maybe a low

blow, but she couldn't stop the words exploding from her mouth. Part of her wanted to hear what he would say. Part of her didn't.

He looked up. His green eyes flashed with anger.

"She is!"

"Well, you don't act like it! If she were my friend, I'd be out there looking for her." She recalled his comment to her back in the tree, about being a mouse in her safe hollow. "I wouldn't be skulking around here."

"You don't know me!" he yelled, leaping to his feet. "You drop in and you think you even know what's going on? Well, you don't. You don't know me, you don't know us, you don't know how we live!" He was so angry now, he was practically spitting.

Lucy felt the rage well up in her belly. Who did he think he was? "I can see that you're all hiding instead of fighting," she threw out. Her hands were shaking.

"You have nothing at stake. These aren't your friends. This isn't your family."

"I'm here right now."

"For how long?"

She froze. Aidan was right. She hadn't exactly decided whether to stay or go yet. She opened her mouth and then closed it again.

He was only a few inches in front of her now. The long muscles in his arms flexed. A vein pulsed in the column of

his neck. She could smell him. Sweat and lemons. He looked like he was going to hit her. She tensed, and then he stepped away. He filled the last two containers in silence, kneeling down and keeping his back to her.

Lucy felt all the annoyance and anger drain away, and it was as if they took her energy with it. All of a sudden she was bone tired.

Aidan was moving. He carried two jugs in each hand, leaving three for her. He was already halfway across the lot, calling to the can-kicking kids to come with him, before she'd even picked hers up. The weight made her wrists ache. She staggered after him, throwing mental daggers into his broad back.

CHAPTER TWELVE
THE RETURN

Two days of hauling and digging helped keep the fear at bay, but the frustration was eating her alive. Lucy had agreed to every duty Grammalie gave her as long as the job took her away from people and the camp. And as far away as possible from Aidan.

She carried in the last load of cabbages and dumped the tub on the floor by the long table. She stretched, easing

the cramped muscles along her spine. She'd been stooped over for so long, her back protested. She stifled a yelp.

Henry looked up from peeling onions. His blackened eye had turned shades of purple and yellow, and tears leaked from between the swollen eyelids. He wiped his nose on the sleeve of his shirt. He smiled at her.

"Thanks. Haven't seen much of you lately."

Lucy jumped up on the edge of the table and swung her legs, noting that her injured ankle had swelled again. "Been out in the fields mostly." She picked gravel out of her knee.

He nodded as if he understood. "Normally I'd do anything before peeling onions, but it feels good to keep my hands busy."

They were both quiet for a few seconds. Slowly, shockingly, things had started to feel like normal again. The busy pace of days spent worrying about food, water, tending fires, and rebuilding shelters didn't leave much room for anything else. Still, Lucy felt guilty every time she thought of Leo and Del.

"Need any help?" she asked.

"You can go check the soup. We're eating at the big tables since the rain's holding off. Grammalie Rose thought we needed a group night."

She nodded and wandered outside. The lanterns were lit, casting wavering shadows and filling the air with stinky smoke. The sun sank down behind the Great Hill in a flush of crimson clouds.

In the square, the awnings were rolled back and tied with lengths of rope. Long tables and benches had been moved out into the open and arranged around the fire pit. It was piled high with roughly chopped wood salvaged from the Sweeper-leveled houses all around, and more fuel lay stacked nearby. The flames roared. The big black cooking pot hung above it in a welded cradle. Even from where she stood, Lucy could smell the mouthwatering combination of root vegetables and browning meat. From the opposite corner of the square she caught sight of Aidan hauling more water. She paused, waiting to see if he would notice her. He responded to a few greetings. His eyes seemed to wander in her direction, but she couldn't be sure. Lucy watched as he crossed over to the pot and set his jugs down. He emptied one in, then another. Steam rose in great billows, his forehead shone with sweat. She moved toward him, a conciliatory smile pasted on her face, but then, before she'd made it to his side, Aidan was gone, crossing quickly and disappearing into one of the side alleys that ran like a maze around the dilapidated houses. She stared after him. He was avoiding her. If she were able to catch up to him, she was pretty sure she'd have punched him in the nose or at least kicked him really hard in the shin. She should have known he was a brooder. They never knew how to fight. Instead of yelling, they kept it all inside. She'd been all ready to continue their discussion from the other day, but he obviously couldn't handle anyone disagreeing with him,

and that's why he was acting like a baby. She felt mad all over again.

Lucy checked the soup. Rabbit. Or was it cat? She couldn't tell. It seemed too thick to her. Aidan hadn't even added enough liquid, she thought, annoyed. She bent to pick up a water container. Her arms, sore from carrying pounds and pounds of vegetables, complained.

Henry hurried over to give her a hand. He reminded her of a puppy. If he'd had a tail, it would have been wagging. She found a smile for him. He grinned back. Between them they poured the rest of the water into the pot. Henry stirred it all together, pushed the lid over the top leaving a gap for steam to escape through, and raked up the embers with the toe of his boot. Almost immediately the water began to simmer.

The heat was strong enough to crisp her eyelashes. Lucy stepped back, looking up at the sky. For the first time in a while, the night was clear and cloudless. She sniffed for the scent of rain but smelled nothing except for the aroma of meat, vegetables, and broth. Henry winked at her.

"Soup's on!" Henry said. "Give it five minutes."

From somewhere nearby came the sound of a pot being pounded with a wooden spoon, and at once people appeared in the alleyways. They hovered in groups of three or four, as if they were family units: one or two small kids and then the older people, the DAs.

Shadows danced across the square. Henry steered her toward the food prep tent, where teetering rows of stacked bowls, platters of the dense, chewy bread, and an assortment of mismatched spoons were laid out. There were saucers of the green olive oil and roughly chopped tomatoes scattered with shredded green leaves. Lucy smelled the pungent scent of basil and her mouth started watering. Yet another thing she'd missed without knowing it—herbs.

Henry grabbed a bowl and a spoon and handed it to her. He ladled tomatoes into her bowl, then filled his own. He shoved a loaf of bread into the front pocket of his sweatshirt, along with a bottle of water. "Can you snag some of that oil?" he asked. She juggled her bowl and picked up a saucer, trying hard not to spill it. They made their way back out past the line of people and sat at a table directly across from the fire. Lucy eased herself down with a sigh. The bench was hard and splintery, but it was the first time all day that she was off her feet. She could feel her toes tingling inside her boots. Her sore ankle throbbed. She closed her eyes and took a deep breath. The fire was like a wall of heat against her face. Slowly the taut muscles in her neck loosened and she opened her eyes. The lanterns were mostly behind her and cast bobbing shadows that made the ground appear to tilt and the dark silhouettes of the shacks on the hill seem to vibrate. It softened edges and concealed the mounds of

rubble. The moon was waning, and the sky was studded with stars. So many stars, she thought.

"Ready for some soup?" Henry asked, distracting her. His bowl was empty already. She drizzled oil on her tomatoes and stuffed them into her mouth. The flavor burst against her tongue. She mopped up the last of the oil with a chunk of bread, then held up her bowl.

The benches beside them were all occupied now. Sue was there. Lucy was glad the girl hadn't been snatched. She couldn't be any older than eleven, but she was mothering a handful of grimy urchins, breaking their bread into small pieces and picking the basil off their tomatoes. Lucy looked around for Grammalie Rose and found her a couple of tables down, warming her swollen fingers before the fire. Lucy checked and saw that she had food. Aidan was nowhere to be seen. She pictured him glowering in the shadows somewhere, breaking sticks or punching walls, or something equally useless. When Henry came back holding a bowl in each hand, she gave him her biggest smile. He stepped backward as if he had lost his balance. Soup sloshed over the edge of the dish.

He sat down, carefully setting her meal in front of her. "Hey," he said, grinning like an idiot.

"Hey," Lucy said, bending her head over her bowl. The soup was thick with chunks of potatoes, onions, and carrots. Droplets of oil swirled on the top with a scattering of fresh

herbs. A few skinny rabbits between thirty people didn't go far, but she figured she had at least a couple of pieces floating in there somewhere. The whole time she was eating, she felt Henry's eyes on her. It made her feel uncomfortable and embarrassed, but good, too, in a way.

She smoothed her hair back, tugging her fingers through the tangles. It fell in wild ringlets around her face. She gave up trying to push it behind her ears, letting it flop. Lucy leaned back in her seat. Her stomach was full. She toyed with another piece of bread, letting the chatter flow over her. The kids were half-in, half-out of their chairs now, wrestling and tagging one another, overtired but full of energy. She was reminded of her brother, Rob, who'd get so wound up at the end of the day, shortly before he crashed. She watched the children run in circles, weaving among the tables, dodging people carrying bowls of soup, ignoring the hands that reached out to grab them or give them a warning shake. And then they were playing in the narrow passages, a complicated game that combined hide-and-seek with tag. One of the boys was blindfolded; the others hooted and catcalled, tossing pebbles around his feet.

The frightening events of the past few days were fading for them already. A couple of the really little kids nestled together, clutching blankets and sucking on their thumbs, already half-asleep.

She figured about a third of the camp dwellers were under the age of thirteen. The rest of them were split equally between young adults and senior citizens. It was weird to be in an environment where she was one of the adults, where her opinion might actually count for something. Of course, she thought irritably, it already sounded like no one would be taking her side.

Lucy watched as a large group of people joined Grammalie Rose's table. She wondered if after all there would be some kind of a meeting this evening. For the most part the people at the table were older, stooped, and gray-haired, but without the fierce strength that radiated through Grammalie Rose. Three of the figures were hooded in thick cloaks. Lucy couldn't stop her nerves from jumping. She caught the flash of Sammy's white mask. The others were masked, too. One was like a cat with huge tufted ears, the other painted blue with silver flourishes. And then she saw that Aidan was there with them, in his bright red sweatshirt, his sun-streaked hair catching the light. He leaned in and slung his arm around Sammy's shoulders. *His brother*, Lucy remembered.

She watched Aidan despite herself, noticing his quick smile, the light rumble of his voice, his graceful movements. She wondered if he was going to come over or if he was still mad. She was sorry, suddenly, that they had argued. But that

didn't mean she was going to walk across the square to him. Lucy tore her piece of bread into little scraps and tossed them onto the table. She turned on the bench so she was looking in the opposite direction.

"Your eyes are the same shade as a stormy sky," Henry said, breaking into her thoughts. He leaned his chin on his steepled hands.

She was just able to keep herself from rolling those eyes. "My sister used to say they were the color of dirty window-panes," she said, trying to laugh it off.

"Oh no," he said, "they're exactly like—"

Lucy interrupted and changed the subject. "So what's going on tomorrow?"

Henry blinked. He looked a little bit like a frog with his big, round eyes. She fought a giggle.

"Tomorrow?" he said.

"Yeah. I was thinking I'd like to go out to the woods and the plateau. Maybe learn how to use a slingshot. Can you show me that?"

He gulped. "Normally we have a rotating schedule every week and people are assigned different chores. So it's the fields one week, and the next, rebuilding or hunting. With everything that's happened lately, we've sort of lost track of who's doing what."

"Great," she said. She grinned at him. "So what do you think?"

"Doesn't Grammalie Rose usually tell you what she needs you to do each morning?" he asked.

Lucy scowled. It was like she'd been drafted.

Henry hurried to say, "But you can already handle a knife, so maybe Aidan will give you some weapons training."

She sat up. "I thought he wasn't much of a fighter?"

"Since Leo and Del are gone, he's the best we've got. He's pretty good with a bow and arrow, and a slingshot. He's on hunting duty more than anyone. You'll have to ask him, though."

She frowned again. Aidan probably didn't want to hear anything she had to say. He was still over there with the S'ans, and he hadn't looked in her direction once.

"Listen, are you mad about something or just hungry?" Henry asked. "Do you want more soup?"

She forced herself to smile at him. "No, I'm fine. Just tired."

Henry got to his feet. He stacked their dishes.

"Okay. I've got to organize the dishwashers, but that'll only take about ten minutes. You'll be here?"

"Sure."

Lucy stretched out her legs and wiggled her toes. Then she leaned forward and cradled her head on her folded arms. The fire smoke tickled her eyeballs. She felt a huge yawn coming.

"Tired, *wilcze?*" Grammalie Rose said in her rough voice. The old lady sat down with a creak and a sigh. "Thank you for all your hard work these past few days."

Lucy straightened her back. "Sure," she said, surprised.

"I see you have befriended Henry, our resident lothario," Grammalie Rose continued. "Has he told you how beautiful you are yet?"

Lucy coughed. "Not exactly."

"He will. He's an eternal optimist."

The old woman beckoned to Connor and Scout, who were walking by with linked hands. "They were responsible for our rabbits today," she informed Lucy. "Did you two have to go out far?"

"A few miles out on the plateau," Connor replied.

Scout frowned. "It took hours. They were really skittish."

"Any trouble?" the old woman asked.

Connor shook his head.

Lucy couldn't help but notice how their fingers clasped and unclasped but never let go, and how they leaned together, as if an invisible string were pulling on them. They walked on, Connor's head bent to hear something that Scout whispered to him. The back of his neck glowed bright red.

"So will there be some kind of meeting tonight?" Lucy asked.

"Not tonight," Grammalie Rose said. She pulled a box from a pocket and opened it. Inside were six or seven of the brown cigarettes and a crumpled book of matches. She lit one, blowing the smoke into the air in a long stream. "Tempers are still too hot tonight." She picked a dried leaf of tobacco from her

lip. "Sammy wanted to storm the hospital." She uttered one of her dry laughs. "He is as foolhardy as his brother."

Lucy was surprised the— Sammy had been thinking along the same lines as she had.

"Aidan doesn't want to go. He wants us to hide here," she said.

"Really? Perhaps he has finally learned to be cautious." Grammalie Rose squashed her cigarette on the sole of her clogs and put the butt into the box, which disappeared again into a pocket. She turned to look at Lucy. "You think Sammy is right, eh?" She patted her on the shoulder and got heavily to her feet. "You wear your emotions on your face, *wilcze*. I understand what you are feeling, but it will help no one if more of us are captured. We need time to plan."

She moved away.

Lucy looked around. The two little kids who'd been at the end of her bench were gone. She imagined them bundled in their blankets under tent cover, a tumble of bodies like drowsy puppies.

Others had pushed their benches closer to the fire pit. From behind her she heard the clatter of dishes and the chime of silverware. Water sloshed into tubs and people talked in low voices. Teams of four and six picked up the long tables and moved them back under the awnings. From the group by the fire she heard the strumming of a guitar, the chords spilling out in a stream of formless music. Someone clapped their

hands, keeping time, and others stamped their feet against the tarmac as the guitar wove around the simple beat. And then a violin came in, a single, sustained note that seemed to climb into the air and hang there, anchoring the guitar. Lucy had never had much time for her parents' classical music. She'd thought it cold and clean and rigid, much like her parents and their friends, and she'd always thought that violins sounded like cats being sawed in half. But this was different. Lucy felt the melody in her chest, as if her heart would explode with fullness. It was the saddest, happiest, wildest, and most human sound she'd ever heard, as if all the yearning in the world had been bottled up and then released in a pure shot of energy. She held her breath, suddenly afraid that she was about to burst into tears.

And then the guitar switched tempo to a folkie reel, speeding up and playing a rippling series of notes wrapped around a repetitive verse and chorus, and the player's hand slapped the body of the guitar at the end of each sequence, speeding up the momentum. The violin came in again and wove around the tune so it seemed as though the two instruments were chasing each other like a dog after a cat. And everyone was clapping in time and stamping their feet.

The younger kids ran around the fire, lit up like little savages. Soon others were up out of their seats and linking hands and dancing. A conga line wound between the benches. It was

so corny, Lucy could only squirm. Sue swept by with her pig-tails bouncing, followed by a dozen people whirling in circles. Connor and Scout stood wrapped around each other, barely moving. Kids she hadn't seen before danced together in groups or couples. Grammalie Rose, her unmistakable hawklike pro-file turned toward Lucy, sat near the fire, nodding her head and tapping her toes.

Lucy was wondering how long the musicians could keep playing when someone tapped her on the shoulder. Her stom-ach flipped.

Oh no! She turned, expecting to see Henry's eager face. It was Aidan.

"Truce?" he said, holding out his hand.

"Sure," she said, shaking it. He didn't let go. His fingers tightened their grip on her own. He pulled her to her feet. She looked up into his face. His green eyes glinted.

"You can't just sit there like a miserable lump."

"I'm not miserable. I was thinking."

"Well, think later." Aidan drew her toward him.

"Oh no, you're kidding!" She dug her heels in.

"Come on. Come with me!"

"I can't dance. I failed dancing in ninth grade. My partner couldn't walk for two weeks afterward."

"I think I can survive it."

"I practically hamstrung the poor guy."

"This isn't really dancing. This is just moving around with another person. You can pretend we're sparring. I'm wearing my motorcycle boots," he added, pointing to his feet.

Lucy hesitated. She could tell that her face was red, but she hoped it was dark enough to disguise the fact.

"Maybe I'll let you take a swing at me later," Aidan said.

She relaxed and let him pull her into the crowd.

The guitarist was playing even faster, a galloping tune, a wild jumble of chords, and the violin soared above it, a high, sweet note. Aidan took both her hands in his and whirled her around, swinging her until it seemed her feet left the ground. Then he brought her closer, one hand clasping her own and the other around her waist. She put her hand on his shoulder, lightly, but she could feel the heat of his body, and they were moving together in a line, up one side of the fire and down the other, and her feet stumbled, but it didn't matter because he was holding her up. Lucy stared at the neck of his sweatshirt, too shy to raise her eyes any higher, conscious of the tickle of her hair against the wet nape of her neck and the sweat sticking her T-shirt to her back, and the drum of his heart. She was out of breath and she couldn't stop laughing.

They whirled and turned, and people's faces came out of the shadows, lit by flickering firelight and tinted by red flames. She caught glimpses as she spun by. The masked S'ans at their table painted a surreal picture, like a photograph of a

Venetian carnival Lucy remembered seeing once long ago. The kids were hysterical, exaggerated in their every movement, heads thrown back, bursting with the giggles. Lucy closed her eyes, feeling giddy. Aidan bent his head to her ear. She felt his warm breath against her cheek. "Lucy," he murmured, "you are so—"

The music stilled. It was abrupt and jolting; the last upstroke of bow on violin sounded harsh and grating. Aidan stopped moving. His hands let go. Lucy stood trying to catch her breath, scraping back the curls clinging to her sweaty face, unsteady on her feet now that the earth had stopped spinning.

From the direction of the road, a figure appeared out of the shadows.

Her face fell into the narrow shaft of light thrown by a lantern.

Lucy recognized the sleek black hair and the silver bangles on tanned arms.

It was Del.

CHAPTER THIRTEEN
BUNNY HUNTING

The tough stalks of grass tickled Lucy's chin. She shifted, earning herself a glare from Aidan. He put a finger to his lips. She scowled back. *I get it! Be silent!* But they'd been lying there on the ridge for over an hour watching the clearing, and nothing had moved in all that time. Her neck was cramped from holding her head at an awkward angle, she had to pee, and the ground was hard and still damp from the morning

rain. Plus, she was lying on her knife and it dug into her hip bone.

The sun beat down. Del lay between them, head lowered, her long black hair tied back in a thick ponytail. She'd taken off the silver bangles she normally wore five inches deep on both arms. No jangling allowed. Lucy studied Del under the cover of her eyelashes. She looked like she was brooding. Even now, three days after she had appeared out of the darkness, she still seemed shaken up and not really present. She didn't say much about what had happened, just a few words at the supper meeting called for the day after the dance. She told them that she'd managed to break out of the waiting room the Sweepers had put her in, but she hadn't been able to provide much detail. The room had been white-walled and stark, and the maze of corridors leading to it were dark and lit only by bare bulbs. A long spiraling stairway rose up the middle of the tower. Del had somehow found a door to the outside, and then, after hours of stumbling around in the pitch-black, she'd been able to orient herself and make her way off the island. She didn't have any idea what had happened to Leo and the others. She'd been separated from them early on, but she did know that Leo had still been unconscious. The Sweepers had had to drag him from the van. At this point she'd started sobbing, and Grammalie Rose had ended the discussion, folded Del in her arms, and taken her from the square.

Aidan had even tried to ask her privately, hoping she could offer some details about the tower, the hospital, ways on and off the island, but she just shook her head and pressed her lips together. "It was dark. I was scared," she'd said, rubbing at her wrists. And the next day, when a bunch of them had been put to work in the pouring rain, shoring up the dikes along the canals with bags full of rubble and old masonry, Del had been even quieter. She'd responded to Aidan's questions and to Henry's flirtation with silence and the smallest of smiles. Only when she'd glanced at Lucy did something flit across her face. It had almost looked like fear.

Lucy had taken her cues from Aidan, and he was definitely concerned. He could barely take his eyes off Del. Lucy thought back to how he'd dropped her hand during the dance and stepped away abruptly. How stricken his face had been, as if he'd woken up and discovered that the girl he'd been dreaming about was not the girl he was with. She got the sense that there was a history between Aidan and Del, but she couldn't figure out if it went beyond friendship.

"The dynamic duo," Henry called them. "Inseparable." Lucy hadn't been completely successful in squashing a twinge of jealousy. She wondered again what Aidan had been about to say to her during the dance. "Lucy, you are so—" he'd whispered.

So what? So strange-looking? So awkward? So annoying? Or maybe, so amazing? It was conceivable, but it didn't seem likely.

And Del hadn't helped matters, either. She'd been distracted, even less friendly than usual (if that was possible), and full of mean looks. Like early this morning, when they were out picking tomatoes and snap peas for their lunch she'd barely said a word, but she seemed annoyed that they were all together again. Lucy had decided to ignore her and focus on her first hunting experience.

Now something stirred in the large thorny bush in front of her. She raised her bow. Aidan's bow, actually. A crescent of smooth red oak rubbed with olive oil until it shone. Aidan put up his hand.

Bird, he mouthed, and motioned downward. That's right. She wasn't to draw on a bird. They were too hard to hit, many of them were sickly, and they couldn't afford to lose arrows. Same went for deer if they were lucky enough to spot one. Just the thought of a whole deer made Lucy's mouth water. She cradled her chin on her arm and watched a column of ants carry tiny white eggs from one hole to another. She stifled a yawn.

When they had first hiked up to the ridge, Aidan had sat beside her on the grass. He'd strung the bow for her, shown her how to check the straightness of her arrows. She had eight slender lengths of springy ash, the tips needle-sharp and hardened in fire, the fletching cut from pliable plastic containers. Lucy had enjoyed Aidan's closeness, liked watching his deft brown fingers and the slight frown that ruffled his

forehead as he explained something to her. There was a lot of physics involved in shooting a bow, apparently.

"So is the wind coming from a good direction?" she'd asked.

Aidan had nodded. "From behind, so we won't be shooting into cross drafts."

Del had snickered. Her hand had swooped down and plucked the arrow from Lucy's grasp. She'd turned it over in her hand, hefting the weight.

"You aim, you point, you fire," she'd said, sitting down between them. There was no room, but she had squeezed in, anyway. She'd flashed Lucy a triumphant look and Lucy had moved over to the right. She wasn't in competition for Aidan. She didn't exactly know how she felt about him. Del had made it pretty obvious what her feelings were, and since then, Lucy had been almost hyperaware of the girl.

Lucy couldn't help noticing that Del's body was right up against Aidan's now. Del's thigh pressed along the length of his leg, her tanned arm inches from his own. She had a purple bruise along one cheekbone and stripes of raw flesh where the Sweepers had fastened her wrists with plastic cuffs, but she was otherwise unhurt. Lucy wondered how she'd gotten away from the Sweepers when no one else ever had. She opened her mouth to ask the question, but stopped when Del leaned forward to whisper something in Aidan's ear. The tip of her ponytail swept across his face and he reached out and

moved it away, his fingers tangled in the shiny black locks. Del laughed, casting a glance over her shoulder at Lucy. Lucy looked in the other direction.

She went over the instructions for shooting an arrow in her head, trying to remember everything Aidan had told and shown her. Months of training were squeezed into a few short hours in between the other work that needed to be done. At first she hadn't been able to hit anything. Just holding the string back without letting her hands shake was harder than it looked. And her fingers always seemed to be in the way when she released, sending her arrow into a wobbly, crooked trajectory that, nine times out of ten, landed it in a bush or the dirt. She'd also had to pretend she felt nothing when Aidan guided her arm or stood behind her with his hands on her fingers and his chest leaning into her back, while Del watched his every move with a glacial stare on her face. Lucy had bitten the inside of her cheek so hard that it bled, but she'd finally made a shot. At least, her arrow had struck the tree the target hung on and had quivered there for a few seconds before falling to the ground with a plop.

"Great!" Aidan had said. "How did it feel?"

"Good." She'd lowered her eyes. Not as good as having his arms around her shoulders, but she thought her fingers were getting used to the cramping grip on the arrow and the pressure of the bowstring and the quick, fluid motion she needed to master to send the arrow off on a straight path.

Now she raised herself on her elbows, positioned an arrow against the bow, and tracked across the glade from left to right, ignoring the soreness in her muscles. Her fingers sweated in the stiff leather glove she had to wear to keep the thin nylon cord from flaying her skin. She had her jacket on to protect the inside of her arm, and the leather was uncomfortably hot in the sun. But at least it wasn't raining. What did they used to call it before the climate went all haywire? *An Indian summer.* They were getting an odd respite from the usual constant heavy storms of the Long Wet. Unfortunately, it seemed as if every flying insect in the world had decided to take advantage of the weather, and they were mating up a storm. Midges, blackflies, and mosquitoes hovered in black clouds. Lucy's legs were clad in cutoff jean shorts, and she'd already counted fifteen bites in rings around her ankles. She squirmed, trying to rub the itches against the stubbly grass, and Del kicked her, then pointed.

Something moved on the sunny slope just beyond the shadows thrown by the spindly trees. It was buff-colored and small. Its pointed head came up and Lucy saw the long ears lying flat against its body. Del looked hard at Lucy with her blazing blue eyes, made sure that she had seen the rabbit, too, and then mimed the action of loosing an arrow. She touched her index finger to a spot just below her shoulder blade to remind Lucy where to aim. The arrow would travel directly to the heart. Death would be quick.

Suddenly, Lucy's fingers felt thick and inflexible. The arrow shaft was slippery and weighted wrong, and she couldn't focus her vision. She pulled the bowstring back. The rabbit's head came up again. It stopped in the middle of chewing a mouthful of grass. Wisps hung from its mouth like a straggly green beard. Lucy felt the scrape of the plastic fletching against her cheek. Her fingers were numb and sweat dribbled into her eyes. She couldn't let go.

Del exhaled, raised her own bow, took aim, and shot. The arrow thrummed, flying straight and true, and hit the rabbit with a force that spun the animal into the air. She was up on her feet and racing toward it before Lucy had lowered her bow. She stared at the ground. She'd caught rabbits, squirrels, and woodchucks before, but in snares. Traps tripped while she wasn't there. This was different, and it was nothing like aiming at a piece of wood.

Aidan touched her arm. "Hey, I puked the first time," he said in a low voice. "Del's always been better at killing things than I am."

Lucy felt her mouth twist. "It just wasn't the same."

"I know. You can try to imagine that it's a tin can or whatever, but it never works. All I can say is try to do it fast and try to do it right."

Del stalked back to them. The bunny swung from her gloved hand. Lucy looked away from the limp head, the eyes like foggy blackberries. A small red hole bloomed on its back.

Del wiped her arrow against a patch of grass and stowed it with the rest in the quiver she wore slung from her shoulders. She hunched down next to Aidan. She danced her fingers up and down his arm.

"If we're lucky," she said, "the other rabbits won't be alerted and we can get a few more. They come out in force just before sunset."

Lucy squinted up at the sky. The sun was behind them now. She rolled over to look at the clouds drifting.

Sure enough, as the sun lowered in the sky, more and more rabbits poked out their quivering noses. They nibbled grass and chased one another, innocent and carefree, reminding Lucy of the camp kids playing kick-the-can.

They're food, she told herself, but it was no good.

She heard the soft whicker as Del notched an arrow.

Lucy watched as Del shot four rabbits in quick succession before finally missing one. Instantly the animal darted to the top of a hillock and drummed the ground with its back foot. The other rabbits vanished into their holes. There was a curious light in the girl's eyes. It wasn't pleasure, but a glint of *something*. Like she was paying the rabbits back for an insult. Lucy was happy that she had finally missed.

"I'll get them," she said, clambering to her feet. She was stiff. The rabbits were hard to find in the long grass. Their soft, brown bodies splayed in awkward positions. They were smaller than she expected. Lucy picked them up, holding

them by their velvet ears, feeling the uncomfortable heavy, boneless quality about them. They were still warm, and they flopped like stuffed toys. One remained lost in the undergrowth despite careful searching.

When she walked back to the others with the dead animals, Del exploded. "I shot four. Where's the other one? Do you think it's easy?"

"Obviously not, seeing as how I couldn't do it," Lucy said. Her cheeks burned, but she met Del's eyes. What was *with* this girl? "I looked for the fourth one. I couldn't find it."

Del snorted. She stripped off her glove and flexed her hand. The abrasions on her wrists looked raw. Again Lucy wondered how she'd escaped the plastic handcuffs by herself.

She unslung her bow and thrust it at Lucy. "Hold this."

She strode off, swishing her quiver back and forth across the long grass, ducking below the branches of a tree that swept the ground.

Lucy held the bow between fingers that didn't seem attached to her hand. The rabbits were cold now and their eyes had filmed over. She felt angry and a little sick.

"Give them to me," Aidan said, standing up and stretching. She looked away from the lean length of him and handed them over. He opened the neck of the canvas bag they'd brought and shoved the bodies in.

"Listen," he said. "She doesn't mean anything by it. That's

just Del. She always says what she thinks. She's been through a lot. . . ." His voice trailed off. He looked uncomfortable.

The kind note in his voice set her eyes prickling. She focused on the scuffed toes of her boots.

"Hey," he said softly. His hand reached out to her arm, fell short, and sort of brushed the air between them. She felt it against her skin, anyway. She took a step toward him.

Aidan pinched her chin and raised her face to his. She'd never seen his eyes so close. They were a deep green with specks of gold. She could smell the sun on his clothes. He smiled and leaned in farther. Lucy felt her head swim. She swore she felt a crackle of electricity. He was going to kiss her. They were going to kiss. His lips looked so soft.

"Crap!" Del yelled. Aidan froze, and Lucy stepped backward so quickly, she tripped over her own feet. Del was a few yards away. She swung a bunny from one hand. She was keeping the weight off of her left foot. "I think I turned my ankle in a rabbit hole." She winced, but Lucy couldn't help noting that the grimace was replaced by a smile as soon as Aidan hurried forward. She slung her arm across his shoulder, hobbled over, and handed Lucy the last rabbit and her quiver. Lucy followed behind, carrying the bag, the bows, and the arrows. She saw how Del clung to Aidan, her sleek head tucked against his chest. Her hand lay over his heart. Lucy quickened her pace until she was ahead of them, and then practically ran back to the camp.

CHAPTER FOURTEEN
MOVING MOUNTAINS

Henry was a whistler. Jaunty little tunes, like that one about working on the railroad, and the other about the hole in the bucket, which totally got on Lucy's nerves after the first hour. If she'd known, she probably wouldn't have requested to be on his team, but the alternatives were even worse: work with Connor and Scout, who were welded together so closely, you couldn't get a thin dime between them, or with

Del and Aidan. Aidan had been in a strange, quiet mood ever since yesterday afternoon. He wouldn't even look at her. And Del had been radiating anger, although Lucy noticed that her sprained ankle was miraculously better. They were working at the other end of the field, over where Sammy and the other two S'ans were raking over the soil. Lucy had discovered that their names were Beth and Ralph, and she was finally able to talk to them without shuddering. Visibly, at least. Inside, she still felt a clench of fear, wondering if some day she would wake up with her skin cracked and oozing and the disease rampaging through her body. She had noticed that they always kept themselves apart from the rest of the scavengers, and that made her feel slightly ashamed of herself.

Henry was all right. He reminded her of her brother, Rob — sort of cute and funny, like a cartoon character — but she'd also found out that he had basically one thing on his mind, with a relentlessness that was almost scary. He was so busy flirting that he had slowed his work to a snail's pace. She looked at the huge pile of stones and chunks of blacktop Henry had yet to load into the wheelbarrow and decided to take a break. They'd already filled the barrow four times and wheeled it out to where the big road entered the camp. Each load seemed pathetically small when they dumped it out. Aidan had been right that it would take time to block it completely, she admitted, but at least they were doing *something*. She wiped sweat from her forehead. Her back was aching,

and she was pretty sure she had some major blisters under her leather gloves.

"The S'ans," Lucy said, leaning on her pickax. "I mean, Sammy, Beth, and Ralph . . ."

Henry looked up and threw down his spade. He stretched with both arms over his head and froze for a minute so she could admire his wiry torso as his T-shirt rode up. She suppressed a laugh. He'd taken every opportunity to show off his biceps. She pointed to the heap and then to the wheelbarrow, and, with a huge dramatic sigh, he began shoveling in the rocks.

Before he could start whistling about the chain gang, Lucy continued. "Are they totally healthy now?"

"Yeah. I mean, their bodies fought off the disease. Normally hemorrhagic smallpox kills in about seventy-two hours."

"And will their skin and their eyes go back to normal?"

He paused. "Hmm. Since there are no documented cases of survivors, I don't know. I mean, the burnt look and the bloody eyeballs are due to bleeding under the skin. I guess it makes sense that eventually the wasted cells will be washed away in the bloodstream, cleansed by the kidneys, and then flushed." Henry frowned and rubbed his nose with his glove, leaving a smudge of dirt. Finally, he said, "Seems likely. Who knows what's going on under those masks? The skin has an amazing ability to rebuild cells." His serious expression was replaced by his usual grin. "You know, I can tell that Beth

might be really pretty. I think she's got those melting brown eyes like dark chocolate and a tight—"

Lucy aimed a punch at him, but he jumped backward and held up his hands in surrender. She lowered her fists, but neglected to tell him about the muddy smear across his face.

"So how'd you get on Lady Del's bad side so fast?" Henry said, pushing the heavy wheelbarrow up a few feet. Lucy glanced over to where Del was working next to Aidan at the other end of the field, and threw a chunk of masonry into the barrow. She pulled her sweatshirt hood forward. It was drizzling, and by the look of the black clouds massing overhead, they were in for a real downpour. The Indian summer was over. The weather matched her mood.

"How do you know it's not you she's throwing those mental daggers at? Maybe she knows you ate the last of those wild strawberries we found."

Henry grinned. "Not me. I'm her go-to man." He stomped on the edge of his shovel, pushing the blade into the iron-hard ground and breaking it into manageable chunks.

"Go-to for what?"

"For whatever she needs. She's Lady Del. Questions, answers, other more urgent needs. You know." His expression was smug. "She seems mightily interested in you, as a matter of fact."

Lucy wrinkled her forehead. "Me? I'm no one."

"That could be argued," he said with a wide grin. "Sometimes it's special favors she's looking for."

"Oh."

"I could be your go-to guy, too, if you like." He waggled his eyebrows.

"Umm, that's okay. I'm good." Lucy peeled her leather gloves from her hands and inspected the blisters across her palms. She looked at the small patch of ground they'd managed to clear. Even the youngest kids were helping—sort of: picking up one pebble every ten minutes and chasing one another around the rest of the time. Grammalie Rose was running lines of rope along what would be the furrows. Connor and Scout were wrestling with a clumsy wooden contraption shaped like a giant V, with two long handles and a thick plate of steel bolted to the underside. It was a plow, Lucy had been told, and it looked as if it would take ten of them to drag it through the ground once they'd gotten rid of as many of the stones as they could.

She eased the gloves back on, wincing as the rough material touched her tender skin. She squinted her eyes against the slanting rain.

"We all have roles. Sammy is her shoulder," Henry said as he piled a scant shovelful of rock into the wheelbarrow.

Lucy stopped in mid-swing. "What does that mean?" she asked. "What's with the high school nicknames?"

"When she needs someone to lean on, or to cry on. He's the shoulder. I'm pretty sure he'd like to be more," he said speculatively. "But for now, that's it."

Lucy turned to look at the small group across the field.

Aidan was talking to Sammy and Del now. Del shook her head at something Aidan was saying. Sammy put his hand on his brother's arm. Aidan shrugged it off and walked away. Lucy wondered what had happened. She saw Aidan hop a low wall and disappear into the jagged terrain left by the orchestrated bombing of what looked like at least three apartment blocks.

Del watched him, too, until Grammalie Rose barked at her. Then she turned around and looked right in Lucy's direction. It was like being targeted by a laser beam. Even from this distance — at least twenty yards — she could sense the anger in Del's deep blue eyes. Henry followed Lucy's gaze. He whistled. One low note.

"Ouch," he said. "You been trespassing on her property?"

"What's that supposed to mean?"

"She laid claim to Aidan a long time ago."

"Oh come on!"

"He's her O-O-H-A," he said.

"Enough already!" Lucy said.

"Object of her affection," Henry said hurriedly.

Lucy snorted. "That's so dumb."

Henry raised his eyebrows. "I've seen the way you look at

him. And he's always staring at you in a sly, undercover way. I bet Del has been scoping out the situation. You guys figured it out yet?"

Her cheeks burned.

"Do you really like him?" He watched her curiously, his voice serious all of a sudden.

She thought about it. "Yeah, I guess I do. But most of the time I'm so mad at him, I could spit."

"The path of love never runs smooth," Henry said, throwing out an arm as if he were declaiming poetry.

"No. I mean, we're friends but . . . nothing has happened. . . . No." Lucy stopped as the rest of what he'd said sunk in. She couldn't help the small smile that spread across her face. She bit the inside of her cheek and tried to look unconcerned. "You've seen him look at me?"

"Yeah. He pretends he's all cool and stuff, but . . ."

She turned her back and savored this information for a moment, and then spun on her heel to face him. "Wait a minute. What business is this of Del's? Doesn't he get any say in this? He's not a trophy, and he's not some pet she can put on a leash."

"I never said he was," Henry said with a roguish grin. "He plays it really well."

"Excuse me?"

Henry put down his shovel. "All I'm saying is, *I* wouldn't mind being caught in the middle of you two."

She whacked him on the arm hard enough that she felt the sting in her fingers. He stood there rubbing the spot, but the grin never wavered. She gritted her teeth.

"Don't be stupid!"

He had the grace to look embarrassed. "I'm just kidding. . . ."

"Well, don't." Lucy grabbed the pickax and hefted it, ignoring the worried look on his face. She worked off some of her annoyance by attacking the ground. After a while, she said, "So are they together?"

"Listen. I shouldn't have shot my mouth off. It's none of my business."

His tone was no longer teasing. She met his gaze. No smirk, no mocking light in his eyes. He looked chastened.

"There aren't too many secrets in camp. I mean, everyone knew Connor was gaga for Scout months before he made his move. But I can tell you honestly, I think it's all on Del's side. Why Aidan would pass up a gorgeous girl like that, I don't know, but that's the truth." He stopped talking all of a sudden. Lucy turned. Grammalie Rose stood behind her. Her clothes were covered in dust. The black leather of her clogs was barely discernible under a thick coating of dirt. Lucy wondered again at the strength of the old woman.

"Is this one talking your ear off, *wilcze?*"

"No, we were just chatting," Lucy said, wiping the sweat out of her eyes.

The old woman glanced at the square they'd cleared, the

half-filled wheelbarrow, the pile of rocks Lucy had collected, and Henry's freckled face, dry and unreddened by exertion.

"Too bad you exercise your tongue more than those strong arms of yours," she said, fixing Henry with a baleful stare. "A couple more hours and some of us can break for lunch. Beth and Ralph found puffball mushrooms in the field this morning."

"Some of us?" Henry asked. He kept the shovel in motion, exaggerating his breathing. He jerked his head at the pile of rubble.

"I believe that Lucy is responsible for that," Grammalie Rose said. She turned so only Lucy saw the way her lips twisted in the beginnings of a smile. "However, if you continue at the pace you are setting now, I think you will be one of the luncheon party."

"Why did Aidan get to slack off?" Henry asked, digging with more enthusiasm than he'd shown all morning.

Grammalie shot him a look. "He works harder than anyone else here. He was in the fields at four A.M. when you were still rolled up in your sleeping bag with those magazines you think I don't know about."

Henry mumbled something and turned away. The tips of his ears were an almost fluorescent red.

"Two hours more, I think," she said, resting her calloused hand for a moment on Lucy's arm. "At least the rain is stopping." The old woman walked away in the direction of the

211

camp. As soon as she was out of sight, Henry put down his shovel. "Man," he said, rubbing his hand along his ribs. "I think I pulled a muscle." He turned to Lucy. "Think you could check it out for me?"

Lucy barely acknowledged him. Del and Sammy were shouting at each other across the field. Del tossed her pickax aside and threw up her hands. She pushed Sammy away from her. Her hair, loosened from its ponytail, swirled around her face. And then she turned.

With a start, Lucy realized the girl was heading toward her. Fast.

"Uh-oh," said Henry. "Lady Del's in a fury." He picked up his shovel again and moved closer to Lucy. She was oddly touched.

Lucy's hand stole to her waist. Her fingers found the comfort of her sheathed knife and then fell away. She was hardly going to stab Del, annoying as she might be.

She took a deep breath and stood her ground as Del stormed up. For a moment, neither of them spoke, and then Lucy said, in as mild a tone as she could muster, "Something going on?"

"Hey, Lady Del," Henry said in a determinedly light voice. "How's the digging going? Your boy Aidan bailed out early, huh? What a slacker!"

Del's deep blue eyes flicked over him and then returned to Lucy's face. Lucy forced herself to stay calm, but it was

difficult. She tried to decipher Del's body language. It was as if she was barely keeping control. And she seemed on the edge of tears.

They had an audience now, too. Beth and Ralph leaned on their rakes. She saw Scout and Connor lower the plow to the ground. Only the youngsters and the older folk seemed unaware of what was going on. Lucy chewed the inside of her cheek, remembering the one and only fight she'd had in grade school, when Gracie Foster had accused her of stealing her heart-shaped pencil eraser. They'd ended up rolling in the dust of the playground with a bunch of kids egging them on. And when Lucy went home that afternoon, she'd had bloody scratches up and down her arms, and her scalp hurt. Gracie Foster had been her best friend after that, all the way up until eighth grade, when they'd gone to different schools. She wondered if Gracie was still alive.

Del got right in her face. "It's all your fault," she snapped. Her hands were clenched in fists. Lucy planted her feet. She felt a surge of anger, and it gave her confidence.

"What are you talking about?"

Del snorted. She paced back and forth, her hair streaming over her shoulders. Her silver bracelets jangled, and she pushed them up her arms impatiently.

"Aidan," she said. Her hand rubbed at her nose, and Lucy saw tears in her flashing eyes.

"What? I didn't . . . What?"

"I think he's either gone to the Wilds or he's trying to find Leo. Either way, you're to blame, Lucy Holloway." She practically spat the words out. Her finger came up and jabbed Lucy in the chest. Lucy slapped her hand away. She didn't understand this at all. She'd thought Del was going to give her grief over almost kissing Aidan, but instead it was these riddles.

"You're crazy!"

Henry moved forward to stand between them. They both glared at him and he stepped back.

"He likes to climb trees in the Wilds. It has nothing to do with me. And he said he wouldn't go after anyone who'd been kidnapped by the Sweepers." *Not even you*, she thought to herself. "He said we weren't prepared." Her toe scuffed the dirt. "And he was right. I get that now. We can't just storm in there with no idea of what to expect."

Del frowned and the accusatory finger rose again. "But that's not what you told him, is it, Lucy Holloway? You said that if he really cared about his friends, he would go anyway, right?"

"I didn't tell him to go anywhere. I just said that, if it were me, I wouldn't wait around for the next bad thing to happen. And besides, I was mad and I was just shooting my mouth off."

Del's shoulders slumped. The anger seemed to drain out of her. She shook her head. "Don't you get it? He can't stand that

you think of him that way. Like he's some kind of a coward. God, don't you know anything about boys?!"

Lucy cleared her throat. She couldn't remember what she'd said to Aidan. She'd yelled. He'd yelled. She tended not to watch her mouth when she was angry. "Are you sure that's where he's gone?" Her voice was hoarse. She grabbed Del's arm. Del stared at her fingers, but she didn't brush them off.

After a moment, Del said impatiently, "No, I don't know. I *think* he's headed to the lake. I hope that's where he's gone. It's where he always goes when he needs space. He won't let me go with him, and he'd never say what he was doing." She looked down at her boots. "I followed him once." Her chin came up, as if she was daring Lucy to say something about it.

"He liked to climb to the top of the elm tree and look north," Lucy said, dropping her hand. "He said he wanted to travel up there one day."

"Yeah, that's what he said." Del sighed. "He probably just needed time to think, but he talked about the argument with you, and Leo and the kids, and then I got so mad at you for picking on him that I wasn't really paying attention to what he was saying." She raised her eyes to Lucy's face. They weren't snapping with fire anymore, but there was still something hard in them.

"Aidan can look after himself," Henry said. "He's been running wild since he was thirteen, right?"

215

Del drew a deep breath. She glanced at Lucy again. "Why did you have to come here?" she said. Her tone was strange. It seemed less angry and more tired. She rubbed her hand over her face and eyes, leaving smeary wet trails mixed with dirt. Her mouth twisted. "We're not friends, Lucy Holloway," she said in a voice that was no more than a whisper.

"Hey now, Lady Del," Henry said softly. He slung an arm around her shoulders. For a moment she collapsed into him, and then, with another shake of her head, she pushed him away.

She strode off. Lucy stared after her. She didn't understand Del at all. One minute she was spitting like an angry cat; the next, she was as emotionless as a robot.

"That was intense," Henry said. "Why'd she keep saying your name in that weird way? It was like she was putting a curse on you or something." He laughed nervously and wiggled his fingers in her face. "Voodoo magic."

Lucy shook her head. How could Del hate her so much?

"Maybe she got too much sun?" Henry said. He rubbed his chin with his gloved hand, staring in the direction Del had taken. "I've got to say, though, all that passion she's bottling up sure makes me wonder. I mean, it's got to find an outlet, right?"

"Oh, for God's sakes, Henry," Lucy said, trying not to laugh.

Sammy tapped her on the shoulder. She turned to face him. His hood was pulled forward, and he was wearing a different mask. This one was painted a glossy red with a broad, upturned smiling mouth and little red horns. Holes were cut out for his mouth and eyes. His red-tinged irises gleamed behind it. "He used to watch you and wonder about you. In your camp. He used to say you were the bravest person he knew."

Lucy raised surprised eyes to his face. "Me? Why?"

"Because you were alone and you were surviving," Sammy said. "Because you just did what needed to be done."

"Half the time I didn't know what I was doing."

"Yeah, but you kept on anyway."

She looked at Sammy properly for the first time. He was shorter than Aidan, probably a shade under six feet. His shoulders were broad, and his gloved hands were wide. He seemed a little clumsy, like he'd recently had a growth spurt. Under his hood she glimpsed a shock of the same dirty blond hair his brother had. And then the gruesome contrast of charred skin. It was smooth, though, not cracked and oozing as she had thought before. The surface was whole. It was underneath that great patches of black and red covered his body, like giant bruises. She thought she could see some hazel in his irises within the bloody whites, and the lobes of his ears were pink, like new skin after a bad sunburn. She

thought about what Henry had said and wondered if he was right about the body fixing itself.

He returned her gaze. "You are brave," he said with the hint of a smile. She caught a flash of very white teeth. "Look at you right now. Not worried I'm going to crack your head like a walnut and eat your brains?"

She blushed. "That was so dumb. I just—"

"—believed the propaganda and the news reports." He nodded. "You're not the first." He shrugged. "They just didn't know what to do with us. Didn't know where to put us. It was like we were nonhumans or something just because we got sick. Ralphie still won't talk to anyone but me and Beth."

Lucy was trying to understand her fear. "I think it's because you survived. You're sort of like a living, walking reminder that there was a plague."

He shrugged. "Yeah, well, I can't do anything about that."

"Hey," said Henry. "Hate to break up the tête-à-tête, but I'm hungry, and if you guys don't help load and dump the rest of this, Grammalie Rose is going to personally see to it that I starve."

Lucy and Sammy exchanged a grin.

"Pretty please?" Henry said.

"With sugar on top?" Lucy said.

"I just can't resist a whiner," Sammy said.

Henry tossed a shovel at him. Sammy caught it.

"I'll give you a kiss," Henry said to Lucy, puckering his lips and opening his arms wide.

"Henry!" Grammalie Rose yelled from the camp.

Suddenly, a jumble of shouts and cries rose from the direction of the square. Something had happened. Lucy's heart started pounding.

"Henry!" And this time they all heard the shrill note of panic in Grammalie Rose's voice.

CHAPTER FIFTEEN
PLAGUE

The shrouded lump lying on the mound of bracken leaves and grasses was scarcely recognizable as human. And the sounds that came from it were more like that of a wounded animal. Lucy and Sammy followed Henry into the open space beneath the awnings. Grammalie Rose crouched at the head of the makeshift bed. A few of the youngest kids huddled together in the corner, and Aidan was there, too. He raised

his head, meeting Lucy's eyes briefly. She was shocked at how pale and drawn his face was. He looked much older than seventeen. Henry dropped down to his knees opposite Grammalie and moved the covers aside. Lucy caught a glimpse of charred skin, a gasping mouth.

"Sue is boiling water for willow bark tea. Aidan soaked some sheets in water. He's bad," Grammalie Rose said. She looked at Sammy and beckoned him closer. "Get the children away," she said in a low voice.

Sammy nodded. He bent his head and pulled the white and gold mask from beneath his cape and switched it for the horned mask. Then he clapped his hands. "Strawberry hunt in two minutes!" The kids clustered about his legs, jabbering in excited voices, and he led them out into the square.

"He's burning up," Henry said, laying a hand on the man's forehead. "The willow bark tea won't bring his temperature down fast enough." He looked miserable. "What else do we have?"

"Elder flower, echinacea for the fever; but if the willow isn't doing any good . . ." Grammalie Rose's voice trailed off. "Valerian, black cohosh for the pain: There may be some motherwort left, but I used most of it when Lottie broke her arm. I have a tincture of rosemary for when he is calmer." Her hand brushed against the man's face for a moment. Then she pulled two small glass bottles from her pocket. One was filled with a gritty brown powder. The other glowed yellow-green.

"He'll probably die," said Henry. He bit his lip, as if ashamed of what he had said.

Grammalie Rose soaked a cloth in a pan of water and dabbed it over the man's face. She muttered under her breath. It sounded like a string of curses. Her eyebrows met in an angry frown over her hawklike nose. She glared at Henry.

"Then we will try to make him as comfortable as we can. Yes?"

Henry lowered his head.

Lucy watched the body writhing beneath the thin covers. She could see blackened skin covering a bald skull and spreading in splotches to the face. His eyes were half-open. The eyeballs were tinged an angry red. She couldn't distinguish his pupils at all. It was as if the sockets were filled with blood. He thrashed and threw off the sheets. Two thick gold hoops dangled from the charred earlobes, and under the skin of the muscular forearms she glimpsed swirls and bands of dark blue. Tattoos, she realized with a shudder of recognition, almost covered by the dusky hue of bleeding beneath the skin. It was Leo.

And now he was ill with the plague.

A feeling of hysteria rose in her throat, and she battled to keep herself under control.

Leo has the plague. She placed her hand over her mouth, ashamed of her weakness, and backed away.

Aidan was on his feet, pacing. The skin above his cheek-bone was reddened and shiny. A new injury. Lucy moved around to join him.

"Did you find him?" she whispered.

"Yeah. I wasn't even looking for him. I was just wandering and there he was, on the big road a couple of miles up. I think they dumped him there." His hands clenched. "He didn't know me. He fought and I had to force him to come with me. He got a few good punches in." He rubbed his cheek. "Luckily he's weak; otherwise he'd have kicked my skull in."

Lucy put a comforting hand on Aidan's arm. He hardly seemed aware of it.

"He's been raving, slipping in and out of consciousness. God knows what he's seeing, or where he thinks he is." His shoulders slumped. "*Monsters*. He kept saying *monsters*. And he screamed like a little kid." Aidan ran trembling fingers through his hair.

"He'll be all right," Lucy said, trying to inject certainty into her voice. "Like Sammy."

"I don't know. Maybe." He swung around and stopped; his arms hung limply at his sides. "We don't have any medicine. Grammalie Rose has a few home remedies. Herbal teas and powders for headaches and minor injuries, but not for something like this!"

"I thought the plague was over," Lucy said. "How can it still be out there when everyone that's left has already survived it?"

223

"I don't think it will ever go away. It hides and it changes. We can't fight it." He groaned and kicked at the ground.

"Aidan and Lucy," Grammalie Rose said. "Come and hold him up. He is too heavy for me." Her raspy voice was calm. Her lips pressed together so firmly, they almost disappeared into the deep wrinkles of her face.

Lucy hesitated. Mentally she screamed at herself to move, but she couldn't. She dug her fingernails into her palms.

"Once the bleeding is visible, the risk of contagion has gone," Grammalie Rose said. "Two days ago he was perhaps a danger, but now he is only a man in pain."

Lucy swallowed her fear and went to the old woman's side.

Sue had returned holding a small steaming saucepan filled with a murky liquid. She held the metal handle with hands shrouded by the long sleeves of her sweater. She was chewing on the end of her pigtail, and her eyes were wet.

"Four heaping teaspoons in two cups of water, Sue?" Henry asked, taking the pot from her. The vapor rising from it smelled dank, like rotting wood.

"Yes," Sue said, taking her pigtail out of her mouth. She pulled her fingers through the wet end.

"Good, my *zabko*," Grammalie said. "This will ease his pain. Go now."

Sue ran from the tent.

Grammalie held her hand up, the vial with the brown powder between her fingers. "Let me add valerian. It may help."

224

Her eyes gleamed like tar. She tipped the opened bottle, tapping in the last few clumps, and then nodded to Henry. Henry swooshed his finger around the liquid, testing the temperature. "We'll do it quickly," he told them. He spared a faint smile. "This stuff tastes like crap and he's not going to like it. He'll struggle."

Lucy took a deep breath and unclenched her fingers. She and Aidan got on either side of Leo and raised his head. Lucy cushioned it on her knees, and clutched his left shoulder and wrist in her hands. Aidan held him still on the other side. Leo screamed at the touch of their hands, as if his skin were being flayed. He bucked, trying to throw them off. His hands twisted and clawed as he attempted to free himself. His fingers were horribly swollen around the knuckles, and the nails were stained a deep purplish red.

Grammalie Rose pinned his legs under the weight of her body. Leo's head flailed from side to side. He tossed the covers off his body. His shirt was torn to shreds. Through it, Lucy saw the dusty blackness creeping across his chest. It was as if he had been beaten all over with steel rods. His eyes rolled back until it seemed like he stared at her through the top of his head. His mouth opened in a long, soundless scream. His tongue was black like a bird's.

"Hold him," Grammalie Rose said.

Henry poured the tea slowly down his throat. After each dose, he pinched Leo's nostrils closed and waited until he saw

his throat swallow convulsively. It seemed cruel and heartless, but Lucy remembered having to give her dog medicine. She'd done it the same way, as quickly as possible and without thinking too much about it. She found herself stroking Leo's broad forehead and mumbling nonsense to him as though he were a baby.

Finally the last of the tea disappeared. They held him for five interminable minutes while Leo fought to free himself. Tears dripped from Lucy's eyes and fell onto his hands.

Eventually, he stopped straining against them and his breathing eased. He seemed to be asleep or unconscious. Henry sat back on his heels. The saucepan fell from his hand. The clanging thud was a signal to relax their hold on him. Aidan let go of his grip and flexed his cramped fingers. He stood abruptly and with his back to the group. Grammalie Rose took Leo's hand and enfolded it between her palms. The skin on either side of her mouth was deeply grooved and looked as delicate as cobwebs around her eyes. She took out the small vial of oil, uncapped it, and began smoothing it into Leo's twisted, lumpy knuckles. The smell was pungent, herbal. It reminded Lucy vividly of the Sunday roast beef dinners her mother had made after church. If she closed her eyes, she could almost see the starched white tablecloth, Rob's eager fingers dipping into the gravy boat, the mountain of buttery mashed potatoes, and the heat from the oven steaming up the windows.

Lucy moved back against the tarp wall and sat with her knees folded and her arms wrapped around her calves. She wanted to cover her head with a soft blanket and rock back and forth. It was something she'd done as a child whenever she was upset and needed to escape an unpleasant situation. She couldn't imagine the pain Leo was experiencing, but she felt she'd shared it somehow. Every spasm he had endured had rocketed through her body, jarred her bones, and made her grit her teeth so hard, the back of her neck was sore. And she was tired, as if she'd just outrun a tsunami again. Her hands shook and her legs felt like rubber. Lucy pulled her sweatshirt down over her knees.

Aidan came and sat next to her. His shoulder pressed against her arm. He was solid and warm. Without looking at him, she shifted slightly, and his hand reached out and found hers. Their fingers interlocked. Aidan watched Leo for a while. Aidan's body was tense, his grip on her fingers was almost painful, but finally his shoulders lowered and he relaxed.

"I was so scared of hurting him," Lucy said. "He screamed." She didn't think she'd ever heard a man scream like that.

"I doubt he knows what's going on. It's like he's trapped in a nightmare," Aidan said.

"He'll need another dose in about four hours," said Henry, wiping sweat from his forehead. His hair was soaking wet. His

face was pale, but there was a stubborn slant to his mouth that Lucy had never seen. His crooked smile had vanished.

Lucy asked, "And all you gave him was the willow bark and the valerian?" She stumbled a little over the unfamiliar name.

"The local Superior Drugs is unfortunately closed for business. Looters took most of the medicine after the first wave. And then flooding took care of the rest. But it's the same stuff that's in aspirin." He nodded toward Grammalie Rose, who was hunched over Leo's still body. She crooned an odd song with guttural words in a different language. "She knows a lot about herbals. There's pretty much a natural alternative to most modern pharmaceuticals. Unfortunately for us, not many grow in New York State these days. There's too much rain."

"Will the tea really bring his fever down?"

"It should, but this disease is tenacious. If his temperature climbs above 103 degrees, it won't have an effect. We can keep him as cool as possible without sending his body into shock, but other than that . . ." He shrugged helplessly.

"Is there something stronger we can try?"

Henry exchanged a glance with Grammalie Rose.

"We have other remedies," she said heavily. "Nightshade. Foxglove."

"Great," Lucy said, and then noticed that everyone was looking grave.

"They're not cures," Aidan said, putting his hand on her shoulder. She reached up and clasped it.

"Not cures," she echoed.

"There are just ways to ease the inevitable," Grammalie Rose said. "They are remedies for the pain. Permanent."

"But Sammy, Beth, Ralph?" Lucy said. "They made it. And Leo, he survived until now."

"Sammy, Beth, and Ralph survived, but not because of any magic pill. Call it God's will, random selection. Luck. But this has gone too far," Grammalie Rose said, patting Leo's hand a final time and getting to her feet. Her hand went to her back as if it pained her. Today her legs were clothed in thick black tights and wrapped tightly at the ankles with bandages, and she wore her heavy woolen shawl tied close around her neck, as if she was cold.

Lucy persisted. "Shouldn't his immune system have kept him safe?"

"Maybe the disease has mutated," Aidan said. The words seemed to hang there.

Del appeared out of nowhere, exploding into the crowded space. Her knees were badly scratched, her boots crusted with mud, and her hair was loose and wild around her face. She dropped to her knees and grabbed Leo's hand.

"*Zabko*," Grammalie Rose said, putting her arm around Del's shoulders and trying to lead her away. "*Zabko*, he is sleeping."

Del shrugged her arm off and leaned forward.

"Leo," she said, her mouth inches from his face. "Leo."

His eyes flickered open. A spasm snaked across his face. "Del," he gasped. His voice was raw and thickened, as if the words were being forced through a closed throat. "They let you go." He reached up and smoothed her hair.

"No. I escaped."

His forehead wrinkled. "How?"

Her fingers tightened on his. The knuckles showed white. He groaned.

"I got away from them," she said again. "Shhhh."

He shook his head. "Let me talk." He tried to rise from the ground but couldn't. "You're okay? They said they wouldn't hurt any of you."

"Oh, Leo." Del's eyes filled with tears. One splashed onto her hand. She rubbed her face against the sleeve of her sweatshirt.

"The kids? Lottie and Patrick. Were they with you?" she asked.

He shook his head. His tongue ran over his blackened lips. "They were kept in the tower. I was put in the hospital with the other adults."

"Hank?" Grammalie asked. "Walter and Olive from the sweep before?"

"I don't know. They were with me, but then—I didn't see them again."

"Why did they put you in the hospital?" Aidan asked.

Leo ran the tip of his tongue over his cracked lips. "Tests. Needles. Dr. Lessing said the secret was in the blood. But which blood . . . the dogs know."

He seemed to be raving again. The muscles along his jaw bulged and jumped. *Dr. Lessing,* Lucy thought to herself. The name was familiar for some reason. Not her family doctor. That had been dear old Dr. Ferguson, who handed out lollipops, and he was dead. Maybe an X-ray doctor at the hospital, though she didn't think that was it, either. But Leo was clearly out of his head. Maybe he was mixing up the past with the present.

"Let him sleep," Aidan said.

Del looked up at him briefly and then back down to Leo. "Did you get sick? Were they treating you?"

Leo's eyes rolled wildly. "Not treating. Infecting."

"What?" said Aidan. He dropped to his knees beside the man. "What did you say?"

Del shushed him. She watched Leo's face. His eyes darted from her to Aidan to Grammalie Rose, who stood behind them, clasping and unclasping her hands. Lucy wanted to leave, but her feet were rooted to the ground.

Leo took a deep breath. It whistled in his chest, as if he were sucking air through a blocked straw. Sweat broke out in huge droplets across his forehead, yet his teeth chattered. Grammalie covered him with the sheets, smoothing them

under his chin as if he were a small child. Del stripped off her sweatshirt and carefully pushed it under his head for a pillow.

"They made me sick," he said between throttled gasps for air. "'Easy,' they said. They'd done it a hundred times. Then they injected me with different serums, vaccines. Looking for the secret in the blood. The blood." His chest rattled as he struggled to breathe. The flesh around his lips was as white as a fish's underbelly. Lucy found herself digging her nails into the palms of her hands. The healed knife wound itched. She remembered how gentle he had been when he bandaged it. She pulled off a loose strip of skin, wondering why it didn't hurt.

After a few torturous seconds, Leo continued. "It didn't work, so the Sweepers dumped me. They promised the kids would be safe if I cooperated. And you," he said, looking at Del.

Del sat up. Her fingers tightened around Leo's hand. "Did they lie about that?" she asked. "What happened to the kids?"

He drew another shuddering breath. His face purpled, visible even through the black blood. His chest seemed to be laboring with no effect. "I don't know. They weren't in the hospital. No one alive there now." His fingers tightened around her hand. "Del, she lied about everything."

CHAPTER SIXTEEN

GOING IN

Lucy stared at the mess of beans and overcooked rice on her plate. She pushed them around with her fork. She had tried to eat the food but it was tasteless, and just the sulfurous smell of it, the look of it, was making her stomach rebel. Across from her, Del was doing the same thing. Shoveling beans up into mountains and smashing them to mush, using her fork as a weapon. Her hair was a tangled mess. She'd

gathered it back into a ponytail, but hanks of it fell forward over her face and trailed over the tabletop. And she'd chewed most of her fingernails off, one finger after another, spitting the half-moon curls into the air. One had ripped into a jagged edge, and a spot of blood welled next to the cuticle. She sat hunched in her seat as if her spine couldn't support her, and her eyes were shadowed and looked huge in her face. Lucy probably looked just as crappy. She knew her hair was one big frizzy knot, but she couldn't make herself care too much. She had the curious sensation that she wasn't really in her body. Beside her, Aidan turned a hunk of bread into a pile of crumbs.

Leo had died half an hour ago, first his breath becoming more and more labored, then the veins on his thick neck standing out like cables. He'd been unable to bear the touch of blankets or wet cloths against his skin. Grammalie Rose and Henry had talked, quietly, and then given him a glass of cloudy water to drink. Leo had gripped Henry's wrist and guided the edge of the glass to his lips. This time the big man hadn't struggled, although it seemed as if most of the liquid had dribbled out of his mouth and onto his chest. He'd kept his hand on Henry's wrist until the water was gone, and then he'd moved the hand to Henry's shoulder before letting it fall limply to his side.

At least, Lucy had *thought* it was water, but afterward Leo had sunk back against Del's pillowed sweatshirt and his eyes

had closed. The lines of pain, the grooves between his blood-filled eyes, had smoothed, and it was only when Lucy had realized that she was holding her breath, waiting for his next breath, which did not come, that the truth became clear.

Now, she felt as if she'd hit her head. Her brain couldn't process everything that had happened in the last few hours, and simple things like eating and talking were beyond her capabilities. She could only sit and stare at the congealed heap of food on her plate. The only thing that felt real and alive to her was Aidan's hand wrapped around her own and the warmth his body gave off. Del had glanced at their interlaced hands and something had passed over her face, but it was so quickly replaced by a glazed expression that Lucy was convinced she had imagined it.

Del shivered now. Her bare arms were goose-pimpled. The wind had picked up, and she, Lucy, and Aidan sat at a far table, not wanting to be close to the people gathered by the fire. They'd made their way there by consensus, although it hadn't been spoken out loud. Lucy didn't think any of them had said more than two words in the last hour. Somehow, though, they had headed in the same direction, in a group, the three of them together.

Sammy had brought over the food. His white mask hung from his neck on a loop of string, his face bare for once. He'd put the plates down, a large bowl filled with beans and rice, bread, some water. He'd pulled Del into a hug. Surprisingly,

she hadn't jerked away from his touch. Instead she'd nestled under his arm, her face turned against his shoulder. A few sobs had escaped from her mouth, and Sammy had stroked her head, murmuring words too faint to catch, before letting her go. And then, after a worried glance at Aidan's face and a nod to Lucy, he'd gone back to the huddled mass of people by the fire.

Lucy pulled her hoodie over her head and pushed it across the table, then shrugged back into her coat. Del glanced at her and put it on without a word. Afterward she went back to picking at her ragged fingernails.

Lucy looked over at the crowd. She thought that all the scavengers were grouped there, the young and the old. Henry was sitting close to Beth. She wore a pearly blue mask that glowed in the light of the flames and, instead of her usual black robes, a light sweater and a pair of jeans. Sammy was crosslegged on a bench. The kids sat in a circle on the ground, away from the flying embers, wrapped in blankets, and Lucy thought Sammy was telling them a story. He was clearly acting something out. She could see his extravagant gestures with his shadow leaping behind him on the hanging tarpaulins, the white mask catching the fire gleam, and she could hear the low hum of his voice. Every once in a while a child shrieked, but it was a joyful sound followed by squeals of laughter.

"Sammy knows how to scare them in just the right way," Aidan said.

236

Lucy searched for Grammalie Rose's unmistakable silhouette but didn't find her. "What will happen now?" she asked quietly. She couldn't tell if Del was listening or not. Her hands were pulled into the sleeves of Lucy's sweatshirt. She had her head down, the hood up. Her silver bracelets sparkled in a stack in front of her. She'd stripped them off as well as the large gold hoops she wore in both ears, pulling so hard she'd ripped the lobe, but Lucy didn't think she'd felt the pain.

"Connor and Scout are collecting wood," Aidan said. "Once dinner is over and the kids go to bed, they'll build up the fire." His voice cracked. "We have gasoline. No cars, but plenty of fuel."

Lucy felt the tears fill her eyes. She knew that was the way of it, but it filled her with horror. It brought back memories of the mattresses piled in the treelined street she had grown up on. She remembered the roar of orange flames and the surprising stink of burning fibers. The billowing black clouds that obliterated the sky like an eclipse.

"Does everyone know that he died of the plague?" Lucy asked.

"Not the littlest kids, but everyone else."

"I can't be here when they—" Del said.

Aidan cleared his throat. He squeezed Lucy's hand and then let it go. He reached over for Del's hand, but she just stared at his fingers until he withdrew them. Lucy noticed the hurt look that flashed across his face, and something in

237

her belly hardened. Aidan cleared his throat again. "We won't be," he said. "We're going to the tower." He looked at each of them in turn. "Right?"

Lucy sat up. "Can we?"

"We know that the kids are in the tower, not the hospital. It probably won't be guarded as well. I mean, they're a bunch of kids, not one over eleven years old." He sounded excited and determined. He looked at Del. "You saw the tower. You remember the basic layout." She nodded slowly and sat up straight.

"I remember the way in. A big winding staircase with lots of rooms coming off it." She frowned. "There was a main entrance and a fire door around the side."

"We know the kids *were* in the tower," Lucy said. "They might not be there anymore."

"We need to try," he said. "Things have changed."

It went unsaid, but the words hung in the air: *They're killing people.*

He shifted on the bench. "We'll have to head west across the plateau and the Great Hill, and then south, and cross the mudflats. Find the bridge to Roosevelt Island."

"What about taking the big road? The way the vans came? Wouldn't it make more sense to go that way?"

Aidan shook his head. "Too exposed. There's nowhere to duck and cover if the vans are out. They could just scoop us up." He met her concerned look.

"I know," he said. "It's miles longer, and over uneven ground, but we have a better chance of getting there unseen."

Del stared steadily at the table, tension visible in the line of her shoulders.

"I've been thinking about it for a while," Aidan continued. "It'll be tough at the beginning, but once we're down on the low ground, it should be pretty straightforward."

"The mudflats are probably still dry enough," Lucy said, feeling a bubble of excitement. "Did you check them out earlier?"

He shook his head. "Not today, but last week. Leo was wandering around that gully where the Grand Canal crosses under the road." His voice roughened and he cleared it. "You'll have to lead us over the Wilds. It'll be dark and you know them better than anyone."

Lucy thought. They were still barely into the Long Wet. When she'd left her camp, the waters had been high, but no higher than the top of the toadstool on the Alice statue. Rainfall had been light and the tsunami had swept through over a week ago. "The ground shouldn't be too bad." She hesitated. "I can find my way around there pretty well, but the island bridge must be half a mile long, exposed, and we'll have to cross it. We'll be easy to spot if they post guards."

Del looked up. "There are no lights on the bridge. If we keep low, we should be okay. There's lighting inside the tower and the hospital. A generator. I could hear it." She swallowed.

A mixture of emotions ran over her face. Lucy had no trouble recognizing one of them. *Fear.* It was on all their faces.

"Good," Aidan said.

Lucy was suddenly nervous. "It could be pretty dangerous." She wasn't sure which was worse, heading in blind or, like Del, knowing what was waiting for them.

"We have to go," said Del. "Otherwise it won't ever stop." She pushed the hood back off her face. Her eyes glittered feverishly. There were dark shadows beneath them, and her face was pale and sick.

"We'll go and we'll bring the kids back home, no matter what," she said quietly.

Slowly the sun went down. The children went to their bedrolls. No one lit the lanterns this night. The light from the great fire and the scattered stars was enough, although there was no moon. Every stick of broken furniture, every scrap of timber gathered for the cold months ahead was thrown onto the blaze. The flames shot up higher and higher, transformed into tongues of orange and red by the gasoline Sammy sloshed everywhere. He had removed his mask again, as had Beth and silent Ralph. In the sporadic flashes of illumination, their features looked deeply etched, swarthy but normal.

From the shadows, Lucy watched the flames climb. It seemed impossible that Leo was dead. She remembered his strength and gentleness. She couldn't see Del's face, but sensed

her overwhelming grief and anger. The girl held herself apart from the others, her gaze fixed ahead, unmoving except for her fingers, which continually worried the red scabs on her wrists. She'd grunted when Lucy told her how sorry she was, and turned her face away when Aidan tried to hold her.

"We should all eat something," Aidan said, after a time. He handed a loaf of bread around. Lucy tore off a hunk and dutifully chewed. Her mouth was dry. She swallowed with difficulty, taking the water bottle from Aidan and washing down the lump that had caught in her throat with a hefty swig. Del ate a tiny bit and shoved the rest into the pocket of Lucy's sweatshirt. She still shivered.

"Let's go now," she said. "I can't stand being here any longer."

"Are you going to be warm enough?" Aidan asked her.

"Once we're moving. Don't worry about me." She sprang up from the bench.

Lucy zipped her jacket and tucked the ends of her hair into the collar. Her legs, clad in cutoffs, were chilled, but she didn't want to change into her jeans in case they went through water. She checked the clasp on her backpack and shrugged the straps over her shoulders. Aidan and Del had retrieved their packs, too. They each carried a short bow and slingshots, and had stuffed their pockets full of sharp rocks. Lucy had her knife. She made sure the sheath was buckled securely at her hip. In her right hand she hefted a long spear. She'd

whittled it out of ash yesterday, and it was similar to the frog spear she'd used at her camp. Five feet long with a three-inch point hardened in the flames of the campfire. She was far better with it than with a bow and arrow, and she had already impressed Aidan by hitting a target four out of five times. Del cast a snotty glance at it, but Lucy ignored her.

"Let's go," Aidan said, getting to his feet. "Slowly, as if we're hunting for rabbits."

Lucy rose from the bench and followed him. The weight of her bag chafed her sore muscles, but it felt good to be moving. Aidan and Del walked ahead, and she was content to let them lead. Maybe Aidan could calm Del down. She could hear the soft murmur of his voice. A short reply from Del— the tone of her voice so musical when she wasn't pissed off. He slung his arm over her shoulders, gave her a quick hug, and then let go.

Their forms were bulked out by the backpacks. Lucy wasn't worried their leaving would give rise to suspicion. Even if someone in the camp saw them, which didn't seem likely, pretty much everyone carried their personal possessions with them at all times. More so since the last Sweeper attack.

If they were lucky, they'd get to the tower before dawn broke.

Lucy concentrated on where she placed her feet, being especially careful while her eyes were adjusting to the dark. Thousands of stars lit up the sky, but the twisting alleys

around the camp were still confusing to her, and they were treacherous, strewn with trash and rubble. She knew they were heading west at first, until they'd crossed the bridge out of the Hell Gate and reached the plateau. Then they would turn toward the south. The terror of her journey across the canal was still fresh in her mind, and she fought to control her breathing. *One foot in front of the other*, Lucy told herself, stubbornly determined not to let Del sense her fear.

Too soon they had reached the suspension bridge and the gorge. The winds seemed stronger here, whistling past like racing cars. A horrid thought occurred to Lucy, and she ran to catch up to the others.

"We're just going to cross one of these things, right?" she asked Aidan in a low voice, darting a glance at Del. She appeared to be distracted. She stood a few feet away, tearing at her raw thumbnail. Her sleek head came up when Lucy spoke.

"Scared?" she said mockingly.

Lucy felt her cheeks redden. She found herself missing the grieving, silent Del. "No." *Yes*, said the voice in her head. "I came this way already, remember?" she reminded the other girl. *Helped along by a tsunami at my back and too panic-stricken to really watch where I was going.*

"Hmm," said Del, like she didn't believe her.

Lucy itched to hit her.

"By herself," Aidan added, putting his arm around Lucy.

Del's face took on a sour expression. She pulled her hair back out of her face and secured it tightly with an elastic.

"Pretty windy tonight. It's going to rock and roll." She touched the thin ropes. They vibrated with the force of the wind.

It looked flimsier than before, Lucy thought, this slender device made of old, braided hemp and recycled planks, which hardly seemed capable of supporting a cat.

Del adjusted her bow and quiver across her back. Then, casting one of her arched-eyebrow, curved-lip smiles at Lucy, she walked out onto the bridge.

Walked wasn't really the correct word. She *danced* her way across before Lucy had even begun to summon up the courage to move forward. Quick, sure-footed, and agile.

Lucy told her feet to move. They ignored her.

In the end, it was Del who provided the impetus. She stood on the other side with her arms folded across her chest, the hood of Lucy's sweatshirt pushed back so that her triumphant face was visible. In another minute she would start prancing back across, just to show how easy it was. Lucy longed to strike Del across her smug mouth.

Lucy tightened her grip on the spear, shrugged her shoulders to center her backpack, and stepped out onto the first wooden slat. After she'd successfully negotiated the jagged hole that had almost killed her the first time, it was just a matter of moving forward. The darkness actually made it a

little easier since she couldn't see the sharp rocks thirty feet below her, and Aidan's presence was comforting. Her breathing calmed, and the panic released her from its grip. She began to relax.

Halfway across, Lucy tripped over a protruding nail and would have fallen if not for Aidan's hand, which shot out and caught the collar of her leather jacket. He pulled her backward with a force that rattled her teeth and squashed her throat. She cracked her hip against a wooden support, bruising the bone against the hard metal knob of her knife as she clutched at the rope. The rough hemp burned her palm and she felt the old wound burst open again. The bridge swayed back and forth, shimmying. Twenty feet away on solid ground, Del watched, her mouth hanging open.

"Lucy!" Aidan yelled, transferring his grip to her arm.

She raised tear-filled eyes. "I'm okay. I just tripped. Bashed my hip. Stupid," she said.

"Didn't drop your spear, though," he said.

"There is that," she agreed. She kept her eyes fixed firmly on her feet for the rest of the way. Aidan's hand remained on her arm.

"Way to jack up the excitement, Lucy-loo," said Del with a sneer.

"It's an adventure, right?" said Lucy, feeling so much better for the hard earth beneath her boots. "It wouldn't be worth much without the terror-fraught moments."

Del laughed.

As soon as Lucy had caught her breath and unzipped her collar from the tender skin of her neck, she removed her knife and scabbard and slipped it into the inside pocket of her jacket. She pressed her fingers against her bruised hip and sucked in a breath. *Ouch!*

"Okay to go on?" Aidan asked her, passing around the water. He winced. Lucy noticed he held his left arm against his body. He must have hurt it when he stopped her from falling.

"Don't be a moron," she said sharply, trying to conceal her concern. His smile lit up his eyes, and the crooked smirk was back, curling his lips. She wondered what it would be like to kiss it off his mouth. *Snap out of it, Lucy!* she told herself. *Stupid notions like this are why you almost took a header off a bridge a moment ago.*

"Is your arm okay?" she asked as they crossed the plateau, moving faster now that the going was relatively easy.

"A little sore," he admitted. "You could come over and make it better."

Del frowned.

Lucy turned away from his grin and checked out the terrain ahead of them.

She walked to the edge of the plateau, using the end of her spear to test how crumbly the ground was. The rain and the unusual heat of the last few weeks seemed to have

solidified it into hard clay. It would get rockier and looser the farther down they climbed. She tried to remember how many gorges she'd crossed on the way here. It had seemed like dozens.

Aidan pointed a few degrees to the right. "That's the easiest way. The way I go." She considered. Easiest but it would take them pretty far out of their way. She could see the thin grove of trees at the crown of the hill where she had first rested beyond the reach of the giant wave. Just out of sight was the meandering rough path the deer used to go down to the water. Before that, though, were several miles of treacherous ground split by crevasses. Slabs of gray granite glimmered in the starlight among deep pockets of shadow where the earth had sunk or cracked. Lucy took a long breath. It would be better once they were in it. The vertigo that seized her at the top of a tree or on a swaying bridge didn't affect her when she was climbing, using her fingers to pull herself up a sharp slope or to steady herself down a hill.

"We going to move, then?" said Del impatiently. She toyed with her slingshot. The pouch of Lucy's sweatshirt gaped with the weight of the pebbles that filled it. Del's face looked even paler in the dim light, and there was a faint sheen of sweat on her forehead.

"Just figuring out the fastest way down," Lucy said easily, wondering why Del looked so sick and nervous now, when she'd been almost giddy just a few minutes before.

247

Aidan peered down the slope and whistled softly. "Faster. Definitely."

Del pushed her hair back and looked over her right shoulder, away from what was left of the city, away from the scree-covered slopes, and Roosevelt Island. *North*. Lucy wondered if the sky was bigger there. It looked bigger, and the stars clustered more thickly, bleaching a wide ribbon of sky that wreathed above the mountaintops.

"Don't you ever just feel like saying 'forget it'?" Del said. "There's just too much . . . responsibility," she said finally. Her mouth clenched around the word. She looked at Lucy, then at Aidan. Her eyes gleamed. "I mean, we're teenagers, right? Aren't we supposed to be getting ourselves into trouble? Having a good time? Sex and drugs and rock 'n' roll! Isn't that what it's supposed to be like?" Her voice lowered and softened until she sounded like a little girl. Wistful and sad.

Del stared across the Wilds with such an expression of rage on her face that it stopped Lucy's breath in her throat. She yelled—it was more of a howl, really—throwing her head back. A clatter of stones rolled beneath her feet and bounced over the rim.

Aidan reached out his hand. "You're standing too close to the edge."

She looked at him. "Isn't that my MO?" Del kicked at the ground viciously with her blunt-toed boots and sent another

torrent of rock over. "Let's go. This way, right?" she asked, throwing a backward glance at Lucy.

For the next few hours, no one could spare the energy for conversation. Lucy went first, followed by Del, and then Aidan. She used her spear for added balance and to prod the earth on the steeper slopes. It seemed that every step caused a mini avalanche. Sometimes Del crowded her, her forward momentum throwing her against Lucy's heels, and Lucy angrily gestured the girl back. Slowly, they made it past the upheaval of old highways and the chunks of tarmac. There was no sound but the crunch of rock, the patter of crumbling soil, and their breathing. Del was oddly quiet except for the occasional cry as she slipped or stumbled. Lucy couldn't help noticing that she wasn't as agile on the uneven ground as she'd been on the bridge. Rocks rolled under her boots; chunks of dirt slid away, shooting past Lucy; and her balance was off. She was stiff and didn't seem to know what to do with her arms.

Lucy paused on a granite outcrop, as if to catch her breath. Aidan had stopped some way back to retie his boot laces. Without looking at Del, she said in a low voice, "Don't stare down at your feet. Look just ahead so your brain can note the slopes and the changes in the ground. Use your hands to grip and for balance. And bend your knees a little."

There was silence.

Out of the corner of her eye, she saw Del bite her lip. She braced herself for the volley of swear words that was sure to come.

"Thanks," Del said finally. She leaned against the cliff wall and pushed her ponytail into the collar of her sweatshirt. She wiped the sweat from her forehead. "I like being up in the air. But all these sheer walls and deep crevices make me feel like I'm being crushed alive."

"Really?" said Lucy. "I feel safer. There are things to grab hold of, to dig my heels into."

Del scanned her face. "We're so different."

"Yeah, we are."

Del paused. "Where is Aidan?" she said in one of her lightning-quick changes of mood.

Lucy looked around. Aidan was still behind a short distance, crouched down and hunched over something. It was too far to see what he was doing. She shrugged. "We're not going anywhere for a minute or two. Catch your breath."

"So, which way next?"

Lucy pointed with her spear. "See the big patches of gray up ahead? It's the beginning of the next plateau. Granite, grass, and earth, and no more of this crumbling cement and blacktop. It'll be easier."

She spun on her heel, considering the best path down. She'd come straight up the hill through the grove, which lay a few hundred yards to their right. It was a narrow, twisting,

ankle-jarring deer track filled with ruts and rocks, and at the end of its sinuous length it would leave them a couple miles out of their way. Lucy dropped to her belly and hunched over to the steepest edge of the escarpment. It dropped about fifty feet at a sharp, almost 45 degree angle. The granite face was wind-roughened and scraped her fingers like sandpaper. She sat up and crossed her legs, considering their options. Her boots were chafing her ankle bones. She untied them and pulled her slouchy socks back up, retied the long laces, and double-knotted them.

"What are you thinking?" Del asked. She was unable to keep the nervousness out of her voice. She picked at the raw skin around her thumbnail, and Lucy wanted to slap her hand away as if she were a little kid with her fingers in her mouth. "A little payback for teasing you on the bridge?" Her chin came up and she looked mulish.

Lucy couldn't get the image of a little kid out of her head. It fit Del perfectly. Her mood swings, her temper, her wanting the whole cake for herself. Lucy disregarded the last statement, noting the frown that appeared when Del didn't get her fight. That was the trick: Ignore the most outlandish remarks. Refuse to play the game. Lucy felt pleased with herself. It would probably drive Del crazy!

"I'm trying to figure out the fastest way. We want to get to the tower while it's still dark, right? Here there are lots of tree roots to grab on to and the slope is gentler after about twenty

yards, although if you fall it'll take some skin off. It's the quickest. So follow me. But not too closely, okay?"

Del nodded.

Aidan jogged up. "Thirsty?" he asked, uncapping the bottle of water. He took a sip and passed it to Del, who drank as if she was parched.

When she'd finished drinking, she held the bottle up to Lucy, who shook her head. "No thanks," she said politely.

"Where to next?" Aidan asked.

"Down," Lucy replied. He immediately sat. She giggled, ignoring Del's over-the-top eye-rolling.

"No, I mean we're going that way." She pointed down the hill.

He peered over the edge. "Looks doable." He shot a glance at Del and then a questioning look at Lucy, who quickly nodded her head.

Del wiped her mouth on her sleeve. "What were you up to?" she asked.

He took his hand out of his sweatshirt pouch. His fist was full of small- and medium-size gray and white pebbles. "Making trail markers. So when we come back it'll be easier to find our way."

"I thought you were just lagging, couldn't keep up." Del had pulled her boots off. Her socks were even worse than Lucy's. The heels were riddled with holes. Del peeled them away from her feet. She had two good blisters going. The skin was

rubbed raw, and bubbles were forming. "Eww," she muttered, digging in her pack for a spare pair of socks.

Lucy felt a sympathetic twinge and turned her attention back to Aidan.

"So, a trail of bread crumbs?" she said.

"A little more permanent." He kneeled down by the edge of the slope and carefully piled two stones, one on top of the other. Lucy sat beside him.

"This identifies a trail." He placed another stone to the right of the pile. "This means turn to the right." He placed a third stone on top of the heap. "And this would be a warning."

"Pretty cool." She bent forward over the stones. "Where'd you learn it?"

"When I was a kid, Sammy and I used to leave each other messages on the road between the foster home and school. But we used tin cans and empty cigarette packs and stuff like that instead of rocks and sticks." Lucy danced her fingers over to where his hand lay. She held her breath.

He brushed her hair away from her cheek and lingered there, and she looked up at him.

Del groaned. Her head was down, her fingers tightening the ties of her backpack with what seemed to be unnecessary force. It might have been the blisters, or it might have been her way of expressing her opinion. In any case, Lucy and Aidan broke apart.

Aidan's eyes twinkled at her as he shouldered his bow.

Casting a quick glance to make sure Del looked ready, Lucy set off down the hill. For a while she was able to step down on a diagonal, zigzagging back and forth. When she reached the sharpest point of the incline, she carefully tossed her spear ahead, flipping it lengthwise and aiming for a patch of grass so as not to blunt the point. She turned to face the pitted rock wall. There were plenty of crannies to fit her fingers into, and the stone was rough enough for her boot soles to grip. The rock face was still warm from the day's sun, and she leaned into it, feeling the heat seep through her clothes. She hadn't realized how chilled she was. The night was not cold, but it was damp and it sank into her bones.

"I don't like this," Del said, just above her. Her boots scrabbled for a hold. A fine spray of dirt was knocked loose and floated into Lucy's eyes. She scrubbed the grit away. Del's limbs were extended like a starfish. She was frozen in place, holding on by determination alone. Aidan was about ten feet above her, but Lucy signaled to him to stay where he was.

The last thing Lucy needed was Del landing on her head. She forced her voice to be calm. "Tree root just past your right foot. If you feel dizzy, press against the wall for two or three breaths, but don't stop moving."

Del inched her foot sideways.

"Okay. Move your hands over, feel that bulge in the rock. You can hold on to that. Now reach down with your left foot. There's a ledge about two feet below you. Good. Grab the tree

root. Now below the ledge is a thick patch of ivy. It's like climbing a rope ladder. You're doing great."

Lucy watched to make sure Del followed her instructions, and then climbed down the last twenty feet. From below she directed the other girl until she stood beside her on the grass. Del collapsed onto her back, her chest heaving, her fingers clutching the ground as if she would never let go.

"I feel like I just got my butt royally kicked," she groaned.

"It's easier going up," Lucy promised, watching Aidan.

Del rolled over onto her stomach. "Oh no. I'm not going back that way. Plus, we'll have the kids with us. We'll go the long way around." She raised her head. "The kids are the important thing." She said this with force.

"Of course," said Lucy, a bit surprised.

Aidan jumped down, brushing his hands on his jeans.

Lucy picked up her spear, checked to make sure the point hadn't been damaged in the fall. It was still sharp enough to draw blood from the pad of her thumb. Then she walked a dozen feet to where the next part of the hill sloped down gently. She caught her breath. Her shortcut had taken them in a straighter line than she'd expected.

Below them and only a mile away was what had once been Lucy's home. If she hadn't known exactly where she was—the southern face of the Great Hill with the giant stone needle, tilted now, and pointing at them like an accusatory finger—she wouldn't have recognized it. Mud was what it was

mostly. An ocean of dry mud, strangely smooth and sculpted into drifts by wind and water. Edged with a white salt crust, like the frosting on a birthday cake. And in places were great troughs and gouges in the earth, where trees had been hurled like javelins by the wall of water. Broken limbs and bushes were tumbled together into rough fences, marking the highest points of the wave. There was an overpowering smell of brine and the stink of organic matter rotting in the sun.

Lake Harlem gleamed in the distance, and on the other side, flanking the land, pressing up against it, the Hudson Sea. Lucy shivered and drew her jacket close. She had never been scared of the water before. She'd loved it. It had fed her and it had offered her protection on two sides, but now she knew it was a huge living thing, and it could be merciless and unpredictable.

She was hardly aware of Aidan and Del as she made her way quickly down the long slope. Stumbling a little, bracing herself with her spear, she slid on the silted, sandy soil unfamiliar to her feet because it was fresh-laid, rootless, and as smooth as a cotton sheet. She stepped over a sodden mess of leaves, disturbing a cloud of small blackflies. This was where her doorway had stood. Two of Lucy's trees had been uprooted and flung far away. Of the two remaining, one leaned over almost flush with the ground, still alive, though, with fresh green growth along the horizontal length of the branch. And

miraculously, the calendar tree was still standing. Its bark was blackened and scoured. The small crown of leaves at the top was curled and shriveled. She ran her fingers over the notches carved in the trunk, counting them silently. Thirteen. It had seemed much longer. Caught in the exposed roots and the drifts of earth she found a few of her pots and pans, dented and crushed.

"Smells like dead fish," Del said, kicking a saucepan lid.

Aidan shushed her.

"What?" she said, then followed his gaze to Lucy. "Oh."

Lucy took one more look around, patted the tree trunk, and faced the lake. They moved faster now, and no longer in single file but spread out. Their boots crunched through the crusty mud and the dried leaves. A low-lying mist wreathed their feet. The hulk of the Alice statue looked black under the stars. The water lapped just below the stiff bronze lace of her petticoats. They skirted the grove of trees where Lucy had first met Aidan. The only indication that the sea had reached this far was the curving tide line of pine needles and the residue of salt. A few of the smaller saplings lay tumbled like pick-up sticks. Lucy felt a chill run up her spine and realized she was braced for the sound of dogs howling, the quick thud of their paws. But it was quiet except for the skittering of small animals in the brush and the constant sound of water. They began the long trek across the mudflats.

Aidan spoke in a whisper. "I can see the tower light." He pointed. The red beam seemed to flicker through the tracery of clouds against the paling sky.

"If we head for that, we should end up at the bridge," Del said.

"You lead," Lucy said.

Del gnawed the tip of her thumb. "Last chance to back out," she said, and then laughed. It wasn't a happy sound, but forced. She adjusted her quiver so it hung within reach of her hand and tapped the string of her bow until it twanged. Feeling suddenly breathless, Lucy unzipped her jacket and felt inside, assuring herself that her knife was still in her pocket. Her grip on her spear was clammy. Her boots felt as if they were filled with concrete. The ground cover was almost nonexistent here. They'd be in the open. Dawn was coming, and the fog was starting to disperse. The red light blinked like the eye she'd imagined it to be. They should be crawling along the ground, not walking three abreast like this, as if they were on a Sunday stroll.

They reached the bridge. It joined up with the road to the left, and then rose out of the bank of mist and curved twenty feet above the lake at its highest point. That was where they would be the most visible even if they kept to the sides. It was wide enough for a vehicle, made of gray concrete with high steel guard rails and a box of welded steel at the end, which supported it. The fog made it appear as

though it were a length of black silk unwinding in space. The three of them would look as if they were walking across the water, Lucy thought, peering ahead, and they would be highly visible.

The stone building, a low and squat block, and the tower, tall and angled, occupied most of the space on the island. A cistern dwarfed by the tower perched on the roof, and some thick pipes jutted out at the side. A whip of black smoke hung in the air. There were no trees, just vast half-moon parking lots in the front, completely empty of cars, and two narrow, rectangular lawns with a dozen park benches. Two or three tall streetlamps burned with a flickering orange light as if they were losing power. There were no lights on behind the windows. Lucy wondered where the white vans were kept. Maybe they were out on a sweep. She remembered the news footage from here. The hospital, with its gleaming floors, bright lights, hordes of doctors in white coats, and smiling nurses, had looked so different from the hospital her family had died in. That had been ill-lit, with gurneys crowded in the halls or pushed into alcoves, the smells of vomit and blood seeping into her nostrils, the floors filthy with soiled bedclothes and pillows piled in heaps in the corners, and rarely a doctor to be seen. Lucy had had to wander for hours searching for her parents, checking charts and toe tags, before grabbing a nurse and forcing her to help. The blood drummed in her head.

Del stepped onto the bridge first. "Let's go," she said. "We've got about an hour of dark left."

"Keep to the sides. Watch for headlights," Aidan said. "Once we're across, we'll make for the side entrance. Right, Del?"

She nodded. "That's the way I came out."

Lucy could hear the suppressed excitement in Aidan's voice. Was she the only one who was scared? She put her foot down hesitantly, as if she were afraid the bridge would crumble under her weight. She had never felt so terrified. They were on their way to a place where people disappeared without a trace. All except for Del, who'd managed to escape, and Leo, who'd basically been murdered. She tried to swallow past the dryness in her throat. Aidan glanced back at her and smiled. She hefted the spear to her left hand, and then switched it back.

Lucy forced herself to move, sliding her hand along the guardrail. She watched the mist swirl around her feet like a net. It reminded her of a nightmare, glue or quicksand trapping her as she tried to run. She looked back. The grove was in shadow. The salt-poisoned pines looked like skeletal fingers. The mudflats were as barren and pocked as the surface of the moon. And still she would rather have been back there than walking across this bridge, the sound of their boots muffled yet loud in the silence. There was a soft, strangling quality to the air. It felt heavy and dank, and

it suffocated her like a tangle of blankets wrapped around her head.

Del had stopped at the point where the bridge began to arch down toward the island shore. When Lucy and Aidan were a couple of paces away she swiveled around to look at them, then turned back and narrowed her eyes. Her arms were wrapped around her body as though she was cold, or in pain. Her face was hidden.

"Why are you stopping?" Lucy whispered.

Del didn't answer her.

Lucy was conscious of an industrial hum coming from ahead. It throbbed, and she could feel it through the soles of her boots.

"Generator," said Del.

"Is the entrance to the left or right of the front door?" Aidan asked.

Del hunched her shoulders. She scrubbed one hand over her mouth. She was very pale. Before Lucy could say anything, she'd crossed to the rails opposite and leaned over the edge. They heard the sounds of her vomiting.

Aidan waited until the heaving had stopped and then walked over to her with the bottle of water in his hand. He held it out to her, standing silently while she drank and splashed her face with a little water. She took a deep breath.

"Are you all right?" Lucy started to say. Aidan shook his head.

"Right or left?" he asked Del again. She stared at him blankly, teeth gripping her bottom lip. Her hand was frozen against her face. Her fingers trembled.

"Right," she said, taking off so fast, her hood blew back.

Aidan and Lucy exchanged worried glances and followed.

Their steps echoed on the concrete. Del walked ahead with her head up, no attempt at concealment. She took a straight line across the parking lot, her shadow stretching ahead of her on the ground. Lucy reached into her jacket pocket and loosened the knife in its sheath. Her eyes darted everywhere looking for a flicker of movement, expecting at any moment to see the Sweepers in their white suits, and the dogs racing like specters toward them. She felt a clamminess grip the back of her neck. Only Aidan's presence by her side gave her the strength to continue.

Now they were crossing the lawn and all was silent again. Pools of darkness thrown by the sides of the building shrouded them. A caged light threw a feeble beam. Moths and mosquito hawks bumbled into it occasionally, combusting with tiny *pop*s against the hot bulb. The door was directly beneath it, a plain steel door with a silver ball handle and a keyed lock above it. Del muttered something. Lucy watched as she reached out for the knob and twisted it. It clicked and the door swung open.

Inside there was a single light. A bare bulb, flickering and emitting an erratic hum like the rest of the lamps outside. A

staircase wound upward like the inside of a conical seashell. Lucy smelled the tang of iodine and some kind of powerful cleaner.

"Three or four floors up," Del whispered, leading the way. Their steps echoed. The light behind them faded to a pinprick and then disappeared. Their breathing sounded as loud as the ocean. The dark, complete now, felt like a pulse against Lucy's skin, it was so thick and impenetrable. She walked with her hands in front of her face, as if she could push it away. They reached a landing, paused, unsure of which way to go. She felt Aidan on one side, Del on the other. She heard the click first, then the buzzing like a hundred angry bees. Powerful incandescent lights flashed on, blinding them.

CHAPTER SEVENTEEN
THE OCTAGON TOWER

I'm sorry," Del said, and stepped away.

"Why—" Lucy started to ask.

More lights blazed, so bright and white they hurt Lucy's eyes. The generator grumbled and then hummed at full roar.

They stood on a large octagonal landing with doors leading off each of the sides. The stairway climbed on upward. At

the very top was a skylight, and through it Lucy could see the last of the stars winking out in the dawn sky.

A woman in a white lab coat stepped through the door opposite them, followed by a troop of hazmat-suited Sweepers. Helmets shielded their faces, and they held Tasers pointed outward.

Aidan notched an arrow and trained it on the closest Sweeper. Lucy swung her spear into position. Del darted forward. Her bow came up and struck Lucy's spear so hard, she felt the vibration in her knuckles. The spear clattered to the ground. Lucy grabbed for it, crouching low, and Del's foot slammed down, crunching Lucy's wrist against the linoleum. With a cry, she pulled loose, ignoring the sting of chafed skin. Still on her knees, she lunged at Del, and the girl stepped back and to the side, easily evading her. Lucy stared up at her face. It was like a mask.

Two of the Sweepers moved closer, pinning Lucy against the stair railing. Blue flames surged and spat from the black boxes they held. One of them kicked her spear across the floor.

Aidan grunted. His bow swept from side to side as he tried to sight on a target and steadied at a point between a helmet visor and the collar of a man's suit. Lucy saw him blink as a drop of sweat trickled into his eye.

Del put her hand on his shoulder.

"It's not you they want. It's her." She faced the woman in the white coat.

"This is her, Dr. Lessing. This is Lucy Holloway."

Aidan moved in Lucy's direction.

Del gripped the hood of Aidan's sweatshirt and yanked him back toward her. He struggled to keep his bow steady. "What are you doing, Del?" he asked through clenched jaws.

"It's complicated," she told him. "But it's for a good reason, I swear. Please, Aidan." Her hands ran up and down his arms.

"No."

"She's just one girl. What does she matter?"

He shook her grip loose, shoved her backward with his shoulder. She hit the steel railing with a thud. Aidan's eyes were furious.

"I don't know you," he said.

A moan of pain escaped Del. She stood apart, rubbing her arm. She looked like she was on the verge of tears. With one last glance, she turned away from him.

"I brought her," she said to the woman. "Now, let the kids go. Like you promised!" She spat the last sentence out.

Dr. Lessing smiled and stepped forward. She swept her gaze over them. Her teeth were very even and small, her soft brown hair was pulled back in a neat bun, and her brown eyes seemed warm and friendly. She laughed. It was a merry sound, and it threw Lucy off balance. This woman reminded her of her favorite fourth-grade teacher.

"You're being overly dramatic, Delfina," she said. "As usual. The children are being well looked after. They've been awaiting your return, in fact." Her glance traveled from Aidan to Lucy. Del made an explosive sound of frustration.

"How quickly you've reverted to savages," the woman said in the same light tone. "There's an article in here somewhere. 'Primitive Response to Traumatic Stress Syndrome,' perhaps?" She sounded amused. "Your weapons are hardly necessary."

"What about the Tasers?" Aidan yelled. His arms were trembling with the effort of keeping his bowstring flexed. Beads of perspiration ran down his forehead.

Three of the Sweepers turned to face him. Dr. Lessing lifted her hand and looked toward the burly man standing to her left. His hands were bare. Lucy noticed the red hairs bristling from his knuckles and his chewed nails — small details that seemed magnified. She tried to see his face, but the visor was too dark. It was disorienting, like trying to see to the bottom of a murky pond. She could tell that the Sweeper standing on the other side was staring at her. A woman, she thought. Medium height, plump, the ends of her blond hair sticking out from under her helmet.

"Simmons," the doctor said. It sounded like an order, though she said no more than the man's name. The Sweeper with the red hair on his fingers jerked his head at the others. The other Sweepers stepped back, holding their semicircular formation.

"Better?" she asked. "Come, now. Surely we can be civilized? You haven't been living in the Wilds for so long?"

Lucy looked at Aidan. She was separated from him by twenty feet of gleaming marble tile floor. Two Sweepers still guarded her, their Tasers primed. They were so close, she could smell ozone frying. Del hovered next to the staircase leading down. Her bow was shouldered, the arrows stowed in her backpack. Lucy stared at her, willing the girl to meet her eyes. Del ducked her head. Her hair hung across her cheek. Tears tracked down her cheeks. Lucy felt no pity. She wondered how Aidan was feeling. One glance at his contorted expression was enough to tell her. He had gone red with anger, but as she watched him, his countenance whitened. She could see the muscles bunch in his jaw as he ground his teeth together.

She turned to face Dr. Lessing. Again the name stirred a memory. "What do you want with me?"

"I wanted to meet you. To talk with you."

"Why?"

Dr. Lessing smiled again. She smiled a lot. "There are things I'd like to ask you, but not here, standing in a foyer. Come to my office. I can make some coffee and we can chat."

Lucy glanced at Aidan, who had not lowered his weapon. "And what about them?"

"If they'd like to join us for coffee, that would be fine. Otherwise, Delfina can go."

"Aidan?"

"Aidan is supposed to come with me!" Del said.

"I'm not going anywhere with you." He cast her such a look of loathing that she backed up.

"We'd like to check Aidan. Make sure he's healthy. I can't help but notice he's favoring his left arm."

"Like you 'checked' Leo?" Lucy said.

Dr. Lessing spread her hands. "Leo was ill. He carried the plague, and it flared up. We tried to help him."

"That's crap and you know it!" Aidan said. "You attacked the camp!"

"We came to the camp to help you. We were attacked before we could explain."

"You brought weapons," Lucy pointed out.

"There are wild animals everywhere. You know that."

"People have been disappearing for months, and it all leads back to this place," Aidan said.

Dr. Lessing transferred her gaze to him. "Is this some kind of conspiracy theory?" she said gently. "Look at us. I am just one doctor. These people are here to keep the hospital and patients safe. Many of my staff lost loved ones. We help people; we don't harm them."

"Leo was healthy until he came here. He was the strongest person I knew."

"The disease lies dormant. In birds, in rats, in people. Sometimes for months. He was already too far gone. We tried

to sedate him, but he fought, injured one of my men and got out of the building. He escaped into the Wilds, and we couldn't find him."

"He died," Lucy said. She had a sour taste in her mouth.

"I am sorrier than I can say," said Dr. Lessing.

"What about the dogs? You use dogs to hunt people," Aidan said. His grip on his bow faltered. With an effort he raised it up to his shoulder. The string pressed against his cheek, and Lucy saw the livid mark there, red against the whiteness around his lips.

"The dogs are a search-and-rescue team. They are trained to find people after a disaster. They track humans by the scent of their blood. It's quite amazing, really," she said with another wide smile. "They can detect the differences."

Lucy shook her head. She was too tired to figure out what was a lie and what was the truth. This woman had an answer for everything, and her voice was calm. She sounded concerned. She looked like someone you confided in.

"There really isn't much choice, Lucy. You're outnumbered, after all." She said this with another broad, white-toothed smile. She was teasing them.

"What'll happen to them if I go with you?"

"Delfina can go home right away. Aidan will be looked after, as I said before. We'll give him a thorough checkup. I'd hate to think the plague was incubating in your camp. There are all those children. Think what a tragedy it would be!"

Lucy hesitated. This rang true. Wasn't she worried that she might be a carrier herself?

Dr. Lessing nodded to Simmons. The Sweepers backed up even farther and lowered their Tasers.

"I just want to talk to you, Lucy," Dr. Lessing said. "You are a very special girl."

"Why do you say that?" Lucy said, suddenly nervous. Could they know that she hadn't been vaccinated?

"I know all about you," she said. "You're a survivor."

"Can I get the kids now?" Del demanded. Her nails were ragged horrors, the tender pads of her fingers torn and chewed.

"Of course," Dr. Lessing said. "You know the way, dear. Your friends will be right behind you. Emi and Jack are on the next floor down. They'll be so excited to see you. They've been ready since six o'clock this evening. So eager!" She laughed again. "Kelly, go with Delfina and help her, won't you?"

The blonde Sweeper stepped forward. She passed by quite close to Lucy, and once again she had the clear sense that the woman was staring at her from behind her dark visor.

Del muttered, "Del, not Delfina. You're not my friggin' mother." She cast one last, pleading look at Aidan, which he ignored, and ran down the stairs. They heard the intake of breath as she stumbled, the click of the door opening and then closing one floor down. Kelly followed at a slower pace.

Lucy didn't want to be separated from Aidan, but it seemed silly to insist on it after Dr. Lessing had shown them to two

rooms adjoining each other. "Simmons is an EMT—one of two on my staff. Kelly is the other," she said, opening the closest door. Inside the small room was an examining table, an IV drip, cabinets, and an armchair. "He can check Aidan's arm. Or is it your ribs?"

"I just wrenched my shoulder," Aidan said. "Could have pulled a muscle," he admitted, opening and closing his fist. A flutter of pain crossed his face.

"He can make you more comfortable. Run a few tests." She looked into his eyes. "Does that sound feasible? It shouldn't take long, and then you can join us for coffee if you'd like. Or I'll send a cup in."

Aidan nodded.

"You can join us later," the doctor continued. "I'll leave the door to my office open."

Aidan shot Lucy a reassuring smile.

She reached out for his hand, moved closer, and spoke in a whisper. "This doesn't seem real. I feel like I'm dreaming. Can we trust her?"

"I'm not sure. See if you can get some answers."

Simmons had removed his helmet. He smoothed his hands over his bushy red hair and slipped his Taser into his pocket. He was younger than Lucy had expected. His face was pale and sweaty. The hazmat suit was zipped up tight under his chin, and the skin above it was red and angry-looking, as if he

had heat rash. It was warm inside the building. She felt the lining of her jacket stick to her skin.

Simmons cleared his throat. "You can put your bow and your backpack just there on the chair, Aidan." And he waved him into the examining room.

"Come along, Lucy," Dr. Lessing said. Lucy entered a room furnished with a large wooden desk, a tall cabinet, and a couple of deep, upholstered armchairs. A thick carpet in rich hues of red and gold covered the floor. It was a comfortable room, but Lucy could smell the strong odor of cleaning fluid and other odors, antiseptic and medicinal. It seemed to permeate everything. And it was chilly, a shock after the humidity on the landing.

"Have a seat," Dr. Lessing said, propping the door open. Her gaze never left Lucy's face, and she frowned as if she were concentrating on a puzzle.

Lucy took the seat closest to the hallway so that she could keep an eye on the closed door of the examining room where Aidan was. She pushed her backpack under the chair. She looked around. The walls were bare and painted white. Floor to ceiling built-in shelves, also painted white, were filled with a collection of wide-spined books covered in red leather. Medical books, Lucy guessed. Off to the side, a door opened onto a closet-sized space with a narrow cot bed. Heavy curtains were drawn over the windows, which she guessed looked

out on the parking lots and the bridge. The thought that Dr. Lessing could have been sitting here in the dark, watching them sneak across it made her feel jumpy.

The doctor seemed nice enough, though. Lucy watched her as she busied herself at the countertop behind her desk. An electric kettle whistled. The air conditioner rattled and wheezed. The air tasted metallic. The drone of the generator was just background noise now and hardly registered. Lucy tried to remember what it would be like to live with electricity, but failed. She wondered if the hospital staff listened to music, had dance parties on Saturday nights. It didn't seem likely.

The two desk lamps felt too bright to her. She was used to the small dancing flames of the lanterns and the steady orange glow of a campfire.

"It's only instant, I'm afraid," Dr. Lessing said and turned around with two steaming mugs. "Artificial creamer?"

Lucy shook her head and accepted the cup.

Dr. Lessing sat down behind her desk. "I miss cows, don't you?"

"I guess," said Lucy. She missed donuts and her family. Mostly her family. And feeling safe.

She took a sip of her drink. It was searingly hot and very sweet. The doctor had added sweetener without asking her. In the past she drank it black and unsweetened, but coffee,

even this chalky, sugary mixture, was coffee. And it was comforting.

She blew on it, watching the woman from behind the rim of the cup.

Dr. Lessing put her cup down on a neatly folded square of tissue paper and opened a drawer to her right. She pulled out a thick folder. Lucy leaned forward. Coffee slopped over the edge of her mug, splashing onto her leg. She yelped. Dr. Lessing looked up momentarily. A little frown creased her forehead and then smoothed itself. Lucy recognized the folder. It was hers, from the nurse's office at school. And now she remembered Dr. Lessing's name from the reports inside. The school nurse, Mrs. Reynolds, had sent all the blood tests here.

"Why do you have that?" Lucy asked. The coffee wasn't waking her up. Just the opposite. She felt like curling up in this soft chair and taking a nap. She forced herself to sit straight. "Did the school send it to you? Why?" She peered at it. There were pages covered in small, precisely written words. It was much bulkier than before.

Dr. Lessing closed the folder and pressed her palms flat against it. She stroked it and smiled. "They did so many tests on you, Lucy. Did you know? A veritable plethora, looking for the usual things: heightened immunity, some kind of increased antibody production, excessive white blood cells,

excessive red blood cells. And then they got creative with it. The most far-fetched possibilities were considered, but there was nothing." Her fingertips caressed the folder as if it were a cat. Her smile didn't waver. "They died without ever finding out. I can't imagine anything more frustrating." Her eyes lingered on Lucy's face. A spasm flickered across her eyelid.

Lucy swallowed the gulp of coffee she'd been holding in her mouth. She sputtered as it went down the wrong way. A tiny thread dribbled down her chin. Dr. Lessing handed her a tissue from the box on her desk.

"Am I sick?" Lucy asked in a whisper.

Dr. Lessing tapped her lip with a pen.

"Your parents didn't vaccinate you."

It sounded like an accusation.

"Yeah, I guess," Lucy said. "I had an older brother who died from an allergic reaction when he was a baby."

The doctor's mouth pursed. Her eyes narrowed. She seemed to be looking at something that was far off in the distance. Lucy shifted in her chair. She finished the rest of her coffee, so hungry she even drank the thick syrup at the bottom, and held the mug in her hands. "You didn't answer me," she finally said. "Am I sick?"

"I didn't believe it at first, but the tests corroborate it completely. You're an anomaly. You shouldn't exist." She slapped the folder so hard, it made Lucy jump. "But you do!"

"What does it mean?"

Dr. Lessing got to her feet in a quick, smooth motion. She walked to the window, pulled the curtain aside. The sun was coming up, flushing the concrete parking lot with pink and gold light. "It means," she said, "I've searched for you for a long time, Lucy Holloway. I almost got you at the Midtown shelter, but you vanished." She frowned. "And then Del mentioned your name while I was asking her a few general questions about the settlement. Such an unbelievable stroke of luck. I don't think she likes you much, by the way. It took some convincing, but she eventually saw that it was the right choice to bring you here."

"She didn't escape," Lucy said, suddenly sure of it. "You let her go."

"She's a capable girl, that one. A little vindictive, but trustworthy, and her heart's in the right place." She swung around. "She'd do anything for the little ones, you know. Quite motherly, although she doesn't look it."

"She's a rat."

Dr. Lessing laughed. "She was stuck between Charybdis and Scylla."

"Whatever."

Lucy didn't care much about Del anymore.

"I don't understand," she said. "The blood tests and all that, that's in the past."

"Somehow, within your body, within your blood, you have

the ability to withstand a disease that killed almost everyone on Earth. I'd say that's still relevant, wouldn't you?"

"Yes, but the plague is over." *But then what about Leo?* She shifted again, pressing her spine to the back of the chair. Her brain was so slow and her eyes felt gritty. She wanted to close them. "I mean, it won't ever come back like before. Will it?" She tried to sit up straighter, but her spine felt like a limp noodle.

"You're missing the point. The answer is what is important. A scientist can't rest until she has the answer."

Rest. That's what she needed. Just a little nap maybe, and then she'd get Aidan and they'd go home.

Dr. Lessing opened the cabinet. It had plain wooden doors on the outside and looked like it belonged in a kitchen to hold plates and dishes, but its interior was more like a refrigerator. Tubes and vials fitted into individual slots and racks. Some were filled with a clear liquid, others with red. There were hundreds of them. She picked up a tube and tilted it. The lamplight turned it into gooey paint.

"What are all those?" Lucy asked. She rubbed her eyes, stifled a yawn. Her eyelids fluttered and then opened again. She was so tired.

"Answers . . . questions . . ." Dr. Lessing murmured. She turned suddenly and stared at Lucy. Her smile was gone. "Every answer fits into a box, and that leads to the next question. That is what is so perfect about science. We can be

methodical about it. Blood. Plasma. Serums. Vaccines. The answer is in the blood."

Lucy had heard that before. It was a creepy phrase and it had stuck in her head. She tried to remember who had said it. Her mind was sluggish. She gripped the arms of the chair, tried to clear the fog. *Leo!* Leo had said the same thing.

"Leo!" she said out loud.

Dr. Lessing was suddenly just above her, so close Lucy could see the large pores on her nose, could hear her breathing, heavy and quick, and smell mint candy. The doctor's soft brown eyes were now hard as pebbles.

"Everything fits, except for you," Dr. Lessing said. "You should want to help. With your blood, I can synthesize a vaccine. A synthetic duplicate. Even if the disease mutates, I'll be able to control it."

"I don't care. I don't want to be a lab rat. It's my choice, not yours."

"It's an opportunity to help so many people and to keep us safe in the future."

Her voice sounded like it was coming from far away.

"What did you put in the coffee?" Lucy said. It was difficult to push the words past her lips. Her tongue felt thick.

Her head snapped back, whacking against the chair. Her eyes flew open. Suddenly, she felt as if she were falling from a great height. She struggled to stay awake, but it was

impossible. She was drowning, so heavy in her body that she couldn't help but be pulled under.

Just before her eyes closed for the last time, she heard Dr. Lessing call out to someone unseen: "Kelly, can you please take this cup of coffee to Aidan?"

CHAPTER EIGHTEEN
IN THE BOX

Lucy woke up. The inside of her mouth felt like it was stuffed with cotton, and her head pounded with a dull pain that started behind her eyes and continued to the base of her neck. She'd felt the same way after her wisdom teeth had been pulled. She pressed her thumbs into the flesh of her temples, and then rubbed her fingers over her forehead. The pain didn't lessen. Her hair felt like one matted clump on top

of her head. Her legs and arms were heavy and almost impossible to move. With an effort, she rolled over and opened her eyes. The faint glow cast by a recessed light showed the white walls of a small room, the bed she was lying on, a small metal nightstand with a plastic pitcher and cup, and a tall bucket in the corner. There was a tiny window high up, and the door was closed.

She swung her legs around, put her feet to the linoleum floor. It was cold. Her arms felt stiff and they hurt. Lucy peeled back her shirtsleeves and stared at a trail of new puncture marks that ran up the undersides of both forearms. There were four or five on each arm, and every hole was circled by bruised skin.

Her head spun. She closed her eyes and bit down on her lip, hard enough to make her eyes tear. She would not faint. She would not vomit. She poured herself a glass of water. It was tepid and tasted unpleasant, but it soothed her dry throat. She stood up. The dizziness rushed back and then ebbed. Her bare feet slapped against the tiles as she walked to the door. She twisted the handle. It was locked from the outside. She pressed her hands against it. It was made of steel and was cold against her palms. She clenched her fists and hammered them against the unyielding metal.

Her boots stood against the wall, her socks balled neatly beside them.

She put on her socks and boots. She kicked the door. Finally she gave up. Her toes hurt, her wounded palm throbbed. It was then she noticed that it had been neatly bandaged. A square, flesh-colored adhesive.

"Dr. Lessing," she yelled. She kept yelling for a few minutes.

Lucy got down on the floor and tried to look underneath the door. It was flush with the linoleum. She ran her fingers along the crack in the doorjamb. She could see the tongue of the bolt lock. Maybe she could jimmy it open. She didn't have anything, but . . . her knife! Was her knife still inside her jacket pocket? She scrambled to her feet and went to the bed. She felt the lump from the outside of the jacket, pulled it out, and ran back to the door. She slid her knife in and eased it down until she felt the top of the bolt, then jiggled it gently. She thought it gave a little. She pushed down harder, wiggled the blade to the side. Metal slid on metal. She twisted and pushed at the same time. With a squeal the knife snapped. She was left with three inches of rough blade, a hilt-heavy thing that felt clumsy and unbalanced in her hand. Her father's knife.

The tears took her by surprise. Hot, they exploded out of her, ripping through her rib cage. When they ceased, she was exhausted. She lay down on the floor, her useless knife clasped between numb fingers. And the door — the door was

closed as tightly as ever. The room seemed too small. It didn't have enough air in it, and her lungs couldn't get a full breath. She felt the walls pressing down on her.

The window. It was at least fifteen feet above her. She could tell that even by standing on the bed she wouldn't be able to reach it, even if she could somehow stack the side table onto the bed and then clamber up on top of it without breaking her neck. And it looked too small to squeeze her shoulders through, anyway.

She paced, feeling the frustration well up in her until she was sure she would explode with it. She sank down onto the bed. It felt weird being so far from the ground. She pulled the covers off and heaped them in the corner. She curled up on top of them, shrugged her arms into her leather jacket, and yanked a rough blanket up to her chin. She turned her knife over and over in her hands. The blade was toothed now, two spikes of metal with a sharp edge. Sooner or later Dr. Lessing would come, and she would jump on her and press the knife to her neck and get out of this box.

She slept fitfully, with her knees tucked in and her sore arms folded across her head. The blanket was scratchy and thin and smelled of detergent. She drifted in and out of sleep. The air conditioner was loud. The rattle of the generator, thrumming far below her as it surged and quieted again, kept her on the edge of wakefulness. And the electric light,

weak though it was, shone down on her. She'd looked for a switch but the walls were bare. She worried about Aidan. What had they done with him? Was he still next door? She scratched at the wall with her fingernail, tapped out a sequence, wishing she knew Morse code or something. Aidan probably knew secret codes, like he knew about trail markers and how to make bows, but it was no good, anyway. Either he couldn't hear her or he wasn't there. She pressed her ear against the wall and slipped into unconsciousness again.

The fumbling noise at the door woke her. She dragged herself upright and then to her feet. Her right hand was behind her back, holding the knife ready. It was still dark outside. She moved forward and to the side of the door, where shadows offered some concealment. It opened outward, and she planned to rush whoever was coming through it, kicking and screaming, punching and stabbing, if that's what it took. The idea crossed her mind that it might be a Sweeper with a Taser. The thought of that bolt of electricity made her shudder with fear. She tightened her grip on her knife. Her eyes were glued to the door handle. She heard the click as the lock disengaged, the handle turned, and the door swung open slowly. Lucy balanced with her weight forward on her toes, ready to spring.

Someone stepped into the room. Her eyes registered black clothing and then she was on him, her weight knocking the person to the floor in the office beyond. They were in

darkness except for a desk lamp. She brought her knife up, ready to plunge it down.

"Tell me what's going on," she said, "or I'll kill you."

The figure beneath her struggled. She pushed her weight down. Her left arm was pressed against what she thought was his neck. The clothes were voluminous, black, his face covered by a hood, and now, as she leaned in closer with the knife, she saw a weird smoothness, an emptiness where the face should be. His legs drummed against the floor. A strangled sputter erupted from his mouth. Never moving the knife, she relaxed her arm somewhat.

"Lucy," he gasped. "You're choking me."

"What?" she said, recognizing Sammy's voice. She rolled off of him, then held out her hand to help him up. "What are you doing prancing around in the dark?"

He pushed his black mask down so it hung around his neck. His red eyes blinked away tears. His hand massaged his throat.

She was so glad to see him, she threw her arms around him and gave him a big hug.

"I wasn't prancing," Sammy said over her shoulder. "Del and I came to rescue you."

Lucy jerked away. She felt the dull thud of anger again.

"Del!" she said. "Rescue us? She's the reason we're here. She led us into a trap."

He shook his head. "Me and Henry tried to follow you guys. It was pretty hard going until I spotted some of Aidan's trail markers. We met up with her by the Needle. She had Lottie and Patrick with her. Henry took them back to the camp, and we continued on. Del insisted on coming back even though she's so exhausted she can hardly walk."

Lucy closed her ears to the note of sympathy in his voice.

"Where is she now? How do you know she isn't raising the alarm?"

"I know her," he said. Oddly, it was the exact opposite of what Aidan had said on the stairs when they realized Del had tricked them.

"If she gets in my way, I'll hurt her," Lucy promised. "You bring weapons?"

He showed her a small knife and a hammer. He grinned. The knife had a curved blade and looked wickedly sharp.

"Nice tiny sickle," she said sarcastically. "You plan the whole look with the robes and the mask and everything?"

He pulled the mask up over his head and stowed it in a hidden pocket under his robes.

"Just working the plague victim—grim reaper angle. In case I run into anyone. You'd be amazed the effect a simple black cloak can have." A broad smile spread across his charred face. "It's a billhook, though. Sickles are those long cut-your-head-off tools. Wish I had one of those."

"All kidding aside. Are you prepared to use it?" she asked.

He looked serious. She saw his throat work and wondered if his mouth was as dry as her own.

"I guess so. You?"

"I will if I have to," she said, realizing it was true.

She slipped her knife into her pocket, then closed and locked the door to the sleeping chamber. The door out to the hallway was shut. Everything was quiet. Gray light leaked through the thick curtains.

"Do you have any idea what time it is?" she asked.

"About eleven thirty, midnight," he said.

"Of the day after we left?"

He nodded. "What's up with you? You seem kind of out of it."

No wonder she still felt groggy. She tried to do the math. The sleeping pills had put her out for about sixteen hours. "I'll be okay," she said. "I'm still lively enough to take you down."

"You jumped me from behind," he said, with a hurt expression.

"I don't think anyone's going to be playing fair here, so be prepared for some dirty fighting." She looked around the room. "See anyone on your way up?" she asked.

He shook his head. "All clear. The dogs were barking up a storm. Maybe they smelled me."

"They're locked up somewhere though, right?"

"Basement, I think. Del said something about kennels."

Lucy hurried over to the desk. Her backpack was still under the chair. She slipped the straps over her shoulders and looked around for her frog spear. It was nowhere to be seen. She remembered how Del had knocked it out of her hands and she knotted her fists.

Her medical folder was still centered carefully on the desktop. Behind it stood the refrigerated cabinet. Lucy stared at the papers—so much information gathered about her without her knowledge. It was weird. And there were probably at least eight new vials of her blood stored in the refrigerator. She felt sick. Although she'd told Dr. Lessing how she felt, the woman had still gone ahead with her plan. She had taken away Lucy's ability to choose. Lucy rubbed her arms, felt the prickle of new scabs.

"Del's getting the rest of the kids. Do you know where Aidan is?" Sammy asked.

"I'm hoping he's still next door. There may be someone with him. But give me a minute, will you? It's important."

Sammy cast a look around. "Listen," he said. "I didn't see anyone on the way up, but this place must have guards, right?"

"There aren't so many of them anymore. I think a few have bailed. Maybe ten total. They won't be expecting a rescue mission." She rested her hand on his arm. "This is seriously important."

He nodded. "Okay, but be quick. This place gives me the creeps."

She stood still, willing her brain to work. It wasn't fair. She wanted to have the choice to decide what to do with her life. But perhaps this was a gift, and it was bigger than she was? She thought of her parents, her sister and brother, of Leo and the terrible pain he had suffered. Maybe if a cure made from her blood had existed they would still be alive. Of course, she argued, if Dr. Lessing hadn't infected Leo in some mad experiment, he'd never have gotten sick. Figuring out the morality of the doctor's motivation was impossible. There was some single-minded craziness going on there, she was sure of it.

But Lucy *could* make a difference.

There were vials of her blood in the refrigerator, and she remembered the doctor saying something about a synthetic duplicate. The question was, what should Lucy do about it?

She moved around to the front of the desk. A white lab coat draped over the chair smelled of Mercurochrome and rubbing alcohol and evoked Dr. Lessing as clearly as if she were standing there. Lucy felt a flutter of fear. Looking increasingly nervous, Sammy followed her over to the desk. Lucy flipped open the front cover of the folder. There was the photo, beginning to fade now. Her hair longer. Her face younger. High school seemed centuries ago. Words jumped out at her.

"Subject shows natural resilience to the highest degree. Possible living source of Mother Vaccine. Risk of death to the subject from controlled blood extraction—97.2%."

"God," Lucy said. Her hand started shaking.

"What's all this medical mumbo jumbo?" Sammy asked, poking his finger at the page.

"It's all about me, Sammy. My blood."

"Yeah, right," he said. "Why would they have a file like this on you? There must be a hundred pages. What's your blood made of? Twenty-four karat gold?"

She shook her head.

"We have to destroy this stuff." She picked up the folder. It was heavy; the papers spilled from it. She kneeled and picked them up. There was a report from when she'd sliced open her calf running through the glass door. There were even the results from the mandatory state physicals all students had to take. Her entire physical history, gathered in one place. *I'll take it with me*, she decided. She pulled opened the rest of the drawers. They slid easily on metal runners. More folders filed neatly. Unfamiliar names. She wondered if any were kids like her, before remembering that the doctor had called her an anomaly. Lucy ignored them, moving on to a thin stack of notebooks covered in Dr. Lessing's neat handwriting. She opened one, scanned the pages, filled with numbers and strange symbols, reams of medical language she couldn't begin to understand, and some diary-like entries, which

seemed oddly personal. Lucy's name leapt off of the pages. Opening her backpack, she stashed them and her medical folder inside. Then she turned to the cabinet holding the samples. It stood as tall as she was. She opened the door and gazed at the rows of glass vials glistening like rubies. There were ten neatly labeled with her name.

She could destroy them; it would be easy. But she hesitated. Insane as Dr. Lessing seemed, she was trying to protect the human race.

"Whoa," said Sammy. He kept an eye on the door.

Lucy spared a glance for him. "This is evil stuff, Sammy. That's my blood in there, and who knows what else."

His teasing expression turned serious.

"Okay. Finish what you've got to do, and then let's find Aidan and get out of here."

Lucy debated. She picked up one of the tubes and held it in her hand. If a cure really did reside in her blood, then it would be wrong not to give that much at least. She tried to see past the emotional, the feeling that she had been violated, and the knowledge that she had been drugged against her will. With a sigh, she closed the cabinet door. Hardly knowing why, she decided to take one and leave the rest.

She opened her backpack and placed the vial inside her tinderbox, padding it with her spare socks. Then she shrugged her arms through the backpack straps and felt the cumbersome weight settle against her back.

"Come on," she whispered.

Lucy opened the door and peered into the hallway. It was empty and quiet except for the weird clicking noises the turned-off air conditioner made. She unlocked, then twisted the knob of the adjacent door. It opened with a creak that set her muscles jumping. The scent of antiseptic was very strong. The room was darker, but she could just make out the shrouded form on the gurney. Plastic IV bags hanging from the stand dripped a viscous liquid, and clear tubes snaked beneath the sheets.

Sammy, close at her heels, flipped on the light. The sudden blaze threw everything into stark relief. Lucy froze, her heart pounding. "God, can you stop doing stuff without warning!" she snapped. "We're supposed to be stealthy," she continued in a furious whisper. The figure on the bed groaned. Lucy sprang forward, tripping in her haste. Her boots squealed on the shiny floor. She caught Sammy's smirk and ignored it.

Aidan lay on his back. His T-shirt was damp with sweat. His eyes were open, but they were bleary. He blinked, shaking his head as if to clear it.

"Aidan," she said, bending over him. A tube ending in a needle ran into the small veins of his hand, another into the larger vein of his forearm. The liquid they carried was clear. They weren't bleeding him. They were doing something else. Lucy frowned. She couldn't think about it now. They would

get him out first. She clawed at the covers. Someone had tucked him in tight.

"Here, let me," Sammy said, putting his arm around Aidan's shoulders and heaving him upright. The blankets fell to the floor. He was still wearing his jeans and socks. Lucy looked around quickly and located his boots and sweatshirt on the chair. His bow and quiver were gone.

Aidan blinked again. "Lucy. Sammy," he said in a rough voice. "I'm feeling a little sick." His head slumped forward. His breathing was labored.

Lucy ripped the needle out of his hand. He groaned again. A trickle of blood leaked from the wounds.

"You going to be sick?" Sammy asked him.

"No."

"Good."

Sammy slapped him across the face. The crack was shockingly loud.

"What are you doing?" Lucy said, trying to get her arms around Aidan. She could feel a bandage of some kind wrapped tightly around his shoulder and ribs.

"He's got to snap out of it," Sammy said, his fingers busy with the tape holding the second, thicker needle in his brother's vein. He ripped it off and slid the needle out.

Aidan's eyes were open now, and they did seem clearer. He swung his legs over the side of the gurney.

"Remind me I owe you one later," he told his brother with a grimace. "What the heck are you doing here, anyway? Didn't I tell you to stay at the camp?"

"Didn't you always tell me to question authority?" Sammy pulled his hood down lower. "Besides, if I hadn't shown up, you guys would still be locked up. So now that I have rescued you, why don't you get a move on so we can get out of here already? Or are you just going to lie around?"

Lucy glared at him. He grinned back at her.

"He's right. We should go. I'm okay," Aidan said to Lucy, squeezing her hand. "Just a little woozy."

"Are you sure?" she asked, smoothing his hair down.

"Yes."

"What were they injecting you with?"

He shrugged. "He took some blood first. After he checked my arm. Pulled muscle, maybe a cracked rib," he said in answer to his brother's querying look. "I think the small IV was a painkiller. The big one. I'm not sure."

Lucy gasped.

"I saw the bottles of medicine," he said. "They were legit. Sealed. Big pharmaceutical names. It could have been an anticoagulant, so I'd bleed quicker. The nurses always had a hard time getting blood out of me. They said my veins were buried too deep. Remind me to ask Henry when I see him next."

"You don't feel like you might be getting sick?" Lucy asked, pressing her hand against his forehead. It was clammy, but not warm. There was no air conditioner, and the room was humid.

"No. I remember drinking some really bad coffee. It must have had six spoonfuls of sugar in it. And then passing out." He rubbed the puncture in his arm. The wounds in Lucy's arm stung in sympathy.

"There were sleeping pills in the coffee," she said. "If you walk around a little you'll feel better."

He took a deep breath, and cautiously probed his ribs. Lucy didn't miss the grimace that flickered across his face.

"Are you sure you're okay?" she asked him again quietly.

With a brief nod, Aidan stood up. "That guy, Simmons, taped me up pretty good." He frowned. "It's weird. I mean, are they bad guys or good guys or what?"

"I vote bad," Lucy said. She brought his boots to him, pushed his fumbling hands away when he tried to lace them, and did it herself. While she was pulling them tight, Sammy brought him up to speed.

"Del came back?" Aidan asked, his face serious. Lucy couldn't read his expression.

"She's getting the other kids out," Sammy said. "Two floors down. Emi and Jack."

The kids who'd been taken in the first raid, Lucy remembered.

"So what's the plan?"

"The plan?" asked Sammy. He rubbed his chin. "To get out of here as fast as possible. Meet up if we can. We didn't have much time to come up with anything." He grinned. "This seems to be working pretty well so far."

"Weapons?"

"I've got my broken knife," Lucy said. "Sammy's got a bill-hook. And a hammer."

Aidan's green eyes opened wide. He looked more awake. His lip curled. "A hammer?"

"It's heavy. It's blunt. It's all we've got," Lucy said. She went to the door, put her ear against it, and listened.

Aidan made a face.

"Well, where's your bow, your slingshot?" Sammy asked him.

"They must have taken them."

"So a hammer doesn't seem like such a bad thing anymore, then, does it?"

"Not if we meet a loose nail or a hanging shutter."

"Stop bickering and get over here," Lucy hissed. "Sammy, give Aidan the hammer."

She flicked the light switch off and eased the door open. The foyer was empty.

"Quickest way out?" Aidan whispered.

"Side door?" Sammy said with a shrug. "That's how Del and I got in."

"Us, too."

"Four floors down," Lucy said.

"Guards?"

She shrugged.

"Likely, then."

She gripped her knife. "Quiet now."

The recessed lights high above them must have been on a dimmer switch. It took a minute for her eyes to adjust to the murk, but she could see the glimmer of the floor tile and the sheen of the metal handrail, which followed the curve of the spiral staircase. She felt Aidan behind her. Sammy, to her right, grumbled to himself, and she nudged him sharply. "Shhhh!"

"I turned the alarm off, but there's a number code for the door lock," a voice said. A shadow on the far side of the corridor peeled itself away and stepped toward them. Lucy froze.

CHAPTER NINETEEN

KELLY

Although the light was dim, Lucy recognized the form of the blonde Sweeper, Kelly. Dr. Lessing's second in command. Lucy sucked in a breath and curled her fingers around the hilt of her knife. Beside her, Sammy and Aidan tensed. Kelly walked forward and showed her hands. They were empty. No Taser. She was wearing regular clothes, a button-down

blouse and jeans. Her hair was tucked behind one ear, and on the other side it hung loose, draping her face.

"Keep your hands where we can see them," Sammy said in a deeper voice than usual. He had the billhook out. His hand trembled.

"You can't stop us," Lucy said. "We'll ... kill you if you try." She eyed the staircase. It was between them. She thought they could tackle Kelly before she could reach the first step. "If you make a sound, you'll be sorry." She pointed the ruined knife and ignored the small voice in her brain that wondered if she had enough blade left to stab someone — and the will to do it. Maybe Kelly would think the tremors shaking her hands were barely suppressed rage.

"Every door has a numeric locking code, and there's a building-wide security check done at midnight, so you won't be able to get back out again without help," the woman said. Somehow her voice was familiar to Lucy. It nagged at her memory. She cudgeled her brain, but her thoughts were still muddied by the drugged coffee.

"We want to leave," said Aidan. "You'll help us get out of here?"

"Yes."

"Why?"

"Lucy."

Lucy forced herself to take a couple of steps. She squinted

into the faint light, trying to see the woman. "Who are you?" she said.

Kelly turned toward them. Her right eye was surrounded by grafts of too-pink flesh, the tint of a pencil eraser. Lucy caught a gleam of a milky pupil and a cheek, cratered and pockmarked and covered in flesh-colored makeup.

The other half of Kelly's face was normal: skin pale and even, her left eye, bright blue. The round-collared cotton shirt she wore was as neat and white as the uniform Lucy had last seen her in; only the wreck of her face spoke of the months that had passed. Time seemed to shift backward. In her mind, Lucy could hear the nurse's measured tones warning of the pinch of the needle, feel the rubber tubing tied tight around her biceps, smell the pine-scented cleanser the school janitors used. Automatically she looked down at the woman's feet, expecting to see the standard issue white brogues, but they had been replaced by gray cross-trainers.

"Mrs. Reynolds!" Lucy said. "I don't understand. What happened to you?"

"Who is that?" asked Aidan, sidling up beside her. He wasn't too steady on his feet. Sammy, one step behind, gripped his elbow.

"The nurse from my high school."

The generator started up its slow grumble again. Frigid air blew from the vents. Lucy felt the skin on her arms rise up in goose pimples. *Equal parts chill and fear,* she thought.

Mrs. Reynolds had moved closer. Now she stood with her good eye facing them.

"What happened?" Unconsciously, Lucy's hand, the one not gripping her knife in a death hold, flew up to her cheek, felt the reassuring smoothness of her skin. Immediately she was embarrassed. The woman's scars were horrifying. In the light, she could see that the nurse's right eye was opaque with a bluish cast. Blind.

"The plague. The risk of nursing sick people."

Lucy's chest contracted in pity. It was awful, but she had to remember the circumstances. Mrs. Reynolds was in this place. Which made her an enemy. She flexed her fingers and tightened her grip on her knife.

"And what are you doing here?"

"I work here," Mrs. Reynolds said.

"For her?"

"It's complicated. Dr. Lessing is . . . she . . . she saved my life. Everyone here owes her a debt of some kind. The work she's doing is important."

"So are you here to convince me to just give up everything?" To her horror, Lucy found that she was crying. Seeing the nurse was a jarring reminder of her life before.

"No, I'm not. You'll have to trust me."

"Crap," Lucy mumbled, swiping her streaming nose against her sleeve. She looked up at Aidan. His face was shocked. She gave herself a mental shake and raised the knife.

"You all right?" he asked.

"Yeah." Lucy transferred her attention back to the nurse. "If that's true and you want to help us, get us out of here now."

Without a moment's hesitation, the nurse said, "Okay."

They crept down the stairs, their boots slapping against the hard marble. Mrs. Reynolds led them quickly, sure-footed in the gloom. Her sneakers were silent. Lucy glanced over the railing. The hallway far below was in complete darkness. It looked like a bottomless hole.

One floor down, then two. She could barely discern the outlines of doors leading to unknown rooms. No light gleamed from the cracks under the doors. She wondered if the other Sweepers were sleeping. The air conditioner had shut itself off again, and the generator was quiet. All was silent but for the puff of their breathing and the faint squeaks of rubber soles.

Slowly, Lucy's eyes adjusted to the dark. The steep stairs became more distinct. There were different grades of black, shades of gray. She quickened her pace, holding on to the railing in case she stumbled. Aidan and Sammy were right behind her.

Lucy could see the front door now. It was massive, steel, with a gleaming column of locks and bolts on one side and a heavy chain looped across it. They should have been suspicious when they'd come and found the place open, should

have known it was a trick. People like this didn't leave their doors unlocked.

"Break to the right once you're outside," Mrs. Reynolds said in a low voice. "The floodlights only illuminate the immediate area around the front entrance. If you stick to the edges, you'll be practically invisible."

Lucy hesitated. She didn't know what to say.

"Just go. Once you're through the door, run as fast as you can."

Lucy hurried toward the door. Her hand reached up to pull the chain free and to click back the first of the deadbolts. The nurse, her shoulder pressed up against Lucy's, tapped a sequence of numbers into a keypad. A red light turned to green. Lucy tried not to stare at the ravaged cheek so closely. She fumbled with the heavy bolt. It was stiff and she needed both hands to pull it back. But what to do with her knife? She didn't dare put it down.

She felt the rush of cold air against her back first. An inner door had opened somewhere. Then the scuffle of heavy steps came out of the darkness behind them. She whirled around. Prepared for the suddenness of blinding lights, she pressed her free hand to her forehead and shielded her eyes. Even so, when the switches were thrown, the glare from the fluorescent tubes was dazzling. Lucy blinked furiously to sharpen her vision. Sammy had fallen back, his hood pulled over his black mask. Aidan braced himself. His eyes

cut left and right looking for an escape route, but they were trapped.

At the corner of her field of vision, Lucy saw Mrs. Reynolds draw away from them. *So much for her help*, Lucy thought. *We should have threatened her. Taken her hostage.* She pressed up against the door, feeling the heavy bolts against her spine, searching for some way out. The foyer split into two corridors, which threaded around behind the steps. She didn't know if they linked up or meandered in opposite directions. She knew the facility was huge, complex.

"Why are you acting like this?" Dr. Lessing said in her calm voice. She rested her hand against the steel banister. Her hair was no longer contained in a neat bun. She had it tucked behind her ears. Her lab coat was unbuttoned as if she had just thrown it on, and she wore blue slippers on her feet.

"Guests don't sneak out like furtive thieves in the night," she continued. "I'm disappointed in you, Lucy and Aidan. I thought we were being honest with one another." The cool gray-painted stone walls picked up her words and threw them back again. The echoes were disorientating. Her gaze lingered on Sammy's form, and a frown appeared momentarily on her forehead. His eyes glittered through the sockets of his mask.

"Who is that?"

"My brother," said Aidan. "He thought he'd come and check up on us."

"At one o'clock in the morning?"

"He couldn't wait."

"And is the costume in keeping with the rest of the theatrics you all seem partial to?"

Sammy pulled his hood lower. Aidan said nothing.

Dr. Lessing broke out into a peal of laughter. It rang loudly. She was bent over double by the force of it. At first the laughter invited them to join in, but it went on for too long, and when she raised her head again, she looked exhausted. Lucy felt a thrill of fear.

Dr. Lessing caught her breath, smoothed her hair down, and buttoned her coat. She uttered a short order, and the figures behind her stepped forward.

"Who are they? Your secret police?" Sammy asked.

"They're here to keep us all safe," she said.

She was flanked by eight Sweepers. All helmeted. All armed.

"Do they sleep in that getup?" Sammy asked. The doctor ignored him.

Lucy recognized Simmons by his bulk and his red hair. *Doesn't he know his boss is crazy?* Lucy raised her knife, feeling foolish as she did so. A broken knife against a bolt of electricity. She wondered how it would feel. A burning sensation, or maybe just a jolt to the heart that stopped it dead? Her palms were instantly slick with sweat. Aidan tried to push her behind him, but she resisted. "You're still weak," she whispered. "I heard you behind me tripping all over your feet."

"Your knife is wrecked," he whispered back.

"It's still sharp."

Dr. Lessing's face softened. Her voice was pitched low. "Aidan, you shouldn't be out of bed. Simmons told me at best you'd torn a ligament in your shoulder. There's considerable tissue damage." She turned toward Lucy with her hands out. "And Lucy, you were so exhausted, you fell asleep in my office between one sentence and the next."

"You drugged us!" Lucy said. "Enough of these lies!"

Dr. Lessing laughed. "Instant coffee, a little stale, but the best that I could offer you. Hardly a drug."

"You know what you did," Lucy spat. "Who are you pretending for? Them? They just do what you tell them, right?" She waved her arm at the Sweepers.

Lines of concern etched themselves onto the woman's face.

"You're confused. You must have had a nightmare of some kind." She beckoned to Lucy. "Come, back to your bed, and we'll talk about it in the morning."

Lucy took a shaking step forward. She felt Aidan's hand on her shoulder, and then it fell away. She stared at the doctor and saw nothing in her eyes but compassion and a wrinkle of worry. It appeared genuine enough. She hesitated.

The scene in front of her seemed too absurd to be real: the Sweepers, silent, invisible faces behind the reflective Plexiglas helmets, grasping their bristling black boxes; the doctor in her lab coat; three kids armed with an assortment of useless

weapons. Even Sammy's curved blade was better suited for slicing off a handful of basil. Lucy's head felt scrambled. She wondered how long it would be before whatever sedatives Dr. Lessing had given them wore off completely. What was real? That was the question that was nagging at her. Dr. Lessing had an explanation for everything. How much of what Lucy was feeling was paranoia seeded in the early days of the plague? The Sweepers, the S'ans. She glanced at Sammy. She'd been totally wrong about the S'ans. What else was she wrong about? Was it true that these people were just trying to make everything better? Safer?

Dr. Lessing moved a step closer, Simmons at her heels, but she gestured him back. The Sweepers kept their positions. A slight smile curved her mouth.

"You're so tired, Lucy. It's been so hard for so long," she said. "But it doesn't have to be this grim struggle for survival, you know. We just want to look after you. And your friends." Her smile broadened. She waited, her eyes intent on Lucy's face.

Lucy shifted from one foot to the other. Her arms were sore. The hilt of her knife was slippery in her fingers. How long could they face off like this? Should they rush the Sweepers? She didn't think any of them was up to a fight.

"I can offer each of you a bath, a bed, and a good hot breakfast in the morning."

Lucy felt her resolve weaken. Her hands dropped a bit. Dr. Lessing came another foot closer. Close enough that Lucy

could see a smear of face cream caught in the fold of skin by her nostril. It reminded Lucy of her mother. She glanced at Sammy, whose face was still masked. Then at Aidan. His expression only mirrored the confusion she felt.

She tried to apply the skills she'd learned living in the Wilds. The ability to judge possible danger, sense predators, and make a quick decision based on a gut feeling. Now, her instincts felt dull. All she could think about was how good a hot bath would feel and the genuine kindness in Dr. Lessing's eyes. And what about Aidan? Aidan was obviously in pain. They could look after him properly here. It was what the staff had been trained for.

"We can leave whenever we want? And no more medical tests without our consent?"

"Of course."

She lowered her knife. Dr. Lessing clapped her hands together.

"Yes?" she said. "I'm so glad."

"Wait a minute," Aidan said. He reached out for Lucy's arm. He pulled her in close, still holding the hammer ready in case of attack. "Are you sure?" he said. "Just like that?"

She couldn't help but notice how he winced as his arm was jarred. His elbow was pressed up against his ribs as if they might shake loose.

"I don't know. I can't tell anymore. It feels like she's telling the truth."

He nodded.

"You talked to her longer than anyone. And your gut reactions are usually pretty right on."

"Really?" She felt herself blushing.

He stroked her arm. "Well, maybe not always," he murmured. "Sometimes it takes you a while to get a clue."

Sammy took a couple of steps back toward them. They stood shoulder to shoulder now. His back was rigid, his knuckles so tight around the hilt of his billhook that the burned skin was white.

"You're not buying this, are you?" he asked. "She reminds me of a teacher I once had. Seemed nice and fair and ready to make a deal, but then as soon as you let down your guard, you'd be hauled into the principal's office."

"That was in second grade," Aidan said.

"Yeah, but she's the same type. If it walks like a duck and quacks like a duck . . . You're forgetting Leo. And, Lucy, she took your blood. And these guys . . . they're not exactly acting friendly, are they?"

He spoke without shifting his glance from the Sweepers. Lucy looked up and noticed that although they still stood in relaxed formation across the foyer, their Tasers were raised and switched on. Dr. Lessing might pretend that everything was safe and civilized, but she still relied on force.

"You're right."

Sammy pushed his way in front of them.

"What about Leo?" he called out. "He was sick when he got out of here. You made him that way. He had no reason to lie to us."

Something flickered across Dr. Lessing's face. The smile on her mouth smoothed away as if it had never been there. Lucy sensed a shift in energy. The Sweepers seemed to stand taller, the blue flames of their weapons leapt and crackled. The hairs on the back of Lucy's neck stood up.

She searched out Mrs. Reynolds. The nurse had made her way around to stand a few feet away from Dr. Lessing. She stared straight ahead, expressionless. Lucy's gaze was drawn to her fingers twisting together. *Run,* she'd said.

The Sweepers inched forward. There was nowhere to go. The main entrance, still locked, was at their backs. The side door they'd come in through was somewhere to the left. Lucy was sure that was locked now, too. The stairs in front of them were blocked by the line of Sweepers. She could see a dark hallway to the left, a few closed doors. She wondered where they led.

Out of the corner of her eye, she saw Dr. Lessing make a gesture.

One of the Sweepers, standing off to the side, suddenly lunged at Lucy. She ducked barely in time to avoid the Taser. She heard the *pop* and *crackle* as it swished past her face. The Sweeper's arm caught her across the skull and sent her reeling backward. Her ear felt instantly swollen and hot. "Not the

Tasers," Dr. Lessing shouted. The Sweeper hesitated, clearly confused, and Aidan aimed a vicious kick at the man's knee. His legs buckled and he fell. And then the others joined in. Lucy heard the dull thud of blows exchanged. She tasted blood at the back of her throat, felt a bruise rising along her jawbone. It was a messy fight. No one really seemed to know what they were doing. And they were packed too close to land a punch with any force. She dug in to a chest with her elbow, kicked out with her heavy boots. Someone's shoulder slammed against her. Lucy fell back. Shaking her head to clear it, she brought her knife up and cleared a little space in front of her. She wondered how long before they realized the blade was broken and charged in again. Beside her, Aidan swung his hammer and connected with a helmet, cracking the heavy plastic. Sammy, his heavy robes swirling around him, wielded his billhook in a broad arc. Around the edge of his mask, she could see the wild grin spread across his face. Through the mouth hole, his teeth were bared like a wolf. He was clearly enjoying himself.

The doctor's choked voice haulted the Sweepers in their tracks. She clawed at something around her throat. "Stop!" she gasped.

Then the Sweepers were falling back. And Lucy, Aidan, and Sammy found themselves in the middle of the floor, with empty space around them. They circled, keeping their tight huddle, wary.

CHAPTER TWENTY

LIGHTS OUT

Del constricted her arm around the woman's throat and stepped out from behind her. Two small kids, wrapped in blankets, shivered at the base of the stairs. Their faces were pinched and gray, like they'd been hungry for a long time. Lucy remembered all the other children who'd disappeared from the shelter, taken away in the white vans. Who knew what had happened to them and the older scavengers who'd

been taken against their will? They'd probably all died here. Somehow she'd forgotten those truths.

Del transferred her grip to Dr. Lessing's arm and held it twisted behind her back. She stepped forward and the Sweepers fell back. Simmons raised his hand and they froze in formation. They were now clustered before the front door. Lucy narrowed her eyes. She was pretty sure there were even fewer of them than before. Had some fled during the night?

"We're all leaving," Del said.

"Delfina. You're safe here," Dr. Lessing pleaded. Her voice was strained and high. "If you go back out into the world, you'll be in danger. You can't do that to the children."

Del's arm jerked and tightened. Dr. Lessing's eyes bulged. "Just shut up," Del screamed. "Stop pretending! Don't act like you know me."

"But you understand the work we're doing here."

"You lied to me. You used me to bring my friends here. You murdered Leo." The tears were pouring down her face. Dr. Lessing writhed, but Del's grip was too strong.

"I'm a doctor," she said, "and a scientist." She looked at Lucy with a pleading expression. "It's your duty. You've been given this gift."

"You were going to bleed me until I died!" Lucy said. Finally her mind felt crystal clear. "I read your report." She stepped forward and lowered her knife. "Did you infect Leo with the plague?"

Dr. Lessing's face contorted with rage. She ripped at Del's hands with her nails, but Del held on with a fierce expression. The doctor whipped her head backward suddenly. Her skull caught Del across the mouth, mashing her lips against her teeth. Del fell backward, her arms pinwheeling as she tried to regain her balance. She dropped to her knees, and blood gushed from her mouth.

The children wailed and shrank against the banister.

Dr. Lessing staggered to her feet. Her lab coat was speckled with Del's blood.

Simmons looked from the doctor to Del. He seemed unsure of how to act. Behind him, the Sweepers shifted uneasily.

"Del!" Sammy shouted. He ran toward her. No one stopped him. He reached her, put his arm around her shoulders, and helped her up.

Uttering a wild scream, Dr. Lessing sprang at him. Her fingers were hooked like claws, her hair streamed over her shoulders. She looked nothing like the calm, composed person who had greeted them hours earlier. Her forward motion threw Sammy off balance and Del was knocked backward. Sammy's boots skidded on the polished floor. His legs gave way and he hit the ground hard. The billhook dropped from his hand and skittered across the floor out of reach. He rolled away, arms wrapped around his stomach.

Del was tearing at her sweatshirt pouch. She pulled her slingshot free, fitted a smooth pebble into the socket, and

extended her arm. She tracked Dr. Lessing as the woman attacked Sammy again with a flurry of blows. Lucy could see the frown of concentration on her face. But Dr. Lessing and Sammy were struggling together, a jumble of arms and legs, and Del couldn't risk hitting Sammy instead.

The doctor's breath came in loud gasps. Sammy tried to protect himself, but Dr. Lessing struck out wildly with the length of her forearm. The savage blow rocked his head to the side. The mask was ripped loose and skidded across the floor. Lucy could hear the murmurs as the Sweepers caught sight of his charred face and flaming eyes, the trickle of blood seeping from a wound on his forehead.

In an instant, Simmons tackled the doctor, pinning her arms behind her body, and dragged her away from Sammy. She struggled, then abruptly went limp. He held her wrists in his broad hands.

Lucy forced herself to move toward her friends. The marble floor stretched ahead of her. Her attention was fixed on the drops of blood, some of which had fallen onto the polished stone of the stairs. She wondered if it was Sammy's or Del's. How badly were they hurt? Dr. Lessing was sprawled, half sitting, on the floor, with Mrs. Reynolds and Simmons bent over her. She seemed really out of it. And the Sweepers. *What were they waiting for?* she wondered.

Suddenly Aidan gripped her arm so hard it hurt, and she heard a *pop pop pop*, and they were plunged into darkness.

"Del shot the lights out with her slingshot," Aidan whispered. He was so close, his breath tickled her ear. "Get up against the wall. She's going to lay down covering fire." Before Lucy had time to ask what that was, something hard whizzed past, inches from her face. She couldn't see it, but she felt the movement of air, and she heard a yelp of pain from someone behind her.

Lucy remembered the small, neat holes Del had made in the rabbits. The speed with which she'd killed four of them. The girl was lethal. She squinted, but the dark was absolute. They could feel their way along the wall, but in what direction?

Aidan pressed her against the wall, shielding her with his body. He whistled, a low warbling sound that was barely audible over the yells of pain and the sharp sounds of impact as stone after stone hit helmet, walls, and, most often it seemed, human flesh. Simmons bellowed orders, but from what Lucy could tell, no one was listening. Someone ran by. She felt clothing brush against her arm.

A second later, Aidan's signal was answered by another whistle. This one more like a trill. "Keep left," Aidan murmured. "Move!"

Lucy could barely see, but Aidan was pushing her into a run toward a deeper darkness, away from the ruckus. She thought they were heading for the short hallway she'd glimpsed before. She stumbled on, and just ahead she could

hear Del and Sammy and the kids. One of the little ones was weeping. Small, feeble cries, like he didn't have the strength to bawl. She ran into a solid body and stifled a gasp. Felt the drape of a cloak — *Sammy* — and heard the sound of him fumbling with a doorknob.

"Locked," he said.

Del's low voice came from farther up the hallway. "This one, too."

They moved as quickly as they could through the darkness. Lucy shuffled her feet, expecting irrationally to fall into a hole at any moment.

And then, a little way past where the corridor made an acute turn, there was a recessed light, and she could see again. She looked back in the direction of the foyer. "They'll be on us in a heartbeat," Aidan said.

"Mrs. Reynolds said the outer doors would all be locked," she said. "Or they'll be rooms with no exit."

"There's a door to the basement somewhere here," Del said. "I remember it from before. Here." She threw it open and groped for the light switch. A bare bulb was set in the sloped ceiling. Old wooden stairs led steeply down, releasing the eye-watering smell of must and mold.

"We're going down there?" Lucy said. She couldn't help thinking of all those old slasher flicks. What was the foremost rule? Don't go into the cellar. . . .

"No choice, right?" Del said.

Lucy reluctantly agreed.

"There's always a way out of a basement," Sammy said. "A window or a coal chute or storm doors—something most people don't think about it." He started going down the narrow steps.

Lucy put out her hand and grabbed hold of his cloak. She looked at the scared kids clinging to Del's fingers and put her lips to his ear so they couldn't hear her. "Aren't the dogs down there?"

She could hear whining, excited yaps. The barks echoed wildly.

"Yeah, I guess—but like Del said, no choice."

Still Lucy hesitated. They didn't know what to expect. It could be a dead end, and they had no way of protecting themselves except for her knife, Aidan's hammer, and Del's slingshot. Aidan pushed urgently against her back. "Hate to tell you, but the Sweepers are coming."

And now she heard hoarse shouts and the scuffling of boots on the hard floors.

She hurried onto the stairs, grabbing a wooden railing, which bent under her weight. Behind her, Aidan pulled the door closed.

"Lock?" Del asked.

"Bolt, but one good kick will break it," he said.

On the first step down, Lucy slipped. The railing pulled away from the wall with a screech of nails. Del's hand shot

out and gripped her elbow, saving her from a nasty spill. As soon as Lucy had regained her footing, the girl released her arm.

"Thanks," Lucy said.

"Don't mention it." She held one kid firmly by the wrist. Lucy thought it was the girl, but she couldn't be sure. The other one stumbled ahead with his arms outstretched. Both of them wore baggy gray pajamas and slippers. Both badly needed their hair washed. *So much for the hot baths Dr. Lessing mentioned,* she thought.

The stairs were steep but short. They found themselves in a large, concrete-floored space. Thick drifts of dust lay on the floor, tracked over by countless footprints. Steel-encased wiring stretched out in a lattice across the low ceilings, as did rusty pipes as thick as Lucy's arm. She could hear the trickle of water pumped down from the cistern. Pink insulation puffed out of crumbling plaster board like masses of cotton candy. Stacks of soggy boxes lined the water-stained walls. It smelled of damp and mushrooms, and overwhelmingly of animals: mouse droppings, but also the close, thick smell of many dogs kept inside, the tang of urine and dander and fur.

Numerous corridors led off in different directions, each poorly lit and dusty. Lucy tried to orient herself, but she'd lost her sense of direction. She thought she could pinpoint where

the dogs were kenneled, even though the echoing barks confused her.

"Any idea what's down here?" Sammy asked Del. He'd
lifted the little girl onto his back. She clung to him, her hair
straggling in her face. Her eyelids drooped.

Del shook her head. "Besides canned goods and bulk food
items? The dogs. A bunch of old boxes. Stuff left over from
before the plague, I guess."

Aidan, who had been hanging back near the stairs listening for the Sweepers, looked excited. "If they were getting big
deliveries of food, then there's probably a loading dock or
something down here. We should head in that direction."

Del shrugged helplessly. "Your guess is as good as mine. It's
as big as a football field. I wandered down here for a couple of
hours before Dr. Lessing—" She broke off, her cheeks red-
dening. *Before Dr. Lessing convinced you to rat me out*, Lucy
thought, and then was a little ashamed of herself. They'd still
be fighting a losing battle if it weren't for Del.

She cleared her throat. She hated to say it, but it seemed
only rational. "If they trucked in mass amounts of dog
food, then they probably stockpiled it near the dog kennels.
We can follow the sound of their barks." She turned slowly,
tracking the sound. They were subdued now, but in her
mind she could see the dogs. She remembered the rottweiler
leaping at her legs as she struggled to climb the tree, the

321

thick froth of spit at the corners of its jaws. Three narrow halls stretched in front of her. They were lit with dim bulbs.

"Look down on the ground," Aidan said. A jumbled trail of muddy footsteps led down the central one. "The middle way gets used a lot." He squeezed Lucy's hand. She moved into the hollow of his arm. A series of sharp *thud*s jolted them apart. Someone was trying to kick in the cellar door.

Exchanging panicked glances, the group crept down the narrow hall, moving as quickly as possible. Del soothed the kids with soft murmurs. The air was very still and dank. The acrid odor of urine and sawdust grew stronger, and the yelps of the dogs increased in volume.

They hurried toward the sound. Sammy ran ahead. His cloaked shadow leapt across the walls. The kid was attached to his back like a monkey.

They'd come a few hundred yards down the passageway, and still Lucy could heard the sound of wood splintering behind them and the buzz of voices. How many were there? Three or four? All of them?

Another volley of barks, louder and more excited.

They can smell us, Lucy thought with a thrill of fear. And then her mouth suddenly turned dry. *They can smell me.*

"Nearly there," Aidan said.

The corridor twisted and then opened up. Wire cages lined one long wall. Dogs of every shape and size pressed against the mesh. Some threw their bodies against the doors or clawed

frantically, hard enough to rip at their paws. The yelping was deafening.

"Can you see a door out?" Lucy yelled. She was transfixed by a large dog that was staring at her. She flicked her eyes away, trying not to challenge the animal. Its black lips lifted away from sharp white teeth and the dog began to howl. At once the rest of the dogs lifted up their snouts and began to howl, too.

Aidan pulled at her arm and she realized she'd been standing still. "Come on," he said.

She tore her eyes away from the dog and moved across the room as quickly as she could, keeping her gaze on the ground under her feet and ignoring the rumble of growls, the clanging of dogs pushing against their metal doors to get to her.

The space narrowed into two corridors. Sammy hurried down one and almost immediately doubled back. "Locked door," he said. They all ran down the other way. The passage was lined with stacks of cardboard boxes. The dogs had quieted again, except for a few excited yips. Lucy heard the dull thud of running feet against the concrete.

CHAPTER TWENTY-ONE
THE BASEMENT

She turned to see Simmons, Dr. Lessing, Mrs. Reynolds, and one other Sweeper who wore his faceguard down and held his Taser in front of him like a sword. Dr. Lessing was sweating and pale. Mrs. Reynolds grabbed her arm. The doctor roughly shook it loose. Lucy stopped, feeling more exhausted than she ever had before. Her hand could barely hold her knife. The generator hummed and then roared into life.

Lucy remembered how she'd thought the light on the roof resembled the gigantic eye of a beast. Now she felt as if she'd been swallowed alive.

Del raised her slingshot. Aidan wrapped an arm around Lucy's shoulders. They backed up as the doctor and Sweepers advanced.

Lucy snuck a look behind her. A shadowy hallway stretched back. More boxes were piled five feet high — rows and rows of them. They were marked with the names of ready-to-eat food, vegetables, precooked meat, and dog chow. There was no outlet that she could see.

"Don't let them force us into a dead end," she said. They spread out in a thin line across the corridor. She noticed that the air was fresher. The scent of dog mingled with something she realized was the smell of rain. Del ordered the two kids to get back as far as they could.

"There's got to be some kind of outside access around here," she said. "How else did they get all these crates in here?" She reached into an open box and pulled out a can of dog food. She tossed it to Aidan, who caught it with his free hand. "Weighty," he said, hefting it.

Sammy helped himself to a couple.

"Just grab the girl, Ross," Dr. Lessing shouted suddenly. "I don't care if the others get hurt."

The Sweeper came toward them at a run. He aimed himself at Lucy. Aidan pelted the can at him, but Ross ducked.

Sammy threw both of his at the same time. One hit the man with a sharp *crack*, fracturing the plastic visor. Mrs. Reynolds shouted out a warning. Attempting to avoid the man's weapon, Lucy threw herself backward so hard she hit the stack of boxes, knocking the topmost one to the ground. The column teetered and came crashing down, splitting the cardboard and spilling tin cans everywhere. Aidan tripped and fell. The Sweeper came on, his Taser dangerously close. He flung his arm out, and the black box skimmed the sleeve of Lucy's leather jacket. She felt a jolt, which seemed to stop her heart for a second, and then her legs turned to water. Her head smashed against the ground, and she felt a trickle of blood edge into her collar. Aidan swept his leg around, felling the Sweeper. He stomped on the man's wrist with his thick-soled boot. There was a *crunch* as the bone broke, and the black box flew from his fingers. Aidan pounced on it quickly.

"Sammy," he said, keeping his eyes on the Sweeper who was curled up, cradling his injured arm. "Help Lucy up, will you?" He stepped toward Simmons. The black box sent out its flickering prongs. Simmons held his hands open in front of him and shook his head. He took a few paces backward.

"Just let me check on Ross, okay?" he said. Aidan nodded. Simmons prodded Ross's wrist. "Broken in about three places," he muttered. He helped the Sweeper to his feet and propped him against the wall.

Lucy's legs still felt like limp noodles. Her heart was pounding, and her head buzzed. It was difficult to fill her lungs with air. She freed herself from Sammy's tight grip. "Where's Dr. Lessing?" she yelled, looking for the white lab coat. The woman was nowhere to be seen. The dogs had started up a crescendo of whining. Then she heard the sound of electronic bolts shooting open. The barking broke out and quickly became a cacophony. A single howl rose. The sound made the hairs on Lucy's arm rise, and she felt cold despite her leather jacket. Mrs. Reynolds's face blanched.

"She's letting the dogs out," she said. "They'll go mad when they scent you. The trainer left a few days ago. If they find you before Dr. Lessing does, they'll tear you apart."

Simmons stepped forward. He spoke hurriedly. "Down that hallway. Green-painted steel door about ten feet on. You can bust the lock. It leads to a dog run with an eight-foot chain-link fence."

"We'll hold them off as long as we can," Mrs. Reynolds said. She looked at Lucy. "Be careful out there. The plague is mutating. It may return. That much is true."

Simmons set his shoulder against a column of boxes and shoved. The heavy boxes came cascading down, partially blocking the narrow corridor. He moved to the next row and heaved. Some split open. Cans rolled underfoot. Slowly the pile grew and wedged against the opposite wall. Mrs. Reynolds

joined him, tugging down crates, and heaping them higher until the lower half of the passage was impenetrable.

Lucy hesitated. The others were already at the door. Sammy was hammering against the lock with a dented tin can.

Mrs. Reynolds met her eyes. "Just run. Run, Lucy!" she said, staggering under the weight of another box. The scars were livid against her flushed skin. Behind the nurse and the growing pile of boxes, she caught sight of Dr. Lessing. She was completely surrounded by furry bodies. The dogs swarmed over one another as they hunted for a scent. Lucy hesitated.

"I took my folder, the notebooks," she said. "They belong to me and no one else. But I left the blood." She turned away, but not before she saw surprise in the nurse's eyes.

Lucy ran to join Aidan. His arm was pressed tight against his ribs again. She saw the pain in the lines of his forehead. Although her skull was still buzzing, she felt surprisingly clearheaded. Sammy threw the can away in disgust. The thin metal was crumpled. Some kind of red sauce leaked out, staining his robes. He pushed his hood back. His blackened forehead was dripping sweat. Aidan set his shoulder against the door and heaved. The lock was battered, but still it held. From behind them, they heard the baying of the dogs.

"They blocked most of the way, but there's still space for the dogs to get through," Lucy said.

She remembered how the animals had propelled themselves halfway up the tree trunk, maddened by her scent. She

pushed Sammy out of the way and slid her knife blade between the lock and the door and slammed it down hard, the impact jarring the old wound on her palm. The lock tore open with an awful squeal, and her knife snapped again. Straight across. An inch from the hilt.

Lucy subdued a stab of grief, shoved it back into her pocket, and thrust the door open. Cool air flooded over her. The dog run was long and concrete, with shallow channels running down each side. It had rained recently, and the cement glistened. Through the links of the fence she could see the shore, and beyond, the stormy surface of Lake Harlem.

"Almost there," she yelled, turning back to grab Aidan's hand.

Two dogs crashed through the barricade of boxes and cans into the hallway. Lucy caught a glimpse of their yellow eyes, the gums pulled back in hideous grimaces. A rottweiler and a pit bull. They leapt, arrowing in at her from two sides. She threw up a defensive arm, and then Aidan pushed her away, yelling. She hit the ground and rolled against the wall, smacking it hard with her head. She shook her head to clear it, scarcely aware of the pain, and dug frantically for her knife before remembering it was useless. Screaming in anger, she threw it at the rottweiler attacking Aidan. The hilt struck it across the skull, but the dog didn't pause. Del stood in front of the door to the outside, shielding the terrified children with her body. Her slingshot was loaded. She raised it, looking

for a clear shot, but everything was happening too fast. Aidan was thrown backward by the weight of the dog. He grappled with it, pushing against its muscular chest and throwing his head around wildly in an attempt to avoid its razor-sharp teeth. He hooked his fingers in the dog's collar and twisted the leather strap, trying to strangle it. The dog's tongue protruded and strands of saliva glistened as the jaws snapped inches from his face.

Sammy hurled himself forward, trying to reach his brother. The pit bull jumped him, seeking the flesh beneath the robes. It clamped down with its jaws and whipped its head from side to side. The heavy black robes tore as Sammy kicked out at the dog. His boot connected with the dog's midriff. Another wild kick, this blow landing on its snout. The dog yelped and released its jaws, falling heavily to the ground. Del took her shot, sending a stone into the meaty part of the dog's thigh. The dog howled and scrabbled at the floor, trying to reach the wound in its leg. Sammy kicked it again in the ribs and ran, panting, to Aidan, who was weakening. He grabbed the rottweiler's collar from behind and yanked it into the air. Scrambling to his feet, Aidan thrust his hand into his pocket, pulled out the Taser, and pressed it against the dog's side. The dog yelped and collapsed, shaking convulsively on the floor. Its pink tongue lolled from its mouth, and then it was still. Aidan limped over to the whining pit bull and Tasered it, too.

He stood looking at the dogs, his face pale and sick, and then he slid down the wall until he was sitting on the ground.

Lucy crawled over to him. His left arm hung loosely. She put out her hand, afraid that she might accidentally hurt him, and settled for stroking him on the cheek. "Are you all right?"

"Painkillers wore off," he said scowling. "I'm pretty sure now that I've pulled a muscle."

"Come on." She helped him stand. They hobbled out the door into the dog run where the others stood waiting.

"Might have brought the Taser out a little sooner," Sammy told his brother.

"Couldn't reach it. You may have noticed the hundred pounds of dog sitting on my chest?"

"Always with the excuses," Sammy said, pulling his shredded robe awkwardly over his head. His forearm was imprinted with four deep tooth marks gushing blood.

Del cursed. "Can you still use it?" She sounded angry.

Sammy looked disappointed. "Yeah, it hurts bad, but—"

"You can climb?" Del asked.

"Of course," he said, watching blood drip onto the floor. "I didn't mean to get mauled by a dog, you know."

Del pressed her lips together. "I know," she said in a softer tone. Stowing the slingshot in her back pocket, she ripped a length of cloth from the discarded robes and tied it tightly around his arm. He gasped.

Giving him a look, she gathered the children to her and faced the fence.

"You go first," she told Sammy. "I'll lift the kids up to you."

He pulled himself up, swung over, and jumped down. Once he was on the ground, he held his arms up for the first of the two children. As soon as they were safely on the other side, Del went up, then Lucy, and lastly Aidan, who favored his left arm and climbed one-handed. He had just reached the top when the fox terrier burst through the door, barking madly. Its toothbrush tail stuck straight up, and the fur on its back stood up in a ridge. It ran back and forth along the fence seeking a way out, and then began throwing itself repeatedly at the chain link as if it were made of rubber.

"Let's go before the poor thing kills itself," Aidan said from the top of the fence.

Lucy turned away.

And then a second dog hit the fence barely a foot below the top. Another rottweiler, even larger than the first. It catapulted itself upward, thick black claws pushing the chain link outward as it tried to find purchase and muscle its way over. Aidan jumped, making no attempt to land gracefully. He staggered and then regained his balance, pulling Lucy back from where she stood almost mesmerized by the animal's single-mindedness. The dog fixed its hot gaze on her and, growling terribly, made another impossible leap into the air, landing

almost on top of the fence before falling heavily back to the concrete.

Aidan hurried her a safe distance away. "It'll be over that in a minute."

The dog was panting heavily, but still it paced and jumped and whined. Lucy's presence was driving it crazy.

"It's not going to stop hunting us," Lucy said. She looked at the exhausted children huddled together in Del's arms and at Sammy trying to smile. "It wants me." She shrugged her arms out of her backpack, carrying it by the strap, and walked toward the fence. The rottweiler's lips inched back from its incisors. Its ears flattened against the bony skull, and an awful snarling rumbled from the barrel chest. Muscles bunched in its back legs as it gathered itself to leap again.

"It can't have you," Aidan said, trying to haul her away. "Get away from the fence, Lucy!" She shook herself loose, jarring his arm. He winced with pain.

"It's okay."

Keeping her eyes on the dog, she opened her backpack and dug around in it, locating the tinderbox.

The dog kept up its continuous growl. "You want my blood?" Lucy shouted, pulling out the vial. She raised it above her head and threw it over the fence. The glass smashed against the concrete. Thick red blood spattered against the wall.

They ran. It wasn't until they reached the parking lot that they paused, looking back at the dark hulk of the building. Lights blazed on the top floors. In an upper window behind heavy curtains, they could see human figures hurrying back and forth. Lucy might have imagined it, but she thought she could hear a single, shrieking note that seemed to go on and on.

"Think they'll follow us?" Aidan asked.

Lucy shook her head, thinking of Mrs. Reynolds. "No. They have what they need."

The sky opened and rain began to fall, a hard-driving shower that soaked them immediately but was as warm as a spring shower. Lucy looked up into the lightening sky and let the rain push her hair off her face. If more dogs did come, the rain would wash away most of their scent. The fog had dissipated and the air smelled fresh and clean. The dull pain in her head subsided to a thump. Sammy carried a kid on his back; Del had linked hands with the other one. Lucy heard her voice, low and soothing, as she urged them to move. It was weird to hear such kindness from the girl.

The empty parking lot glittered like an ice rink. They ran through the rain, slowing down again once they came to the bridge. Lucy looked back at the tower. The red light was dark.

Aidan slipped his arm around her waist. She leaned into him, careful of his wounded ribs and arm.

"Which way?" he said.

She looked ahead. Del and Sammy and the children were walking slowly. They'd reached the bridge. With the sleepy kids and the exhaustion that she was sure everyone was feeling, it would take them hours to get home. She scanned the horizon. The long bridge curved above the wind-whipped waves of Lake Harlem. Beyond that lay the Wilds, as familiar to her as the lines on her palm. Lucy could close her eyes and in her mind navigate over the flats, the grove, past the salt marsh and the blighted pines, the remains of her camp and the Great Hill. And then onward up the shifting terrain of the gorges and the escarpment and the suspension bridges swinging wildly with the slightest breath of wind. They'd have to carry the children, or haul them up the granite cliff face somehow. Aidan and Sammy were injured. Her own body hurt so much, everywhere, that it was almost funny.

Lucy turned away from the thicket of tall trees and the gleam of the restless sea she glimpsed between the black trunks, and toward the broad, solid road that snaked north for five miles before entering the Hell Gate. The road the Sweepers had taken. "That's simple," she said with a grin. "For once we'll take the easy way."

She held hands with Aidan as they crossed the bridge, walking into him when he suddenly stopped.

"What?" she said, startled. "Do you hear something?" He put his finger to her lips.

"Shh," he said. "C'mere." He pulled her to the side where the shadows concealed them. His voice sounded thick.

Del, Sammy, and the kids had almost reached the road.

Lucy moved closer.

Aidan traced his finger to her cheek and then to her chin. He tilted her face upward.

And then she was looking only at him, his bright eyes shadowed, the messy fall of his hair over his forehead, his wide mouth with that infuriating curl in the corner. His hand moved to the side of her face, he leaned forward, and, letting her breath leave her in a sigh, she rose onto her tiptoes to meet his lips. Her fingers tangled themselves in his hair and she pressed against him, feeling the warm solidity of his body, the crushing strength in his arm as he pulled her against him, and the doubt draining away from her, leaving nothing but happiness.

After a long minute, Aidan pulled back a little. Her lips felt bruised. She was flustered now, conscious of the tingling sensation left on her mouth, the need to keep touching him. Her mouth hardly felt as if it belonged to her anymore. He kissed the tip of her nose and, linking his hand with hers, drew her toward home.

EPILOGUE

ABOVE THE WORLD

Do you think the rain will ever stop?" Lucy asked Aidan.

He shrugged and she gripped his arm more tightly. "Oww," he said. "You think you could relax that death grip?"

"Did I hurt you?" she asked.

"No, I'm healed."

"Well, you know I don't like being up so high," she said. Aidan shifted his back against the tree trunk.

"Come here, then," he said, prying Lucy's fingers from his arm and guiding her forward so that she was nestled against his shoulder. She still didn't understand how he could be twenty-five feet off the ground and act like he was lounging on a couch, but she settled into the crook of his arm and crossed her feet on top of his legs. "So," she said, "The rain?"

"It's been, what? About two weeks?"

She thought back. "Ever since . . . you know . . . that night." The night they'd escaped. The night they'd first kissed.

He yawned, stretching like a cat. Her hand tightened around his arm again. She pushed the drift of her hair away from her mouth. The canopy of this elm was so thick that the raindrops ran out of steam before they reached their branch.

"It'll clear up any minute," he said lazily. "Either that, or it'll go on for months."

She pressed her palm against his forehead. It was cool and smooth.

"I'm not sick."

"I know, I just have to check."

"Every day?"

"Just until I'm sure that Dr. Lessing didn't do something to you."

He exhaled deeply.

"Are you falling asleep?"

"Maybe. I was up at dawn hunting bunnies," he said. "Ever since Del and Sammy left, I've been the guy. At least until you learn to handle a bow as well as you handle a spear."

His lips hovered near her ear. She felt the soft shushing of his breath. A shudder went up her spine. She snuggled closer. She could hear the dull roar of the waves, the rustle of the wind. As long as she didn't think about the ground, it was nice being up high, cradled and surrounded by thick, green leaves.

Aidan had picked his favorite tree, the elm, and his favorite place in it. At the very top. When he stood up, he said he could see fifty miles in every direction. Lucy had to take his word for it, because there was no way she was going to balance on a branch that dipped up and down under her weight, with nothing to hold on to but whiplike stems. He liked to be here at dusk, when the bullfrogs started their nightly warblings and the broken string of beacon fires along the northern route became visible.

Del had left a week ago. She'd taken the Geo Wash Bridge west, before heading due north to find the settlement up there. And surprisingly, Sammy had opted to go with her. Actually, Lucy amended, not so surprisingly. They had spent a lot of time together after they'd gotten back from the island, and Lucy had seen something in Sammy's eyes. Del had kept herself apart from the jostle and bustle of the camp. She'd hunted, she'd helped shore up the dikes now that the canals

were filled with roaring cascades of water, she'd harvested tomatoes and squash, she'd worked like she was possessed, but after the work was done, she'd disappeared to places only she knew about.

Lucy had been anxious. Mostly for Aidan. She knew what Del's friendship meant to him.

"We talked. It's cool," Aidan had said. "She made a mistake. And you know, maybe I . . ." He'd stopped and looked at her carefully then. "Maybe I wasn't straight with her. About you. How I felt about you. That was wrong of me."

Lucy had dropped her eyes, suddenly shy.

"You'll make it up to me, then?" she'd said, teasing to break the tension.

"Lucy, you know how I feel about you, right?" he whispered now.

She was breathless. "Why don't you tell me?"

He tilted her face up. "Why don't I show you?"

"How?" she said, fighting the urge to giggle. If she started laughing, she'd probably fall out of the tree.

"Like this," he said, kissing the lobe of her ear. She closed her eyes, looking at him through her lashes. Her hands tightened on the tree limb. She felt dizzy all of a sudden.

"This," he breathed, planting more kisses along her hairline. His fingers tangled with hers. She was holding on to nothing but him. He was kissing her eyelids now. Each movement of his lips made her shiver. Aidan murmured her name.

She was melting. She couldn't feel the hard bark against her hip. Nothing existed but his gentle hands and his warm lips.

"You smell like blackberries in the sun," he said. "You taste like honey."

He could do this for hours. Kiss every inch, every centimeter of her neck and face except for her mouth. It drove her crazy, made her want to scream *Enough!*

"Aidan. Aidan! I'm going to fall!"

"Hmmm," he said, against her neck. He opened his eyes. They were sleepy, but she saw the glint that hovered in them.

Lucy leaned back in his arms and rested her head against his shoulder. Rocked in the elm's broad branches, she felt safe. The fires weren't visible through the heavy screen of leaves, but up above where the branches thinned were the stars. Aidan had shown her the North Star, tracing its path from the handle of the Big Dipper, which was pretty much the only constellation she could identify with any certainty. It hung low, not the brightest star, but special now after so many nights of picking it out together, as though it somehow belonged to them. "Are you sorry you didn't go?"

He took a moment to reply. "Some day, when we're ready. If you want to," he said, drawing his eyebrows together. The crooked smile was still there, dancing in the corner of his mouth, but he looked serious.

Lucy followed his gaze northward.

ACKNOWLEDGMENTS

Writing a book is often a solitary endeavor but making it good takes many people.

A million heartfelt thanks to: Silvia Rajagopalan, Charise Isis, Alison Gaylin, Jennifer May, and Charity Valk, who were there at the beginning. Without all your enthusiasm and help it is very possible there never would have been a book.

ACKNOWLEDGMENTS

My family and friends, in particular, Arnaldo Treggiari, the Rajagopalans, Gail Parris, Lesley Sawhill at the Woodstock, NY Children's Library, and all the kids who have workshopped with me.

I'm indebted to my agent, Garrett Hicks, who digs deep; my über-editor, Lisa Sandell, who knows her way around words and then some; and the Scholastic team, especially Jody Corbett, Starr Baer, and Elizabeth B. Parisi.

To Milo, who understands when Mommy has to shut herself away for a few months, and still thinks the whole author thing is cool.

As always, to my husband, Marcus, who gives good advice and delivers mega cappuccinos to order.

And special gratitude to my dearest, brave friend Sacha McVean, who tried to outrun a tsunami and inspired a heroine.

THE WORLD HAS ENDED. . . . WHAT COMES NEXT?

Smallpox epidemics, floods, droughts — for sixteen-year-old Lucy, the end of the world came and went, stealing with it everyone she ever loved. Even the landscape of her beloved New York City is ever-shifting and full of hidden dangers. As the weather rages out of control, she survives alone in the wilds of Central Park, hunting and foraging for food and making do with the little she has, while avoiding roving scavengers and thieves. But when an unrelenting pack of vicious hounds begins to hunt her, Lucy is not sure she can continue on her own. Then, suddenly, she is swept to safety by a mysterious boy named Aidan, who helps her escape the hounds and urges her to join a band of survivors. Reluctantly, she finds him after her home is destroyed; however, new dangers await her.

An army of Sweepers terrorizes the camp, carting off innocent people and infecting them with the plague. Lucy and Aidan realize that it's up to them to save their friends, but Lucy doesn't know that the Sweepers have laid a trap — for her. There is something special about Lucy, and the Sweepers will stop at nothing to have her in their clutches.

Jo Treggiari spins a thrilling tale of adventure, romance, and one girl's unyielding courage through the darkest of nightmares.

SCHOLASTIC

www.scholastic.com

ISBN 978-0-545-43656-4

$6.99

596°

EAN

9 780545 436564